Lecture Notes in Electrical Engineering

Volume 823

The book series *Lecture Notes in Electrical Engineering* (LNEE) publishes the latest developments in Electrical Engineering - quickly, informally and in high quality. While original research reported in proceedings and monographs has traditionally formed the core of LNEE, we also encourage authors to submit books devoted to supporting student education and professional training in the various fields and applications areas of electrical engineering. The series cover classical and emerging topics concerning:

- Communication Engineering, Information Theory and Networks
- Electronics Engineering and Microelectronics
- Signal, Image and Speech Processing
- Wireless and Mobile Communication
- Circuits and Systems
- Energy Systems, Power Electronics and Electrical Machines
- Electro-optical Engineering
- Instrumentation Engineering
- Avionics Engineering
- Control Systems
- Internet-of-Things and Cybersecurity
- Biomedical Devices, MEMS and NEMS

For general information about this book series, comments or suggestions, please contact leontina.dicecco@springer.com.

To submit a proposal or request further information, please contact the Publishing Editor in your country:

China

Jasmine Dou, Editor (jasmine.dou@springer.com)

India, Japan, Rest of Asia

Swati Meherishi, Editorial Director (Swati.Meherishi@springer.com)

Southeast Asia, Australia, New Zealand

Ramesh Nath Premnath, Editor (ramesh.premnath@springernature.com)

USA, Canada:

Michael Luby, Senior Editor (michael.luby@springer.com)

All other Countries:

Leontina Di Cecco, Senior Editor (leontina.dicecco@springer.com)

**** This series is indexed by EI Compendex and Scopus databases. ****

More information about this series at https://link.springer.com/bookseries/7818

Ashwani Kumar · S. C. Srivastava · S. N. Singh
Editors

Renewable Energy Towards Smart Grid

Select Proceedings of SGESC 2021

Springer

Editors
Ashwani Kumar
Department of Electrical Engineering
National Institute of Technology Kurukshetra
Kurukshetra, India

S. C. Srivastava
Department of Electrical Engineering
Indian Institute of Technology Kanpur
Kanpur, India

S. N. Singh
Department of Electrical Engineering
Indian Institute of Technology Kanpur
Kanpur, India

ISSN 1876-1100 ISSN 1876-1119 (electronic)
Lecture Notes in Electrical Engineering
ISBN 978-981-16-7471-6 ISBN 978-981-16-7472-3 (eBook)
https://doi.org/10.1007/978-981-16-7472-3

This Springer imprint is published by the registered company Springer Nature Singapore Pte Ltd.
The registered company address is: 152 Beach Road, #21-01/04 Gateway East, Singapore 189721, Singapore

Preface

Renewable energy systems are becoming popular due to high oil prices, growing energy demand, and environmental concerns. These energy systems can replace the fossil fuel-based sources and lowering the carbon emissions in addition a supply to the rural areas. The future power systems need to be a smart network with the mix of generation and efficient transmission and distribution of energy with higher efficiency, better quality, and high reliability and security. Smart Grid Technology will provide a solution to these challenges of increasing electric demand, ageing utility infrastructure, environmental impact on greenhouse gas emissions due to fossil fuel-based generation, high reliability of energy supply, better quality power, and stable and secure network. Integrated distributed generation can improve the efficiency and reliability of the electric grid and benefit utilities and customers. Grid Support DG can also contribute to the provision of providing an ancillary service. These include services necessary to maintain the sustained and stable operation of the grid, and DG plants can help to stabilize a frequency. Furthermore, most government policies aiming to promote the use of renewable energy will also result in an increased impact of the integration of DG technologies for sustainable development of smart grid. Information and communication technology with computer software and hardware integration will provide a solution to the complete automation of the renewable energy integrated systems.

The book presents advanced research in the emerging fields of renewable energy, smart grid, micro grids, distributed generation with solar, wind and other renewable sources, ancillary services and demand-side management, electricity markets, and energy efficiency improvements from experts working in these areas over the last decade. Most of the authors of the papers have research contributions towards the development of smart grid, improvement in the renewable energy efficiency, ancillary services, electricity market, dispatching of the mix of generation, processes or algorithms, and applications of computational techniques in these areas. The book is a unique collection of chapters from the mentioned areas with a common theme and will be of immense use to academic researchers and practitioners in the industry who work in this field.

Kurukshetra, India Dr. Ashwani Kumar
Kanpur, India Dr. S. C. Srivastava
Kanpur, India Dr. S. N. Singh

Contents

About the Editors

Dr. Ashwani Kumar is a Professor in the Department of Electrical Engineering, National Institute of Technology (NIT) Kurukshetra, India. He received his B.Tech. in Electrical Engineering from Pant Nagar University in 1988 and Masters' degree in Power Systems from Panjab University, in 1994. He obtained his Ph.D. degree in Electrical Engineering from IIT Kanpur, India. He did Post-Doctoral from Tennessee Technological University, USA. His research areas are power system restructuring, distributed generation, integration of renewable energy systems into the deregulated power system, energy management, demand-side management, ancillary services, micro-grid, and smart grid. He has over 29 years of teaching and research experience, published 160 papers in various prestigious journals, conferences. He has executed two research projects sponsored by different funding agencies including DST and TTU. He has also carried out few consultancies assignments for power companies. Many of his research students have been awarded the prestigious POSOCO Power System Award (PPSA) for best thesis on all Indian levels.

Prof. S. C. Srivastava is a Professor in the Department of Electrical Engineering, Indian Institute of Technology (IIT), Kanpur, India. He completed his Ph.D. from IIT Delhi, India. His research areas are power system dynamics and stability studies, optimal power dispatch, security analysis and control, wide-area monitoring and control of power systems, power system restructuring and issues in the electricity market, renewable integration, DC and AC microgrid. He has published over 100 papers in several journals.

Prof. S. N. Singh obtained his M. Tech. and Ph.D. in Electrical Engineering from the Indian Institute of Technology Kanpur in 1989 and 1995. Presently, he is Vice-Chancellor of the Madan Mohan Malviya University of Technology Gorakhpur, and on leave from the position of Professor (HAG), Department of Electrical Engineering, Indian Institute of Technology Kanpur, India. His research interests include power system restructuring, FACTS, power system optimization and control, security analysis, and wind power. He has published over 440 papers in international and national journals and conferences. He has also written two books on Electric Power Generation, Transmission and Distribution, and Basic Electrical Engineering, published by PHI, India. Prof. Singh has completed over 20 projects in India and abroad.

Impact of OLTC on Nodal Pricing of Distribution System and Comparison with DSTATCOM

Banothu Sridhar and Ashwani Kumar

Abstract The advent of deregulation in the electricity sector introduced a lot of changes in the operations of the distribution systems. The addition of distributed generation in the conventional distribution system has changed the passive network to active. There is a need to remunerate the distributed generation connected at a particular node and to determine the 'node Price' for accurate quantification. In this paper, OLTC and DSTATCOM are modelled and their impacts on nodal pricing are analysed. This study is carried out on a standard IEEE 69-bus Radial Distribution System (RDS) using General Algebraic Modelling System (GAMS) software.

Keywords Deregulation · Electricity market · Nodal pricing · OLTC · D-FACTS · DSTATCOM · RDS · Load flow · GAMS

1 Introduction

The analysis of the distribution system has been changed greatly in recent times. Nodal pricing is a cost-effective pricing method commonly practiced in transmission systems and which can also be used in the distribution network for analysis according to current national plans [1]. Nodal pricing is also called Locational Marginal Pricing (LMP). This technique is used to evaluate the price at any particular node of a given distribution system. As we know, the nodal pricing methodology is dependent on various parameters such as total power loss, active power injection, reactive power injection, sending and receiving end voltages at the nodes, and voltage angles. Therefore, voltage improvement and loss minimization are very essential in order to evaluate the nodal pricing in the distribution system. Currently, voltage control in low voltage distribution networks is performed by making use of manual no-load tap changers whose position is only calibrated and changed in case of expansion or change of the network, possibly with seasonal variations. Some utilities have already implemented distribution transformer prototypes with on-load tap changers

B. Sridhar · A. Kumar (✉)
National Institute of Technology, Kurukshetra 136119, India

A. Kumar et al. (eds.), *Renewable Energy Towards Smart Grid*, Lecture Notes in Electrical Engineering 823, https://doi.org/10.1007/978-981-16-7472-3_1

(OLTCs), taking into account the transition to Active Network Management (ANM) strategy and technology [2].

OLTC is a device for changing the tap connection of a winding suitable for operation when the transformer is energized and the load condition adjusts the voltage. The tap selector can be located in the transformer oil. By choosing the tap position optimally, the transformer's downstream bus voltage can be adjusted. In this paper, the impact of on-load tap changer is studied for potential or voltage development and loss reductions. To change to the new tap position [3], the total time is about 2–10 s. The number of unique tap positions of the OLTC is generally standardized on equipment of various manufacturers. In this paper, as shown in Fig. 1, we will consider 7 steps (6 positions) from −5 to +10% in 2.5% steps. Under normal operating conditions, the tap position is set to the nominal value. At nominal tap position, the test system minimum bus voltage is 0.9091 P.U. This violates the statutory requirements. Lowering the tap position increases the number of primary taps, which increases the transformer secondary voltage and thus the downstream voltage. As you increase the tap position, the downstream voltage decreases. If the bus voltage is kept within tolerance, it means that the KVAR load requirements are met by the generator, transformer, and reactive power sources in the system. In this scenario, OLTC operation is not expected. However, during high load conditions, the bus voltage may deviate from the acceptable voltage band limit, thereby adjusting the tap position to maintain the bus voltage as needed. Tap positions can be calculated optimally with ECS program logic or manually controlled by the system operator via the transformer RTCC panel.

As concluded in [4], the proposed control technology can exacerbate the voltage imbalance in the grid due to the independent OLTC control of each phase. In addition, the OLTC can only partially reduce the over-voltage limit (1.1 P. U.) violation due to the finite response of the device. Thus, the proposed voltage control

Fig. 1 OLTC winding diagram

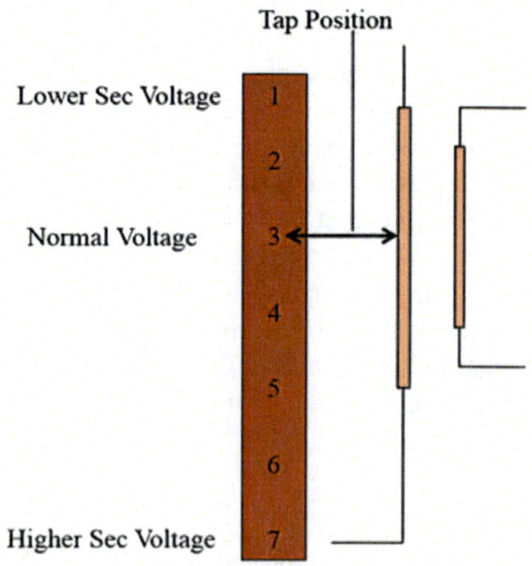

algorithm is required to be combined with additional active network management techniques aimed at solving the identified problems. FACTS devices are widely used around the world and are themselves quite mature technologies. Innumerable control functions include voltage regulation, system damping, and power flow control. The use of those devices at the LV level is almost non-existent due to cost, awareness of complexity, and possibly some concerns about their reliability and technological advancement. The concept of distributed FACTS (D-FACTS) has recently been proposed as an alternative approach to realizing the function of FACTS devices that eliminates the above-mentioned barriers. DSTATCOM is suitable for improving the power quality of distribution systems, which alleviates various problems such as voltage fluctuation and flicker, voltage imbalance, and current distortion. Compared to SVC, DSTATCOM allows more flexibility, and reactive power is more independent of the actual voltage at the junction.

In this paper, the nodal pricing of a 69-bus radial distribution system with only OLTC and OLTC with DSTATCOM has been analysed.

2 Nomenclature

i, j, l & t	Index variables for buses, lines and time
U_i^{max}, U_i^{min}, U_i	Maximum and minimum voltage limit
P_{gi}^{max}	Active power generation maximum limit
P_{gi}^{min}	Reactive power generation minimum limit
$Q_{D_{STATCOM(z)}}^{min}$	DSTATCOM reactive power minimum limit
$Q_{DSTATCOM(z)}^{max}$	DSTATCOM reactive power maximum limit
P_{gi} & Q_{gi}	Active and reactive powers generation
P_{di} & Q_{di}	Real and reactive power load demand at each bus i
P_i & Q_i	Real and reactive powers injected at bus i
ρ_{pi}	Active power Marginal Loss Coefficient (MLC)
ρ_{Qi}	Reactive power Marginal Loss Coefficient (MLC)
G_{ij}	Branch conductance
B_{ij}	Branch susceptance
θ_i	Voltage angle
T_i^{max}&T_i^{min}	Upper and lower limits of OLTC tap positions
λ	Electricity price(USD/MWh)
DSTATCOM	Distribution static synchronous compensator
N_{Pi} & N_{Qi}	Active and reactive power nodal price (USD/MWh)
RF	Reconciliation Factor
TPL	Total real power loss in branch ij
θ_{ij}	Angle of node voltages ij

3 Problem Formulation

3.1 Modelling of OLTC and DSTATCOM

Mathematical modelling of OLTC and DSTATCOM [5] has been developed in GAMS software to minimize total power loss by making use of below-given equation,

$$\mathrm{TPL} = \sum_{\mathrm{ij}} \mathrm{P}_{\mathrm{lossij}} \tag{1}$$

Power injection equations,

$$P_i = P_{\mathrm{gi}} - P_{\mathrm{di}} \tag{2}$$

$$Q_i = Q_{\mathrm{gi}} - Q_{\mathrm{di}} \tag{3}$$

Power flow equations,

$$P_{\mathrm{ij}} = T_i^2 * U_i^2 * G_{\mathrm{ij}} - T_i * U_i \sum_{j=1}^{n} U_j \left(G_{\mathrm{ij}} * \cos\theta_{\mathrm{ij}} + B_{\mathrm{ij}} * \sin\theta_{\mathrm{ij}}\right), i \in S_b \tag{4}$$

$$Q_{\mathrm{ij}} = -T_i^2 * U_i^2 * \mathrm{B}_{\mathrm{ij}} + T_i * U_i \sum_{j=1}^{n} U_j \left(\mathrm{G}_{\mathrm{ij}} * \sin\theta_{\mathrm{ij}} - \mathrm{B}_{\mathrm{ij}} * \cos\theta_{\mathrm{ij}}\right), i \in S_b \tag{5}$$

System constraints are as follows:

$$U_{\mathrm{i}}^{\mathrm{min}} \le U_i \le \mathrm{U}_{\mathrm{i}}^{\mathrm{max}}, i \in S_b \tag{6}$$

$$\delta_{\mathrm{i}}^{\mathrm{min}} \le \delta_{\mathrm{i}} \le \delta_{\mathrm{i}}^{\mathrm{max}}, i \in S_b \tag{7}$$

$$\mathrm{T}_{\mathrm{i}}^{\mathrm{min}} \le T_i \le \mathrm{T}_{\mathrm{i}}^{\mathrm{max}} \tag{8}$$

The modified power injection equation,

$$Q_i = Q_{\mathrm{gi}} + \sum_{z} Q_{\mathrm{DSTATCOM(z)}} - \left(Q_{\mathrm{di}} + Q_{\mathrm{loss(ij)}}\right) \tag{9}$$

$$Q_{\mathrm{DSTATCOM(z)}}^{\mathrm{min}} \le Q_{\mathrm{DSTATCOM(z)}} \le Q_{\mathrm{DSTATCOM(z)}}^{\mathrm{max}} \tag{10}$$

Minimum and Maximum limits of DSTATCOM are taken as (30–40) % of total active power load and 2/3 of total reactive power load of the given system [6].

3.2 Determination of Nodal Pricing

Active and reactive power nodal pricing [7, 8] at every node can be found by using the following equations [9]:

$$N_{\mathrm{Pi}} = \lambda * (1 + \mathrm{Rf} * \rho_{\mathrm{Pi}}) \tag{11}$$

$$N_{\mathrm{Qi}} = \lambda * \mathrm{Rf} * \rho_{\mathrm{Qi}} \tag{12}$$

4 Results and Discussions

In this section, a case study of IEEE 69-bus RDS base of 100 MVA, 12.66 kV is chosen. The total load on the network is $3.8013 + j * 2.6936$ MVA [10, 11]. Programing codes have been developed in a MATLAB 7.0.4 environment and GAMS 23.4 CONOPT solver; with the help of interfacing, results are obtained. For analysis, bus-4 is considered.

Tables 1 and 2 represent the nodal prices of Real power and Reactive power for each tap position with OLTC individually and OLTC coordinated with DSTATCOM for a 24 h time duration. So, it is clearly observed from the above tables that the active and reactive power nodal prices are gradually increasing at lower taps (tap positions 0.95 and 0.975) and decreasing at upper taps (tap positions 1.025 and 1.05) which means OLTC lower taps are used when the system load is under light load condition, and the OLTC upper taps are used when the system load is under heavy load conditions. Therefore, for this particular work, upper taps of OLTC tap positions are utilized.

Nodal prices of all the 69 buses are represented in the 3-D graph as shown (Figs. 2, 3, 4, 5, 6, 7, 8, and 9).

Voltage and power loss profiles are as shown in the Figs. 10, 11, 12 and 13.

Figures 10 and 11 give the bus voltage magnitudes for each bus of a given 69-bus RDS. It is observed that the voltages in the system are improved by using on-load tap changer as well as distributed static synchronous compensator. So, due to the improvement of voltages in the system, the nodal prices at each bus are greater in reduction.

5 Conclusion

Considering the fact that OLTC can partially improve the over-voltage indicator while the phase-independent tap changes increase the voltage imbalance, this article discusses coordinated voltage control using OLTC and DSTATCOM devices. It was

Table 1 Active power nodal prices

Hours	Tap position 0.95		Tap position 0.975		Tap position 1.0		Tap position 1.025		Tap position 1.05	
	OLTC	OLTC & DSTATCOM	OLTC	OLTC & DSTATCOM	OLTC	OLTC & DSTATCOM	OLTC	OLTC & DSTATCOM	OLTC	OLTC & DSTATCOM
1	25,939.56	4551.999	38.28079	39.72207	242.3127	39.72414	40.11554	39.74558	40.12699	39.73387
2	41,124.7	6142.905	38.33532	39.72003	249.6644	39.75122	40.08641	39.73314	40.1275	39.73175
3	10,395.08	8103.196	134.1399	39.71834	255.7942	39.73982	40.09105	39.74057	40.12792	39.73
4	7596.555	9299.434	38.39304	39.71768	258.1738	39.69471	40.09285	39.70409	40.12808	39.72932
5	47,486.58	6058.039	221.5631	39.72011	249.532	39.72462	40.08616	39.73115	40.12748	39.73184
6	1781.739	12,446.05	38.47514	39.71385	40.10544	39.69075	40.10359	39.72419	40.12903	39.72535
7	2156.419	3126.21	38.74261	39.69616	40.10906	39.695	40.15565	39.68626	40.13336	39.71031
8	2718.938	39.38181	38.81424	39.68938	40.11551	39.67925	40.17664	39.66924	40.135	39.7057
9	2716.396	39.38221	38.81412	39.68941	40.10986	39.66897	40.17654	39.66927	40.13499	39.70573
10	2690.569	39.38635	38.81084	39.68975	40.10982	39.67957	40.17548	39.67869	40.13491	39.7061
11	2519.086	39.41326	3509.794	39.69192	40.11076	39.68014	40.08219	39.68172	40.13439	39.70852
12	1900.317	2911.562	2340.153	39.6987	40.11193	39.67957	40.14774	40.06264	40.13274	39.72938
13	2403.238	39.43071	38.77543	39.69333	40.11051	39.68026	40.08135	39.68276	40.13405	39.71009
14	2340.556	3299.914	38.76786	39.69406	40.11341	39.68006	40.08091	39.68527	40.13387	39.71091
15	48.2427	3025.699	38.73215	39.69731	40.10893	39.694	40.15198	39.68781	40.13308	39.70822
16	48.23694	3003.511	38.72843	39.69759	40.1089	39.68873	40.15116	39.68818	40.13301	39.73404
17	48.18605	2802.157	2969.531	39.70007	323.851	39.68838	40.14362	40.06428	40.13241	39.71108
18	87.55369	13,361.97	38.54845	39.71017	40.10809	39.72632	40.45833	39.70704	40.12994	39.72154
19	19,142.17	4192.562	38.26872	39.72259	240.3861	39.73361	40.07941	39.74931	40.12686	39.7344
20	5072.507	27.13997	37.75508	39.73291	40.11426	39.71046	40.08658	39.71528	40.12435	39.7451

(continued)

Table 1 (continued)

Hours	Tap position 0.95		Tap position 0.975		Tap position 1.0		Tap position 1.025		Tap position 1.05	
	OLTC	OLTC & DSTATCOM	OLTC	OLTC & DSTATCOM	OLTC	OLTC & DSTATCOM	OLTC	OLTC & DSTATCOM	OLTC	OLTC & DSTATCOM
21	6385.891	601.168	437.8249	39.72819	40.10692	39.73406	40.08556	39.7545	40.12546	39.7402
22	6421.542	24.00452	38.08547	39.72813	40.10677	39.75273	40.08554	39.71142	40.12548	39.74014
23	6871.164	1176.09	425.7436	39.72746	40.10525	39.74788	40.08541	39.71629	40.12565	39.73944
24	9464.052	2671.288	38.18982	39.72511	231.4848	39.73974	40.0727	39.72147	40.12623	39.73701

Table 2 Reactive power nodal prices

Hours	Tap position 0.95		Tap position 0.975		Tap position 1.0		Tap position 1.025		Tap position 1.05	
	OLTC	OLTC & DSTATCOM	OLTC	OLTC & DSTATCOM	OLTC	OLTC & DSTATCOM	OLTC	OLTC & DSTATCOM	OLTC	OLTC & DSTATCOM
1	42,982.29	9037.35	4.173482	0.207671	13.2881	0.180154	0.025276	0.191919	0.128146	0.264047
2	97,283.15	9110.669	4.041757	0.209064	11.1498	0.16673	0.048085	0.198561	0.128779	0.265675
3	33,232.67	8752.514	1859.357	0.210219	9.224647	0.172689	0.061738	0.195037	0.129303	0.267023
4	27,441.35	8364.498	3.902268	0.210666	8.442014	0.195756	0.067033	0.213787	0.129505	0.267545
5	110,565.3	9115.555	1665.412	0.209008	11.63952	0.180155	0.047348	0.199566	0.128752	0.265609
6	15,613.58	13,393.67	3.703726	0.213288	0.253065	0.198264	0.098494	0.204071	0.130689	0.270601
7	8852.764	1966.613	3.055377	0.225419	0.261748	0.198288	0.249325	0.226285	0.136066	0.282343
8	8245.729	0.592639	2.881142	0.230079	0.277215	0.207249	0.309419	0.236168	0.138086	0.287037
9	8248.234	0.592294	2.881439	0.23006	0.263668	0.212545	0.309133	0.236147	0.138077	0.287013
10	8273.625	0.588702	2.889437	0.229827	0.263572	0.207039	0.306099	0.231189	0.137976	0.286731
11	8447.917	0.565381	4359.015	0.228333	0.265829	0.206462	0.026511	0.229283	0.137328	0.284919
12	9171.024	1827.258	1697.178	0.223668	0.268386	0.205893	0.226549	0.070207	0.135289	0.268959
13	8571.646	0.550259	2.975622	0.227366	0.265224	0.206222	0.024339	0.228531	0.136908	0.283741
14	8640.77	2081.955	2.994015	0.226863	0.272147	0.206232	0.023204	0.227118	0.136688	0.283127
15	6.243446	1900.961	3.080841	0.224628	0.261435	0.198657	0.238767	0.225309	0.13571	0.283771
16	6.240251	1886.554	3.089872	0.224436	0.261353	0.201327	0.236392	0.225072	0.135628	0.265941
17	6.211748	1757.475	2917.389	0.222729	21.44523	0.201187	0.214683	0.069499	0.134878	0.281571
18	12,374.88	17,867.78	3.526325	0.215808	0.259356	0.173022	0.112215	0.213426	0.131809	0.273535
19	28,828.9	8973.702	4.202638	0.207318	13.81769	0.175292	0.027417	0.189949	0.12798	0.263634
20	196.4804	7.978131	5.440672	0.200206	0.274331	0.185693	0.034143	0.205641	0.125181	0.255398

(continued)

Table 2 (continued)

Hours	Tap position 0.95		Tap position 0.975		Tap position 1.0		Tap position 1.025		Tap position 1.05	
	OLTC	OLTC & DSTATCOM	OLTC	OLTC & DSTATCOM	OLTC	OLTC & DSTATCOM	OLTC	OLTC & DSTATCOM	OLTC	OLTC & DSTATCOM
21	2668.846	7812.303	1121.457	0.2035	0.256678	0.174371	0.029509	0.186455	0.126233	0.259166
22	2738.034	9.773977	4.644937	0.203537	0.256298	0.164959	0.029423	0.208437	0.126252	0.259209
23	3617.066	8014.085	1156.608	0.203997	0.252662	0.167499	0.028531	0.20604	0.126463	0.259747
24	8826.764	8550.933	4.393111	0.2056	16.10392	0.171876	0.007577	0.203746	0.127199	0.261624

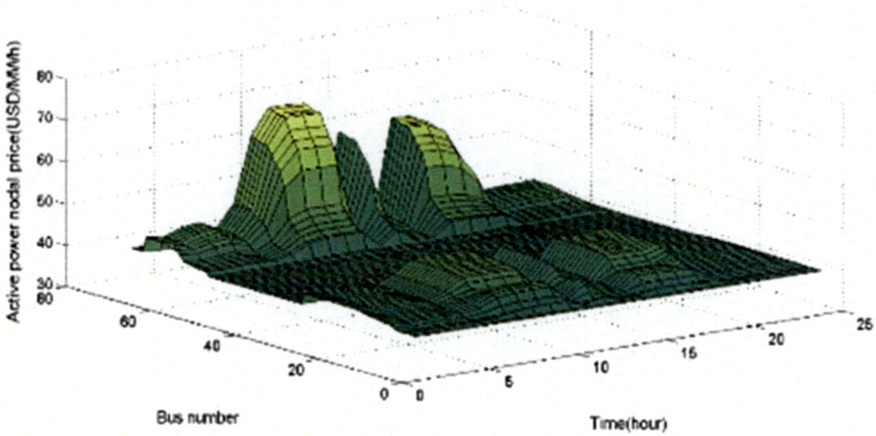

Fig. 2 Active power nodal price with OLTC tap 1.025

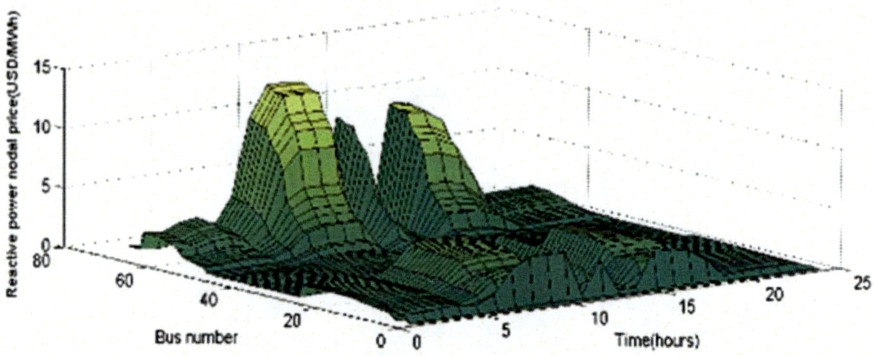

Fig. 3 Reactive power nodal price with OLTC tap 1.025

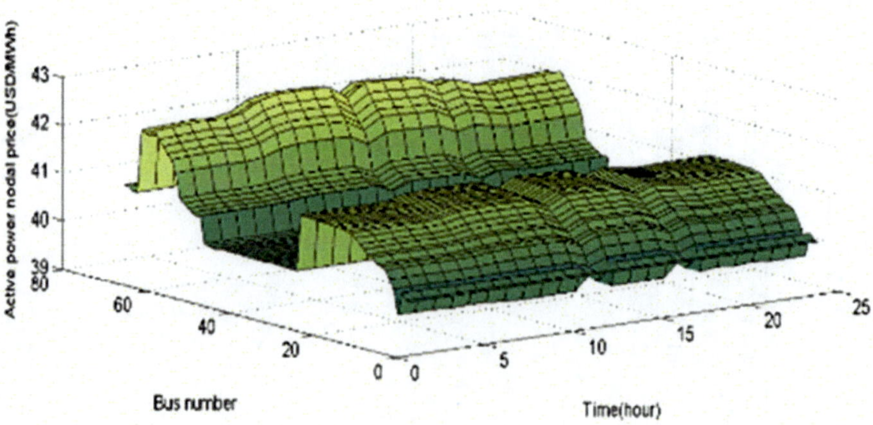

Fig. 4 Active power nodal price with OLTC and DSTATCOM tap 1.025

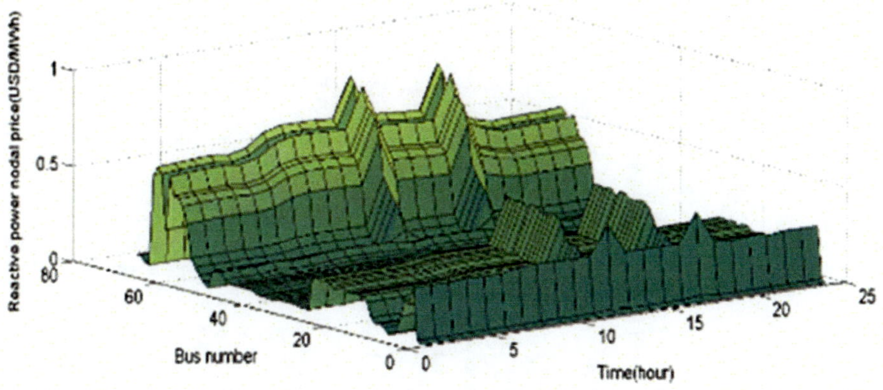

Fig. 5 Reactive power nodal price with OLTC and DSTATCOM tap 1.025

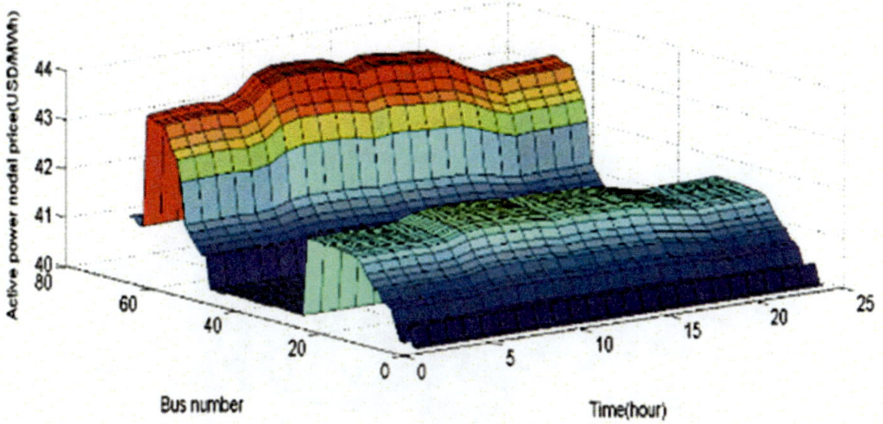

Fig. 6 Active power nodal price with OLTC tap 1.05

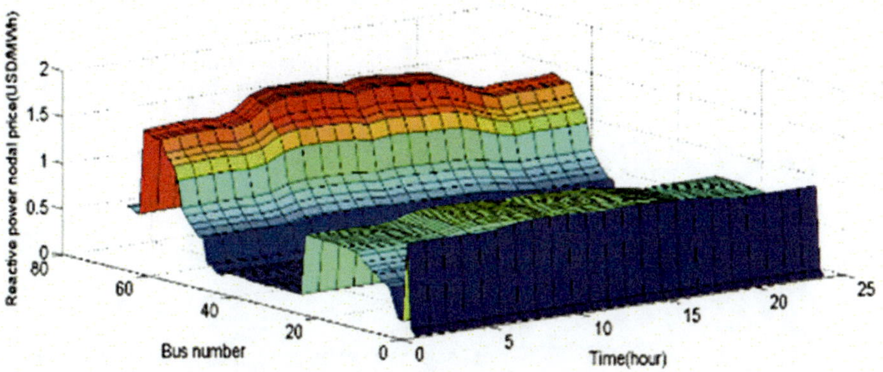

Fig. 7 Reactive power nodal price with OLTC tap 1.05

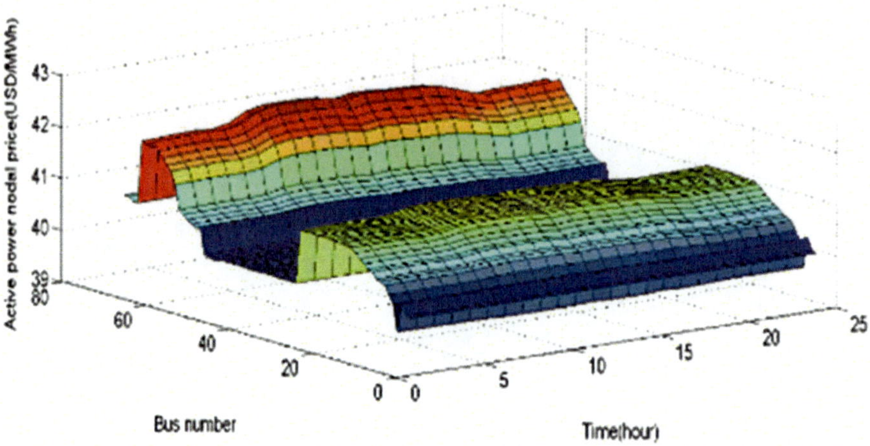

Fig. 8 Active power nodal price with OLTC and DSTATCOM tap 1.05

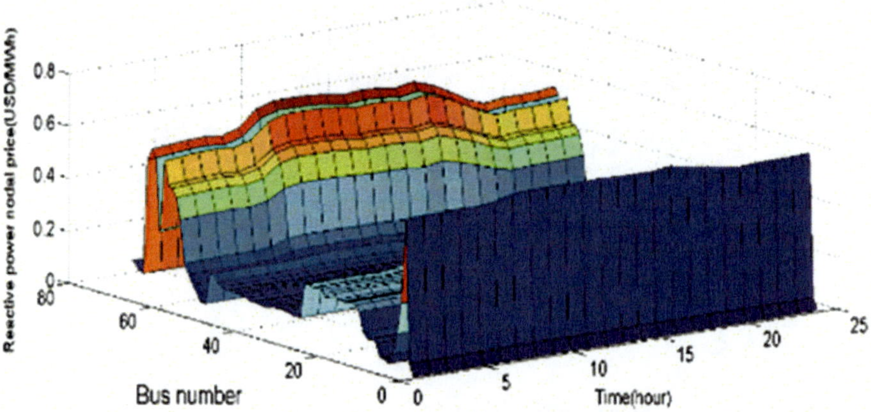

Fig. 9 Reactive power nodal price with OLTC and DSTATCOM tap 1.05

Fig. 10 Voltage profile for tap position 1.025

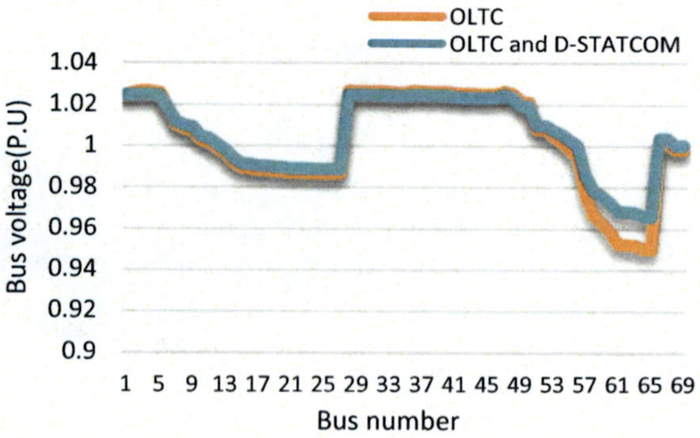

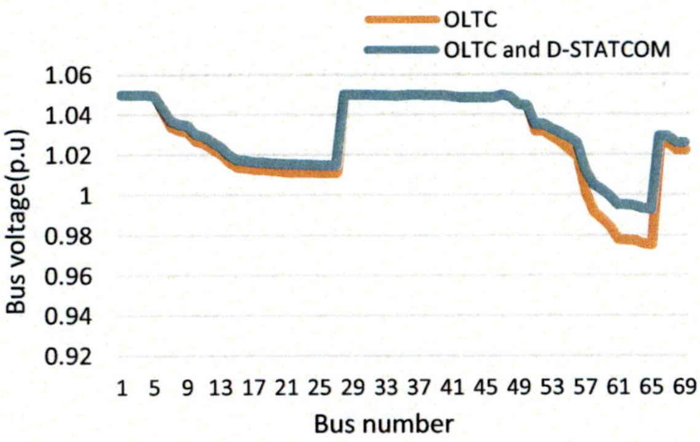

Fig. 11 Voltage profile for tap position 1.05

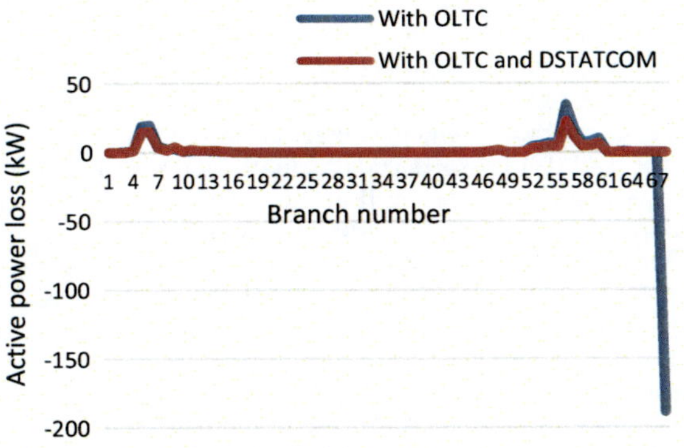

Fig. 12 Power loss for tap position 1.025

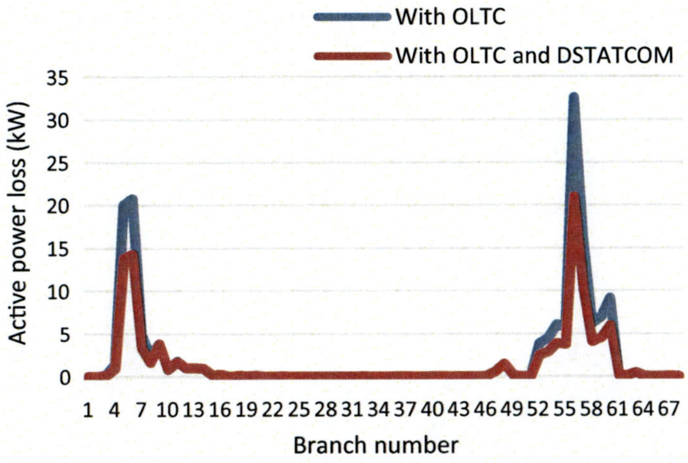

Fig. 13 Power loss for tap position 1.05

concluded that the proposed method could completely eliminate the actual power over-voltage indicator violation, and it has been found that controlling the reactive power can significantly reduce the voltage regulation. Also, active and reactive power nodal prices are obtained for each tap position step-wise (in steps of 2.5%) for a 24 h duration with only OLTC and OLTC with DSTATCOM.

References

1. Ghayeni M, Ghazi R (2011) Transmission network cost allocation with nodal pricing approach based on ram say pricing concept. Gener Transm Distrib (IET) 5
2. SIEMENS, FIT former REG, the adaptable distribution transformer. Tech
3. IEEE std C57.131–2012, IEEE standard requirements for tap-changers
4. Efkarpidis N, Gonzalez C, Wijnhoven T, Dommelen DV, Rybel TD, Driesen J (2013) Technical assessment of on-load tap-changers in flemish LV Distribution Grids. In: International workshop on integration of solar power into power systems
5. Murty VV, Sharma AK (2018) Optimal coordinate control of OLTC, DG, D-STATCOM, and reconfiguration in distribution system for voltage control and loss minimization
6. Sridhar B, Kumar A (2019) Loss reduction in distribution system with wind energy and D-STATCOM considering uncertainty. In: IEEE 2019 1st international conference on energy, systems and information processing (ICESIP2019). Kanchipuram
7. Narayan KS, Kumar A (2015) Distribution system nodal pricing analysis with realistic ZIP load and variable wind power source. In: IEEE India international conference on (INDICON-2015)
8. Polisetti K, Kumar A (2016) Distribution system nodal prices determination for realistic ZIP and Seasonal loads: an optimal power flow approach 25:702–709
9. Sridhar B, Kumar A (2019) Nodal pricing analysis of distribution system with D-STATCOM and wind power for realistic ZIP and RIC loads. ICSC 2019,THDC-IHET,Tehri, Springer
10. Parasher R (2014) Load flow analysis of radial distribution network using linear data structure. ArXiv14034702
11. Zimmerman RD, Murillo-Sánchez CE, Thomas RJ (2011) MATPOWER: steady-state operations, planning and analysis tools for power systems research and education. IEEE Trans Power Syst 26(1):12–19

Allocation of Wind-Based Distributed Generation Using Bacterial Foraging Optimization Algorithm in Electrical Distribution System

Ashish Verma and Atma Ram Gupta

Abstract Wind-based DG is used by DNO for the production of power by maintaining the voltage profile and minimization of power losses in the power system and as DG has many techno-economic as well as environmental advantages. The proposed work shows the reduction in power losses using WG in the system. Whenever load in the power system increases, the power losses increase and voltage at the bus also reduces. In the proposed work, the electrical load is taken as different types like CP, CI, CZ, and ZIP load. Also, the behavior of electrical parameters like power losses and voltage profile at 20% overloading and underloading conditions has been studied. Optimal allocation of WG and DSTATCOM for the fulfillment of active and reactive powers, respectively, with different types of load using BFOA for Portuguese 94-bus RDS is simulated and studied. The proposed model results in a very large amount of loss reduction and energy cost savings for the system.

Keywords DG · ZIP load · BFOA · DNO · DSTATCOM · Distribution system

1 Introduction

The power system is moving toward restructured power systems and also electrical energy consumers' willingness has made small production units more important to receive energy in their residence. Previously, Distribution Network Operator (DNO) had the responsibility of planning for the future behavior of the power system. With the introduction of Distributed Generations (DGs), energy storage, renewable sources, and micro-grids, the classical model of DNO is changing. So, DNO shifts to Distribution System Operator (DSO) for active system operation. DNO has the responsibility to perform voltage regulation, minimize power losses, ancillary services in the transmission, and local generation, that is, DG, and implement Distribution Static Compensator (DSTATCOM) [1]. DGs are small power generating units that are allocated at strategic points of the system installed on the distribution

A. Verma (✉) · A. R. Gupta
Department of Electrical Engineering, NIT Kurukshetra, Haryana, India
e-mail: ashish_31910102@nitkkr.ac.in

A. Kumar et al. (eds.), *Renewable Energy Towards Smart Grid*, Lecture Notes in Electrical Engineering 823, https://doi.org/10.1007/978-981-16-7472-3_2

side. DGs may include wind power plants, fuel cells, solar power plants, etc. [2]. In this proposed paper, Wind Generation (WG) is used on the distribution side as DG. WGs can help to supply different levels of power for micro-grid or whole grid to provide loads for desired consumers separately. DGs are used to provide electricity for load centers in the open and closed positions and as connected to the distribution side. WG provides network support, peak shaving, and standby power. The use of non-conventional energy resources is increasing day by day worldwide due to the atmospheric problems, energy crisis, technological advancement, technical problems, power outages, and economic temptation. Reactive power compensation in a power system plays an important role in improving voltage profile, power quality, power factor improvement, and reducing voltage sag in the grid [1]. Voltage source converter (VSC) is one of the adjustment tools which gives a fast response to the demand of reactive power, so DNO uses VSC to improve voltage regulation and power factor. There are different types of VSC-based controllers that are called DSTATCOMs which are the best alternative for SVC. DSTATCOMs are used to improve power quality and to manage voltage fluctuations, unbalanced load, reactive power, voltage imbalance, and voltage drop that occur in a short time within milliseconds [3]. Smart Energy Management (SEM) is the best way to manage energy in different methods and how the systems work most efficiently. Allocation of WG and DSTATCOM is the method of SEM. Some of the easy methods of SEM are the use of energy-saving air conditioners and smart appliances. Some of the smart energy systems are the installation of wind generation, installation of solar panels, multi-building automation utility systems, etc. There are different methods to control the system parameter and increase efficiency to save energy cost which depends on types of DG. The first step in SEM is to analyze and forecast electrical energy losses and existing power quality problems in the Transmission and Distribution (T&D) network. After the installation of the SEM system, DNO can use smart technology to identify electrical theft, times of high usage, and especially electrical energy losses [4]. Energy usage has always been a concern for consumers. For DNO and consumers without SEM, it is hard to find the resources because the electrical cost and load demand continuously increasing may help to use energy efficiently. T&D losses always increase with load demand. So, SEM helps DNO and consumers to understand in which period they can use the most energy and the best techniques that can be used to manage energy with minimum losses. SEM takes the methods that consumers already use in the next level and promotes the highest level of energy-saving by minimizing losses. Electricity Act provides guideline to DNO for effective management of distribution system. Previously, researchers have developed the allocation of DG and DSTATCOM by using the Bee Colony algorithm [5], Particle Swarm Optimization algorithm [6], Honey Bee Mating Optimization algorithm [7], and Bat algorithm [8]. In the proposed paper, Bacterial Foraging Optimization Algorithm (BFOA) is used to analyze the behavior of system parameters toward DG and DSTATCOM.

2 Wind Generation Allocation

2.1 *Wind Generation*

Installation of WGs usually gives continuous power, and less investment compared to other power plants in T&D lines [9]. WG is an eco-friendly renewable energy sources with less requirement of land area and provide more economic energy solutions compared to other sources of energy.

2.2 *Wind Generation as DG*

Distributed WG includes modeling and simulation, siting, resource characterization, and technology development. WG on the distribution side can be used as DG. DGs can be classified into four categories based on their abilities to inject active power (P) and reactive power (Q): Type A (injects P only), Type B (injects both P and Q), Type C (injects P but absorbs Q), and Type D (injects Q only) [9]. In this proposed paper, DG that injects real power only with different types of loads is used for analysis.

2.3 *Mathematical Modeling of WG*

Wind energy conversion is the process by which wind is converted into electricity. Wind turbines convert kinetic energy to mechanical power; after that, mechanical power is converted into electricity which acts as a source of energy in the power system.

Kinetic energy exists in the motion of an object and when the wind is in motion, kinetic energy in moving wind can be determined by Eq. (1):

$$E = \frac{1}{2}\mathrm{mV}^2 \tag{1}$$

Wind introduces lift forces and drag forces on the blade, resulting in the movement of these forced blades, and wind energy is transferred to the blade power. The blade power P is the rate of change of energy [10]. Wind power can be calculated by using Eq. (2) for different velocities at different times:

$$P = \begin{cases} 0 & V < V_{\mathrm{ci}} \\ \frac{1}{2}\rho \mathrm{AV}^3 & V_{\mathrm{ci}} \leq V < V_{\mathrm{rated}} \\ P_{\mathrm{rated}} & V_{\mathrm{rated}} \leq V < V_{\mathrm{cutoff}} \\ 0 & V > V_{\mathrm{cutoff}} \end{cases} \tag{2}$$

where V_{ci} = minimum speed (usually 3 m/s), V_{rated} = rated speed, V_{cutoff} = cutoff speed, P = power (kW), ρ = air density (= 1.23 kg/m), A = Area of blades (= $\pi * 50^2 m^2$), and V = Wind velocity; here, the proposed paper takes different velocities of different months as per the online weather report.

There are various wind power parameters that control the wind power generation such as power coefficient, effective power output and total power conversion coefficient, Lanchester-Betz limit, power curve, tip speed ratio, and wind turbine capacity factor. But the general expression of active power generation is given in Eq. (2). Wind is a highly intermittent energy source which may cause fluctuation in the power. So, it is required to determine expected velocity and load demand in the power system. In terms of cost-effectiveness, wind energy is much cost saving because wind turbine lifetime is approximate 20–30 years [11].

2.4 Cost Analysis

Allocation of WG at the distribution side helps to improve different parameters (voltage profile, energy losses, and active power demand) of T&D which cause economic savings, life, and efficiency of the T&D equipment. The cost of Energy losses is given by Eq. (3):

$$\text{CEL} = (\text{Total Real Power loss}) * (\text{E}_c * T)\ \text{INR} \tag{3}$$

$where E_c$ = Energy cost rate (= 4.4664 INR/kW h), and T = Time duration (= 8760 h).

So, the cost of energy losses for a period of one year can be written as shown in Eq. (4) and savings due to WG or DSTATCOM can be written as shown in Eq. (5):

$$\text{CEL} = (\text{Total Real Power loss}) * (39, 125.664)\ \text{INR} \tag{4}$$

$$\text{Savings} = \text{CEL} - \text{CEL}' \tag{5}$$

where CEL = Cost of Energy losses without WG, and CEL' = Cost of Energy losses with WG or DSTATCOM.

Cost of WG depends on wind velocity, installation, operation, maintenance, and tax. Cost of WG varies according to manufactures but the specific cost of WG for more than 200 KW rated power can be taken as 85,600 INR/kW [12]. Per unit cost of Electricity for a wind power plant is very less as compared to other sources of electricity. Wind energy cost per unit is the ratio of accumulated value cost (AVC) to the total energy generated ($\text{E}_{out} = 8760 P_r C_F$) by WG which is given by Eq. (6):

$$C = \frac{\text{AVC}}{E_{\text{out}}} = \frac{C_I}{8760n}\left(\frac{1}{P_r C_F}\right)\left\{1 + m\left[\frac{(1+I)^n - 1}{I(1+I)^n}\right]\right\} \quad (6)$$

where C_I = total initial investment cost, C_F = Capacity factor, P_r = Rated power, m = annual operation and maintenance cost, n = lifetime (in years), and I = real discount rate.

2.5 LSF for WG and DSTATCOM

LSF is a sensitivity factor for the allocation of WG and DSTATCOM as WG is used for real power injection; so LSF for allocation of WG depends on real power load demand as shown in Eq. (7). DSTATCOM decides reactive power injection, so LSF for allocation of DSTATCOM depends on reactive power load demand as shown in Eq. (8);

$$\text{LSF}(k, k+1) = \frac{\text{dP}_{\text{lineloss}}}{\text{dP}_{k+1,\text{eff}}} = \frac{2P_{k+1,\text{eff}} R_{k,k+1}}{|V_{k+1}|^2} \quad (7)$$

$$\text{LSF}'(k, k+1) = \frac{\text{dQ}_{\text{lineloss}}}{\text{dQ}_{k+1,\text{eff}}} = \frac{2Q_{k+1,\text{eff}} X_{k,k+1}}{|V_{k+1}|^2} \quad (8)$$

where $P_{k+1,\text{eff}}, Q_{k+1,\text{eff}}$ is real and reactive power load at the bus $(k+1)$ $R_{k,k+1}, X_{k,k+1}$ is resistance and reactance between bus (k) and $(k+1)$, and V_{k+1} is voltage of bus k.

3 DSTATCOM

DNOs are essential for the effective governance of the power system on the distribution side. Previously, Shunt capacitors were used in T&D lines for the compensation of the reactive power in the radial distribution system (RDS). Resonance and bulky weight are drawbacks of a shunt capacitor. So to resolve drawbacks, Static Compensators (STATCOMs) are used to compensate the reactive power requirements in T&D lines in order to manage the reactive power continuously [9].

STATCOM is a shunt Flexible AC Transmission Systems (FACTS) device that can be used in T&D lines. When it is used on the distribution side, it is known as DSTATCOM. It is a device which is connected in parallel with a network to inject or absorb the reactive and active current into the buses of a performance-oriented congestion (PCC) network. With the help of a DSTATCOM, real and reactive powers can also be exchanged. But in the proposed paper, DSTATCOMs are used to inject reactive power which improves voltage profile, efficiency, and reliability and reduces energy losses of the distribution system. Within a short period of time,

DSTATCOM injects reactive power to compensate parameters of sensitive buses, and active power injection should be done by WG. The modeling effects of DSTATCOM for the compensation of power quality of sensitive load problems are considered and dynamic effects of DSTATCOM will be studied in a short period of time. The energy storage system can't inject active power into the system for a long time due to the limited capacity and cost of energy storage systems [13, 14]. Usually, DSTATCOM injects suitable compensation current to the point of load connection to the network at steady state and faulty conditions or in case of heavy load conditions. Due to this, the bus voltage and T&D losses will regulate and can be placed in the allowed range. The major components of a DSTATCOM device are coupling transformer, DC-link capacitor, PWM control strategy, AC filters, and inverter modules. Reactive power depends on voltage and current, so DSTATCOMs can work in two different modes that are voltage and current control modes.

4 Wind Speed with Different Geographical and Environmental Conditions

Wind is non-uniform for different geographical locations and environmental conditions. Due to variation in speed, WG generates variable electrical power. Due to this power quality, transient problems occur. But expected wind speed at different locations can be determined from different weather experts and previous data. DNO can predict wind speed by using an artificial neutral network [15]. Type of WG decides power quality characteristics. Wind generators use sensitive power converters which require monitoring, and FACTS devices use power electronics converters. Wind force or wind speed per day in Mumbai from January 2019 to December 2019 as per the online weather report is given in Table 1.

Table 1 Wind force per day

Wind force per day (January 2019–December 2019)						
Month	Jan	Feb	Mar	Apr	May	Jun
[Km/h]	7.4	8.2	9.0	10.1	11.1	13.0
Data availability	100	89	74	100	100	100
Month	Jul	Aug	Sep	Oct	Nov	Dec
[Km/h]	13.7	15.4	12.0	7.8	7.3	7.3
Data availability	100	100	100	100	100	100
Averaged value (January 2019–December 2019): 10.2 km/h						

5 Load Modeling

As the integration of renewable energy and smart energy metering is increasing, the importance of load modeling in power systems is increasing day by day [16]. In the power system, load may be of different types that can affect the power system differently. It is required to know the variation of power demand with voltage for better performance of the power system. Depending upon the voltage, load can be classified into four parts—these are CP load, CI load, CZ load, and the combination of these three types of load which is known as ZIP load [17]. In the case of constant power load, active and reactive powers are independent of voltage and that load is always constant. In the proposed paper, it is the base case when the power load is taken as the same as load (P_i, Q_i) for thePortuguese 94-bus RDS as shown in Eqs. (9) and (10):

$$P = P_i \tag{9}$$

$$Q = Q_i \tag{10}$$

In constant current load, active and reactive powers are proportional to the voltage of an individual bus. In the proposed paper, active and reactive power loads at any bus is taken as actual power of that bus multiplied by voltage of that bus as shown in Eqs. (11) and (12):

$$P = P_i V \tag{11}$$

$$Q = Q_i V \tag{12}$$

In constant impedance load, active and reactive powers are proportional to the square of the voltage of an individual bus. In the proposed paper, active and reactive power loads at any bus is taken as actual power (P_i and Q_i) of that bus multiplied by the square of the voltage of that bus as shown in Eqs. (13) and (14):

$$P = P_i V^2 \tag{13}$$

$$Q = Q_i V^2 \tag{14}$$

Most of the loads can be represented as a combination of these loads; that load model is also called ZIP load. Z stands for impedance, I represents current, and P refers to power [18]. The polynomials for active and reactive powers are given in Eqs. (15) and (16):

$$P = P_i[0.333V^2 + 0.333V + 0.333] \tag{15}$$

$$Q = Q_i[0.333V^2 + 0.333V + 0.333] \quad (16)$$

6 Allocation of WG Using BFOA

BFOA is an algorithm which is based on the nature-inspired optimization algorithm. BFOA estimates objective function; in the proposed paper, minimum Real power loss function is used as an objective function as shown in Eq. (17). Locomotion is achieved by a set of tensile flagella during foraging of real bacteria. There are two operations in this algorithm, the first is 'tumble' and the other is 'swim'.

After each iterative step of the program, as the execution proceeds and leads to better fitness of power loss function. The parameters to be optimized represent coordinates of the bacteria. One bacterium is present at each point after each progressive step the bacteria move to new positions. At each position, the power loss function is calculated and then with this calculated value of power loss function, further movement of bacteria is decided by decreasing the direction of the power loss function. So, this will lead to a position with minimum power loss [19].

Before implementation of BFOA for calculation of the optimal time of operation of wind generation with minimum power loss, it is required to know about the buses for allocation of wind. And it can be calculated by using the power loss sensitivity factor which can be given by Eqs. (7) and (8). LSF can be calculated from the load flow analysis, and values of LSF are arranged in descending order for all buses of the given system. Descending order of LSF decides the sequence in which the bus is selected for WG installation and DSTATCOM [20]. According to LSFs, the three most sensitive buses are 90, 40, and 88 for the Portuguese 94-bus RDS for WG. The three most sensitive buses are 58, 40, and 88 for the Portuguese 94-bus RDS for DSTATCOM. The optimal size of WG at the candidate buses is calculated by using BFOA.

Steps to perform BFOA:

Step 1. Set parameters.

Number of wind turbines (p) = 3, since three most sensitive buses are selected,
Total number of wind power samples (S) = 100,
Swimming length (N_s) = 4,
Elimination-dispersal probability (P_{ed}) = 0.5, and
$C(i)$: Size of the array for the step taken in the random direction specified by the tumble = (0.05*ones(S,1)) Allocation.

Step 2. Start elimination-dispersal loop (l = 1:2).

Step 3. Start reproduction loop (k = 1:4).

Step 4. Start chemotaxis loop (j = 1:4).

- For $i = 1{:}S$, take a chemotactic step for bacteria i as follows.
- Compute Real power loss fitness function, PL(i, j, k, l):

$$\mathrm{PL}(i, j, k, l) = \mathrm{Min(Real\ power\ losses)} \tag{17}$$

- Save this value in $\mathrm{PL}_{\mathrm{last}} = \mathrm{PL}(i, j, k, l)$ to find better value via a run.
- Tumble to generate a random vector with each element $\Delta_m i$, $m = 1{:}S$, a random number on $[-1, 1]$.
- Move:

$$P^i(j+1, k, l) = P^i(j, k, l) + C(i, k)\frac{\Delta(i)}{\sqrt{\Delta^T(i)\Delta(i)}} \tag{18}$$

Equation (18) results in a step of size $C(i, k)$ in the direction of the tumble for bacteria i.

- Calculate $PL(i, j+1, k, l)$ with $P^i(j+1, k, l)$.
- Swim:

(i) let $m = 0$ (counting variable of swim length).
(ii) while $m < N_s$ (if not climbed down too long).
  (a) let $m = m + 1$.
  (b) if $PL(i, j + 1, k, l) < PL_{\mathrm{last}}$, let $\mathrm{PL}_{\mathrm{last}} = \mathrm{PL}(i, j + 1, k, l)$, Then, another step of size $C(i, k)$ in the same direction will be taken as $P^i(j+1, k, l)$; use the newly generated $P^i(j+1, k, l) = \mathrm{P}^i$ to compute the new $PL(i, j + 1, k, l)$.
  (c) else let $m = N_s$.

- Go to next bacterium, i.e. $(i + 1)$: if i is not equal to S then go to (b) to process the next bacteria.

Step 5. If $j < 4$, go to Step 3. In this case, continue chemotaxis since the life of the bacteria is not over.

Step 6. Reproduction.

- For the given k and l, and for $i = 1{:}S$, let

$$\mathrm{PL}^i_{\mathrm{health}} = \sum_{j=1}^{N+1} \mathrm{PL}(i, j, k, l) \tag{19}$$

Equation (19) shows the health of the bacteria. Sort bacteria in order of ascending values ($\mathrm{PL}_{\mathrm{health}}$).

- The S_r bacteria with the highest $\mathrm{PL}_{\mathrm{health}}$ values die, and the other S_r bacteria with the best values split, and the copies that are made are placed at the same location as their parent.

Step 7. If $k < 4$ go to Step 2. In this case, the number of specified reproduction steps is not reached; start the next generation in the chemotactic loop.

Step 8. Elimination-dispersal: for $i = 1{:}\ S$, with probability p_{ed}, eliminate and disperse each bacterium, which results in keeping the number of bacteria in the population constant. To do this, if a bacterium is eliminated, simply disperse one to a random location on the optimization domain. If $l < 2$, then go to Step 2, otherwise, end [21].

7 Results and Discussion

7.1 *Input Data for Different Bus Systems*

The proposed paper uses the Portuguese 94-bus RDS. Data for this bus system is taken from reference papers [22, 23].

7.2 *LSF at Different Buses*

LSF helps to select sensitive buses that are suitable for the allocation of WG and DSTATCOM. LSF for different buses for the 94-bus system is as shown in Fig. 1a for WG. LSF for different buses for the 94-bus system is as shown in Fig. 1b, respectively, for DSTATCOM.

7.3 *Bus Voltages at Different Loading*

Load is variable throughout the year so the proposed paper is also showing bus voltage for different loading (20% overloading, 20% underloading) at all buses as shown in Fig. 2. From the figure, it is clear that the voltage profile is better for 20%

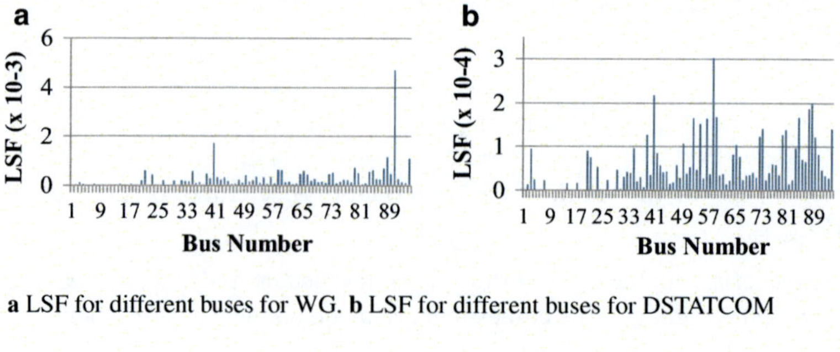

Fig. 1 **a** LSF for different buses for WG. **b** LSF for different buses for DSTATCOM

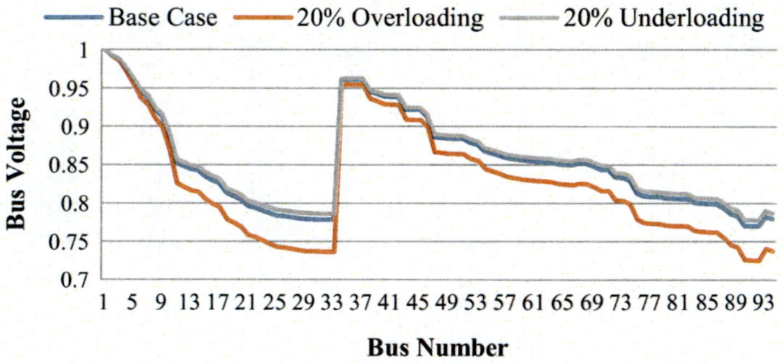

Fig. 2 Per unit bus voltage for different loading for 94-bus system

underloading than 20% overloading. So DNO has to manage the load demand by using WG and DSTATCOM.

As shown in the diagram, there is a very small difference between the base case and 20% underloading as compared to the difference between the base case and 20% overloading.

7.4 *Different Types of Loads*

As discussed earlier, there may be different types of loads in the power system. Figures 3, 4, 5, and 6 show bus voltage for different types of loads, real power loss for different cases, bus voltage with WG, and bus voltage with DSTATCOM for the Portuguese 94-bus system, respectively.

Fig. 3 Per unit bus voltage for different types of load without wind

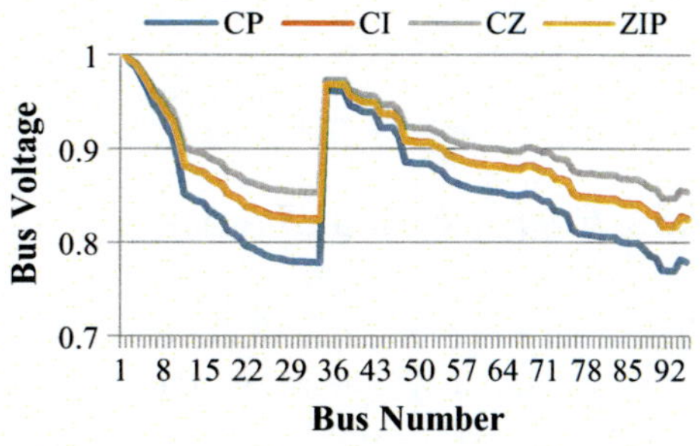

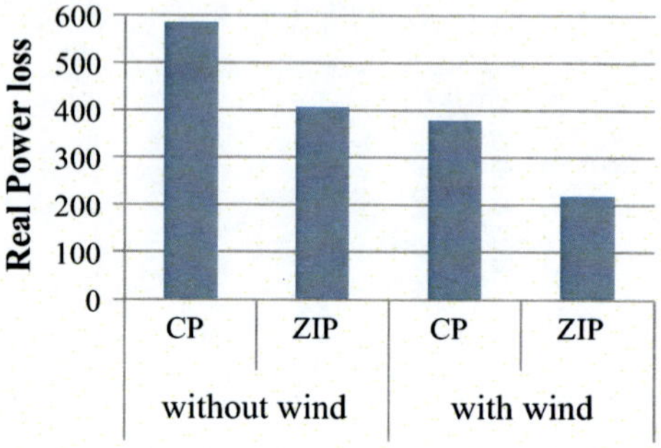

Fig. 4 Real power losses (in kW) for different types of load

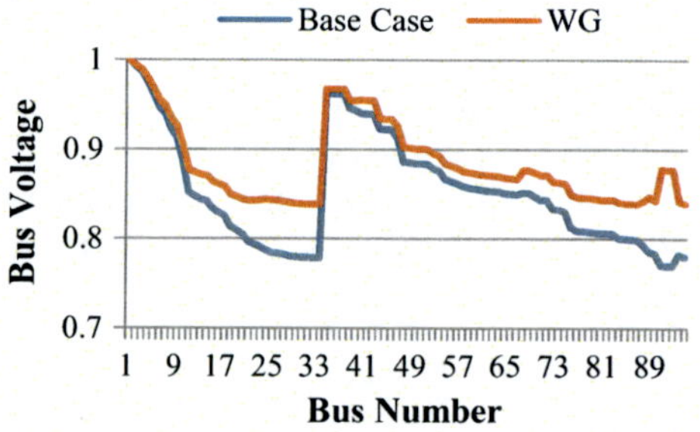

Fig. 5 Per unit bus voltage with WG

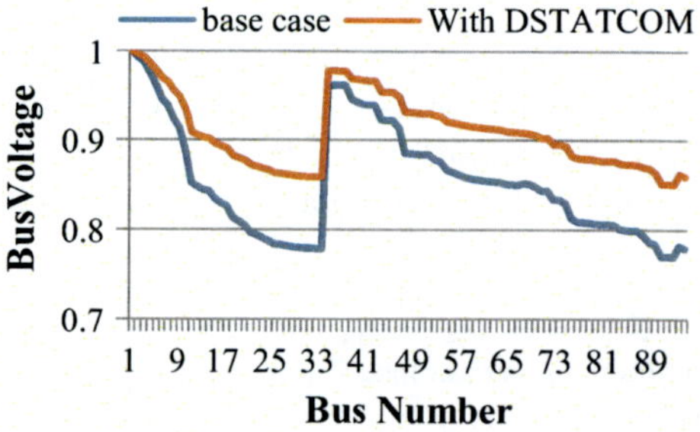

Fig. 6 Per unit bus voltage with DSTATCOM

7.5 Energy Cost and Savings

The earlier proposed paper shows the better voltage profile and less power losses with the allocation of WG and DSTATCOM. But DNO should ensure that allocation of WG and DSTATCOM is economical. Figures 7 and 8 show energy cost and savings for wind and energy cost for DSTATCOM in the Portuguese 94-bus system.

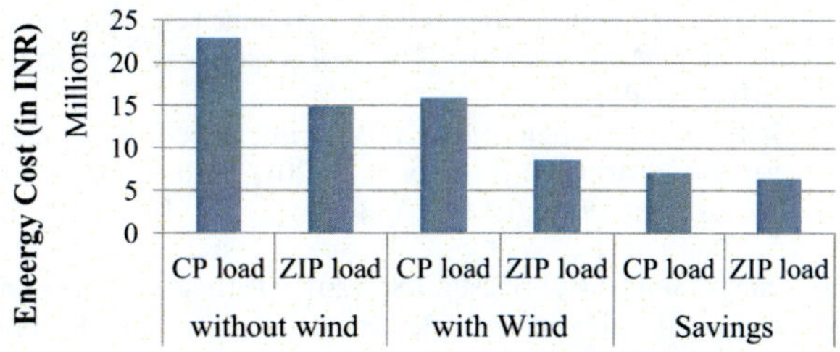

Fig. 7 Energy cost with and without WG (in INR)

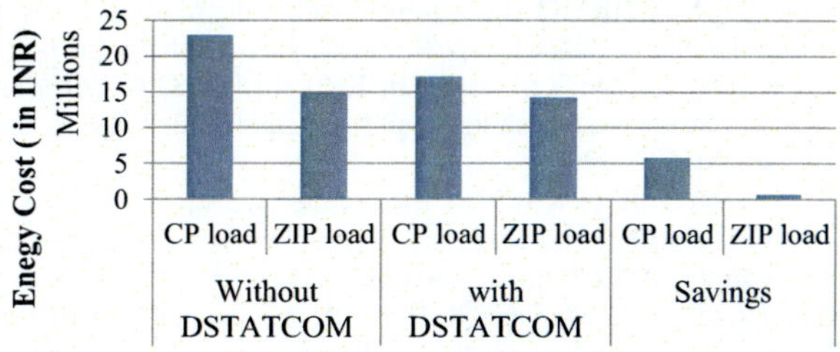

Fig. 8 Energy cost with and without DSTATCOM (in INR)

8 Conclusion

In this proposed paper, the allocation of WG and DSTATCOM with different types of load proposed using BFOA has been studied. Base value of voltage and power is taken as 12.66 kV and 10 MVA, respectively. Firstly, the most suitable bus for allocation of WG and DSTATCOM is selected from the load sensitivity factor. After that, the behavior of bus voltage at 20% overloading and underloading at each bus is analyzed. The behavior of bus voltage for different types of load is discussed. WGs are playing a key role in power loss minimization. By placing WG and DSTATCOM, power losses reduce and the voltage profile gets improved. It is necessary for DNO that allocation of WG and DSTATCOM is economic, so the proposed paper is also showing a large amount of savings in INR.

References

1. Ali S, Mutale J (2015)Reactive power management at transmission/distribution interface. In: 2015 50th international universities power engineering conference (UPEC), Stoke on Trent, pp 1–6. https://doi.org/10.1109/UPEC.2015.7339816
2. Payasi RP, Singh AK, Singh D (2012) Planning of different types of distributed generation with seasonal mixed load models. Int J Eng Sci Technol 4. https://doi.org/10.4314/ijest.v4i1.13S
3. Bahrman MP, Johansson JG, Nilsson BA (2003)Voltage source converter transmission technologies: the right fit for the application. In: 2003 IEEE power engineering society general meeting (IEEE Cat. No.03CH37491), vol. 3. Toronto, Ont, pp 1840–1847. https://doi.org/10.1109/PES.2003.1267441

4. Loganathan N, Mayurappriyan P, Lakshmi K (2018) Smart energy management systems: a literature review. In: MATEC web of conferences, vol 225. https://doi.org/10.1051/matecconf/201822501016
5. Seker AA, Hocaoglu, MH (2013) Artificial bee colony algorithm for optimal placement and sizing of distributed generation. In: 2013 8th international conference on electrical and electronics engineering (ELECO), Bursa, pp 127–131. https://doi.org/10.1109/ELECO.2013.6713817
6. Jain N, Singh SN, Srivastava SC (2010) Particle swarm optimization based method for optimal siting and sizing of multiple distributed generators. In: Proceedings of 16th national power systems conference, pp 669–674
7. Afzalan M, Taghikhani MA, Sedighizadeh M (2012)Optimal DG placement and sizing with PSO&HBMO algorithms in radial distribution networks. In: 2012 Proceedings of 17th conference on electrical power distribution, Tehran, pp 1–6
8. Yuvaraj T, Ravi K, Devabalaji KR (2015) DSTATCOM allocation in distribution networks considering load variations using bat algorithm. Ain Shams Eng J 8. https://doi.org/10.1016/j.asej.2015.08.006
9. Yuvaraj T, Ravi K (2018) Multi-objective simultaneous DG and DSTATCOM allocation in radial distribution networks using cuckoo searching algorithm. Alex Eng J 57(4):2729–2742. ISSN 1110-0168. https://doi.org/10.1016/j.aej.2018.01.001
10. Manyonge A, Manyala R, Onyango F, Shichika J (2012) Mathematical modelling of wind turbine in a wind energy conversion system: power coefficient analysis. Appl Math Sci 6:4527–4536
11. Malik MZ, Kumar M, Soomro AM, Baloch MH, Farhan M, Gul M, Kaloi GS (2020) Strategic planning of renewable distributed generation in radial distribution system using advanced MOPSO method energy reports 6:2872–2886. ISSN 2352-4847. https://doi.org/10.1016/j.egyr.2020.10.002
12. Mohammadi K, Mostafaeipour A, Dinpashoh Y, Pouya N (2014) Electricity generation and energy cost estimation of large-scale wind turbines in jarandagh, Iran J Energy 1–8. https://doi.org/10.1155/2014/613681
13. Iqbal F, Khan M, Siddiqui A (2017) Optimal placement of DG and DSTATCOM for loss reduction and voltage profile improvement. AEJ—Alex Eng J. https://doi.org/10.1016/j.aej.2017.03.002
14. Gupta A, Kumar A (2019). Deployment of distributed generation with D-FACTS in distribution system: a comprehensive analytical review. IETE J Res 1–18. https://doi.org/10.1080/03772063.2019.1644206
15. Malik H, Savita (2016) Application of artificial neural network for long term wind speed prediction. In: 2016 conference on advances in signal processing (CASP), Pune, pp 217–222. https://doi.org/10.1109/CASP.2016.7746168
16. Arif A, Wang Z, Wang J, Mather B, Bashualdo H, Zhao D, Load modeling—A review. United States. https://doi.org/10.1109/TSG.2017.2700436
17. Hatipoglu K, Fidan I, Radman G (2012)Investigating effect of voltage changes on static ZIP load model in a microgrid environment. In: 2012 North American power symposium (NAPS), Champaign, IL, pp 1–5. https://doi.org/10.1109/NAPS.2012.6336407
18. Sairam S, Daram S, Venkataramu PS, Nagaraj M (2018) Analysis of ZIP load modeling in power transmission system. Int J Control Autom 11:11–24. https://doi.org/10.14257/ijca.2018.11.7.02
19. Imran A, Kowsalya M (2013) Optimal size and siting of multiple distributed generators in distribution system using bacterial foraging optimization. Swarm Evol Comput 15. https://doi.org/10.1016/j.swevo.2013.12.001
20. Teng J-H (2003) A direct approach for distribution system load flow solutions. IEEE Trans Power Deliv 18(3):882–887. https://doi.org/10.1109/TPWRD.2003.813818
21. Chen H, Zhu Y, Hu K (2011) Adaptive bacterial foraging optimization. abstract and applied analysis. In: Abstract and applied analysis. https://doi.org/10.1155/2011/108269

22. Kashem MA, Ganapathy V, Jasmon GB, Buhari, MI (2000)A novel method for loss minimization in distribution networks. In: DRPT2000. International conference on electric utility deregulation and restructuring and power technologies. Proceedings (Cat. No.00EX382), London, UK, pp 251–256. https://doi.org/10.1109/DRPT.2000.855672
23. Malik MZ, Kumar M, Soomro AM, Baloch MH, Farhan M, Gul M, Kaloi GS (2020) Strategic planning of renewable distributed generation in radial distribution system using advanced MOPSO method. Energy Rep 6:2872–2886. ISSN 2352-4847. https://doi.org/10.1016/j.egyr.2020.10.002

Reactive Power Requirement for Operating Wind-Driven Micro Grid in the Presence of Several Proportions and Classes of Static Load

Nitin Kumar Saxena, Varun Gupta, Raj Shekhar Rajput, Ashwani Kumar, and Atma Ram Gupta

Abstract Squirrel Cage Induction Generators (SCIGs) are the most common electro-mechanical energy conversion devices used in wind-driven micro grids. In most of the available studies, researchers have focused on the (i) type of induction generators, (ii) reactive power compensation for induction generator's excitation, (iii) control techniques for voltage control, (iv) method of excitation using FACTS device, etc. In this paper, the authors tried to identify the effect of load rating as well as characteristics on reactive power demand in the presence of SCIG. It has been verified how the reactive power requirement for SCIG depends upon the proportions and classes of static load connected with it. Fixed Capacitor as static compensator is fulfilling the demand of reactive power for the system analyzed in this paper.

Keywords Squirrel cage induction generator · Static compensator · Static load · ZIP and exponential load mode · Reactive power compensation

1 Introduction

In decentralized or stand-alone plants such as micro-hydro plants and wind farms where locally available renewable resources are exploited for power generation, the induction generators are very popular. Decentralized power generation has been considered as a viable alternative to the power grid. This has become crucial in those areas where the erection of transmission lines is very difficult. By using appropriate technologies, such a system can be easily installed and maintained by local personnel [1]. These systems consist of prime mover, generator, controller and distribution system to feed local loads. Hydro electric, fossil fuels and nuclear energy are the

N. K. Saxena (✉) · V. Gupta · R. S. Rajput
KIET Group of Institutions, Delhi-NCR, Ghaziabad, UP, India

V. Gupta
e-mail: varun.gupta@kiet.edu

A. Kumar · A. R. Gupta
National Institute of Technology Kurukshatra, Haryana, India
e-mail: argupta@nitkkr.ac.in

A. Kumar et al. (eds.), *Renewable Energy Towards Smart Grid*, Lecture Notes in Electrical Engineering 823, https://doi.org/10.1007/978-981-16-7472-3_3

popularly used conventional energy resources for the generation of electrical power. However, these energy resources are seriously damaging the earth's environment, and these fossil fuels cannot be regenerated, and their cost will be very high. Hence, renewable energy resources like solar and wind will prove to be a great alternative [2]. Since the cost of wind power is coming down very fast to the point where its cost is almost equal to conventional electric power generation [3], the use of induction machines as a generator is becoming very popular for renewable resources. Reactive power consumption accompanied by poor voltage regulation under varying speed are the two major limitations faced by induction generators, but the development of a static power converter has facilitated the control of output voltage of the induction generator. The reactive power required can be supplied by static capacitor bank, static VAR compensator (SVC), static compensator (STATCOM), etc. [4].

In a decentralized power system, reactive power procurement involves the additional payment from the users in lieu of reactive power supplied by the distributors or gencos as one of the ancillary services out of total six ancillary services. The load rating and characteristics will ask for reactive power from the compensating source connected with SCIG. Also, the behavior of SCIG will be different for different loads depending upon the load rating and characteristics.

In the literature, load is broadly categorized as (i) static load, (ii) dynamic load and (iii) composite load [5]. Out of these, static load used in most of the steady state studies, as it can be modeled mathematically either as a ZIP model or an Exponential model [6]. This paper mainly focuses on how the fixed capacitor bank participates to supply reactive power to the generator for different static loads using the exponential model.

The main contribution of this paper can be summarized into the following sections: in Sect. 1, an introduction to the theme of this present work is given. In Sect. 2, induction generators in terms of their type, voltage build-up requirement, static compensation and cost are discussed. In Sect. 3, the static load model is illustrated. In Sect. 4, stepwise algorithms for this work are presented. In Sect. 5, results and discussions are presented. In Sect. 6, the paper is concluded. Required parameters and flowcharts are presented with the help of figures and tables. Relevant references are also cited at the end of this paper.

2 Descriptions about Induction Generators

Induction generators are either wound rotor type induction generators or squirrel cage induction generators based on rotor configuration. Generating schemes chosen for the operation of stand-alone or grid on induction generator depend on the use of prime movers. They can be classified as [7–9]

Constant Speed Constant Frequency, CSCF;
Variable Speed Constant Frequency, VSCF;
Variable Speed Variable Frequency, VSVF.

Constant Speed Constant Frequency: The prime mover speed is kept constant by following generator characteristics continuously. We can operate an induction generator at a slip of 1–5% above the synchronous speed on an infinite bus. Also, induction generator operation, control and maintenance are easy, and it does not have any synchronization problem as in the case of a synchronous generator.

Variable Speed Constant Frequency: With variable speed, wind-operated electric system executes higher output in low as well as high wind speeds. The result is higher annual energy yields per rated install capacity. Under variable speed operations, this gain is exhibited by both horizontal and vertical axis wind turbines.

Variable Speed Variable Frequency: The performance of a synchronous generator can be affected by variable prime mover speed. The generator scheme focused on in this paper is constant speed constant frequency.

2.1 Reactive Power for Voltage Build-Up in SEIG

The induction machine works as an induction generator over synchronous speed, i.e. at negative slip mode, and the induced torque polarity gets reversed at this moment. From the analysis in Ref. [10] with the help of a circle diagram of the induction generator, it has been explained that the machine draws a lagging current with respect to voltage by more than 90° in the negative slip region. This means that the machine supplies the real power and draws the reactive power. Voltage can be built with the help of external capacitor-based excitation. The induction generator can work in two modes: grid-connected and isolated-connected modes.

1. Grid-connected mode: In this mode, it draws reactive power by a grid. While drawing reactive power from the grid, it will place a burden on the grid [7, 9].
2. Isolated-connected mode: In this mode, the induction generator draws reactive power from a capacitor bank. Hence, a suitable capacitor bank is connected across the generator terminal. This process is called capacitor self-excitation, and the induction generator is called self-excited induction generator SEIG [9].

2.2 Static Compensation Using Fixed Capacitor Bank

Reactive power demand mainly originates from the inductive load connected to the system and for the excitation of the induction generator. The inductive loads are mostly electromagnetic circuits of electric motors, induction furnaces, fluorescent lightings, air conditioners, etc. The reactive power should be supplied adequately, so that the voltage is maintained constant throughout the system. The most popular and easiest method for supplying reactive power is through fixed capacitor banks. Since it is a static device, this method of reactive power compensation is called static compensation. A fixed or static capacitor is used to supply only a fixed amount of reactive power to the system [11]. Static capacitors are of two types: (i) shunt capacitor and (ii) series capacitor.

Shunt Capacitor: The shunt capacitor draws a fixed amount of leading current which superimposes the load current and thus a reactive component of load reduces. The advantages of using a shunt capacitor are

a. Line current of the system reduces.
b. Voltage level of the load improves.
c. System losses reduce.
d. Power factor of the source current improves.
e. Capital investment per megawatt of the load reduces.

Series Capacitor: Series capacitor does not have any control over the flow of current. These are the capacitors which are connected in series with the load. The reactive capacitance of the series capacitor reduces the effective reactance of the line by neutralizing its inductive reactance. Therefore, the voltage regulation of the system gets improved. But during a fault condition, the capacitor voltage may rise up to 15 times the rated value, which is its major disadvantage [12]. Due to this, we use a shunt capacitor for supplying reactive power.

2.3 Cost of Static Capacitor

Compensation cost from static compensator, i.e. capacitor is formularized as a linear function with supplied reactive power. A static capacitor's fixed cost is expressed as [4, 13]

$$\text{SCFC} = r \times Q_{\text{FC}} \text{ in } \$/h \tag{1}$$

where r is the depreciation rate, and Q_{FC} (in MVAR) is the reactive power from the static capacitor.

The fixed cost for 15 years' life span is given as [4, 13]

$$\text{SCFC} = 0.132 \times Q_{\text{FC}} \text{ in } \$/h \tag{2}$$

3 Understanding of Static Load Model

Static load depends only on the present value of voltage and frequency. It is independent of their previous values. Static load model is used for steady-state calculations and so is a better choice for steady-state simulation conditions in a power system

[14]. Mathematically, a static model of load can be modeled in different way as described in subsection below.

3.1 ZIP Model or Polynomial Type

In this model, the relation between voltage and power is represented in the form of a polynomial equation expressed as

$$P = P_0[k_1 + k_2(V/V_0) + k_3(V/V_0)^2] \tag{3}$$

$$Q = Q_0[k_4 + k_5(V/V_0) + k_6(V/V_0)^2 \tag{4}$$

Bus voltage, real and reactive powers at any time are quantified as V, P and Q, respectivelys while initial operating powers are P_0 and Q_0, respectively. As this model consists of constant impedance (Z), constant current (I) and constant power (P) component, it is referred to as the ZIP model too [15]. The proportion of each component is defined by the model parameter coefficient as in the equation of P and Q.

3.2 Exponential Type

The ZIP model explains the characteristics of the load model individually like constant impedance (Z), constant current (I) and constant power (P) load [6]. But in the real world, loads are unpredictable and so it is not possible to identify them using the ZIP load model. Hence, an exponential formulation is used to quantify the most accurate load model.

$$P = P_0(V/V_0)^{\mathrm{np}} \tag{5}$$

$$Q = Q_0(V/V_0)^{\mathrm{nq}} \tag{6}$$

where P, Q, V, P_0 and Q_0 have the same nomenclature as in the last section. np and nq are parameters of this model referred to as exponential coefficients. With these exponents, constant power, constant current or constant impedance characteristics can also be represented by putting them equal to 0, 1 and 2, respectively. To model the load in the Simulink model, a dynamic load block is available in MATLAB. However, several types of loads can be developed simply by fixing the values of np and nq according to the load characteristics. In Refs. [16, 17], exponential model coefficient parameters for various industrial loads are given as illustrated in Table 1.

Table 1 Exponential coefficient parameters for various industrial loads

S. no.	Load component	np	nq
1	Air conditioner	0.5	2.50
2	Resistance space heater	2.00	0.00
3	Fluorescent lighting	1.00	3.00
4	Pump, fan, other motors	0.08	1.60
5	Large industrial motors	0.05	0.50

4 Development of Simulink Model

As explained in the preceding sections, the induction machine is used to work as an induction generator first, and then several loads with ratings and characteristics are examined on it in the presence of static/fixed capacitor as static compensation. For this, the whole work is further classified into some subsections, and every step is elaborated with proper explanations with the help of a flowchart, algorithm and figures.

4.1 Selection of Induction Machine

For this work, a 4 kW squirrel cage induction machine is used. Table 2 gives the important parameters that are evaluated with the first reference method as explained in Ref. [5] and mechanical parameters as in Ref. [18].

Table 2 Parameters of induction machine used in Simulink block

S. No.	Parameters	Value
1	Reference frame	Synchronous
2	Nominal power	4000/0.9
3	Voltage (line-line)	400
4	Frequency	50
5	Stator resistance	1.405 Ω
6	Stator inductance	0.005839 H
7	Rotor resistance	1.395 Ω
8	Rotor inductance	0.005839 H
9	Mutual inductance	0.1 H
10	Moment of inertia	0.0131 kg m^2
11	Friction factor	0.002985 N m s
12	Pole pairs	2

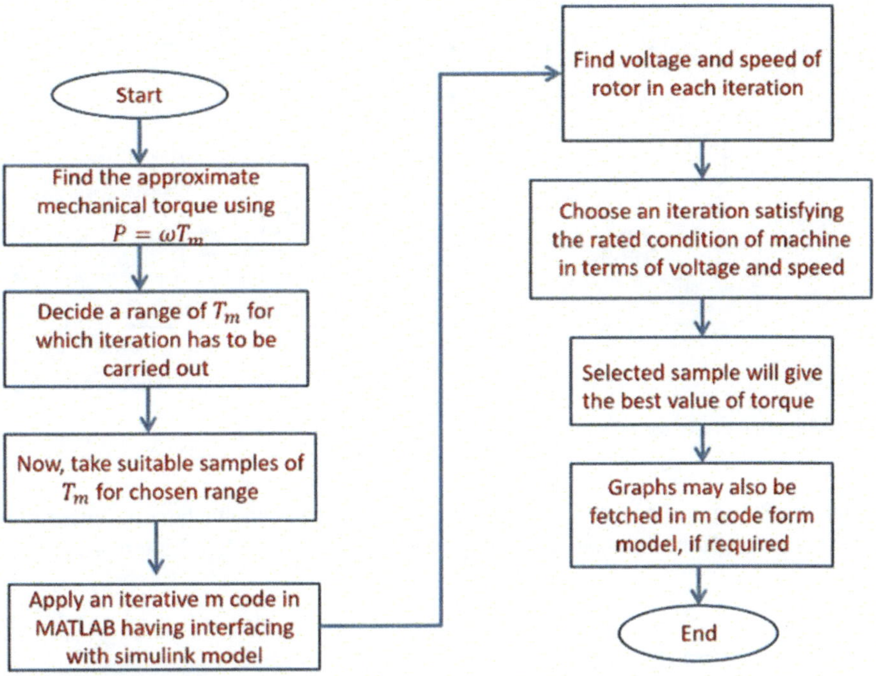

Fig. 1 Flowchart for torque estimation

4.2 Estimation of Torque

It has already been explained that an induction machine will work as an induction generator if negative torque would be applied to it as input of the machine. A simple expression $P_{\text{mech}} = \omega T$ can be used to get a rough idea about the approximate value of torque. Now, an iteration procedure may be adopted to find the actual value of this machine for getting a specified speed. It is assumed that the induction machine is operated at 4% slip and so, its rated speed would be 1500 rpm for this 2-pole pair machine. As a generator, its speed would be 1560 rpm at 4% slip. The flowchart depicting the stepwise procedure for the estimation of torque (T) is shown in Fig. 1.

4.3 Development of Simulink Model Including Load and Compensator

For this study about load effect on reactive power demand from SCIG, the induction machine model is connected with the load and shunt capacitor model. For the full idea about the model developed in MATLAB 2016a software, Fig. 2 is designed as below.

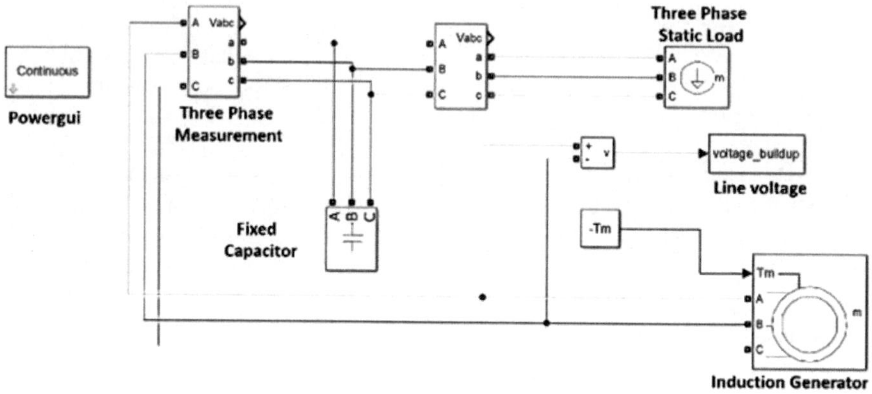

Fig. 2 Simulink model for SCIG with load and compensator

All the measurement blocks like scope, display, etc. are not shown here because their connections might be changed according to individual researchers' interests.

4.4 Estimation of Reactive Power

Reactive power required for the excitation of the induction generator and to supply the load decides the capacity of the Static Capacitor Bank. In this study, the authors are elaborating how reactive power is changing with load applied to the induction generator. This estimation can be performed as shown in the flowchart in Fig. 3.

5 Results and Discussion

As discussed earlier, this paper gives a method to estimate the reactive power demand for different load conditions so that the overall behavior of generators with different load demands can be investigated. Table 1 gives an idea about the load categories. It can be understood from Table 1 how the load proportions and classes have changed with the value of np and nq. A Simulink model presented in Fig. 2 shows the complete Simulink model except for output blocks. An induction machine of 4 kW is used to run as an induction generator at 4% negative slip. The parameters for induction machines used in the Simulink model are given in Table 2, and the procedure to investigate the required mechanical input in terms of the negative value of torque at the induction generator's shaft is explained with the help of the flowchart as in Fig. 1. Finally, reactive power is calculated for a variety of loads at several ratings and characteristics. For this, a code is generated in the MATLAB 2016a environment. This m code has an iterative program to evaluate the optimum value of reactive power

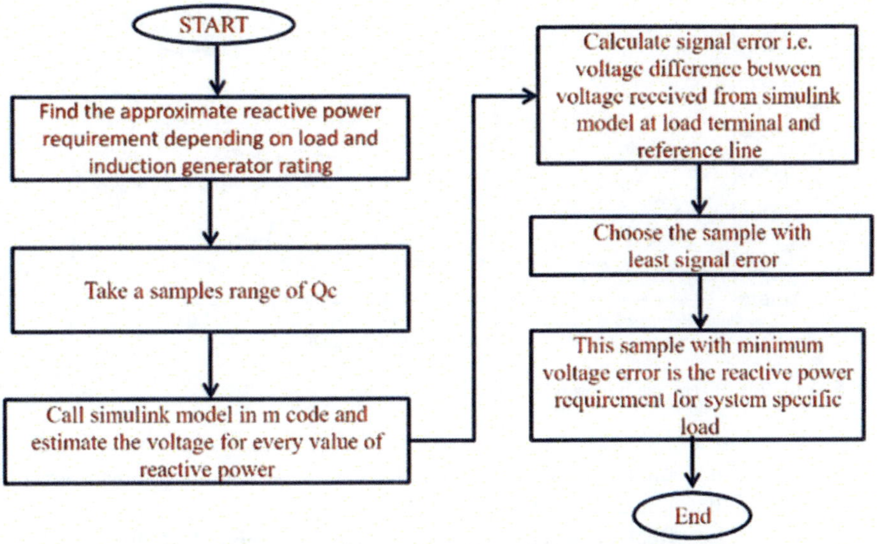

Fig. 3 Flowchart for reactive power estimation for specific load parameters

required for specific rating and specification of load. The Simulink model presented in Fig. 2 is called again and again in m code developed for getting the system performances and finally, the m code chooses an optimum solution of reactive power for this particular load.

Figures 4 and 5 represents the reactive power demand from static compensator when, either np or nq, only one parameter is being changed for the system at full load rating. It is quite interesting to analyze that reactive power demand is increasing with an increase in np in Fig. 4. In Fig. 5, it can also be seen that till nq = 0.5, reactive

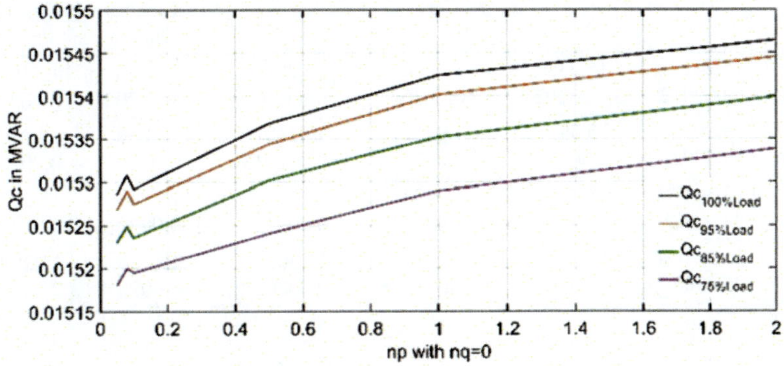

Fig. 4 Variation of reactive power with np

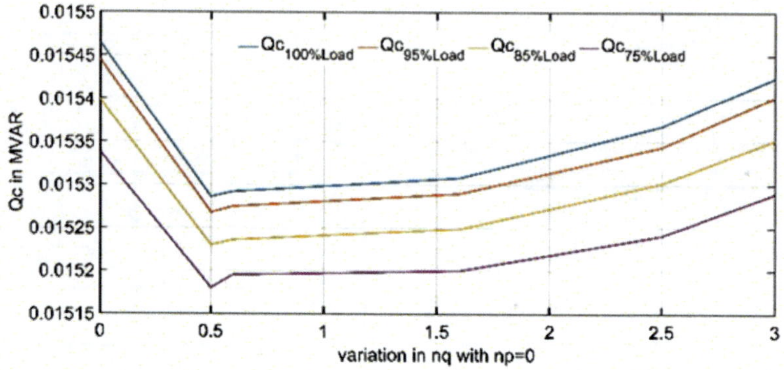

Fig. 5 Variation of reactive power with np

power demand is decreasing; after that, reactive power demand is almost constant till nq = 2 and then it is again increasing.

To analyze the effect of several loads categorized in Table 1 at different load percentages, a tabular comparison is presented in Table 3. Since the consumers have to afford the cost of reactive power as an additional service required from distribution companies or generating companies in case of a decentralized system, the cost of reactive power production is also given in this table. This table helps to understand the behavior of different consumers' tariffs on the basis of their load category.

Table 3 Pattern of reactive power demand and its cost of compensation

	Air conditioner	Resistance space heater	Fluorescent lighting	Pump, fan, other motors	Large industrial motors	Small industrial motors
Load %	Qc (in VAR) Cost (in $/hr)	Qc (in VAR) Cost (in $/hr)	Qc (in VAR) Cost (in $/hr)	Qc (in VAR) Cost (in $/hr)	Qc (in VAR) Cost (in $/hr)	Qc (in VAR) Cost (in $/hr)
75% of load	15,241 0.002012	15,338 0.002024	15,290 0.00202	15,200 0.00201	15,180 0.002004	15,195 0.00200
85% of load	15,302 0.00202	15,399 0.002032	15,352 0.002026	15,249 0.002013	15,230 0.00201	15,236 0.00201
95% of load	15,344 0.002025	15,445 0.002038	15,402 0.002033	15,290 0.00202	15,268 0.002015	15,275 0.002016
100% of load	15,368 0.00203	15,465 0.002041	15,424 0.002036	15,308 0.002021	15,286 0.002018	15,292 0.00202

6 Conclusions

The estimation of reactive power for excitation of SEIG has been done using an exponential load model for different types of commercial loads. Also, the cost in each case has also been calculated. From the results, it can be concluded that as load percentage increases, more and more reactive power is consumed by IG to maintain constant voltage and torque and so does the cost of the static capacitor. So, a fixed capacitor is more suitable to supply light load connected with a wind-driven micro grid for keeping the low cost of reactive power. This study can help in estimating the reactive power requirement on the basis of load behavior and characteristics. The distribution companies and generating companies in the case of decentralized power plants will be able to decide the tariff of individual consumers based on their load components and characteristics. In future studies, dynamic load conditions can also be examined with advanced compensating devices connected with distributing generators.

References

1. Bhamu S, Pathak N, Bhatti TS (2019) Power control of wind-biogas-PV based hybrid system. In: 2018 international conference on computing power and communication technology GUCON 2018, pp 421–426. https://doi.org/10.1109/GUCON.2018.8675057
2. Hinz F, Most D (2018) Techno-economic evaluation of 110 kV grid reactive power support for the transmission grid. IEEE Trans Power Syst 33:4809–4818
3. Bhamu S, Bhatti TS (2018) Automatic power control of a wind-hydro-grid based interconnected system for rural electrification. In: India international conference on power electronics IICPE 2018-Decem, pp 1–5
4. Saxena NK, Kumar A (2015) Electric power components and systems analytical comparison of static and dynamic reactive power compensation in isolated wind—Diesel system analytical comparison of static and dynamic reactive power compensation in isolated wind—Diesel, pp 37–41. https://doi.org/10.1080/15325008.2014.993777
5. Hajiabbas MP, Optimization of power system problems
6. Saxena NK, Sharma AK (2015) Estimation of composite load model with aggregate induction motor dynamic load for an isolated hybrid power system. Front Energy9
7. Taylor P, Singh B (2007) Electric machines & power systems induction generators-a prospective, pp 37–41
8. Osheiba AM, Rahman MA (1991) Performance analysis of self-excited induction and reluctance generators. Electr Mach Power Syst 19:477–499
9. Bansal RC (2005) Three-phase self-excited induction generators: an overview. IEEE Trans Energy Convers 20:292–299
10. Kothari DP, Nagrath IJ (2006) Electric machines. Tata-McGraw-Hill, India
11. Saxena NK, Kumar A (2016) Electrical power and energy systems reactive power control in decentralized hybrid power system with STATCOM using GA. ANN ANFIS Methods 83:175–187
12. Saqib MA, Saleem AZ (2015) Power-quality issues and the need for reactive-power compensation in the grid integration of wind power. Renew Sustain Energy Rev 43:51–64
13. Qu B, Zhuan X, Cui X (2014) Optimal sizing and allocation of fixed reactive power compensation. IFAC Proc 19:10778–10783
14. Kundur P (2006) Power system stability and control. Tata-Mcgraw-Hill, India

15. Saxena NK, Kumar A (2018) Electric power components and systems dynamic reactive power compensation and cost analysis for isolated hybrid power system dynamic reactive power compensation and cost analysis for isolated hybrid power system. Electr Power Compon Syst 45:2034–2049
16. Ram SS, Daram SB, Venkataramu PS, Nagaraj MS (2018) Analysis of ZIP load modeling in power transmission system. Int J Control Autom 11:11–24
17. Murty VVSN, Kumar A (2013) Comparison of optimal capacitor placement methods in radial distribution system with load growth and ZIP load model. Front Energy 7:197–213
18. Saxena NK, Kumar A (2019) Analytical approach to estimate mechanical parameters in induction machine using transient response parameters. Int Trans Electr Energy Syst 29:1–14

Renewable Energy Resource Availability and Supply Guide in India

Parveen Kumar, Manish Kumar, and Ajay Kumar Bansal

Abstract Renewable Energy Resources (RERs) due to their environmental friendliness and sustainability are adopted by the countries all over the world. As the increasing pressure due to cost and pollution, various agreements are signed by different countries like Kayota Summit and Paris Agreement. Wind, solar, biomass, hydro, geothermal are some of the numerous RERs which can be harnessed without compromising with environment. This paper puts forward the characteristics of different RERs across the world and particularly in India. Their share and increasing capacity are represented by different graphics. In the end, their comparison in terms of capacity and cost is also presented. India has a tremendous capacity of renewable power assets, and it has one of the most important programs within the world for deploying renewable power products and systems. Indeed, it's far the only USA within the world to have a ministry for renewable strength improvement, the Ministry of Non-Conventional and Renewable Energy (MNRE). Since its formation, the Ministry has released one of the international biggest and most ambitious applications on renewable strength. This paper offers an outline of various renewable electricity resources inclusive of solar energy, wind electricity. This paper comprehensively elucidates why we are going closer to RES their economic, social, and environmental impact, demanding situations associated with Renewable Energy System (RES).

Keywords Renewable energy source · Renewable positional · Wind and Solar energy

1 Introduction

Continuous economic as well as exponential population growth is responsible for increased global demand of power and a large portion of this demand is met by carbon-fossil-based energy sources, which have limited capacities and adversely

P. Kumar · M. Kumar (✉) · A. K. Bansal
Department of Electrical Engineering, School of Engineering and Technology, Central University of Haryana, Mahendragarh, India
e-mail: manish.kumar@cuh.ac.in

A. Kumar et al. (eds.), *Renewable Energy Towards Smart Grid*, Lecture Notes in Electrical Engineering 823, https://doi.org/10.1007/978-981-16-7472-3_4

affect the environment. The concern about the environment has increased all across the world. Due to air quality issues, cost and climate impact, emphasis has been laid on the deployment of low carbon intensive and more sustainable power resources. The production units of DG have progressed in the direction of power sector development. Utility of distributed generation resources (photovoltaic, fuel cells, wind energy, biomass, geothermal, small hydropower plants, and tidal, etc.) in distribution grid is increasing due to its various techno-economical advantages. Regulatory power industry provides economic opportunities for investors and provides many potential benefits for utilities (peak shaving, loss reduction, asset use, etc.), which encourage further tendency for addition of DG [1, 2].

The emergence of wind, solar and other renewable technologies and their integration into the power system can be attributed to the power sector reforms, policy support of the Govt., and direction-oriented guidelines towards the market in the last decade. It has created new business opportunities for Independent Power Producers (IPPs), private investors, non-institutional, for supplying the electricity to the grid, which results in a huge flow of capital in the electricity sector.

Renewable Energy Source (RES) integration has led many technical and economic challenges to the power producers in smooth operation of the power system. Major challenges associated with RES integration in both transmission and distribution network systems include impact on power system operating costs and losses, power imbalances (scheduling and dispatch), transmission planning (congestion), nodal pricing with DG in distribution system, etc. Thus, RES integration in the power system is now an important issue to optimize resource usage and to increase the installation of renewable capacity, in order to achieve the sustainability and security goals of the supply [3].

In this review paper, a detailed analysis of renewable energy in India and across the worldwide has been provided. Abundant amount of renewable energy is present in India which can be utilized to meet the future demand for electricity generation. In this review paper, different regions and different states of India to study renewable energy are considered. The Solar and Wind energy are the main focus area of this paper.

2 Global Scopes and Status in Renewable Energy

Energy is a fundamental requirement in the world of economic development and in every sector of the economy. In this way, it is essential that the nations have looked at new, clean, and sustainable sources of energy around the world and implemented new Renewable Energy Promotion and Energy Conservation Act. According to annual energy report (MNRE 2019) capacity of renewable energy production has become 85,908.37 MW.

The major part of total energy capacity is Wind Power (37,505 MW) then 33,712 MW of Solar Power and above 14,533 MW of Hydropower. Its shows that Solar PV becomes in a lead role of energy. The major part of total energy capacity

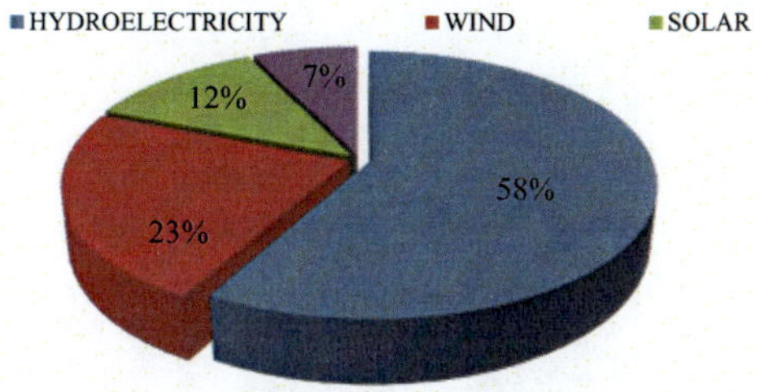

Fig. 1 World electricity generation in year 2019 by MNRE source

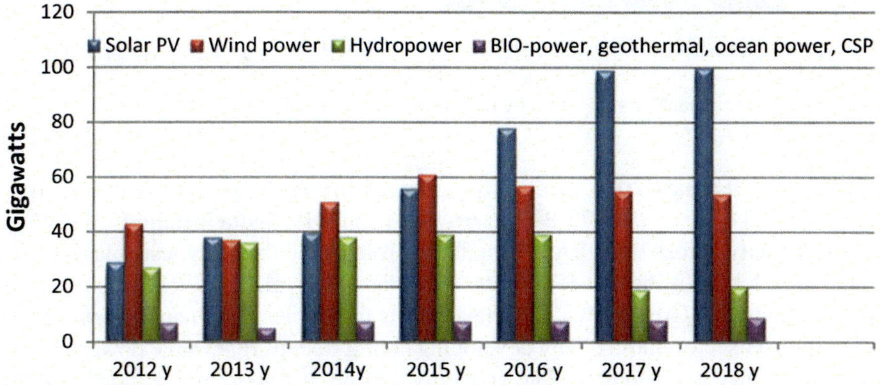

Fig. 2 Renewable power capacity and yearly growth rates, 2012–2018

is hydroelectricity (58%), then 23% Wind energy, 12% solar energy, and around 7% other energy sources like bioenergy [4]. It shows that Solar PV becomes in a lead role of energy.

The world-wise electricity generation in year 2019 has been and the overall yearly growth rate of 2012–2018 has been shown in Figs. 1 and 2, respectively. It is observed that overall renewable energy growth increased year by year and contribution of solar energy more increased as compared to the other renewable energy resource.

Total investment in RE worldwide has been shown in Fig. 3. In this, including early stage and corporate-level funding as well as the financing of new capacity was $288.3 billion in 2018. This was 11% down on 2017 record $325 billion [5].

3 India Status of Renewable Energy

India is at the 3rd rank for renewable energy sector in US clever in line with EY Report May 2019 [7]. As of February 29, 2020, the established R.E renewable energy capability stands at 86.75 GW out of which 34.40 GW is SOLAR (sun) strength and 37.66 GW is wind strength. The small hydro strength is 4.67 GW and biomass strength is 9.86 GW. By December 2019 15,100 MW of wind initiatives had been

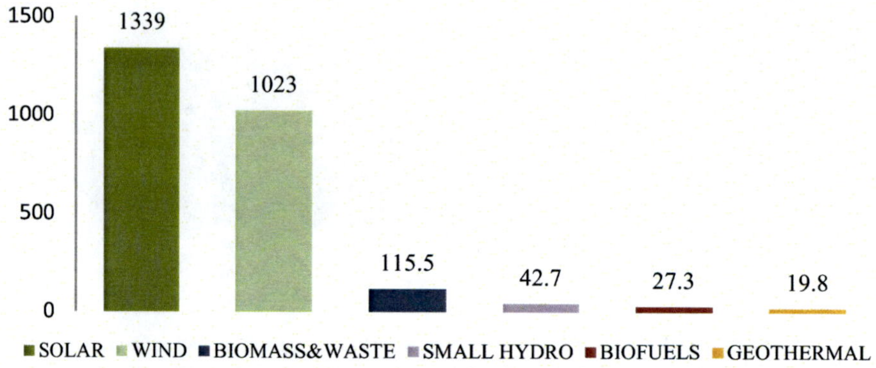

Fig. 3 World investments in renewable by technology, 2010–2019

issued. As of February 2020, electricity generated from sun, biomass, wind stood at 39.40 BU, 2.34 BU, and 57.72 BU, respectively. In 2019, India installed 7.3 GW of solar electricity across the USA organizing its role as the third-biggest solar market within the global. In the last few years, the solar power of India is increased more because the Indian Govt. provided some benefits in policies and financial promotion in solar power sector. India or any developing country needs renewable power source for their country development. The economy will be increased and demand will be fullfilled by extra power source.

The Government of India is leading active part in adopting RES through different ways/incentives such as capital and interest subsidy, viability gap funding, Generation-Based Incentives (GBI), concessional finance, financial incentives, etc. Renewable Energy Development for the year (Annual Report 2019) [8] is shown in Fig. 4 and the growth rate has increased in the last five years.

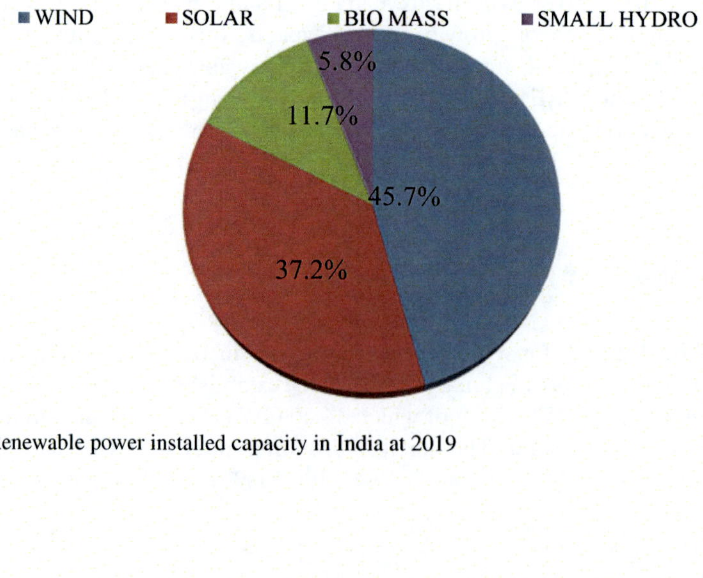

Fig. 4 Renewable power installed capacity in India at 2019

3.1 *Region-Wise Status of Renewable Energy in India*

The region-wise thermal, nuclear, hydro, and renewable energy source was installed capacity (MW) in India as shown in Fig. 5. In Northern region, thermal power was (57,721.46 MW), nuclear power was (1620 MW), hydropower was (197.707 MW), renewable was (14,199.02 MW), and total grand power for northern region installed was (93,248.25 MW) as on 31.3.2019. In Western region, the thermal (85,155.11 MW), hydro (7547.50 MW), nuclear (1840 MW), renewable (23,078.94 MW), and total grand power for western region was (117,621.55 MW). In Southern region, the thermal (53,217.26 MW), hydro (11,774.83 MW), nuclear (3320 MW), renewable (38,620.18 MW), and total grand power for western region was (1.6932.27 MW). In Eastern region, thermal power was (27,563.84 MW), hydropower was (4942.12 MW), renewable was (1401.48 MW), and total grand power for Eastern region installed was (33,907.24 MW). In North-East region, thermal power was (2581.83 MW), hydropower was (1427 MW), renewable was (324.29 MW) and total grand power for North-East region installed was (4333.211 MW). The Island only for renewable power was (17.73 MW) and Thermal Power was (40.05 MW) and grand total power in Island was (57.78 MW) [6].

The role of center, state, and private sectors is very important to fullfill the power demand. In India, the sector-wise installed capacity (mw) are shown in Fig. 6. The region-wise (Northern Region, Southern Region, Western Region, Eastern Region, North-East Region, Island) total installed power capacity was 93,248.25 MW, 106,932.27 MW, 117,624.55 MW, 33,907.24 MW, 4333.11 MW,57.78 MW till March 31, 2019 respectively. In the private sector, total installed power region-wise was 38,953.29 MW, 62,267.43 MW, 55,211.70 MW, 7902.37 MW, 85.54 MW, 7.38 MW and recently private sector was interested in the renewable power sector. Center installed power in region-wise 25,424.65 MW, 23,909.92 MW, 18,346.5 MW, 15,851.84 MW, 3058.62 MW, 5.10 MW, respectively.

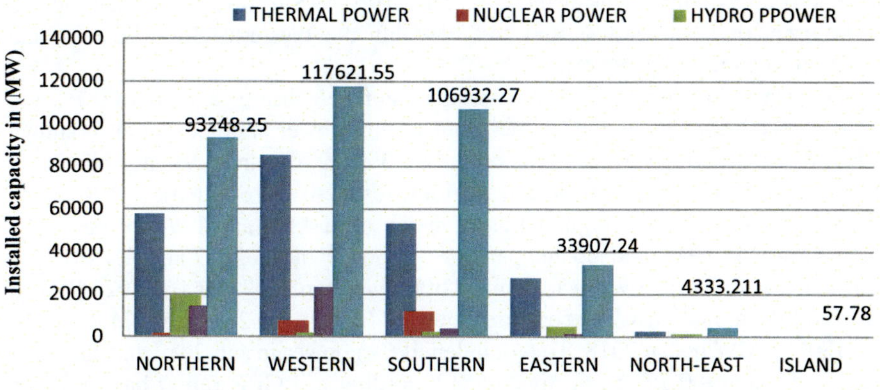

Fig. 5 Region-wise all India installed power capacity (MW) with renewable and non-renewable. *Source* as on 31 March, 2019

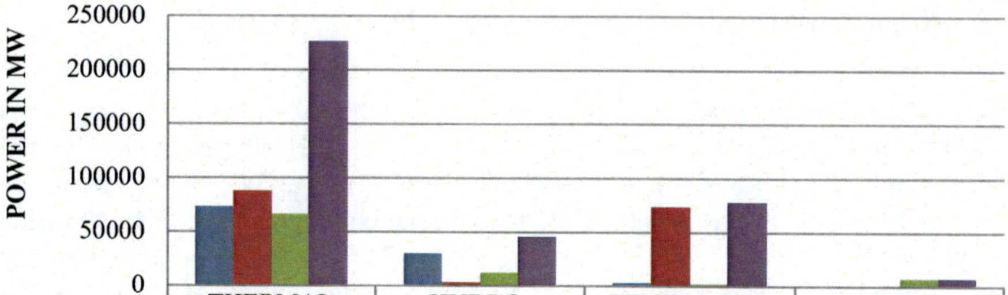

	THERMAL	HYDRO	RENEWABLE	NUCLEAR
STATE	72849.13	29878.8	2347.93	0
PRIVATE	87372.3	3394	73661.4	0
CENTER	66057.91	12126.42	1632	6780
TOTAL	226279.34	45399.22	77641.63	6780

Fig. 6 All India installed ability (MW) sector-wise as on 31, March 2019

State-wise installed power capacity 28,870.31 MW, 31,444.21 MW, 33,374.07 MW, 10,153.03 MW, 1188.95 MW, 45.30 MW respectively. The renewable power till March 31, 2019 was 73,661.40 MW installed by private sector and thermal power was 87,372.30 MW, and hydropower was 3394 MW and renewable power installed by states till March 31, 2019 was 2347.93 MW, thermal power 72,849.13 MW, and hydropower 29,878.80 MW and Center installed renewable power till March 31, 2019 was 1632.30 MW, thermal power 66,057.91 MW, hydropower 12,126.42 MW and Nuclear power was 6780 MW, respectively [7] and its shown in Fig. 7.

During the year 2017–18, overall ex-bus energy provided improved by way of 6.0% over the preceding year and the height met improved by 2.0%. The power requirement registered increase of 6.2% at some point of the year in opposition to the projected growth of 7.6% and Peak demand registered a boom of 3.0% against the projected boom of 6.0%. During the 12 months 2018–19 Surplus energy is anticipated of the order of 1.9%, 22.9%, and 14.8% within the Western, North-Eastern, and Northern Regions, respectively.

Southern and Eastern areas are possibly to stand electricity scarcity of 0.7% and 4.2%, respectively, which may be met from surplus power in other regions is shown in Fig. 8. The peaking surplus of the order of 9.3%, 12.6%, and 4.9% is predicted in Western, North-Eastern, and Eastern Regions, respectively. Southern and Northern regions are likely to face height deficit of the order of 4.5% and 1.2%, respectively. The region-wise peak demand and peak met of power in Jan 2017 and Jan 2018 are shown in Fig. 8. The all over peak demand of India in Jan 2017 was 165,292 MW and 169,130 MW in Jan 2018. The peak met power of India in Jan 2017 was 171,440 MW and 180601 MW in Jan 2018. The total deficit (%) power in Jan 2017 and Jan 2018 was −0.7% and −0.7%, respectively, [8] according to LGBR report has shown in Fig. 9.

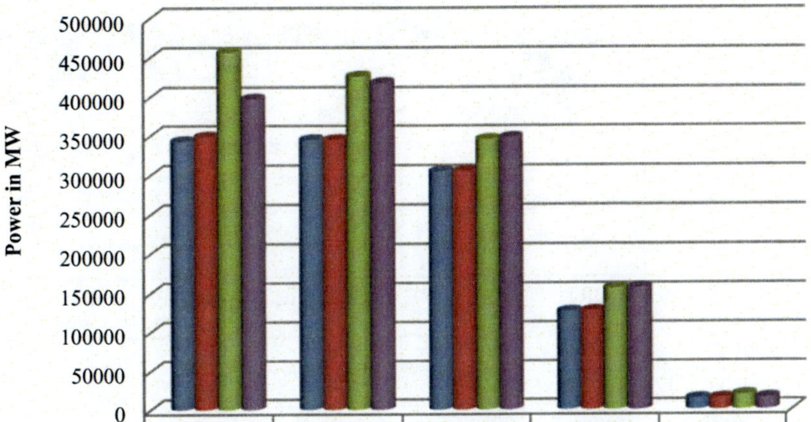

	NORTHERN	WESTREN	SOUTHERN	EASTERN	NORTH-EAST
AVAILABILTY 2017-18	343513	345127	305107	126868	14720
REQUIREMANT 2017-18	349172	345247	305586	127783	15140
AVAILABILITY 2018-19	456,855	426401	345708	156192	19550
REQUIREMANT 2018-19	398020	418323	348077	156703	15914

Fig. 7 Region-wise energy (MU) requirement and availability in (2017–2019)

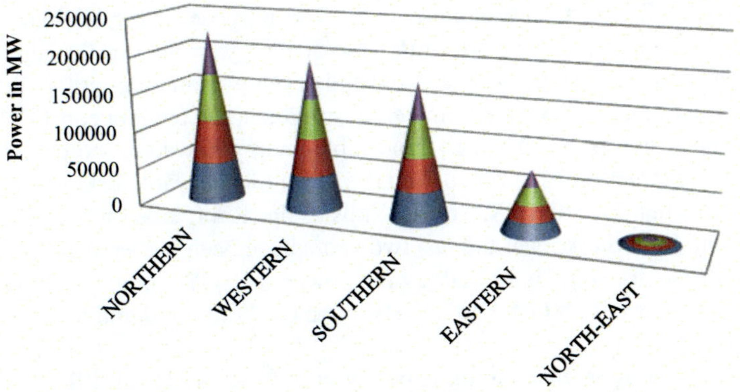

	NORTHERN	WESTERN	SOUTHERN	EASTERN	NORTH-EAST
PEAK MET 2018	58448	50085	47210	20485	2520
PEAK DEMAND 2018	60749	50477	47385	20794	2629
PEAK MET 2017	56612	48313	42232	18788	2475
PEAK DEMAND 2017	53372	48313	42232	18908	2487

Fig. 8 All India installed ability (MW) sector-wise as of March 31, 2019

	NORTHERN	WESTREN	SOUTHERN	EASTREN	NORTH-EAST
DEFICIT 2018	-3.8	-0.8	-0.4	-1.5	-4.1
DEFICIT 2017	-1.6	0	-0.2	-0.7	-2.8

Fig. 9 Region wise deficit (%) energy for Jan 2017 and Jan 2018

3.2 Solar Power in India

National Solar Mission has set a goal to increase the development of solar power for energy production and also use it in different ways, so that solar energy can compete with fossil-based energy alternatives. The purpose of the National Solar Mission is to minimize the overall price rate of solar energy production through large-scale deployment targets, aggressive research and development, long-term policy and domestic production of important raw materials, products and components [9]. Fossil Fuel-based generation is competed by renewable energy nowadays comparatively. To achieving the target of 227 GW (earlier175 GW) of renewable energy by 2022, major projects/schemes, Solar PV, Solar Pumps, Solar Roof, etc., on the top of the Canal Bank and Solar Park, over the last two years have been started during. Out of 227GW R.E the solar energy is 113 GW, wind power is 66 GW, biomass power is 10 GW, small hydropower is 5 GW and 31 GW from offshore wind and floating solar power [10, 11].

Top 10 States in Solar Installation capacity in MW as on 31-12-2019 is shown in Fig 10. It went to the introduction of various policy measures for 2022 new and renewable energy schemes implemented by the Ministry (MNRE) targeted for renewable energy capacity of 227 GW (earlier175 GW) to achieve have been taken additional steps to provide financial assistance to Renewable Purchase Obligation (RPO) strengthening and install Renewable Generation Obligation (RGO) to provide the appropriate amendments to tariff policy and the Electricity Act, exclusive solar park, through the Green Energy Corridor power transmission network development rooftop identify large buildings projects/government complexes, solar roof and 10% renewable energy up Interfere with the mission guidelines and statement for the development of smart cities. Providing free solar bonds to raise long-term loans to banks/solar make up the roof as a part of the housing loans granted by NHB,

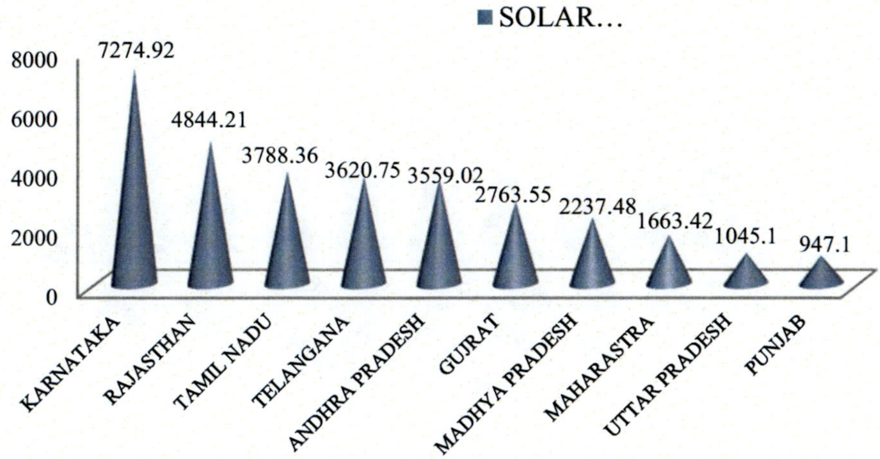

Fig. 10 Top 10 States in solar installation capacity in MW as on 31-12-2019

encouraging distribution companies to achieve the targets, of making pure metering mandatory and increase funds [12, 14].

3.3 *Wind Power in India*

In India the development of wind energy began in 1986, with 55 KW of wind turbine in the coastal areas of Gujarat (Okha), Maharashtra (Ratnagiri), and Tamil Nadu (Tirunelveli) in the first wind farm. These performance projects were supported by the MNRE as shown in Fig. 11.

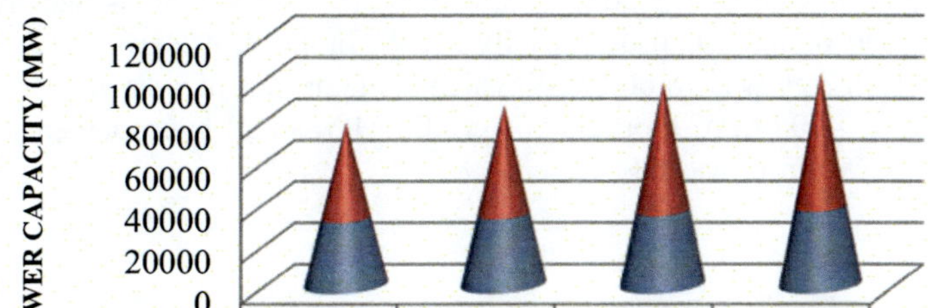

	2016-17 Y	2017-18 Y	2018-19 Y	2019-20 Y
GENERATION (GWH)	46011	52666	62036	64485
INSTALLED CAPACITY (MW)	32280	34046	35626	37669

Fig. 11 Yearly installed wind power capacity (MW) in India

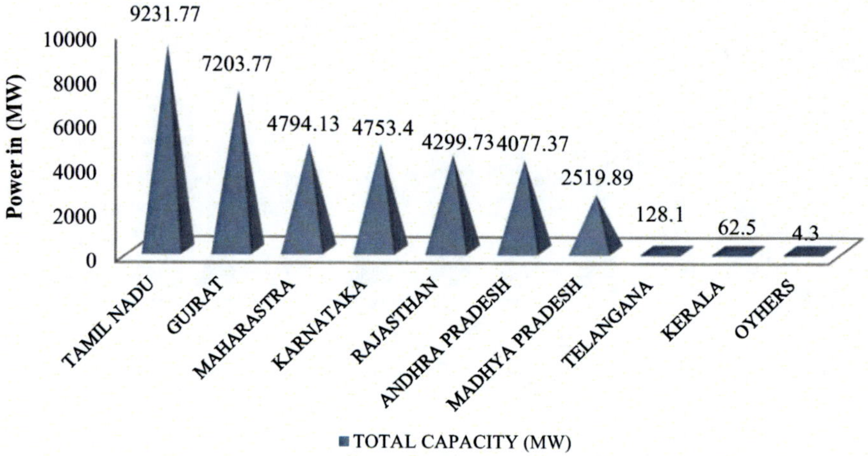

Fig. 12 State-wise Wind potential in India as on 2019

Wind power generation ability in India has increased significantly in the last few years and by 2018 the installed ability of wind power was 35,626 MW, mainly spreading south, west, and northern areas. In 2015, MNRE had set a target of 60,000 MW for wind power generation capacity by 2022 [13]. In year 2016, the installed capacity wind power was 28,665 MW and after two years in 2018 the installed wind power capacity was 35,626 MW which means it's increased by 6961 MW.

The state-wise wind potential of India is shown in Fig. 12 and observed that Tamil Nadu is the maximum wind potential state. The month of July is more wind potential in India and the southern region are more wind potential as compared to other regions during this month. The Indian Government set a target of production 227 GW (earlier175 GW) of renewable energy capacity till the year 2022. The target having part of wind energy 66 GW, solar energy 113 GW, biomass electricity 10 GW and rest part of small hydroelectricity 5 GW. In the year 2016–2017, the target of renewable energy was 16,660 MW which contains part of solar power 12,000 MW, wind power 4000 MW, biomass power 400 MW, small hydropower 250 MW, and 10 MW power from waste [14].

4 Conclusions

In this paper, the importance of RERs for maintaining greenhouse gases in limits, CO_2 emissions, replacing fossil fuels, and reducing the price of energy is described. Firstly, the types and nature of different RERs are discussed and then their presence in different countries is presented. Thereafter, the statue of RERs in India is described in detail. This paper mainly focuses on the state-wise and region-wise growth of renewable energy on yearly basis. In India, wind and solar play a very important role in the renewable energy power sector market as compared to other renewable resources. Lot of investor is interested to invest money in the solar energy sector.

References

1. REthinking Energy 2017, Accelerating the global energy transformation. In: International renewable energy agency (IREA 2017), Abu Dhabi. www.irena.org
2. International Energy Agency (IEA) (2019) Headline energy data, Paris. https//www.ira.org/newsroom/news/September/worldenergy-inverstment-2019-html
3. IEA (2019) World energy outlook 2019, IEA Paris
4. REN21 (Renewable Energy Network for the 21st Century) (2019) Renewable global status report Paris. www.ren21.net/gsr
5. Warren D (2019) Renewable energy country attractiveness index (RECAI) (47). http://www.ey.com/Publication/vwLUAssets/EY-RECAI-47-May-2016/$FILE/EY-RECAI-47-May-2016.pdf
6. Govt. of India Ministry of power center electricity authority New Delhi report Jan (2020). www.cea.nic.in/reports/monthly/executivesummary/2020/exe_summary-01.pdf
7. India Renewable Energy Development, Annual Report (2019). http://mnre.gov.in/file-manager/annual-report/2018-2019/EN/.../chapter_1.htm
8. Ministry of New and Renewable Energy (MNRE), Annual reports. http://mnre.gov.in/mission-and-vision-2/publications/annual-report-2
9. Shah V, Booream-Phelps J (2019) F.I.T.T. for investors—Solar. Deutsche Bank. United States, Deutsche Bank Markets Research, p 5
10. This just became the world's cheapest form of electricity out of nowhere. Fortune. Accessed 05 Feb 2019
11. Grid connected SPV with VGF under JNNSM (PDF). Accessed 25 Dec 2019
12. Solar auction companies seeking lowest state support to win
13. Record capacity addition of wind power of 5400 MW in last fiscal (2020). Ministry of New and Renewable Energy, Govt. of India
14. Press information Bureau Government of India Ministry of New and Renewable Energy. http://pib.nic.in/newsite/PrintRelease.aspx?relid=155612

Cloud Computing Data Security Techniques—A Survey

Mayanka Gaur and Manisha Jailia

Abstract From past few years cloud computing has become an IT slang. It is the grassland of computing that is spreading briskly day-by-day in scholastic and business in form to accomplish conditions of final users. Cloud computing in real sense is achieving any duty by building use of assistance that are granted by cloud providers. It empowers an ample dimension of customer to access Shared, Ascendable, and Virtualized resources on top of the World Wide Web. It is a chunk of shared computing. It is currently arriving grassland by dint of its act, towering opportunity, and small price. With the brisk rise of internet computing machinery there is a big requirement for data storage security on cloud. There endure many cloud data storage security techniques now a days. In this paper, we are declaring some of these current techniques, literature review, and comparison between them. From the given comparison table of various current cloud computing data security approaches, it is concluded that the best approach for cloud data storage security must be taken and considered. We believe that there exist many more approaches for cloud data storage security and among them the best approach must be chosen.

Keywords Cloud computing · Encryption · Decryption · Cryptography · Steganography · Data security

1 Introduction

"The Smart Grid is an advanced digital two-way power flow power system capable of self-healing, adaptive, resilient, and sustainable with foresight for prediction under different uncertainties" [14].

"To achieve the targeted aim, it is recommended to integrate IoT technology with cloud technology. This integration is called CloudIoT, and it exhibits magnificent benefits in alignment with the aims of smart grids in addition to posing new challenges" [8].

M. Gaur (✉) · M. Jailia
Banasthali Vidyapith, Vanasthali, Rajasthan, India

A. Kumar et al. (eds.), *Renewable Energy Towards Smart Grid*, Lecture Notes in Electrical Engineering 823, https://doi.org/10.1007/978-981-16-7472-3_5

"By assimilating concepts of advanced communications and future control technologies, Smart Grid has become the next generation power system. Due to its greater robustness, efficiency, and flexibility over conventional power system, it is gaining importance. As in modern electrical power system, need of resources and storage is increasing which can be dealt with cloud computing. It is a promising technology with functionality of using computing resources in scalable and virtualized manner. Cloud computing integrates the electrical power system resources through internal networks, thus improvement in robustness, load balancing, and storage capacity is observed" [26].

"The use of emerging technologies such as cloud computing, Internet of Things, and Big Data is increasing as tools to assist the management of data and information related to energy systems grow. This allows for greater flexibility, scalability of solutions, optimization of energy use, and management of energy devices" [34].

Now-a-days improvement in computer science and Internet technology contrived computing on actual cloud a big appeal. Cloud computing in actual touch is doing any assignment by formulating usage of assistance that is granted by cloud suppliers. It is the grassland of gauge that is developing briskly gradually twain in scholastic and business in the form to accomplish conditions of final customers. It is a chunk of split computing. Its ambition is to give shared, fundamental, and malleable assets as assistance to customers.

From the past few years, cloud computing has become an IT slang. It acquires all the computing assets and administers them by some program. There are tons of customers using cloud assistances all at once. Customers do not have to fear regarding how to purchase the assistances or programs for a prolonged aspect rather they can straight use or purchase evaluating assets from the cloud by making use of world wide web (WWW).

2 Cloud Computing Definition

Cloud computing in actual touch is doing any assignment by formulating usage of assistance that are granted by cloud providers. Cloud computing is the field of gauge that is developing briskly day-by-day both in scholastic and business in form to accomplish conditions of final users. Cloud computing is a chunk of split computing. Ambition of cloud computing is to give shared, virtualized, and malleable assets as assistance to customers.

From the past few years cloud computing has become an IT slang. It acquires all the computing assets and administers them by some program. There are tons of customers using cloud assistances all at once. Customers need not to fear about how to purchase the assistances or programs for a long-time aspect rather they can straightly use or purchase computing resources from the cloud using internet. Customers broadly concern that the cloud computing providers can harm their essential info present on cloud. The only approach is to use the best technique for cloud data security. Now a days there exist many cloud data security techniques some of which are listed as follows:

- Spread Spectrum and Adaptive method
- LSB Insertion
- New Stego Key Adaptive LSB (NSKA-LSB)
- Convolution
- Randomized and indexed word dictionary
- Auditing protocol for data integrity
- Deoxyribonucleic Acid (DNA) computing
- Challenge Handshake Authentication Protocol (CHAP) and Rivest Shamir Adleman (RSA) Algorithm
- Encryption technique based on identity
- Encryption technique based on attribute
- Proxy-re-encryption (PRE).

3 Existing Approaches

3.1 Spread Spectrum and Adaptive Method

In [17], there are two methods which are as follows:

Spread Spectrum Method: It is a technique that is using pseudo code to spread energy signal in a communication line (bandwidth).

Adaptive Method: Insertion of LSB on a bit of pixel image and modifying level of intensity of digital image pixel block adaptively.

3.2 LSB Insertion

In [3], conceals word using Sequence Mapping procedure in pixels of a cover image which allows steganalyst to recover word due to clarity of algorithm.

3.3 New Stego Key Adaptive LSB (NSKA-LSB)

In [1], there are four stages: (a) random selection of pixel is done, (b) compression and encryption of secret data is done, (c) adaptive hiding of chaotic secret data within cover images is done, and (d) extracting secret data from stego image at the receiver terminal is done.

3.4 Convolution

In [4], convolution technique has three steps: (a) conversion of the color image to greyscale and then to black and white image, (b) feature extraction is done, and (c) hiding selected plain text in the image.

3.5 Randomized and Indexed Word Dictionary

"There are two steps: (a) Stego key is generated using two approaches, one approach uses date field of forwarding email cover template to generate stego key. Another approach uses system date to generate stego key (b) Stego key transmission using the RSA algorithm" [24].

3.6 Auditing Protocol for Data Integrity

In [18], minimizing the calculation complications of the buyer during the organization arrangement stage of the auditing agreement.

3.7 Deoxyribonucleic Acid (DNA) Computing

"1024—bit secret key is generated based on Deoxyribonucleic Acid (DNA) computing that allows the organization to save in opposition to several safety hacks" [28].

3.8 Challenge Handshake Authentication Protocol (CHAP) and RSA Algorithm

In [23], CHAP improves security of Authentication and RSA is used for encryption and decryption.

Authentication request is being send by the user to cloud service provider confirms the approval using CHAP and dispatch the acceptance back to the user. User encrypts the information by making use of RSA algo and dispatches the encoded information to the cloud. Customer apply RSA algorithm for decryption whenever he/she downloads the data from the cloud and retrieves it in encrypted form. Execution of CHAP is shown in Fig. 1.

Fig. 1 Execution of CHAP

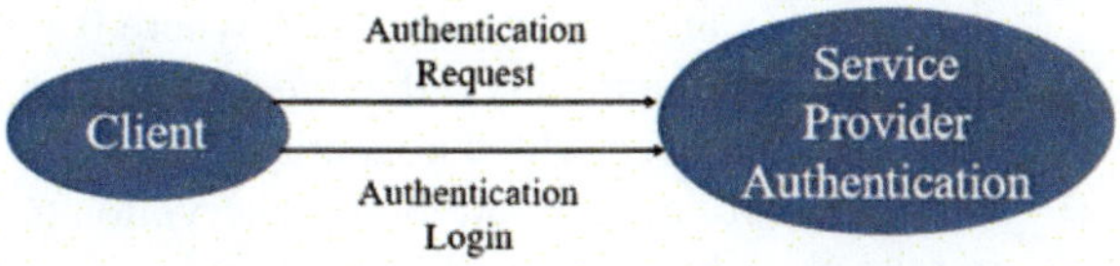

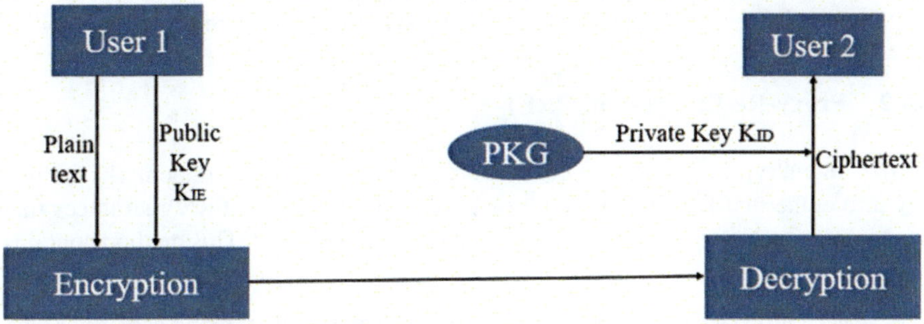

Fig. 2 Encryption technique based on identity

3.9 Encryption Technique Based on Identity

"This method (as shown in Fig. 2) authorize some set of customers to talk security and to confirm every other trademark beyond swapping public and private key and beyond putting key storages. This method allows users to pick a random line of words that gives a special name for him to the other group as a public key. The customers' private key is produced by making use of private key generator (PKG).

This method consists of 3 important stages:

(1) Encryption: The email is encrypted by Bob's email address as the public key when alice dispatches email to bob.
(2) Identity authentication: Bob authenticates himself and acquires private key generated by private key generator when bob receives encrypted email.
(3) Decryption: Bob decode the encoded email and gets the original message" [43].

3.9.1 Encryption Technique Based on Attribute

In [43], the ciphertext and secret key are associated to attributes. Decryption operation can be executed when the no. of customers and ciphertext attribute set comes to the threshold parameter. "This technique is divided into two types: Key-policy-attribute-based-encryption and Ciphertext-policy-attribute-based-encryption.

(1) Key-policy-attribute-based-encryption (KPABE): In this, nonentity text is labeled with sets of feature and private label are related with entry tree structure. The interior nodes of tree are threshold policies. Interior nodes consist of

flag vertex x and threshold value k where $0 < k \leq x$. Customer decrypts the nonentity text if and only if the tree related with a private key is pleased by the attribute set related with a nonentity text.

(2) Ciphertext-policy-attribute-based-encryption (CPABE): In this, nonentity text is related with entry tree structure and private keys are labeled with sets of attributes. User decrypts the nonentity text if and only if entry tree related with nonentity text is pleased by the attribute set related with the private key" [43].

3.9.2 Proxy-Re-Encryption (PRE)

"Proxy-re-encryption is used to convert ciphertext (message or trademark) for one key into ciphertext for another key by making use of a proxy. PRE guarantees that the proxy cannot get some analogous messages with plain text. This method contains 4 important stages:

(1) Encryption: Alice encrypts the original information by making use of her own public key E_A. Alice produces the first layer nonentity text C_1 and then transfer to the proxy.
(2) Generation of re-encryption key: Alice gets the bob's public key E_B. Alice encode $\mathrm{E_A}$ below $\mathrm{E_B}$ to give rise to the re-encryption key $E_{A \to B}$ and Alice then transfer it to the proxy.
(3) Re-encryption: Proxy encodes the first layer nonentity text by making use of re-encryption key when proxy obtains $\mathrm{C_1}$ and $E_{A \to B}$ and produces the second re-encrypted nonentity text C_2.
(4) Decryption: Bob acquires the re-encrypted nonentity text $\mathrm{C_2}$ from the proxy. Bob decode it with his own private key SB" [43].

Data sharing technique based on PRE is shown in Fig. 3.

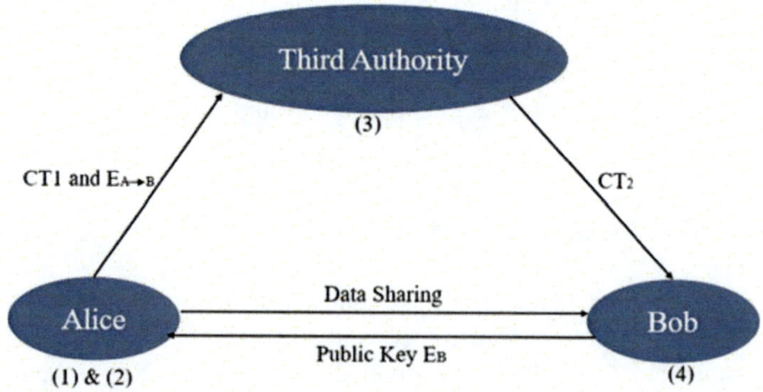

Fig. 3 Data sharing technique based on PRE

4 Literature Review and Comparison Between Existing Approaches

See Table 1.

Table 1 Literature review and comparison of some of the different existing approaches

S.no	Title	Methodology	Description	Drawback
1	Analysis of steganography on TIFF image using spread spectrum and adaptive method Febryan et al. [17]	Spread spectrum and adaptive method	Spread Spectrum: It is a technique that is using pseudo code to spread energy signal in a communication line (bandwidth) Adaptive method: insertion of LSB on a bit of pixel image and modifying level of intensity of digital image pixel block adaptively	Time-consuming and too lengthy
2	A Symmetric key based steganography calculation for anchored information [3]	LSB insertion	This technique conceals word using Sequence Mapping procedure in pixels of a cover image which allows steganalyst to recover word due to clarity of algorithm	Time-consuming and too lengthy
3	An effective and secure digital image steganography scheme using two random function and chaotic map [1]	New stego key adaptive LSB (NSKA-LSB)	Four stages: (1) random selection of pixel is done (2) compression and encryption of secret data is done (3) adaptive hiding of chaotic secret data within cover images is done (4) extracting secret data from stego image at the receiver terminal is done	Very time consuming and too lengthy

(continued)

Table 1 (continued)

S.no	Title	Methodology	Description	Drawback
4	Text in image steganography hiding based convolution technique [4]	Convolution technique	Three steps: - (1) conversion of the colour image to greyscale and then to black and white image (2) feature extraction is done (3) hiding selected plain text in the image	Time-consuming and too lengthy
5	A forward email-based high-capacity text steganography technique using a randomized and indexed word dictionary [24]	Randomized and indexed word dictionary	Two steps: (1) Stego key is generated using two approaches, one approach uses date field of forwarding email cover template to generate stego key. Another approach uses system date to generate stego key (2) Stego key transmission using the RSA algorithm	Very time consuming and too lengthy
6	An efficient data integrity auditing protocol for cloud computing [18]	Auditing protocol for data integrity	Minimizing the calculation complications of the buyer during the organization arrangement stage of the auditing agreement	• Data Confidentiality is a major challenge • Very time-consuming protocol • Too lengthy
7	Towards DNA based data security in the cloud computing environment [28]	Deoxyribonucleic Acid (DNA) computing	1024—bit secret key is generated based on Deoxyribonucleic Acid (DNA) computing that allows the organization to save in opposition to several safety hacks	• Very time-consuming • Too lengthy • Method complexity is high
8	Data security protection in cloud computing by using encryption [23]	CHAP and RSA	CHAP is used to improve the security of authentication and RSA is used for encryption and decryption	• Very time-consuming • Too lengthy • Other algorithms can also be used to protect data

(continued)

Table 1 (continued)

S.no	Title	Methodology	Description	Drawback
9	Data security and privacy-preserving in edge computing paradigm: survey and open issues [43]	Identity-Based Encryption (IBE)	This method enables any pair of users to communicate securely and to confirm each other's signatures without exchanging private or public key and without keeping key directories	• Very time consuming • Too lengthy
10	Data security and privacy-preserving in edge computing paradigm: survey and open issues [43]	Attribute-based encryption (ABE)	In this method, the cipher text and confidential key are related to attributes	• Very time-consuming • Too lengthy • Other approaches can also be used
11	Data security and privacy-preserving in edge computing paradigm: survey and open issues [43]	Proxy re-encryption (PRE)	PRE is used to convert ciphertext for one key into another key by using a proxy	• Very time-consuming • Too lengthy • Other approaches can also be used

5 Conclusion

Cloud-computing is the new emerging technology that faces a lot of security challenges. In this paper we had briefly described some of the current cloud computing data security techniques and comparison between them. From above literature review and comparison table of various current cloud computing data security approaches, it is concluded that the best approach for cloud data storage security must be taken and considered. We believe that there exist many more approaches for cloud data storage security and among them the best approach must be chosen.

References

1. Abdulwahed MN (2020) An effective and secure digital image steganography scheme using two random function and chaotic map. J Theor Appl Inf Technol 98(1):78–91
2. Ahmed M, Hossain MA (2014) Cloud Comput Secur Issues 6(1):25–36
3. Al-Halabi YS (2020) A symmetric key based steganography calculation for anchored information. J Theor Appl Inf Technol 98(1):103–123
4. Al-Tuwaijari JM, Ismael HA (2020) Text in image steganography hiding based convolution techniques. Int J Adv Sci Technol 29(4):372–387

5. Alabdulatif A, Khalil I, Yi X (2020) Towards secure big data analytic for cloud-enabled applications with fully homomorphic encryption. J Parallel Distrib Comput 137:192–204. https://doi.org/10.1016/j.jpdc.2019.10.008
6. Albugmi A, Alassafi MO, Walters R, Wills G (2016) Data security in cloud computing. In: 5th international conference on the future generation communication technologies FGCT 2016, pp 55–59. https://doi.org/10.1109/FGCT.2016.7605062
7. Assanelli D, Maifredi G, Manessi E, Giovannini G, Canclini S, Malossi C, Mertins V, Zywietz C (1995) Portable cardiological workstation designed for acute coronary events. Comput Cardiol 361–364. https://doi.org/10.1109/cic.1995.482660
8. Bagherzadeh L, Shahinzadeh H, Shayeghi H, Dejamkhooy A, Bayindir R, Iranpour M (2020) Integration of cloud computing and IoT (Cloud IoT) in smart grids: benefits, challenges, and solutions. Int Conf Comput Intell Smart Power Syst Sustain Energy CISPSSE. https://doi.org/10.1109/CISPSSE49931.2020.9212195
9. Banerjee I, Bhattacharyya S, Sanyal G (2012) Text steganography through quantum approach. Commun Comput Inf Sci 292:632–643. https://doi.org/10.1007/978-3-642-31686-9_74
10. Bender W, Gruhl D, Morimoto N, Lu A (1996) Techniques for data hiding. IBM Syst J 35(3–4):313–335. https://doi.org/10.1147/sj.353.0313
11. Bernsmed K, Jaatun MG, Meland PH, Undheimy A (2011a) Security SLAs for federated Cloud services. In: Proceedings of the 2011 6th international conference on availability, reliability and security ARES 202–209. https://doi.org/10.1109/ARES.2011.34
12. Bernsmed K, Jaatun MG, Undheim A (2011b) Security in service level agreements for cloud computing. CLOSER 2011—Proceedings of the 1st international conference on cloud computing and services science, pp 636–642. https://doi.org/10.5220/0003391606360642
13. Deore MP, Chaudhari MT (2008) Hybrid encryption for database security. Int Res J Eng Technol 9001(3):1264
14. Dileep G (2020) A survey on smart grid technologies and applications. Renew Energy 146:2589–2625. https://doi.org/10.1016/j.renene.2019.08.092
15. Ebedon MM (2015) Ensure data security in cloud computing by using cryptography
16. Eloff J, Dlamini MT, Zielinski MP, Inter net of people, things and services-the convergence of security, trust and privacy
17. Febryan A, Waluyo Purboyo T, Erfa Saputra R (2020) Analysis of steganography on TIFF image using spread spectrum and adaptive method. J Eng Appl Sci 15(2):373–379. https://doi.org/10.36478/jeasci.2020.373.379
18. Garg N, Bawa S, Kumar N (2020) An efficient data integrity auditing protocol for cloud computing. Futur Gener Comput Syst 109:306–316. https://doi.org/10.1016/j.future.2020.03.032
19. Gaur M (2015) A new PDAC parallel encryption with digit arithmetic of cover text based text steganography approach for cloud data security. Int J Recent Innov Trends Comput Commun 3(3):1344–1352. https://doi.org/10.17762/ijritcc2321-8169.150397
20. Juels A, Oprea a new approaches to security and availability for cloud data
21. Kandukuri BR, Ramakrishna PV, Rakshit A (2009) Cloud security issues. In: SCC 2009–2009 IEEE International Conference on Service Computer, pp 517–520. https://doi.org/10.1109/SCC.2009.84
22. Kumar PR, Raj PH, Jelciana P (2018) Exploring data security issues and solutions in cloud computing. Procedia Comput Sci 125(2009):691–697. https://doi.org/10.1016/j.procs.2017.12.089
23. Mahmood G (2019) Data security protection in cloud computing by using encryption Kirkuk University. J Sci Stud (KUJSS) Data Secur Protect Cloud Comput Using Encr
24. Maji G, Mandal S (2020) A forward email based high capacity text steganography technique using a randomized and indexed word dictionary. Multimed Tools Appl. https://doi.org/10.1007/s11042-020-09329-z
25. Mehra N, Aggarwal S, Shokeen A, Bura D (2018) Analyzing cloud computing security issues and challenges. Adv Intell Syst Comput 710(March):193–202. https://doi.org/10.1007/978-981-10-7871-2_19

26. Mishra N, Kumar V, Bhardwaj G (2019) Role of cloud computing in smart grid. In: 2019 international conference on automation, computational and technology management ICACTM 2019, pp 252–255. https://doi.org/10.1109/ICACTM.2019.8776750
27. Modi C, Patel D, Borisaniya B, Patel A, Rajarajan M (2013) A survey on security issues and solutions at different layers of Cloud computing. J Supercomput 63(2):561–592. https://doi.org/10.1007/s11227-012-0831-5
28. Namasudra S, Devi D, Kadry S, Sundarasekar R, Shanthini A (2020) Towards DNA based data security in the cloud computing environment. Comput Commun 151:539–547. https://doi.org/10.1016/j.comcom.2019.12.041
29. Petitcolas FAP, Anderson RJ, Kuhn MG (1999) Information hiding—A survey. Proc IEEE 87(7):1062–1078. https://doi.org/10.1109/5.771065
30. Purushothaman D, Abburu S (2012) An approach for data storage security in cloud computing 9(2):100–105
31. Ramachandra G, Iftikhar M, Khan FA (2017) Science direct the 3rd international workshop on cyber security and digital investigation (CSDI 2017) a comprehensive survey on security in cloud computing. Procedia Comput Sci 110(2012):465–472. https://doi.org/10.1016/j.procs.2017.06.124
32. Rao RV, Selvamani K (2015) Data security challenges and its solutions in cloud computing. Procedia Comput Sci 48(Iccc):204–209. https://doi.org/10.1016/j.procs.2015.04.171
33. Rasheed H (2014) Data and infrastructure security auditing in cloud computing environments. Int J Inf Manag 34(3):364–368. https://doi.org/10.1016/j.ijinfomgt.2013.11.002
34. Rees D, McBain R (2004) Management: challenges and opportunities 53(9):1689–1699
35. Rodero-Merino L, Vaquero LM, Caron E, Muresan A, Desprez F (2012) Building safe PaaS clouds: a survey on security in multitenant software platforms. Comput Secur 31(1):96–108. https://doi.org/10.1016/j.cose.2011.10.006
36. Rong C, Nguyen ST, Jaatun MG (2013) Beyond lightning: a survey on security challenges in cloud computing. Comput Electr Eng 39(1):47–54. https://doi.org/10.1016/j.compeleceng.2012.04.015
37. Streit K, Garfinkel Y (2015) Horned figurines made of stone from the neolithic and chalcolithic periods and the domestication of sheep and goat. Palest Explor Q 147(1):39–48. https://doi.org/10.1179/1743130114Y.0000000006
38. Subashini S, Kavitha V (2011) A survey on security issues in service delivery models of cloud computing. J Netw Comput Appl 34(1):1–11. https://doi.org/10.1016/j.jnca.2010.07.006
39. Tari Z (2014) Security and privacy in cloud computing. IEEE Cloud Comput 1(1):54–57. https://doi.org/10.1109/MCC.2014.20
40. Yesilyurt M, Yalman Y (2017) New approach for ensuring cloud computing security: using data hiding methods. Sādhanā 41(11):1289–1298. https://doi.org/10.1007/s12046-016-0558-8
41. Yesilyurt M, Yalman Y (2016) New approach for ensuring cloud computing security: using data hiding methods. Sadhana Acad Proc Eng Sci 41(11):1289–1298. https://doi.org/10.1007/s12046-016-0558-8
42. Yesilyurt M, Yalman Y, Ozcerit AT (2013) A new DCT based watermarking method using luminance component. Elektron ir Elektrotechnika 19(4):47–52. https://doi.org/10.5755/j01.eee.19.4.2015
43. Zhang J, Chen B, Zhao Y, Cheng X, Hu F (2018) Data security and privacy-preserving in edge computing paradigm : survey and open issues 4(c). https://doi.org/10.1109/ACCESS.2018.2820162

Review on Power Restoration Techniques for Smart Power Distribution Systems

D. Sarathkumar, Albert Alexander Stonier, M. Srinivasan, and L. Sahaya Senthamil

Abstract When an abnormal condition happened in a power delivery networks, it must be segregated. The power delivery system should be restructured to recover the not reachable region after fault isolation. The reconfiguration can be executed through various breaking operations on the feeder lines. Recent days restoration of power system brings to more attentiveness also made more progression. This article primarily reviews the various power restoration techniques for power distribution systems, and it needs communication tools. Based on the communication architecture the power restoration techniques should be classified into three types namely hierarchical approaches, centralized approaches, and distributed approaches. The several optimization techniques can be used for the power restoration systems such as stochastic programming, expert systems, heuristic algorithms, and multi-agent systems. Research developments and future areas of research in the distribution restoration techniques are discussed in this paper.

Keywords Service restoration · Distribution networks · Fault identification · Fault isolation · Smart grid · Self-healing

1 Introduction

Smart power grid provides sustainability, reliability, efficiency through incorporating smart metering, and Information Communication Technology (ICT) tools in already available conventional power systems [1]. United States of the National Energy technology Laboratory was derived seven important aspects of smart power systems [2], out of this one of the main features of smart grid is self-healing. The power

D. Sarathkumar (✉) · A. A. Stonier · M. Srinivasan
Department of Electrical and Electronics Engineering, Kongu Engineering College, Erode District 638 060, Tamilnadu, India

L. S. Senthamil
Department of Electrical and Electronics Engineering, PSNA College of Engineering and Technology, Dindigul, India

A. Kumar et al. (eds.), *Renewable Energy Towards Smart Grid*, Lecture Notes in Electrical Engineering 823, https://doi.org/10.1007/978-981-16-7472-3_6

system is entitled as a self-healing network in few reviews [3, 4]. It is a capability of power delivery networks to restore naturally after the abnormal conditions occurrence [3, 4].

Smart power grid designed to separate the fault affected networks also re-configure the power systems with less manpower interruption, based on the automation level implemented. After identifying the fault, optimum power restoration strategy reduces the amount of power outage areas as satisfying the operational restraints. An only completely automated self-healing strategy can be applied in the areas of smart grid context [5], to isolate the faulty section remote-controlled switching equipment was used, to restore the faults, and changing in the network topology. Wide range blackout problems are still continuing and are unavoidable, even though a large range of research task has been executed to construct the electrical systems resilient across outages [6]. An appropriate power restoration scheme can efficiently alleviate the negative consequence on the economy, society, and power networks itself. The more research needed for fast and efficient power system restoration after the abnormal conditions is of essential task.

The objective of power restoration is to restore the loading after the abnormal events by varying the network topology as satisfying the operational constraints [7]. Several studies already accomplished to execute the power recovery issue through decentralized and centralized ways. All of its studies individual advantages also limitations [8] (i.e., heuristic rules in paper [9], Stochastic programming in paper [10], artificial neural networks in paper [11], expert systems in paper [12], etc.). Systematic outputs of self-controlled smart power grids were shown in [13]. In paper [14], to identify the faults software control implementation of real time inputs was presented.

Restoration of power system after a sectional or entire disruption is totally a complicate activity. Several parameters requirement to be taken in to account such as functioning level of the network, availability of equipment, fault service restoration period, and the attainment rate of performance. Restoration requires not only wide range of interpretation and authentication but also the decisions making handled by executing team. Restoration is a multi-variable, multi-constraint, and objective; it is complete uncertainty and non-linearity. It can be expressed as a classic semi-structured decision-making also it is complex to receive an entire solution. The prosperous establishment of self-healing power network based on the extensive deployment of fault position, separation, and power restoration. As an important characteristic of smart grid self-healing is, power restoration has taken out significant attentiveness. So this paper surveys the already available power restoration techniques of power distribution systems.

The remaining part of the article is written as follows. Section 2 presents the context of location of fault, isolation of fault and power restoration of electrical networks. Section 3 demonstrates an overview and strategies of power restoration techniques for distribution systems. Sections 4 presents the various power restoration methods is explained. Section 5 represents the conclusion (Fig. 1).

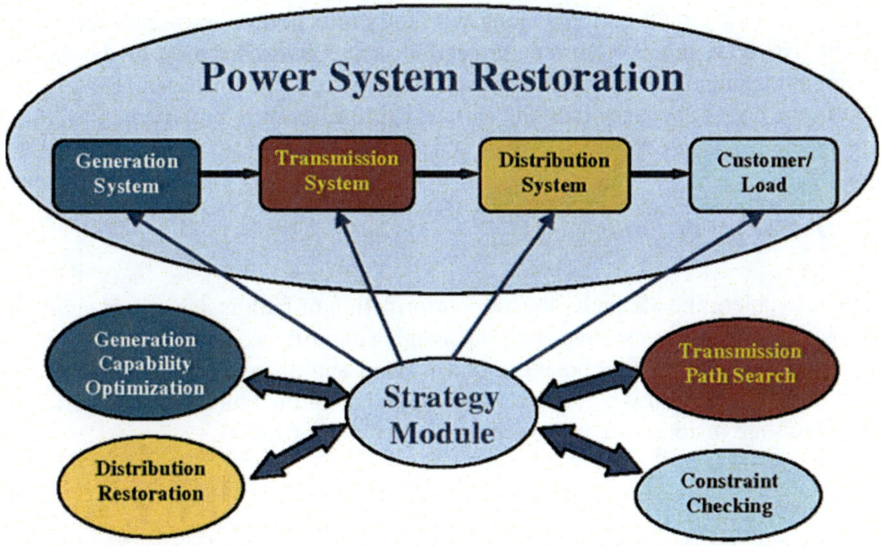

Fig. 1 Restoration for smart power distribution systems

2 Fault Location, Isolation, and Service Restoration

2.1 Location of Fault

Power distribution networks certainly dominated to diverse kind of short circuit faults. The faults can be identified through alerts depending on the over currents or minimum voltages. In illustration, if a fault happens, protection devices like Fault Passage Indicators (FPIs) and feeder circuit breakers (FCBs) detected beyond the short circuit fault current route can perform. If a fault current value reaches above the predetermined current value FCBs will trips. The time response is analyzed based on calculated value of current [15]. The universal operational scale applied to protective relay of a feeder circuit breaker is close to double of magnitude in the largest supplying load. This instrument instantly notifies to distribution automation system (DAS) regarding the abnormal currents by its responsible transmission linkage [16]. The physical position of fault occurred point necessarily to be computed after tripping the faulty feeder.

Several kinds of faults occurred in power distribution systems. Based on the survey, greater than 80% of faults are one line-to-ground fault occurred on the distribution systems. Based on the method of grounding the neutral, the one-line-to-ground fault having diverse attributes and requires various identification methodologies [17].

For appropriately grounded neutral systems, the short-circuit, route is created when a one-line-to-ground fault happen. So, the more short circuit fault current entering into faulted lines and protective systems which are turn on instantly to disconnect short circuit currents along with the protection of zero-sequence over-current. Anyhow, the over-current protection is inefficient because of the vague properties belonging to the states of high impedance fault. In paper [18, 19], to identify the high impedance fault the various approaches were proposed.

2.2 Isolation of Fault

In order to determine the faulty location, information of fault is naturally examined through data acquisition system applying suitable algorithms. Then the least path of the network is separated by breaking downstream and upstream switches from all directions from fault to separate the fault. Immediately the faulty part is separated, the systems are restored via closing of the Field Circuit Breaker.

2.3 Restoration of Fault

Power restoration is the very essential control scheme for Distribution Automation (DA) systems. The faulty network was separated, whenever the short circuit fault take place in power distributed networks. In this moment, few of active portion of network become de-activated. Power recovery is applied for maintaining its continuity of supply of de-powered parts. In previous days schemes and regulations were little to utility systems because of minimum size of distribution networks, i.e., no systematic algorithms were framed to reconfigure the power delivery networks.

When the complexity of the power system was increased, the issues facing by engineers increased and restoring system in minimum interval is the major challenge. For those complex power systems additional powerful restoration strategies were essential. To execute the effective restoration, the schemes and rules were framed by the IEEE Power and Energy Society (PES). Many areas were surveyed such as overvoltage, pick up the load in system, etc.

3 Overview of Power System Restoration

In previous days, almost all the industries have established their individual power restoration schemes and guidelines. Those schemes were depending on the working conditions relevant to the operation of the power system. Analytical tools based power restoration schemes were derived by Potomac Electric Company (PEC) [20]. Still the common scheme was not developed to restoration of power systems. It is essential to develop the centralized schemes for the restoration. In paper [14] it was explained about the prime mover characteristics after the fault occurrences. In [21] the impact on security by applying two levels power recovery scheme was reviewed.

During initial level comprises of reconfiguration a present group user's also next stage extra group of customers is restored through manual operations.

The various methodologies were presented to explain the service restoration issues. Based on the execution, it should be classified into two methods, such as centralized approaches and decentralized approaches. In centralized approaches, every network function details were gathered through intelligent electronic devices (IEDs) also transmitted into the central room via transmission medium. Behind the segregation of affected area, the control room will execute the restoration techniques and establish the continuous switching functions. Anyhow, in decentralized approaches the scheme of switching functions is collected through coordination and communication among the Intelligent Electronic Devices (IEDs). The types of centralized and decentralized approaches are explained in Sect. 4.

The centralized and decentralized structure, the service restoration issues is resolved through various methods, out of these expert systems, metaheuristic optimization etc., was applied to work out the issues based on its defined algorithms, guidelines computed from heuristics and domain specific knowledge, the stochastic programming and heuristic rules resolve the issues depend on the mathematical optimization systems.

Followed by the several restoration techniques, the successful implementation of service restoration is depending upon the hardware systems. The recently used instruments can be modernized to manage with restoration techniques. For illustration, manually operated switches can be repossessed by automatic regulated switches, and advanced intelligent systems, such as intelligent electronic devices and remote terminal units are needed. Furthermore, the reliable and effective bi directional transmission medium systems were needed to assure secure distribution of information details. The flow chart for power system restoration is indicated in below flowchart (Fig. 2).

4 Methods for Service Restoration

This part explains the various existing centralized and decentralized optimization methods of power restoration schemes.

Centralized schemes of power restoration.

4.1 Knowledge Based Approach

Knowledge based technique for power restoration is presented [22]. This method is milestone and initiating work developed for power restoration. This method is assumed to be executing the meta-heuristic approach to perform power restoration issues in electrical distribution system. Here sixteen rules are framed for restoration.

Fig. 2 Flow chart for restoration scheme

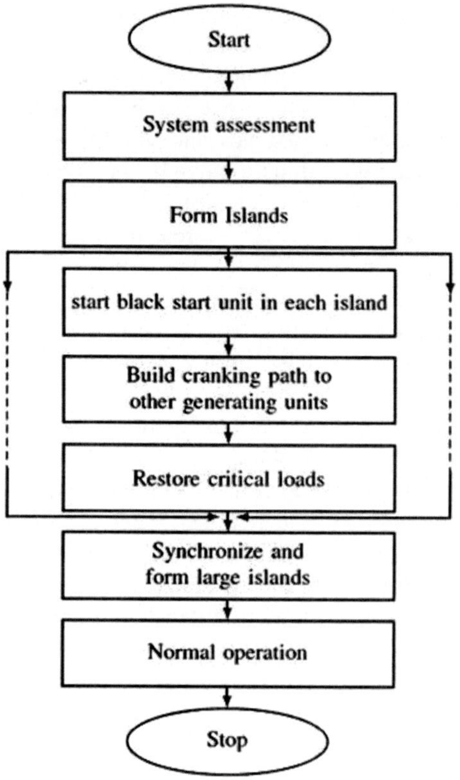

Main objective of this method is to perform and support full in developing rules for the service restoration issues. In paper [23], power system planning, restoration, and degradation problems are described for stabilization of power systems during restoration. In this paper, the prominence is on combination of conventional method along with knowledge based system used for restoration of power system. Paper [24] describes on the consequences and the study of important constraints, specification during power restoration, managing of constraints variables, generator start-ups sequencing, finding of transmission path, simulations and design of software. To restore power system, knowledge based approach was used in this paper.

In paper [25] knowledge based approach was proposed. It is exchange of knowledge of expert with the help of few set of algorithms (if and else rules) also make use of a presumption mechanism to infer from those algorithms [25]. In paper [26] framed knowledge support with 180 rules obtained through survey and deliberation with electrical engineers. The framed expert systems should be implemented load transfer, zone restoration, and group restoration.

The knowledge systems is established in paper [27] along with algorithms divided by the power outages rules which divides the outage zone into individual grouping or multitudinous grouping depend on tie switches, branch edges, and feeder margins.

The expert systems were established on paper [27] which depends on inference pattern and its attributes.

4.2 Heuristic Algorithm

This algorithm uses domain based heuristics to lead the process of searching. In paper [28–31], various procedures of service restoration were formed depending on heuristic algorithms and converted into software programming to work out the issue. The heuristic process was converted into a heuristic algorithm in paper [28]. To restore the continuity of supply the supportive feeders are initially applied, if the entire restoration falls, loads should be fully restored through supporting laterally was restored continued by the loads but not be entirely restored through supporting systems.

In paper [29], the typical structure for framing the rules for service restoration in electrical system and configuration of network was elaborated. The problem for the restoration is developed in this paper based on the discrete methods. Heuristic based fault restoration of the distribution systems was elaborated in this paper. In paper [30] to solve the restoration issue the set of fuzzy rules was framed to address the imprecise system variables.

In paper [32] in comparison analysis of various advanced heuristic optimizations (AHO's) and reactive tabu search optimization techniques used for recovery of power in power delivery networks. Tabu search optimization algorithm provided better results in those mentioned algorithms because of quickest computing and provides best outputs. The newest optimization to reduce the line losses depending on nodal power for reconfiguration of feeder is presented [33]. The reconfiguration of sequential switching operation is presented depending upon the power flow of branch rather than flow of current. In paper [34], meta-heuristic techniques were presented to resolving the service restoration issue for the unbalanced three phase distribution network.

Selection of the switching is depending upon two types of indices. The issue framed was inhibited multi-objective function contains the following objectives; minimizing the impact of faulted area, minimizing amount of transient operations, also the priority of customer's.

4.3 Fuzzy Control System

Paper [35] provides the optimization technique for reduce the amount of switching functions. The technique depending on set of fuzzy rules was presented for to compute adjacent system parameters. Paper [36] advanced methodologies is formed to restoration of power supply. This scheme contains of two components such as fuzzy decision making and candidate set generation. It makes the candidate position of feasible restoration scheme achieved by continuation six fundamental restoration approaches explained in this paper. To improve the manufacturing of software object oriented programming language was applied.

4.4 Genetic Algorithm

In paper [37] to restoration of power supply, Parallel Genetic Algorithm (PGA) was applied. Algorithm framed in this paper to reduce the cost of hardware components and increases the speed of computation. The algorithmic technique used in paper [38] gives the objective function for best load dropping. The strategy was applied for this algorithm is integer permutation encoding. The scheme every chromosome indicates group of regulated switching systems also conditions in every switching are described with the help of best optimization techniques in consideration with radility of power network.

4.5 Petri Nets

The paper [39] presents about the common process of power restoration actions using petri net method. This method is applied to indicate the Generic Restoration Action. Also petri net applied to sample the conditions as well as changing an energy level during restoration. The moment to recovering of the power networks should be evaluated to receive the balanced system.

4.6 Artificial Neural Network (ANN)

The paper [40] is applied for power restoration with help of presents Artificial Neural Network (ANN). Because ANN has more performing speed and capacity. Supervised learning with multi-layered perceptron is applied in this paper.

4.7 Ant Colony Optimization Technique

This technique is a type of probabilistic technique. This is applied to compute the combinational optimization task. This optimization is derived through via artificial swarm intelligence. This approach is applied to minimize the power problem during the process of restoration [41].

4.8 Multi Agent System

In paper [42] to achieve the power restoration the multi Agent System was presented. Multi agent system provides the advanced level of autonomous software conception. Load shedding with partial and entire restoration and substituted restoration direction was illustrated by applying multi agent system. Advanced distributed optimization algorithm for power restoration is presented in paper [43] along the support of distributed energy storage systems. Artificial intelligent agent is applied for this distributed algorithm. The various operations were executed through these agents such as identification of fault, determining the fault location, and isolation of fault place, and de-energizing the restoration zone was executed through switch agents. Islanding and grid integration of power system the renewable energy systems will help to delivering continuous power to this.

4.9 Particle Swarm Optimization (PSO)

PSO was applied to restoration of power supply in the distributed energy system in paper [41]. This method presents the application of particle swarm optimization acting as decision supporting device. This technique provides a group of switches to receive better outputs based on what power maximization is applied, minimum amount of switching functions, and also no overloading.

5 Conclusion

This paper presented the various research work given to resolve service restoration issues in previous decades. The various restoration schemes, issues such as knowledge Based System, heuristic rules, stochastic programming, and multi-agent systems, are also given in this article. There are enormous changes in techniques applied for restoration of power, from traditional techniques to artificial intelligence (AI) based optimization techniques. By applying AI methodology for power recovery was shown better results such as reduction of loss, minimum restoration time, and security enhancement.

References

1. Moslehi K, Kumar R (2010) A reliability perspective of the smart grid. IEEE Trans Smart Grid 1(1):57–64
2. Nguyen CP, Flueck AJ (2012) Agent based restoration with distributed energy storage support in smart grids. IEEE Trans Smart Grid 3(2):1029–1038

3. Bhaskar MA, Sarathkumar D, Anand M (2014) Transient stability enhancement by using fuel cell as STATCOM. In: International Conference on Electronics and Communication Systems (ICECS). IEEE, Coimbatore, India, pp 1–5
4. Massoud Amin S, Wollenberg BF (2015) Toward a smart grid: power delivery for the 21st century. IEEE Power Energy Mag 3(5):34–41
5. Aguero JR (2012) Applying self-healing schemes to modern power distribution systems. In: Proceedings IEEE power energy society general meeting. San Diego, CA, USA, pp 1–4
6. Massound A (2015) A smart self-healing grid: in pursuit of a more reliable and resilient system. IEEE Power Energy Mag 12(1):108–112
7. Cavalcante PL et al (2016) Centralized self-healing scheme for electrical distribution systems. IEEE Trans Smart Grid 7(1):145–155
8. Parikh P, Voloh I, Mahony M (2013) Fault location, isolation, and service restoration (FLISR) technique using IEC 61850 GOOSE. In: Proceedings IEEE power energy society general meeting. Vancouver, BC, Canada, pp 1–6
9. Kleinberg MR, Miller KM, Chiang H (2014) Improving service restoration of power distribution systems through load curtailment of in-service customers. IEEE Trans Power Syst 26(3):1110–1117
10. Golshani WS, Zhou Q, Zheng QP, Tong J (2017) Two-stage adaptive restoration decision support system for a self-healing power grid. IEEE Trans Ind Inform 13(6):2802–2812
11. Chen SJ, Zhan TS, Huang CH, Chen JL, Lin CH (2015) Nontechnical loss and outage detection using fractional-order self-synchronization error-based fuzzy petri nets in micro-distribution systems. IEEE Trans Smart Grid 6(1):411–420
12. Lee H-J, Park Y-M (2016) A restoration aid expert system for distribution substations. IEEE Trans Power Delivery 11(4):1765–1770
13. Dai W, Riliskis L, Wang P, Vyatkin V, Guan X (2018) A cloud-based decision support system for self-healing in distributed automation systems using fault tree analysis. IEEE Trans Ind Inform 14(3):989–1000
14. Zidan et al (2017) Fault detection, isolation, and service restoration in distribution systems: state-of-the-art and future trends. IEEE Trans Smart Grid 8(5):2170–2185
15. Milioudis N, Andreou GT, Labridis DP (2012) Enhanced protection scheme for smart grids using power line communications techniques–Part I: detection of high impedance fault occurrence. IEEE Trans Smart Grid 3(4):1621–1630
16. S. Kazemi, Millar RJ, Lehtonen M (2014) Criticality analysis of failure to communicate in automated fault-management schemes. IEEE Trans Power Delivery 29(3):1083–1091
17. Liu J, Dong X, Chen X et al (2016) Fault location and service restoration for electrical distribution systems. Wiley, Beijing, China
18. Elmitwally, Elsaid M, Elgamal M, Chen Z (2015) A fuzzy multiagent service restoration scheme for distribution system with distributed generation. EEE Trans Sustain Energy 6(3):810–821
19. M. Aucoin and R. H. Jones.: High impedance fault detection implementation issues. IEEE Transactions on Power Delivery 11(1), 139–148(2016).
20. Coelho, Rodrigues AB, Da Silva MG (2014) Distribution network reconfiguration with reliability constraints. In: Proceedings of international conference on power system technology. Singapore, pp 1600–1606
21. Latare NA, Bhat SS, Srivastava I (2017) Literature review of service restoration in distribution system. In: 2th International Conference on Electrical, Computer and Communication Technologies (ICECCT), Coimbatore, pp 1–6
22. Albert Alexander S (2017) Development of high performance solar photovoltaic inverter with advanced modulation techniques to improve power quality. Int J Electron (Taylor and Francis)104(2):174–189
23. Srinivasan SD, Trivedi A (2015) A decentralized multiagent system approach for service restoration using DG islanding. IEEE Trans Smart Grid 6(6):2784–2793
24. Sanches S, London Jr. JBA, Delbem ACB (2014) Multiobjective evolutionary algorithm for single and multiple fault service restoration in large-scale distribution systems. Elect Power Syst Res 110(1):144–153

25. Srivastava S, Butler-Burry KL (2016) Expert-system method for automatic reconfiguration for restoration of shipboard power systems. IEE Proc Gene Transm Distrib 153(3):253–260
26. Liu C, Lee SJ, Venkata SS (2018) An expert system operational aid for restoration and loss reduction of distribution systems. IEEE Trans Power Syst 3(2):619–626
27. Lee SJ, Kim KH, Kim HY, Lee JK, Nam KY (2012) Expert system-aided service restoration in distribution automation. In: Proceedings of IEEE international conference on systems, man and cybernetics, vol 1(1). Chicago, IL, USA, pp 157–161
28. Chen C-S, Lin C-H, Tsai H-Y (2012) A rule-based expert system with colored petri net models for distribution system service restoration. IEEE Trans Power Syst 174(4):1073–1080
29. Tsai MS (2018) Development of an object-oriented service restoration expert system with load variations. IEEE Trans Power Syst 23(1):219–225
30. Hsu YY et al (2012) Distribution system service restoration using a heuristic search approach. IEEE Trans Power Delivery 7(2):734–740
31. Miu KN, Chiang H-D, McNulty RJ (2014) Multi-tier service restoration through network reconfiguration and capacitor control for large-scale radial distribution networks. IEEE Trans Power Syst 15(3):1001–1007
32. Kleinberg MR, Miu K, Chiang HD (2011) Improving Service restoration of power distribution systems through load curtailment of in-service customers. IEEE Trans Power Syst 26(3):1110–1117
33. Shirmohammadi (20178) Service restoration in distribution networks via network reconfiguration. IEEE Trans Power Delivery 7(2):952–958
34. Morelato L, Monticelli AJ (2019) Heuristic search approach to distribution system restoration. IEEE Trans Power Delivery 4(4):2235–2241
35. Sarathkumar D, Srinivasan M, Stonier AA, Samikannu R (2021) A research survey on microgrid faults and protection approaches. In: IOP conference series: materials science and engineering, vol 1055(012128), pp 1–15
36. Sarathkumar D, Srinivasan M, Stonier AA, Samikannu R, Vijay Anand D (2021) Design of intelligent controller for hybrid pv/ wind energy based smart grid for energy management applications. In: IOP conference series: materials science and engineering, vol 1055(012129), pp 1–15
37. Sarathkumar D, Srinivasan M, Stonier AA, Samikannu R, Dasari NR, Raj RA (2021) A technical review on classification of various faults in smart grid systems. In: IOP conference series: materials science and engineering, vol 1055(012152), pp 1–11
38. Sarathkumar D, Srinivasan M, Stonier AA, Samikannu R, Dasari NR, Raj RA (2021) A technical review on self-healing control strategy for smart grid power system. In: IOP conference series: materials science and engineering, vol 1055(012153), pp 1–15
39. Albert Alexander S, Lehman B (2018) An intelligent based fault tolerant system for solar fed cascaded multilevel inverters. IEEE Trans Energy Convers 33(3):1047–1057
40. Albert S, Manigandan T (2015) Optimal harmonic stepped waveform technique for solar fed cascaded multilevel inverter. J Electr Eng Technol 10(10):742–751
41. Albert Alexander S, Manigandan T (2014) Power quality improvement in solar photovoltaic system to reduce harmonic distortions using intelligent techniques. J Renew Sustain Energy 6(4):1–19
42. Gnanavel C, Albert Alexander S (2018) Experimental validation of an eleven level symmetrical inverter using genetic algorithm and queen bee assisted genetic algorithm for solar photovoltaic applications. J Circuits Syst Comput 27(13):1850212–1850223
43. Shanmuga Aravind P, Albert Alexander S (2013) Harmonic minimization of a solar fed cascaded h bridge inverter using artificial neural network. In: 2013 international conference on energy efficient technologies for sustainability, pp 163–167

Feasibility Analysis of Standalone Hybrid Renewable Energy System for Kiltan Island in India

Mohammad Shariz Ansari

Abstract The net present cost (NPC) and the cost of energy (COE) are very high in the remote islands because energy is mainly generated from diesel generators. Pollution is also increasing due to the usage of diesel generators. So, the objective in this work is to minimize the NPC, COE, and pollution by using renewable energy sources such as solar and wind in addition to diesel. Kiltan Island of Union Territory of Lakshadweep in India has been considered for the study. The average wind speed and average daily radiation at Kiltan Island are 5.51 m/s and 5.76 kWh/m^2/day respectively. The modeling, simulation, and optimization have been performed by Hybrid Optimization Model for Electric Renewable (HOMER) software. The model that gives an optimal solution includes Diesel-PV-Battery. This model gives the least NPC and COE. For sensitivity analysis, the wind speed (m/s) and the fuel rate ($) have been considered as sensitivity variables. The optimal system is obtained for the sensitivity variable of 5.21 m/s of wind speed and 1$/L of diesel price. This optimal system also reduces pollution by 90%.

Keywords Optimal system · Net Present Cost (NPC) · Cost of Energy (COE) · Renewable energy · HOMER

1 Introduction

Energy is the primary and most universal need for all kinds of work by human beings and nature. Then again, energy is additionally and still to be the greatest emergency to individuals, since right now, most of the energy utilized on earth originates from customary petroleum products, and some of them will be depleted in a very long while as per the ongoing investigating and devouring rate [1]. Also, there are still about 1.5 billion occupants overall as yet have no accesses to power [2]. For these remote islands and towns, energy is being regularly supplied by diesel generators. In any case, they felt progressively increasingly tense since they as often as possible

M. S. Ansari (✉)
KIET Group of Institutions, Delhi-NCR, Ghaziabad, India
e-mail: shariz.ansari@kiet.edu

A. Kumar et al. (eds.), *Renewable Energy Towards Smart Grid*, Lecture Notes in Electrical Engineering 823, https://doi.org/10.1007/978-981-16-7472-3_7

face inflated fuel costs as a result of the gigantic climb in the cost of diesel and shipment cost [3]. Additionally, the negative natural effects from crafted by diesel hurt the nearby biological system and cause water, noise, air, and soil contamination. In the issue, they experience the ill effects of energy shortage or power outages often. Luckily, remote regions are typically rich in locally accessible sustainable power resources [4]. Because of the increasing expense of diesel fuel and the quickly decreasing expense of sustainable power resources, the supply of energy by renewables is presently getting to be more focused with conventional energy, accordingly encouraging broadly use of sustainable power sources for stand-alone system, for example, PV–battery, wind–battery, or hybrid system [5]. Until now, look into on RESs is normally done in the area of modeling of the system, sizing of the element, simulation, economic evaluation, and especially optimization of the system. For doing such type of research, the computer tools and simulation models are normally required. 37 PC tools for understanding the RESs have been looked into in [6], and an audit of the strategy of optimizing the hybrid renewable energy sources was done in [7, 8]. Out of these simulation devices, HOMER software is a standout among the most broadly utilized for standalone RESs [9]. Utilizing HOMER for RES techno-economic analysis, simulation and modeling have been the topic of considerable prior investigations, for instance, the likelihood of accomplishing energy sovereignty in an island utilizing wind turbine, PV, biogas generator, and battery was assessed in [3]. A techno-economic investigation of Renewable Energy System for Rural Electrification in South Algeria has been done in [10].

2 Profile of Kiltan Island

Kiltan is situated 51 km north-east of Amini Island. The distance from Kiltan Island and Kochi is 213 nautical miles (404 km). It is shown in Fig. 1. The area of Kiltan Island is 2.20 km^2. The Island largest length is 3.4 km and breadth is 0.6 km at its broadest point. There are high storm beaches on the southern and northern ends of the Island. It is situated between latitude 11°28′ and 11°30′ N and longitude 72°59′ and 73°01′ E. Kiltan is having a lagoon of length 1.76 km [11].

The power is generated mainly from wind, solar, and biomass in renewable category.

3 Optimal Hybrid Model for Kiltan Island

3.1 Meteorological Data for Kiltan Island

Average monthly Solar Global Irradiance (SGI) Data for Kiltan Island is downloaded from NASA data base. Average monthly global horizontal radiation has been

Fig. 1 Location of Kiltan Islands of UTL

recorded by NASA for duration of 22 years from 1983 to 2005. The monthly average solar global irradiance for Kiltan Island is show in Table 1 and in Fig. 2. Table 1 gives the index of clearness and daily radiation in kWh/m^2/day.

This table shows that the maximum clearness index is 0.682 and minimum clearness index is 0.449. This maximum index of clearness occurs in February and March, and the minimum index of clearness occurs in July. The maximum and minimum daily radiation shown in Table 1 is 6.93kWh/m^2/day and 4.67kWh/m^2/day respectively. The maximum solar radiation at selected location falls in March and April, and the minimum falls in July. The annual average radiation is 5.76kWh/m^2/day.

Table 1 Average monthly Solar Global Irradiance (SGI) Data for Kiltan Island[12]

Month	Index of clearness	Daily solar radiation (kWh/m^2/day)	Month	Index of clearness	Daily solar radiation (kWh/m^2/day)
Jan	0.660	5.74	Jul	0.449	4.67
Feb	0.682	6.42	Aug	0.505	5.28
Mar	0.682	6.93	Sep	0.561	5.74
Apr	0.658	6.93	Oct	0.551	5.29
May	0.575	6.04	Nov	0.599	5.29
Jun	0.459	4.77	Dec	0.644	5.43

Annual average radiation = 5.76 kWh/m^2/day

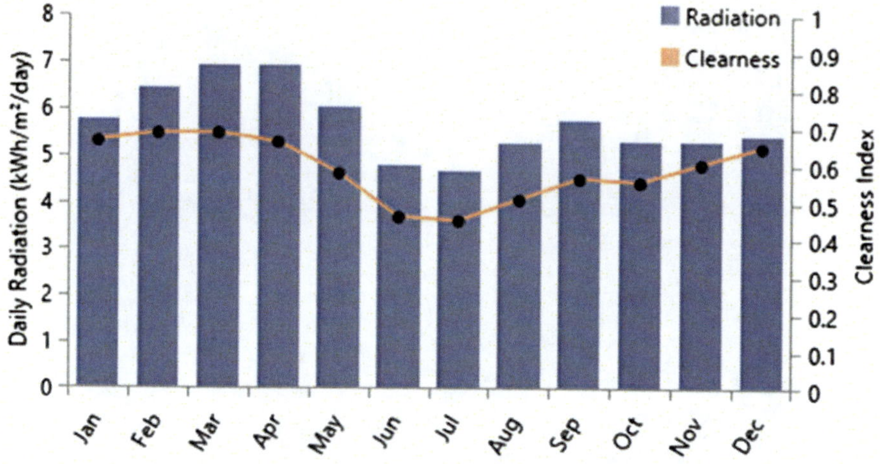

Fig. 2 Average monthly Solar Global Irradiation (SGI) Data at Kiltan Island

Table 2 Average monthly wind velocity data at Kiltan Island in m/s[12]

Month	Average wind speed (m/s)	Month	Average wind speed (m/s)	Month	Average wind speed (m/s)
Jan	4.12	May	5.2	Sep	6.08
Feb	3.66	Jun	8.95	Oct	4.22
Mar	4.15	Jul	8.3	Nov	4.1
Apr	4.4	Aug	7.92	Dec	5

Average value of wind velocity = 5.51 m/s

Average monthly wind velocity data at Kiltan Island in (m/s) is downloaded from NASA data base. Wind speed is noted at 50 m above the earth's surface for territory same as airports. Monthly average wind speed has been recorded by NASA for duration of 10 years from 1983 to 1993. Average monthly wind velocity is shown in Table 2 and in Fig. 3.

Maximum wind velocity is 8.95 m/s, and it is available in June. Minimum wind velocity is 3.66 m/s, and it is available in February. The average wind speed is 5.51 m/s.

3.2 *Kiltan Island Load Profile*

The profile of load for the selected Island is an important input data for the optimization of proposed model in HOMER, because for an optimal sizing, different energy sources depend on the load profile. Figure 6.6 shows the load profile of

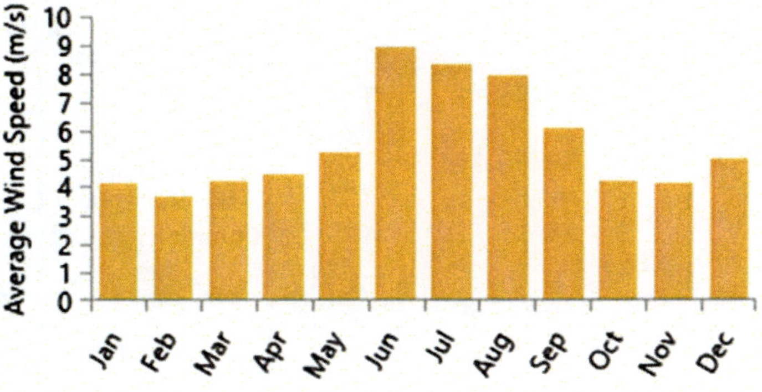

Fig. 3 Average monthly wind velocity data at Kiltan Island

Kiltan Island. Figure 4a, b shows the month wise and daily residential load shape respectively. Figure 4c, d shows the month wise and daily commercial load shape. In Kiltan Island, generated energy is mainly used to meet the residential and commercial loads only. Hence, we will consider only these two types of load in HOMER for optimization and simulation.

The consumptions of energy annually of Kiltan Island by the residential and commercial load are 1.87 MU and 0.41 MU respectively in 2016 [11]. These data are collected from Indian Government and Administration of UTL. These data then converted into daily energy consumption as 7,342.47kWh/day for residential load and 1,561.64kWh/day for the commercial load. The peak loads of residential and

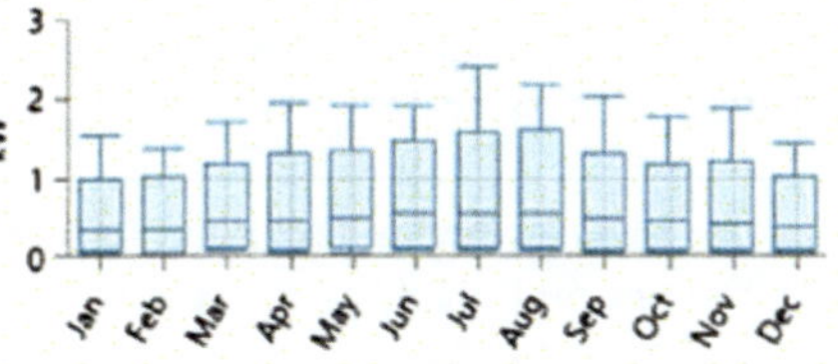

(a) Monthly residential load profile

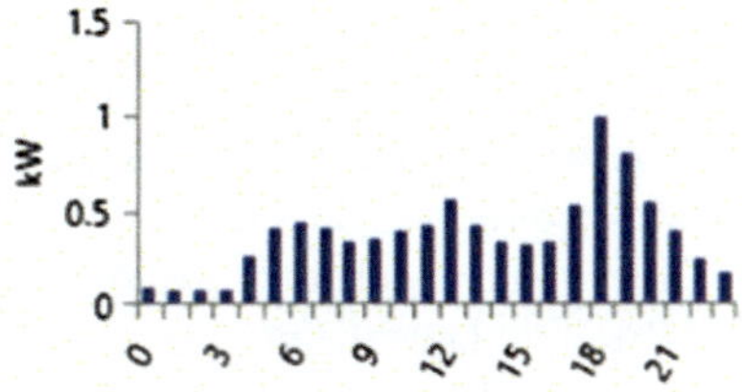

(b) Daily residential load profile

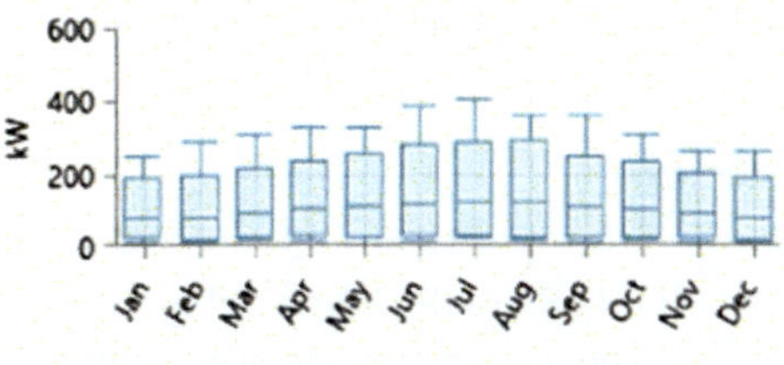

(c) Monthly commercial load profile

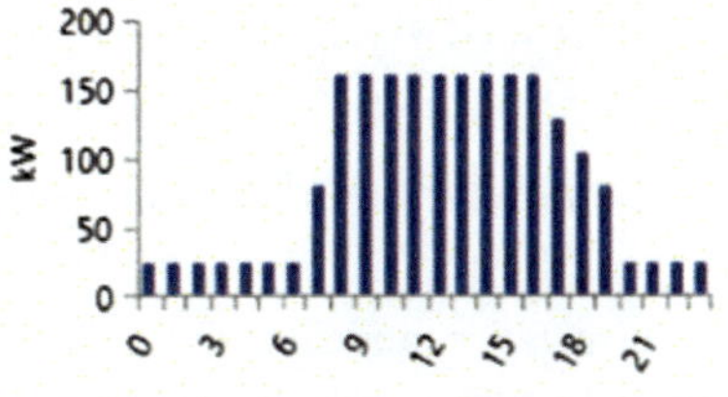

(d) Daily commercial load profile

Fig. 4 Load profile of Kiltan Island

Fig. 5 Proposed system model for Kiltan

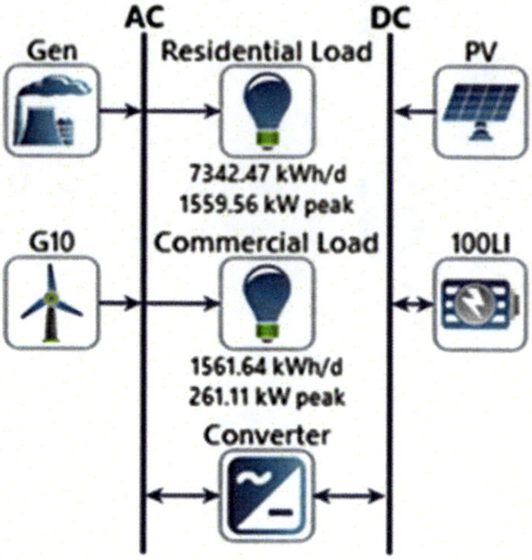

commercial loads are 1,559.56 kW and 261.11 kW respectively. These daily load data and peak load data are utilized for simulation and optimization.

4 System Architecture of Proposed Model for Kiltan Island

In Kiltan Island power is generated mainly from diesel generators but there are huge potential of solar PV and wind is also available. Since diesel is costly and it is depleted very fast so the aim is to reduce the use of diesel and make the most of the usage of RES such as wind energy and SPV. I have proposed a model shown in Fig. 5 that consists of wind turbines, solar PV, and diesel generators for emergency use, batteries, converters, and loads.

The rating and cost of these components have been described in previous chapter. HOMER simulates the proposed model to give the optimal result.

5 Results of Proposed Model for Kiltan Island

5.1 Sensitivity and Optimization Results

The optimization and sensitivity results from HOMER are displayed in Fig. 6. The upper part of the figure shows the result of different sensitivity variables and lower part gives the optimization result for a particular selected sensitivity variable. The sensitivity variables for diesel price have been taken as 0.9$/L, 1$/L, and 1.1$/L. The sensitivity variables for wind speed have been taken as 3.0 m/s, 8 m/s, and 5.51 m/s.

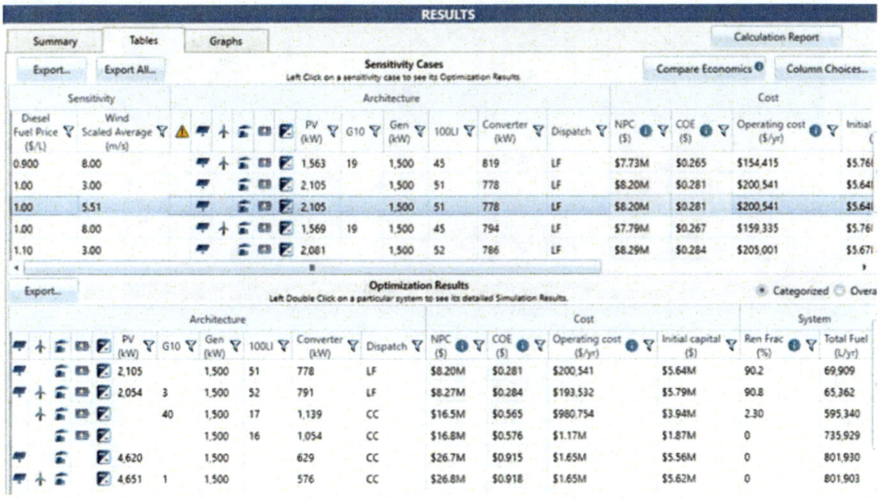

RESULTS

Summary | Tables | Graphs | Calculation Report

Export… | Export All… | **Sensitivity Cases** Left Click on a sensitivity case to see its Optimization Results | Compare Economics | Column Choices…

Diesel Fuel Price ($/L)	Wind Scaled Average (m/s)	PV (kW)	G10	Gen (kW)	100LI	Converter (kW)	Dispatch	NPC ($)	COE ($)	Operating cost ($/yr)	Initial
0.900	8.00	1,563	19	1,500	45	819	LF	$7.73M	$0.265	$154,415	$5.76
1.00	3.00	2,105		1,500	51	778	LF	$8.20M	$0.281	$200,541	$5.64
1.00	5.51	2,105		1,500	51	778	LF	$8.20M	$0.281	$200,541	$5.64
1.00	8.00	1,569	19	1,500	45	794	LF	$7.79M	$0.267	$159,335	$5.76
1.10	3.00	2,081		1,500	52	786	LF	$8.29M	$0.284	$205,001	$5.67

Export… | **Optimization Results** Left Double Click on a particular system to see its detailed Simulation Results. | Categorized | Overa

PV (kW)	G10	Gen (kW)	100LI	Converter (kW)	Dispatch	NPC ($)	COE ($)	Operating cost ($/yr)	Initial capital ($)	Ren Frac (%)	Total Fuel (L/yr)
2,105		1,500	51	778	LF	$8.20M	$0.281	$200,541	$5.64M	90.2	69,909
2,054	3	1,500	52	791	LF	$8.27M	$0.284	$193,532	$5.79M	90.8	65,362
	40	1,500	17	1,139	CC	$16.5M	$0.565	$980,754	$3.94M	2.30	595,340
		1,500	16	1,054	CC	$16.8M	$0.576	$1.17M	$1.87M	0	735,929
4,620		1,500		629	CC	$26.7M	$0.915	$1.65M	$5.56M	0	801,930
4,651	1	1,500		576	CC	$26.8M	$0.918	$1.65M	$5.62M	0	801,903

Fig. 6 HOMER sensitivity and optimization result for Kiltan Island

The wind average velocity for the considered island has been taken from NASA by HOMER, and it is 5.51 m/s. HOMER has used all the combination of these sensitivity variables for simulation and optimization. Total 26,254 solutions were simulated, and all of them are feasible. Out of them, total 3,900 were omitted in which 2,595 omitted for lacking a converter and 1,161 omitted for having an unnecessary converter. The actual value of sensitivity variables at the considered island is 5.51 m/s for wind average velocity and $1/L for price of diesel. Using these sensitivity variables for the optimum solution, HOMER uses only solar PV, diesel generators and storage systems. The optimum solution for this combination comes out to be Total NPC as $8,201,914.00 and levelized COE as 0.281 $/kWh.

Optimal system type plot for the proposed model of the Kiltan Island is shown in Fig. 7. It shows that which combination of hybrid system is optimal at a particular value of wind velocity and price of diesel. Green color shows the Diesel/Wind/Battery/PV hybrid system and red color shows the Diesel/Battery/PV hybrid system. From the optimal system type plot, it is clear that for slow wind speed at Kiltan Island Diesel/Battery/PV hybrid system gives optimal solution and if wind speed is high then Diesel/Wind/Battery/PV hybrid system gives optimal solution. Table 2 displays that the wind average velocity is 5.51 m/s. For this velocity of wind, the optimal system consists of Diesel/Battery/PV systems only. Interpolated values of hybrid system cost are shown in Fig. 8. It gives the Interpolated values of hybrid system cost at a particular point on the optimal system type plot. Figure 9 shows the surface plot for the proposed model. It gives the NPC for a particular value of wind velocity and price of diesel. In this figure the purple color shows the least NPC and red color shows the maximum NPC.

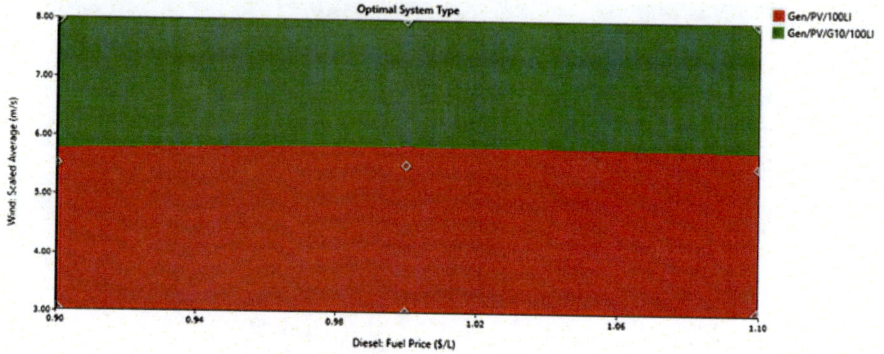

Fig. 7 Optimal system type plot for Kiltan Island

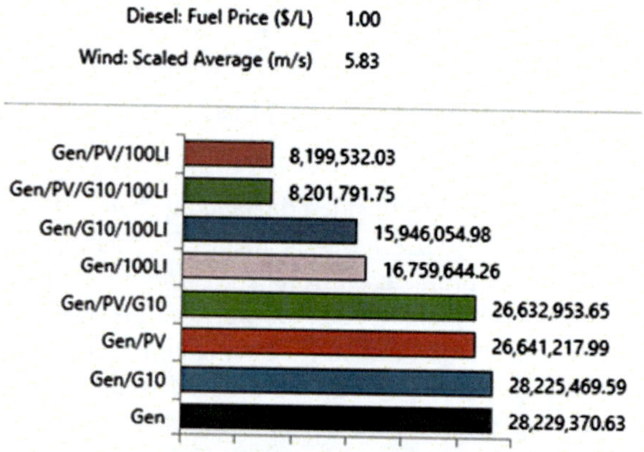

Fig. 8 Interpolated values of hybrid system cost for Kiltan Island

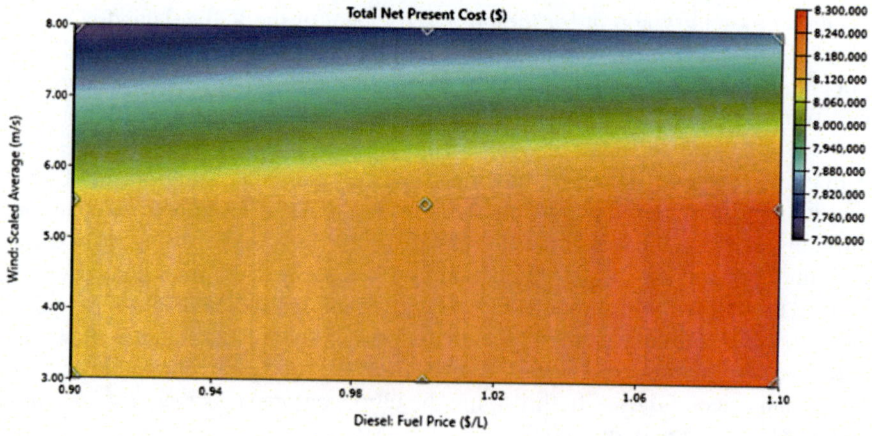

Fig. 9 Surface plot for Kiltan Island

5.2 Net Present Cost (NPC)

Figure 10 and Table 3 give the category wise NPC in bar chart form and in tabular form respectively.

Total NPC comes out to be $8,201,914.00 that consists of cost of capital, operating cost, cost of replacement, salvage value, and cost of resource for the considered model. Total NPC includes $5,640,000.00 as capital cost, $838,406.00 as operating cost, $978,844.00 as replacement cost, -$147,322.00 as salvage value, and $893,666.00 as resource cost. Among these, capital cost is supreme and it is $5,640,000.00 which includes Gen sets capital cost ($ 750,000), Li-Ion batteries capital cost ($2,550,000.00), PV capital cost ($2,110,000.00), and converters capital cost ($233,251.00). The total operating cost is $838,406.00, total replacement cost is $978,844.00, total salvage value is −$147,322.00, and total resource cost is $893,666.00.

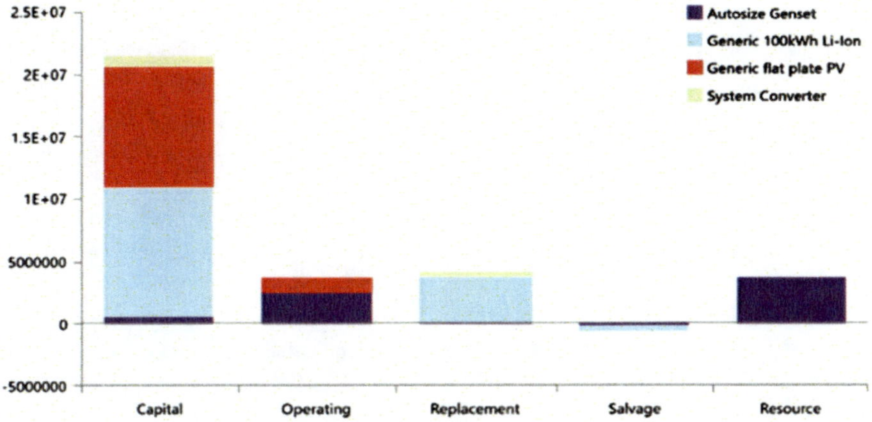

Fig. 10 Cost summary of proposed model at Kiltan Island

Table 3 Net Present Cost of proposed model at Kiltan Island

Name	Capital	Operating	Replacement	Salvage	Resource	Total
Autosize Genset	$750,000	$562,787	$0.00	−$3,786	$893,666	$2.20 M
Generic 100kWh Li-Ion	$2.55 M	$6,520	$881,516	−$125,420	$0.00	$3.31 M
Generic flat plate PV	$2.11 M	$269,099	$0.00	$0.00	$0.00	$2.37 M
System Converter	$233,251	$0.00	$97,327	−$18,116	$0.00	$312,463
System	$5.64 M	$838,406	$978,844	−$147,322	$893,666	$8.20 M

5.3 Output Power of the Different Resources in the Optimal System

The optimal hybrid system for the actual value of sensitivity variable, i.e., diesel price 1$/L and wind speed 5.51 m/s at the Kiltan Island consist of auto size genset, generic 100kWh Li-ion battery, generic flat plate PV, and system converter. Figure 11 shows monthly electrical energy output of the optimum system. From Fig. 6.71 it is clearly seen that the most of the energy is generated by RES, i.e., in this case PV system. For the period of Jan, Feb, Mar, Apr, and Dec renewable fraction is 100% which means PV generation itself is enough to meet the load demand. Figure 12 shows the diesel generator output power of the optimal system. This figure shows that for every hour of a day and throughout the year what is the generation of the diesel generator. In this figure black color shows the minimum generated power and red color shows the maximum generated power. The maximum generated power from diesel generator is 1,000 kW, and minimum power generation from diesel generator is 0.0 kW. This

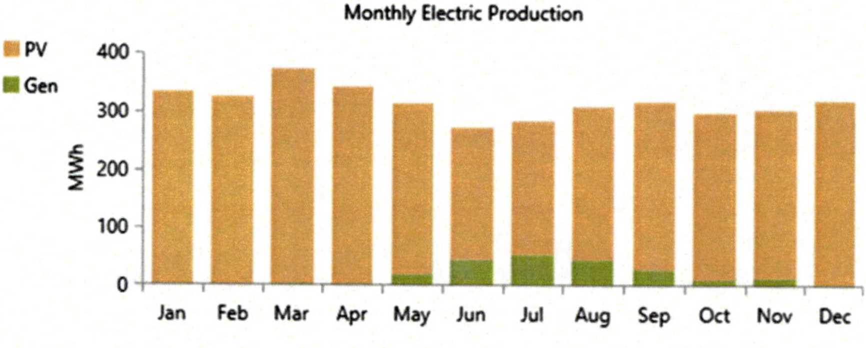

Fig. 11 Monthly electrical power output of the optimal system

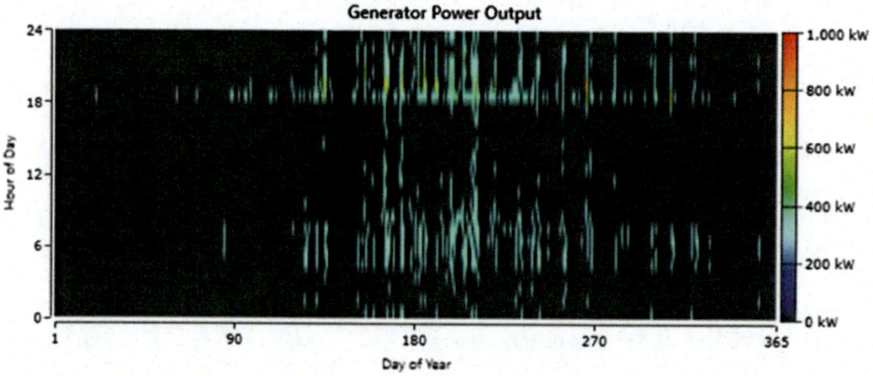

Fig. 12 Diesel generator power output of the optimal system

figure shows that most of the time, diesel generator is in off position and only PV will supply to the load. Figure 13 displays the PV output power of the optimal system. This figure shows that for every hour of a day and throughout the year what is the output power of the PV system. In this case black color shows the minimum power output and yellow color shows the maximum output power. The maximum output power from PV system is 2,500 kW, and minimum are 0.0 kW. From Fig. 13 it is clearly seen that since no sun in available during 0 to 6 h and again from 18 to 24 h so no power output is available by PV system during that periods of the day. Figure 14 displays the converter output power of the optimal system. The upper gives the inverter output and lower figure shows the rectifier output. This figure shows that for every hour of a day and throughout the year what is the output power of the

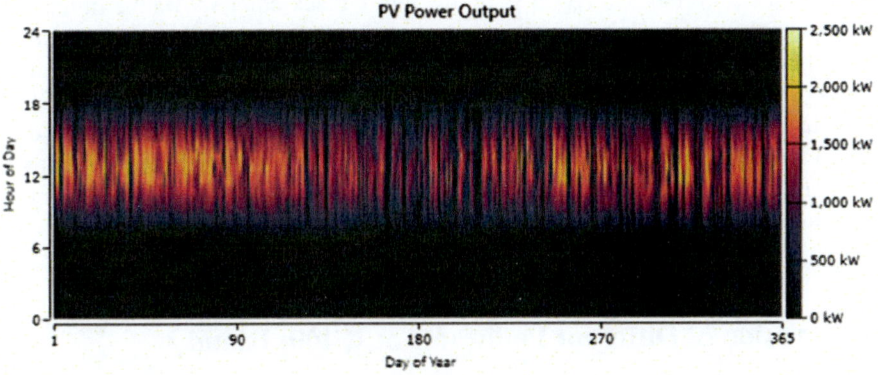

Fig. 13 PV power output of the optimal system

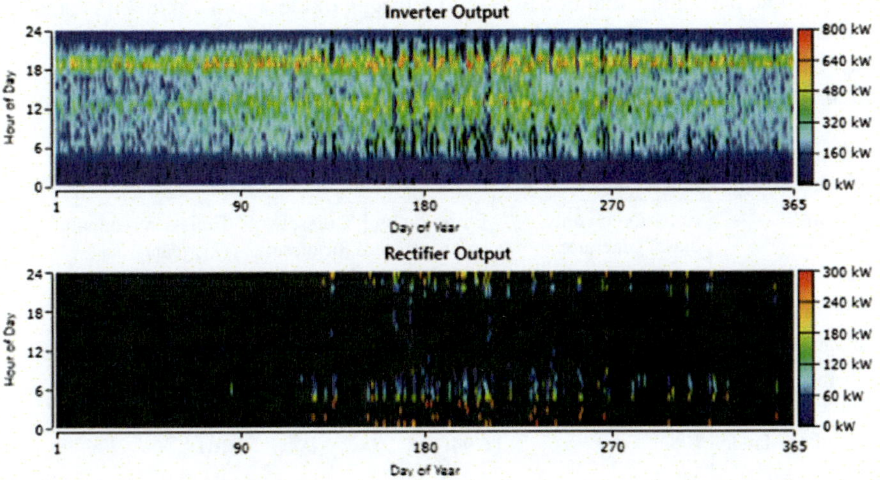

Fig. 14 Converter power output of the optimal system

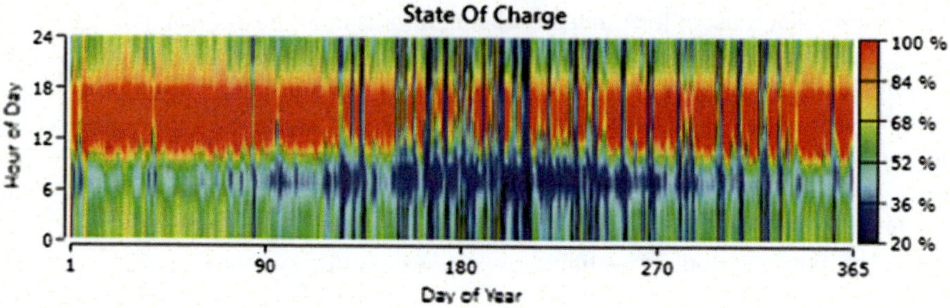

Fig. 15 Battery state of charge for the optimal system

converter, i.e., inverter and rectifier. In this figure black color shows the minimum power output and red color shows the maximum power output. The maximum output power is 800 kW and 300 kW for inverter and rectifier respectively, and minimum output power is 0.0 kW for both inverter and rectifier. Figure 15 displays the SOC of the battery for the optimal system. This figure shows that for every hour of a day and all through the year what is the SOC of the battery storage system. Here again black color shows the minimum, i.e., 0.0% state of charge and red color show the maximum, i.e., 100% state of charge.

6 Emission of Different Pollutants at Kiltan Island

Table 4 shows the emission of different types of pollutant materials by only diesel system and by a hybrid system. These pollutant materials are CO_2, CO, unburned hydrocarbon, particulate matters, SO_2, and nitrogen oxide. From Table 4 it is clearly seen that a huge amount of pollution is reduced if we use a hybrid system in place of a diesel only system. In a diesel-only system the CO_2 emission is 2,827,596 kg/year and in the hybrid system the CO_2 emission is 181,716 kg/year. It means a reduction in

Table 4 Emission by diesel only system and hybrid system

Pollutant	Emission by only diesel generator system (kg/year)	Emission by PV-diesel-battery system for diesel price $1/L (kg/year)	Emission reduction (kg/year)
Carbon Dioxide	28,27,596	1,81,716	26,45,880
Carbon Monoxide	17,824	1,145	16,679
Unburned Hydrocarbons	778	50	728
Particulate Matter	108	6.94	101
Sulphur Dioxide	6,924	445	6,479
Nitrogen Oxides	16,743	1,076	15,667

CO_2 emission is 2,645,880 kg/year which is a great reduction. The reduction of other pollutant materials such as carbon monoxide, sulfur dioxide, and nitrogen dioxide are mentioned in Table 4.

7 Comparative Analysis of Hybrid System and Diesel Only System

Table 5 gives Comparative analysis of Hybrid and Diesel only system for different value of sensitivity variables, i.e., for different value of diesel price and wind speed. This table shows that NPC of $7.73 M and COE of 0.265 $/kWh are minimum for a hybrid arrangement consisting of Diesel/Wind/Battery/PV for diesel price 0.9 $/L and wind speed 8 m/s. But the wind average velocity is limited to 5.51 m/s, and the actual price of diesel as of today is $1/L at the considered island. That is why we are considering wind velocity of 5.51 m/s and price of diesel as $1/L for the actual value of NPC and COE that comes out to be $8.2 M and 0.281 $/kWh respectively for a hybrid arrangement comprising of Diesel/Battery/PV systems. Table 5 also shows that the COE and NPC for the same sensitivity variable of diesel only system and that comes out to be $28.3 M and 0.970 $/kWh respectively. Comparing these two values for hybrid and diesel only system at the same wind speed and diesel price, i.e., 5.51 m/s and $1/L respectively it is clearly seen that the NPC is reducing from $28.3 M to $8.2 M and COE is reducing from 0.970 $/kWh to 0.281 $/kWh. It means if we use a hybrid system in place of only diesel system then there will be a great reduction in NPC and COE. If we use a hybrid system instead of diesel only system then then CO_2 emission will also be reduced from 2,827,596 kg/year to 182,994 kg/year which is again a great reduction.

8 Conclusions

Sensitivity analysis and optimization results are shown in Fig. 6. In order to get the optimal result HOMER simulates the proposed model for thousands of combinations. The optimal solution includes Diesel/Battery/PV storage and converters. For this optimal system, the NPC is $8,201,914.00 and Levelized COE is 0.281 $/kWh which is minimum. The wind speed in (m/s) and fuel rate in dollars ($) are taken as variables for sensitivity analysis. The above NPC and COE have been calculated for diesel rate $1/L and wind speed 5.51 m/s. Figure 7 displays the optimal system type plot. This plot shows that for different value of wind speed and for different value of diesel price which type of system is economical, i.e., Diesel/Battery/PV system is economical or Diesel/Wind/Battery/PV system is economical. From the optimum system type plot, it is clearly visible that for a higher value of wind velocity, Diesel/Wind/Battery/PV system is economical and for a lower value of wind velocity,

Table 5 Comparative analyses of diesel only system and Hybrid system (Kiltan Island)

Sensitivity variables		Hybrid system					Diesel only system		
Diesel fuel price ($/L)	Wind speed (m/s)	System	NPC (M$)	COE ($/kWh)	Renewable fraction (%)	CO2 Emission (kg/year)	NPC (M$)	COE ($/kWh)	CO2 Emission (kg/year)
0.9	3	PV-Diesel-Battery	8.11	0.278	90.2	1,81,716	26.9	0.922	28,27,596
0.9	5.51	PV-Diesel-Battery	8.11	0.278	90.2	1,87,705	26.9	0.922	28,27,596
0.9	8	Wind-PV-Diesel-Battery	7.73	0.265	93.7	1,17,080	26.9	0.922	28,27,596
1	3	PV-Diesel-Battery	8.2	0.281	90.2	1,82,994	28.3	0.97	28,27,596
1	5.51	PV-Diesel-Battery	8.2	0.281	90.2	1,82,994	28.3	0.97	28,27,596
1	8	Wind-PV-Diesel-Battery	7.79	0.267	93.6	1,18,546	28.3	0.97	28,27,596
1.1	3	PV-Diesel-Battery	8.29	0.284	90.3	1,81,138	29.6	1.02	28,27,596
1.1	5.51	PV-Diesel-Battery	8.29	0.284	90.2	1,81,705	29.6	1.02	28,27,596
1.1	8	Wind-PV-Diesel-Battery	7.84	0.269	94.1	1,10,844	29.6	1.02	28,27,596

Diesel/Battery/PV system is economical. Figure 10 and Table 3 display the cost summary of NPC that contains cost of capital, operating cost, cost of replacement cost, salvage value, and resource cost. Among these, capital cost is maximum, and it is $5,640,000.00 that includes Gen sets capital cost ($ 750,000), Li-Ion batteries capital cost ($2,550,000.00), PV capital cost ($2,110,000.00), and converter capital cost ($233,251.00). The total operating cost is $838,406.00, total replacement cost is $978,844.00, total salvage value is -$147,322.00, and total resource cost is $893,666.00.

References

1. Legros G, Havet I, Bruce N, Bonjour S (2009) The energy access situation in developing countries. USA. https://doi.org/10.1503/cmaj.160656
2. Elsen C, Darses F, Leclercq P (2011) An aanthropo-based standpoint on mediating objects: evolution and extension of industrial design practices. Paris, France
3. Manaullah AMS (2019) Feasibility analysis of renewable energy options for the union territory of Lakshadweep Islands. Int J GlobEnergy Issues 42:63–80 https://doi.org/10.1504/IJGEI.2019.100690
4. Nasiruddin I, Khatoon S, Jalil MF, Bansal RC (2019) Shade diffusion of partial shaded PV array by using odd-even structure. Sol Energy 181:519–529. https://doi.org/10.1016/j.solener.2019.01.076
5. Dawoud SM, Lin X, Sun J, Mohsin QK, Flaih FM, Long P (2015) Reliability study of hybrid pv-wind systems to isolated micro-grid. In: Sixth international conference on intelligent control and information processing. Wuhan, China, pp 431–437
6. Manaullah AMS (2019) Techno-economic feasibility of PV-wind-diesel battery hybrid energy system for Lakshadweep Island in India. Indian J Power River Val Dev 69:141–148
7. Ansari MS, Manaullah JMF (2017) Investigation of renewable energy potential in union territory of Lakshadweep islands. In: CIPECH 2016. IEEE, Ghaziabad, India, pp 209–213. https://doi.org/10.1109/CIPECH.2016.7918768
8. Anoune K (2018) Sizing a PV-Wind based hybrid system using deterministic approach Sizing a PV-Wind based hybrid system using deterministic approach. Energy Convers Manag 169:137–148. https://doi.org/10.1016/j.enconman.2018.05.034
9. Manaullah AMS (2014) Solar photo voltaic power generation in union territory of Lakshadweep island: projected level dissemination using technology diffusion models. In: CIPECH 2014. IEEE, Ghaziabad, India, pp 448–452. https://doi.org/10.1109/CIPECH.2014.7019036, Electronic ISBN: 978-1-4799-5871-9.
10. Chermat F, Khemliche M, Badoud A, Latreche S (2018) Techno-economic feasibility study of investigation of renewable energy system for rural electrification in South Algeria 8:3421–3426
11. A Joint Initiative of Government of India and Administration of UT of Lakshadweep: 24 X 7 POWER FOR ALL (Lakshadweep Islands) (2016)
12. Stackhouse P, Whitlock C, Zhang T (2004) Solar renewable energy data sets from NASA satellites and research 279–283

Impact of Wind Power Participation on Congestion Management Considering Seasonal Load in Pool Electricity Market While Ensuring Loadability Limit

Rahul Sagwal and Ashwani Kumar

Abstract Recent shift towards Renewable Energy Sources, backed by encouraging regulatory structure of Govt. and increased awareness regarding their clean energy, greatly affects power network's operations due to intermittent behaviour of their sources. Be it wind, hydro or solar, all are intermittent in nature due to change in wind velocity, water discharge and radiation intensity respectively. It necessitates the pre-planning or day-ahead scheduling of the dispatch for optimal operation of the power network as a whole. Impact on congestion management needs to be observed on hourly basis for power system's stable operation. Current work examines the issue of congestion management in pool electricity market in presence of hourly varying wind power and seasonal load variation including 24 h load model for winter, spring, summer and fall seasons, while respecting line security limits and ensuring system voltage stability through maximization of loadability limits. Market clearance is done for the day-ahead market by submitting 3-block bid structure to the ISO. Linear curve modelling has been done for three block bid structure as a function of increment/decrement components of rescheduling. Impact of wind and seasonal load variations on congestion costs payable by ISO has been studied for IEEE 24-bus RTS system.

Keywords Congestion Management (CM) · Wind Power Plants (WPPs) · Generation rescheduling · Block bid function · Seasonal load

Nomenclature

k	refers an hour
t	refers block of generator bid function

R. Sagwal (✉)
Madhav Institute of Technology & Science, Gwalior 474005, MP, India

A. Kumar
NIT Kurukshetra, Kurukshetra, Haryana 136118, India
e-mail: ashwani.k.sharma@nitkkr.ac.in

A. Kumar et al. (eds.), *Renewable Energy Towards Smart Grid*, Lecture Notes in Electrical Engineering 823, https://doi.org/10.1007/978-981-16-7472-3_8

t_{max}	maximum no. of bid blocks i.e. 3
g	refers to thermal power plants
w	refers to wind power plants
$P_g(i,t,k)$, $P_w(i,t,k)$	base case generation plan for k hour, t bid block, i bus for thermal & wind respectively
$P_{gn}(i,t,k)$, $P_{wn}(i,t,k)$	new generation plan for k hour, t bid block, i bus for thermal & wind respectively
$P_d(i, k)$	base case demand in k hour for i bus
ΔP^{up}, ΔP^{down}	increment and decrement in active power generation
ΔCP^{up}, ΔCP^{down}	up and down regulation prices offered by generators
k1n(i,t,k)	price coefficient for linear bid curve ($/MWh) by generator for rescheduling generation
R_g^{up}, R_g^{down}	price offered ($/h) as fiexd part of linear bid curve by generators for rescheduling
P_{ij}, Q_{ij}	active & reactive power flows in line from bus i to bus j
P(i,k), Q(i,k)	active and reactive powers injected at bus i in hour k as per load variation in hour k

1 Congestion Management

1.1 Introduction

Restructuring of Electric Supply Industry (ESI) is taking place across the globe in different rates and shapes. Restructuring includes both privatization and deregulation. Deregulation of Electricity sector brought a competitiveness, amongst its so-called players, that includes GENCOs, TRANSCOs and DISCOMs. It proved good for the growth of Electricity sector as a whole. But at the same time buyers, due to now available capability to choose amongst different sellers, want to get the cheapest electricity. This natural tendency to get cheapest electricity actually resulted in magnifying the problem of transmission line's congestion. Furthermore, in today's era where Renewable Energy Sources (RESs) are gaining more and more attention due to their clean and subsidized power, the congestion problem becomes even more complicated. Intermittent nature of Renewable Energy Sources makes the job of the Independent System Operator (ISO) more complex. ISO has the responsibility to ensure secure, reliable and economic power dispatch, for which ISO makes plans for power dispatch beforehand for day-ahead market. But, again presence of intermittent RES in the system makes it an important aspect to study their impact on congestion management. In this paper, same has been done with Wind power plant in the system.

1.2 Literature Review

In literature, transmission line congestion has been tackled with different approaches. These may be classified based upon the market structure; Singh, Padhy and Sharma presented congestion management approach in pool electricity environment in [1]. Verma and Kumar tackled congestion while securing bilateral transactions amongst buyers and sellers in [2]. Bilateral model has been included in [3] as well. Open market model has been considered for study in [4, 5]. Fattahi and Ehsan [6] have considered sensitivity wise redispatching technique to manage congestion in pool market model. Redispatching of generators to alleviate congestion of transmission lines has always been the most effective technique and so has attracted a lot of researchers, the same method has been utilized in [2, 7–9].

Ultimate goal in congestion management (CM) model is to achieve optimal generation schedule in this approach. This can be achieved by creating and solving an optimization problem. This optimization problem has been solved by researchers by means of heuristic optimization techniques or by help of non-linear programming (NLP) or mixed integer non-linear programming (MINLP) [2, 10] models in General Algebraic Modeling System (GAMS) [11]. Particle swarm optimization (PSO) [8], multi-objective PSO [12] and cat swarm optimization [13] based approaches are also present in literature. These were mainly focused on generation side participation; some researchers have also focused on load side participation in congestion management by load shedding [9] and demand side management [14]. DG placement for congestion management has been explained in [15, 16]. Zonal congestion management by ac CDFs have been presented by [17]. Voltage stability limits have also been taken care of in [18–21] while removing congestion from system. Esmaili has presented stochastic approach for CM in [22]. Use of FACTS devices has also been done for congestion management as given by [7, 14, 23]. Comparative analysis of various techniques and a survey on congestion management can be seen in [24, 25]. Pricing based approach has been proposed in [26].

1.3 Major Contributions

- The current work examines how wind power source's presence on the system can impact congestion management. For this, results for optimal generation rescheduling are obtained in the presence of wind considering 24 h variation in wind power and seasonal load consisting of winter, spring, summer and fall seasons.
- Optimal rescheduling based approach has been proposed incorporating voltage stability by means of loadability limit's maximization.
- Linear block bid structure consisting of three bid blocks has been considered for market clearance in a pool electricity market environment. Optimal power flow

study has been carried out with the help of CONOPT solver by using GAMS-MATLAB interfacing. [27].

2 Mathematical Model

2.1 Congestion Management Model with Bid Function

GENCOs submit their bids to ISO beforehand, and it is ISOs duty to clear the market economically and securely. Three block bid structure is considered to be submitted by generators to ISO in this study. Congestion of transmission lines is a short term phenomenon and hence needs to be studied on hourly basis. The congestion management model for 24 h with known block bids has been described as follows: [7]. Flowchart of the methodology is represented in Appendix A.

$$\text{Min TCC} = \sum_{k=1}^{24} \text{CG1}(k) \text{ and Max loadability limit } \rho \tag{1}$$

Where

$$\begin{aligned} CG1 &= \sum_{i=1}^{ng} \sum_{t=1}^{t\max} \left(\Delta CPg^{up}(i,t,k)\right) + (\Delta CPg^{down}(i,t,k)) \\ &+ \sum_{i=1}^{nw} \sum_{t=1}^{t\max} \left(\Delta CPw^{up}(i,t,k)\right) + (\Delta CPw^{down}(i,t,k)) \end{aligned} \tag{2}$$

$$\Delta CPg^{up} = k_{1n}(i,t,k) * \Delta P_g^{up}(i,t,k) * MVA_{base} + R_g^{up}(i,t,k) \tag{3}$$

$$\Delta CPg^{down} = k_{1n}(i,t,k) * \Delta P_g^{down}(i,t,k) * MVA_{base} + R_g^{down}(i,t,k) \tag{4}$$

$$\Delta CPw^{up} = k_{1n}(i,t,k) * \Delta P_w^{up}(i,t,k) * MVA_{base} + R_w^{up}(i,t,k) \tag{5}$$

$$\Delta CPw^{down} = k_{1n}(i,t,k) * \Delta P_w^{down}(i,t,k) * MVA_{base} + R_w^{down}(i,t,k) \tag{6}$$

Subjected to inequality constraints:

$$\Delta P_{g\min}^{up}(i,t,k) \le \Delta P_g^{up}(i,t,k) \le \Delta P_{g\max}^{up}(i,t,k) \tag{7}$$

$$\Delta P_{g\min}^{down}(i,t,k) \le \Delta P_g^{down}(i,t,k) \le \Delta P_{g\max}^{down}(i,t,k) \tag{8}$$

$$\Delta P^{up}_{w\min}(i,t,k) \le \Delta P^{up}_{w}(i,t,k) \le \Delta P^{up}_{w\max}(i,t,k) \quad (9)$$

$$\Delta P^{down}_{w\min}(i,t,k) \le \Delta P^{down}_{w}(i,t,k) \le \Delta P^{down}_{w\max}(i,t,k) \quad (10)$$

$$(P_{ij}(i,j,k))^2 + (Q_{ij}(i,j,k))^2 \le (S_{ij}(i,j,k))^2 \quad (11)$$

Equality constraints:

$$\begin{aligned}&\sum_{i=1}^{ng}\sum_{k=1}^{24}\sum_{t=1}^{tmax}(\Delta P^{up}_{g}(i,t,k)-\Delta P^{down}_{g}(i,t,k))\\&+\sum_{i=1}^{nw}\sum_{k=1}^{24}\sum_{t=1}^{tmax}(\Delta P^{up}_{w}(i,t,k)-\Delta P^{down}_{w}(i,t,k))=0\end{aligned} \quad (12)$$

By consideration of loadability limit, ρ, power balance equation becomes:

$$\sum_{t=1}^{tmax} P_{gn}(i,t,k) + \sum_{t=1}^{tmax} P_{wn}(i,t,k) - \rho P_d(i,k) = P(i,k) \quad (13)$$

$$\sum_{t=1}^{tmax} Q_{gn}(i,t,k) - Q_d(i,k) = Q(i,k) \quad (14)$$

$$P_{gn}(i,t,k) = P_g(i,t,k) + \Delta P^{up}_{g}(i,t,k) - \Delta P^{down}_{g}(i,t,k) \quad (15)$$

$$P_{wn}(i,t,k) = P_w(i,t,k) + \Delta P^{up}_{w}(i,t,k) - \Delta P^{down}_{w}(i,t,k) \quad (16)$$

Here, Vi and Vj are the voltages at bus i and bus j respectively. Bij and Gij are susceptance and conductances of line ij. No. of thermal and wind buses are given by ng and nw; gmin, gmax, wmin, wmax are minimum and maximum limits of thermal and wind buses. The bids submitted by GENCOS are arranged in ascending order with generation to obtain market clear price and to settle the market. The bid values which are higher than market settle price have been set to market clear price. In this paper, the congestion cost has been evaluated for 24 h on a particular day of each season.

2.2 Wind Model

Hourly wind variation for 24 h duration has been considered in this paper. For this purpose, historical wind speed data of a particular geographical location has been taken and since wind power generated is directly linked with wind speed as given by

(17). [28, 29]. Here, vci, vco, and vr are cut-in, cut-out and rated wind velocities, and Pr and Pw are rated and generated wind power outputs of wind turbine respectively.

$$P_w(v) = \begin{Bmatrix} 0,\, v \le v_{ci} \cup v \ge v_{c0} \\ P_r\left(\dfrac{v - v_{ci}}{v_r - v_{ci}}\right),\, v_{ci} \le v \le v_r \\ P_r,\, v_r \le v \le v_{c0} \end{Bmatrix} \tag{17}$$

Different wind turbines with great variations in their hub heights are commercially available for the purpose of electricity production. Hub height under consideration might not be the same as the height at which wind speeds have been measured. As the wind speeds observed at different hub heights vary considerably, it necessitates considering this important factor before using the wind turbine power output model for calculating wind power output.

$$\frac{v_2}{v_1} = \left(\frac{h_2}{h_1}\right)^m \tag{18}$$

Dependence of wind speed on hub height is shown in relation (18). [30, 31]. Where, v_1 and v_2 indicates the wind velocities at h_1 and h_2 hub heights respectively. Atmospheric stability and surface roughness also affects the wind speeds, and this is taken care of by the factor m. Value of this exponential factor is usually approximated to 0.14. m is taken as 0.14 in this research work as well.

2.3 Load Model

Hourly load model for 24 h has been considered in this paper for IEEE 24 bus RTS system. To make the load model more realistic, seasonal load profiles have been considered, which includes winter, spring, summer and fall season's hourly load variation for a day. Load profiles for these seasonal loads [32] have been depicted in Fig. 1. There are no loads connected at buses 11, 12, 17, 21, 22, 23 and 24.

3 Case Study

Proposed methodology has been tested upon modified IEEE 24 Bus RTS system, which includes 11 thermal buses namely buses 1, 2, 7, 13, 15, 16, 18, 21, 22, 23 and 1 wind farm at bus 17, consisting of 1000 VESTAS-110 wind turbines each of specifications: 2 MW rated power, 3 m/s cut-in speed, 11.5 m/s rated speed, 20 m/s cut-out speed. System contains 17 load buses. Optimization study to get optimal

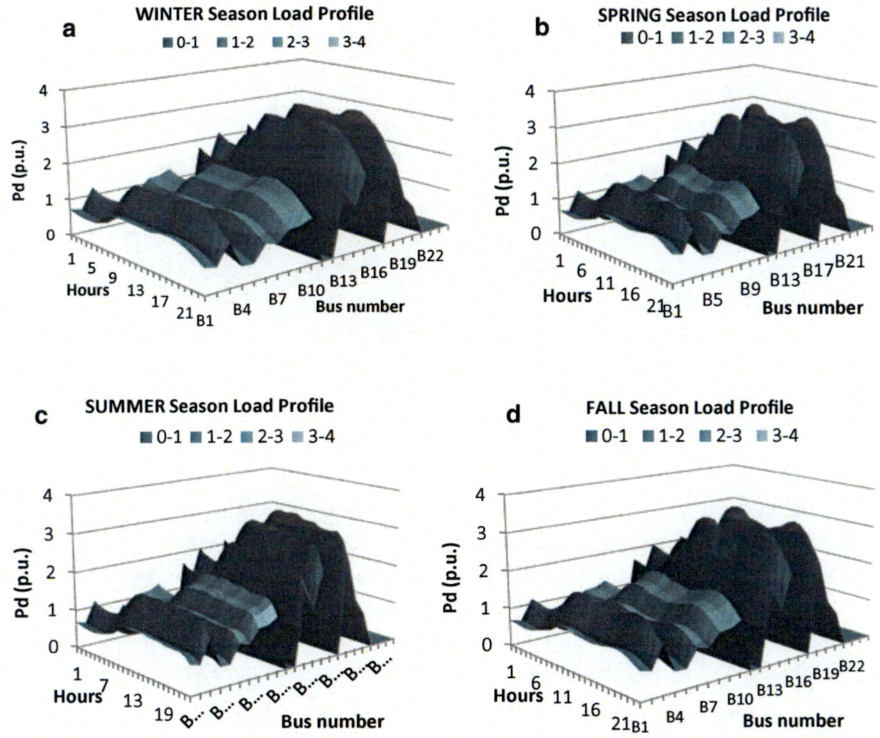

Fig. 1 Demand curve for 24 h for **a** Winter **b** Spring season. Demand curve for 24 h for **c** Summer **d** Fall season

generation plans for minimal congestion cost has been performed using CONOPT solver in GAMS by means of Non-Linear Programming (NLP).

Wind power has been generated based on historical wind velocity data of geographical location (27.15, 78.05) for 24 h. For this, wind velocities have been taken for 11th January, 2014, due to availability of moderate to large velocities [33]. Since impact of wind on congestion for different seasonal loads has been studied in this work, so base case generation schedule for day-ahead market has been considered as uniform in each bid block, depicted in Fig. 2, instead of taking economically dispatched schedule, which would have been different for each load and a fair comparison for various seasons would not have been possible. Practically that approach would give economical results; authors have worked in that direction as well. Congestion has been considered on three transmission lines: 6–10, 14–16 and 15–16 by limiting their maximum power transmission capability to 100, 300 and 150MVA from 175, 500 and 500 MVA respectively [2]. Details on bid structure can be referred in [7].

Winter Season: For the considered 24 bus system, hourly variation of loads for 24 h span has been considered for various seasons. It has been observed from load profiles

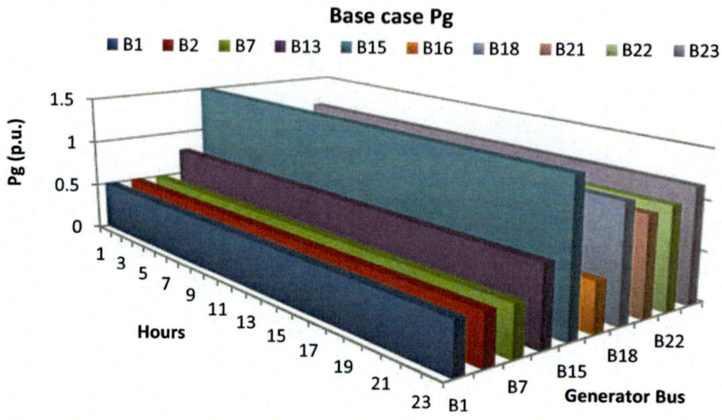

Fig. 2 Base case active power generation schedule, Pg

that for the system winter and fall seasons are having equal and maximum total load, though hourly variations are different for both, followed by spring and least total load is found to be in summer season. Nature of load variation and peaks is different in each season. Figure 3a–f depicts the rescheduled generation plans for all three bid blocks with and without wind participation in winter season. To comply with the limitation (number of pages), summer and fall season curves are not shown.

Spring Season: Requisite up and down regulations, done by generators for rescheduling in first bid block of this season, has been plotted and is depicted by Figs. 4 and 5 respectively. It has been observed from these figures that rescheduled generation plans are different for different seasons, as the load profiles of various seasons are different and are also different in different bid blocks, as bid values are higher in block 2 and 3 in comparison to block 1, so generators are found to be more willing to reschedule their generation plans in bid block 1. Different generation plans in various seasons resulted in different congestion costs that ISO has to pay to generation side in virtue of congestion removal from the lines. Comparison of congestion cost in various seasons has been presented in section-5 of this paper.

Another interesting observation made from these figures is the difference in participation from thermal generators in presence of wind power. In presence of wind power in the system, more number of generators have to change their generations though generator bus 23 rather got relieved, but as a whole intermittent nature of wind forced thermal generators to undergo change to maintain power balances in the system, this resulted in higher congestion costs in the case of wind participation. It is also worth noticing that, in bid block-1, generations required are less with wind but in other bid blocks different behaviour has been observed. In spring season, in some of the hours during 24 h period, congestion costs have been found to be lesser with wind. For analysis purpose, up and down generation regulations for spring season without and with wind are explicitly shown by Figs. 4 and 5. On careful observation of the same for 22nd hour, it has been found that due to presence of wind need of up generation

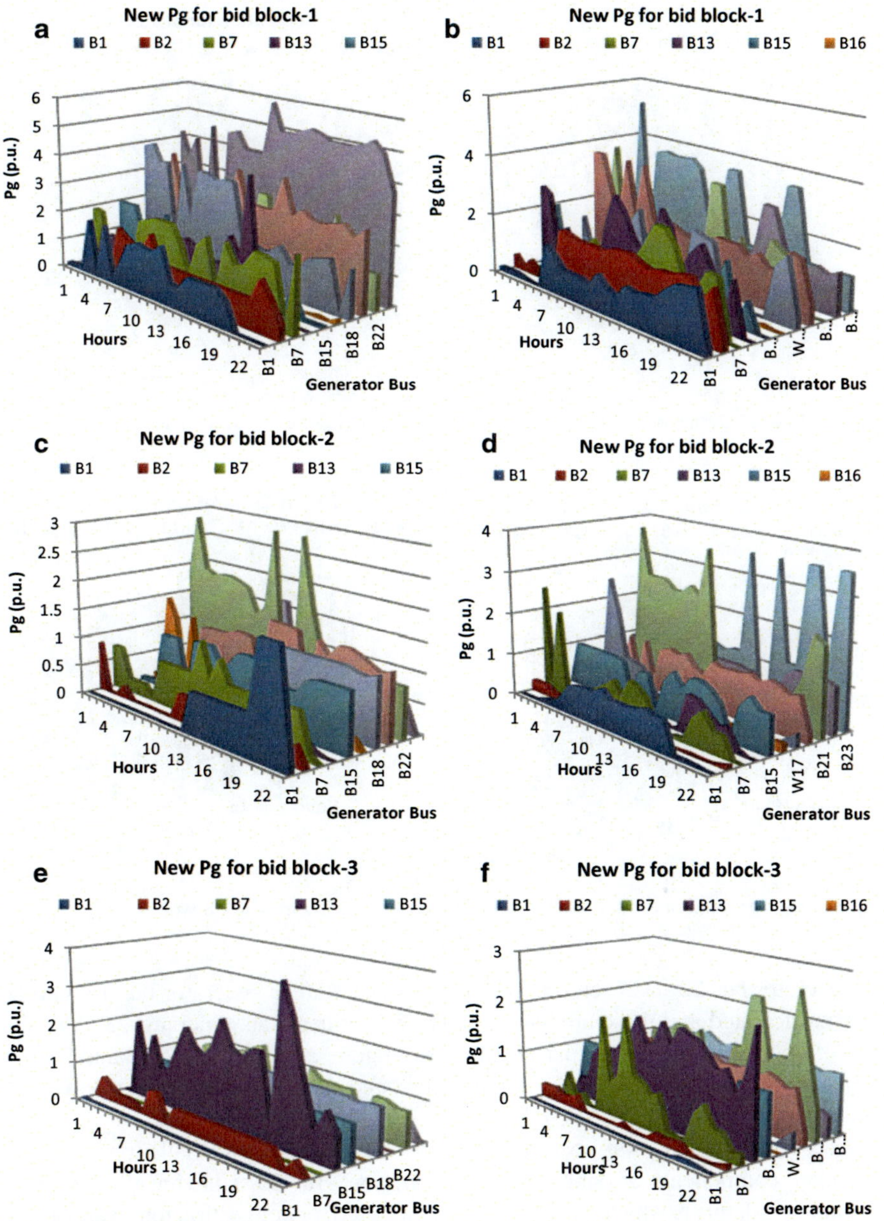

Fig. 3 New generation plan for bid block-1 in winter **a** without wind; **b** with wind. New generation plan for bid block-2 in winter **c** without wind; **d** with wind. New generation plan for bid block-3 in winter **e** without wind; **f** with wind

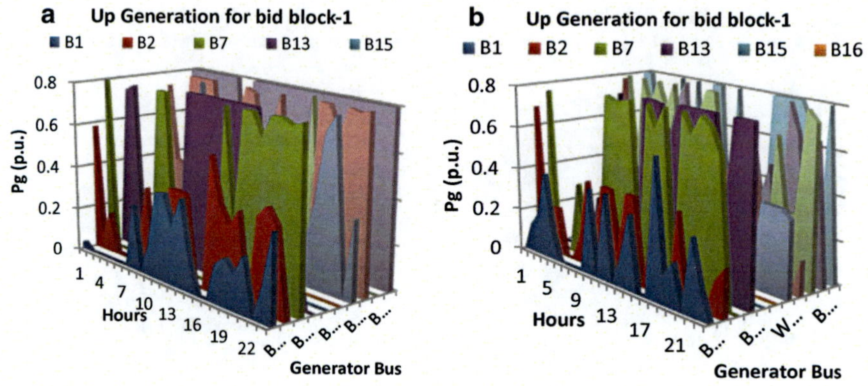

Fig. 4 Up generation for bid block-1 in spring season **a** without wind; **b** with wind

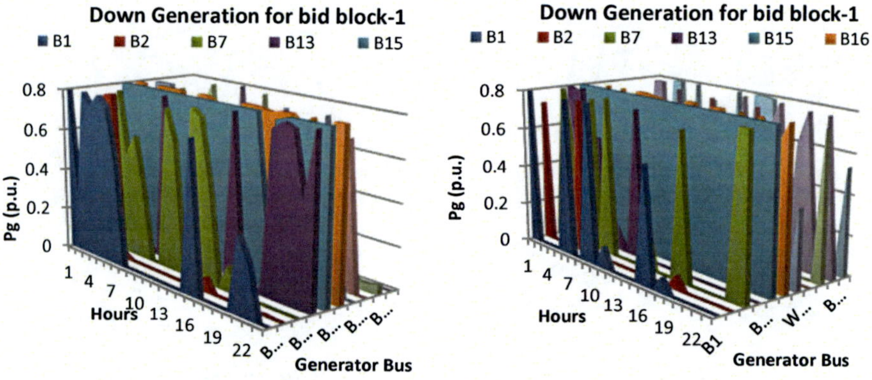

Fig. 5 Down generation for bid block-1 in spring season **a** without wind; **b** with wind

by generator bus 23 has been omitted in this hour. Rather generator bus 13 and 22 have participated and these are having lower bid values than generator 23. Also, in case of down regulation, less participation from generator buses 1, 2, 13 and 22 has been seen in presence of wind; on the other hand, bus 7, having lesser down-bid value, participated more.

As a result, congestion cost got reduced as up and down regulations directly formulate components of congestion cost function. Bus 15 is having lowest down regulation bid, this is why it can be seen to participate for congestion removal regardless of the season and hour. In general for all the seasons, if area under up and down regulation curves is taken as a measure of congestion cost, it can be stated that this area has been found to be higher with wind participation and so is the congestion cost higher with wind. Bus 7, 13 and 23 have been found to be most affected by presence of wind power plant (WPP) at bus 17.

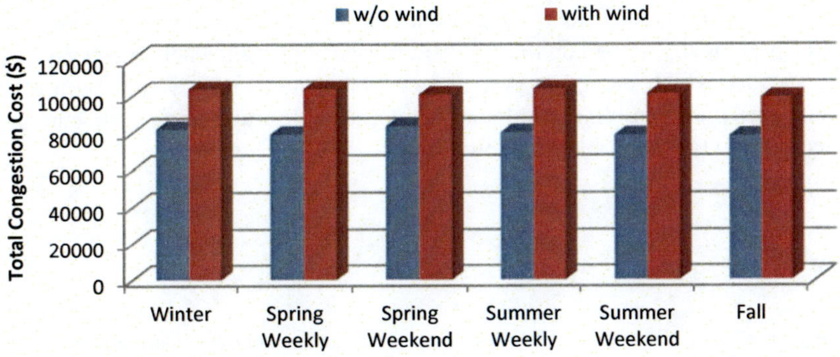

Fig. 6 Comparison of total congestion cost for various seasons on different days without and with wind

4 Comparison of Congestion Cost Per Hour for Various Seasons

For the considered IEEE 24 bus RTS system hourly seasonal loads, winter and fall seasons have the highest total loads followed by spring and summer has the lowest total load. Comparison of total congestion costs in different seasons in absence and presence of wind has been presented by Fig. 6. Without wind, winter season is found to have highest congestion cost, followed by spring weekend and fall is having least costs. With wind, winter is still having highest cost, a considerable rise in spring weekday has also been observed, fall is having least costs. In each season a rise in cost is there with wind participation, reason for this has been explained in section-4. Weekday and weekend loads are slightly different which causes difference in costs.

5 Conclusions

In this work, authors have used rescheduling based approach ensuring bus loadability limits for system security, to relieve congestion of transmission lines. Block bid function considering 3 bid blocks has been implemented in the hourly congestion management model for 24 h duration. Impact of wind power participation in congestion management has been analysed in presence of hourly load model. To make the situation more realistic seasonal load namely winter, spring, fall and summer with 24 h load variations have been included in the study. Although a direct relation in variation of cost and variation of wind is hard to obtain as intermittent nature of wind, variable seasonal loads and different bid blocks have simultaneous impact on congestion cost. It can be confidently concluded that ISO would have to pay more for congestion management in presence of wind power in the system in almost all the seasons. Hourly variation of congestion cost without and with wind has been presented. Comparison of total congestion cost for various seasons has also been depicted graphically. Based on the study, in presence of Renewable Energy Sources in the system, which indeed is going to be the scenario of power system everywhere, a need has been felt to implement other possible ways along with generation rescheduling to tackle problem of congestion.

References

1. Singh K, Padhy NP, Sharma J (2011) Congestion management considering hydro-thermal combined operation in a pool based electricity market. Electri Power Energy Syst 33(8):1513–1519
2. Verma YP, Sharma AK (2014) Congestion management solution under secure bilateral transactions in hybrid electricity market for hydro-thermal combination. Electri Power Energy Syst 64:398–407
3. Cheng JW, Galiana FD, McGillis DT (1998) Studies of bilateral contracts with respect to steady-state security in a deregulated environment of electricity supply. IEEE Trans Power Syst 13(3):1020–1025
4. Fang RS, David AK (1999) Transmission congestion management in an electricity market. IEEE Trans Power Syst 14(3):877–883
5. Barbulescu C, Kilyeni S, Cristian DP, Oprea DJ (2006) Congestion management using open power market environment electricity trading. IEEE 45th international UPEC, pp 1–6
6. Fattahi A, Ehsan M (2005) Sensitivity based redispatching method for congestion management in a pool model. Int J Emerg Electri Power Syst 3(2):1077
7. Sekhar C, Kumar A (2013) Congestion management with FACTS devices in deregulated electricity markets ensuring loadability limit. Electri Power Energy Syst 46:258–273
8. Dutta S, Singh SP (2008) Optimal rescheduling of generators for congestion management based on particle swarm optimization. IEEE Trans Power Syst 23(4):1560–1569
9. Shandilya A, Gupta H, Sharma J (1993) Method for generation rescheduling and load shedding to alleviate line overloads using local optimisationIEE Gener, Transm Distrib 140(5):337–342
10. Borghetti A, D'Ambrosio C, Lodi A, Martello S (2008) An MILP approach for short-term hydro scheduling and unit commitment with head-dependent reservoir. IEEE Trans Power Syst 23(3):1115–1124
11. GAMS (General Algebraic Modeling System) software package. http://www.gams.com.
12. Hazra J, Sinha AK (2007) Congestion management using multiobjective particle swarm optimization. IEEE Trans Power Syst 22(4):1726–1734
13. Kumar GN, Kalavathi MS (2014) Cat swarm optimization for optimal placement of multiple UPFC's in voltage stability enhancement under contingency. Electri Power Energy Syst 57:97–104
14. Kumar A, Sekhar C (2012) DSM based congestion management in pool electricity markets with FACTS devices. Energy Procedia 14:94–100
15. Afkousi-Paqaleh M, Fard AAT, Rashidinejad M (2010) Distributed generation placement for congestion management considering economic and financial issues. Electri Eng 92(6):193–201
16. Singh AK, Parida SK (2013) Congestion management with distributed generation and its impact on electricity market. Electri Power Energy Syst 48:39–47
17. Kumar A, Srivastava SC, Singh SN (2004) A zonal congestion management approach using ac transmission congestion distribution factors. Electri Power Syst Res 72(1):85–93
18. Esmaili M, Shayanfar HA, Amjady N (2009) Congestion management considering voltage stability of power systems. Energy Convers Manag 50(10):2562–2569
19. Conejo AJ, Milano F, Bertrand RG (2006) Congestion management ensuring voltage stability. IEEE Trans Power Syst 21(1):357–364

20. Amjady N, Velayati MH (2009) Evaluation of the maximum loadability point of power systems considering the effect of static load models. Energy Convers Manag 50(12):3202–3210
21. Bonab SMM, Rabiee A, Ivatloo BM (2015) Voltage stability constrained multi-objective optimal reactive power dispatch under load and wind power uncertainties: a stochastic approach. Renew Energy 85:598–609
22. Esmaili M, Amjady N, Shayanfar HA (2010) Stochastic congestion management in power markets using efficient scenario approaches. Energy Convers Manag 51(11):2285–2293
23. Singh SN, David AK (2001) Optimal location of FACTS devices for congestion management. Electri Power Syst Res 58(2):71–79
24. Bompard E, Correia P, Gross G, Amelin M (2003) Congestion-management schemes: a comparative analysis under a unified framework. IEEE Trans Power Syst 18(1):346–352
25. Kumar A, Srivastava SC, Singh SN (2005) Congestion management in competitive power market: a bibliographical survey. Electri Power Syst Res 76(1–3):153–164
26. Jafariana M, Scherpenb JMA, Loeffb K, Mulderc M, Aiellod M (2019) A combined nodal and uniform pricing mechanism for congestion management in distribution power networks. Electri Power Syst Res 180:1–8
27. A User's Guide (1998) GAMS software. In: Brooke A, Kendrick D, Meeraus A, Raman R, Rasenthal RE (eds) GAMS development corporation
28. Chang T, Liu F, Ko H, Cheng S, Sun L, Kuo S (2014) Comparative analysis on power curve models of wind turbine generator in estimating capacity factor. Int J Energy 73:88–95
29. Sheen J, Tsai M, Wu S (2013) A benefit analysis for wind turbine allocation in a power distribution system. Int J Energy Convers Manag 68:305–312
30. Kavak Akpinar E, Akpinar S (2005) An assessment on seasonal analysis of wind energy characteristics and wind turbine characteristics. Int J Energy Convers Manag 46:1848–1867
31. Singh SN, Erlich I (2006) Wind power trading options in competitive electricity market. IEEE Power Eng Soc General Meet
32. The IEEE Reliability Test System-1996 (1999) Report prepared by the Reliability Test System Task Force of the application of Probability Methods Subcommittee. IEEE Trans Power Syst 14(3):1010–20
33. Renewable Resource Data Centre (RRDC), National Renewable Energy Laboratory (NREL). Available: www.rredc.nrel.gov/solar/new_data/India/about.html.
34. Shahidehpour M, Alomoush M (2001) Restructured electrical power systems, operation, trading and volatility. Marcel Dekker, Inc., New York

Research and Analysis of the Efficiency of Power Consumption in Tunneling Sections

V. Petrov, A. Sadridinov, and A. Pichuev

Abstract The operation efficiency of the mining industry is determined by the technical level of mechanical equipment and automation of industrial mineral extraction processes. The article provides a method of research and results of the statistical analysis of indices defining the energy efficiency of mining complexes by the example of the Severnaya coal mine. The authors defined the energy and technological profiles for a quantitative and qualitative assessment of the energy efficiency of mining works. Recommendations on ensuring sustainable operation of mining complexes during a whole period of mining section tunneling are given. To research and analyze the energy characteristics of mining sections of a coal mine, the authors applied the methods of correlation and regression analysis. The theoretical performance of a tunneling machine is defined by its technical characteristics while the level of performance (technical and operational)—by relative time factors of scheme perfection, availability, and the partial operation continuity factor. With a change in the scope of works performed by tunneling machines, the level of power consumption is correspondingly changed; therefore, the change in specific indicators is insignificant and within the limits of a minimum statistical error. Therefore, for the research of the energy efficiency of tunneling, it is necessary to analyze shift averages for each section, i.e., to take into account the peculiarities of the technological process of mine tunneling for each tunneling machine type.

Keywords Power supply · Coal mine · Energy efficiency · Mining complexes

V. Petrov (✉) · A. Sadridinov · A. Pichuev
National Research Technological University "MISiS", Leninsky prospect 4, 119049 Moscow, Russian Federation
e-mail: petrovv@misis.ru

A. Sadridinov
e-mail: sabrtima@mail.ru

A. Pichuev
e-mail: pvlmsmu@mail.ru

A. Kumar et al. (eds.), *Renewable Energy Towards Smart Grid*, Lecture Notes in Electrical Engineering 823, https://doi.org/10.1007/978-981-16-7472-3_9

1 Introduction

The operation efficiency of the mining industry is determined by the technical level of mechanical equipment and automation of industrial mineral extraction processes. The key requirements for mining equipment are increased operation efficiency and safety, reduction of the metal intensity of mining machines, and energy intensity of rock mass destruction, decreasing the environmental damage from mining works [1, 2, 2–4]. Various processes in the mining industry determine the specifics of studying the energy consumption of machines and equipment that are involved in the implementation of technology [4–7]. This confirms the need to use special methods for analyzing energy consumption, taking into account mining technology factors. The used methods of mathematical modeling and measurement instruments are important [7–9]. The mining industry is characterized by the special nature of energy supply systems, which is also significant for assessing the level of energy consumption [10–12]. The article discusses issues related to the analysis of electricity consumption in the tunneling complexes since the level of tunneling along the leading coal mines varies from 72 to 98% [1, 7].

2 Basic Power Consumption Models

The methodology for studying the energy efficiency of mining operations is based on the integrated use of methods for collecting, processing, and analyzing data from energy technology flows, which allow solving this scientific problem.

To create a database on energy technology flows, a software-analytical complex for energy resource management was used [7]. The power consumption modes of tunneling complexes in coal mines were modeled according to the methodology developed by the authors [7–12]. Microsoft Excel and Statistic Neural Networks were used as licensed software for processing the energy consumption database. The research object of the paper was 6 tunneling sections at the Severnaya mine. The mining works at the Severnaya mine are conducted by tunneling combines MB670, JOYR75, 12CM30, and KP-21. The reference data are shift performance indices and energy efficiency of the tunneling section in the period from November 2016 to October 2017.

To define the interdependence between the average statistical indices of shift performance of separate sections, the volume of the extracted rock mass Q_1, and passed running meters of the mine Q_2, as well as the shift consumption of power W and its specific consumption ω_1 and ω_2, the authors used the energy and technological indices of tunneling sections for a shift.

Correlation ratios for corresponding functions studied are provided in Table 1.

The analysis of correlation ratios demonstrated the following:

The interconnection of power consumption W (kWh) and the mass of extracted rock Q_1 (tons) from the mine is feeble or even absent $RWQ_1 = 0.071–0.164$. It

Table 1 Correlation ratios

Function	Correlation ratio R		
	Section no. 1	Section no. 8	Section no. 9
$W = f(Q_1)$	0.121	0.114	0.071
$W = f(Q_2)$	0.008	0.072	0.008
$\omega_1 = f(Q_1)$	−0.68	− 0.67	− 0.7
$\omega_2 = f(Q_2)$	− 0.92	− 0.74	− 0.75

can be explained by the fact that a tunneling machine unloads the rock in the same volume which corresponds to the section and length of the mine but its density and, consequently, the rock mass change. Thus, an increase or decrease in the rock mass is directly associated with the power consumed for its extraction. It is confirmed by the fact that the specific power consumption $\omega 1$ for all tunneling sections is virtually unchangeable and is equal to 2.41–2.57 kWh/tons. The dependence $W = f(Q_1)$ can be presented in the form of ф correlation ellipsis [2, 2].

There is no interdependence between power consumption W (kWh) and the length of tunneling Q_2 (rm) $RWQ_2 = 0.002–0.008$. The tunneling speed is directly related to the rock hardness and the cutting effort of the tunneling machine control device or a drilling assembly. This is confirmed by the fact that tunneling sections during 4 shifts consume approximately the same amount of energy for 1 running meter of tunneling. At this, the increase in the tunneling length for a shift is directly related to power consumption. The dependence $W = f(Q_2)$ can be also presented by a simple correlation ellipsis.

Specific power consumption ω_1 (kWh/tons) and the mass of extracted rock Q_1 (tons) from the mine have a robust correlation $R\omega_1 Q_1 = -0.67 \div -0.7$. This proves the fact that the increase in the volume of the extracted rock mass at virtually constant power consumption leads to a decrease in energy capacity ω_1 ($R\omega_1 Q_1 < 0$).

The dependence $\omega_1 = f(Q_1)$ can be presented by the equation of regression of a hyperbolic form.

Specific power consumption ω_2 (kWh/rm) and the tunneling length Q_2 (pm) also have a robust correlation $R\omega_2 Q_2 = -0.74 \div -0.92$. This means that with the increase in the tunneling length during the shift at virtually constant power consumption, the energy capacity ω_2 decreases. The dependence $\omega_2 = f(Q_2)$ also can be presented as the regression equation of a hyperbolic form.

Regression dependencies of the changeable specific power consumption on the extracted rock mass and specific power consumption on tunneling running meters on each section are provided in Table 2. The analysis of the dependencies $\omega_1 = f(Q_1)$, $\omega_2 = f(Q_2)$ shows that the nature of change in specific power consumption virtually coincides for the shifts in general at each separate section.

Table 2 Dependencies of specific power consumption on the extracted rock mass and running meters at mining works

Section	Shift	Regression equation	R	Regression equation	R
No. 1	1	$\omega_1 = 0.06 + 106.4/Q_1$	−0.53	$\omega_2 = 0.95 + 104/Q_2$	− 0.87
	2	$\omega_1 = 0.18 + 101.6/Q_1$	− 0.51	$\omega_2 = 0.85 + 106.7/Q_2$	− 0.91
	3	$\omega_1 = -0.02 + 109.7/Q_1$	−0.54	$\omega_2 = -2.37 + 115.1/Q_2$	− 0.92
	4	$\omega_1 = -0.14 + 115.6/Q_1$	−0.57	$\omega_2 = 1.73 + 104.4/Q_2$	− 0.92
No. 8	1	$\omega_1 = 0.38 + 179.2/Q_1$	− 0.62	$\omega_2 = -16.7 + 273.9/Q_2$	− 0.84
	2	$\omega_1 = 0{,}231 + 192.5/Q_1$	−0.61	$\omega_2 = -19.6 + 283.1/Q_2$	− 0.8
	3	$\omega_1 = 0{,}03 + 201.8/Q_1$	− 0.63	$\omega_2 = -4.51 + 214.1/Q_2$	− 0.76
	4	$\omega_1 = 0.19 + 216.4/Q_1$	− 0.71	$\omega_2 = 7.49 + 180.7/Q_2$	−0.72
No. 9	1	$\omega_1 = -0.137 + 452/Q_1$	− 0.69	$\omega_2 = -3.23 + 450.9/Q_2$	−0.75
	2	$\omega_1 = -0.03 + 409.9/Q_1$	−0.72	$\omega_2 = -5.64 + 458.4/Q_2$	−0.81
	3	$\omega_1 = -0.42 + 485/Q_1$	−0.75	$\omega_2 = -4.34 + 441.9/Q_2$	−0.71
	4	$\omega_1 = -0.62 + 516.2/Q_1$	− 0.7	$\omega_2 = -0.81 + 428.9/Q_2$	−0.76

3 Results and Discussion

The analysis of the dependencies of specific power consumption on the shifts and separate actions for a whole observation period showed the following.

1. For the entire section, the indices of specific power consumption ω_1 (kWh/tons) and ω_2 (kWh/rm) are virtually the same for all shifts.
2. The difference in the indices is explained by the deviations from industrial process modes and, to some extent, also depends on the qualification of the team staff.
3. The control over energy efficiency indices at the level of shift teams allows accurately registering the compliance with the regulations for mining works and operation modes of the technological complex, timely identifying the causes of their violation, and taking corresponding measures on their correction.
4. The regression analysis of the indices of day power consumption on the sections involved also the data for Q_1 of the observed period.
5. Correlation ratios for the functions $W = f(Q_1)$, $W = f(Q_2)$, $\omega_1 = f(Q_1)$, $\omega_2 = f(Q_2)$ are provided in Table 3. The analysis of correlation ratios demonstrates that the indices of specific power consumption and performance of the section in terms of the volume of the extracted rock mass, as well as passed running meters of the mine, have a robust reverse correlation ($R = -0.69 \div -0.789$). There is no correlation between power consumption and performance in terms of the volume of rock mass and running meters ($R \approx 0$).

Therefore, the regression dependence equations $W = f(Q_1)$ and $W = f(Q_2)$ can be represented by correlation ellipses while the dependencies $\omega_1 = f(Q_1)$ and $\omega_2 = f(Q_2)$—by hyperbolic or polynomial functions.

Table 3 Correlation ratios

Function	Correlation ratio R		
	Section no. 1	Section no. 8	Section no. 9
$W = f(Q_1)$	0.052	0.08	−0.095
$W = f(Q_2)$	−0.027	−0.15	0.051
$\omega_1 = f(Q_1)$	−0.69	−0.703	−0.756
$\omega_2 = f(Q_2)$	−0.93	−0.755	−0.789

The dependence between the specific energy consumption and the shift productivity of the tunneling section according to the passed linear meters of the mining is represented by the energy and technological characteristic $\omega_2 = 163.12 - 19.78 \cdot Q_2 + 0.78 \cdot Q_2^2$ (Fig. 1).

The specific flow rate of ω_2 varies from 44 to 75 kWh/pm (41.3%) with a change in performance in the range $Q_2 = 7.8 \pm 2$ linear meters.

An increase in the tunneling rate leads to a decrease in the specific consumption of electrical energy. However, if the rate exceeds 9 linear meters per shift, this leads to increased wear of technological equipment, accidents, disruption of the technological process, and an increase in the downtime of the site for repair and restoration work.

The relationship between the specific power consumption and the shift productivity of the tunneling section in terms of the volume of the extracted rock is represented by the energy and technological characteristic $\omega_1 = 4.411 - 0.01 \cdot Q_1 - 0.62 \cdot 10^{-5} \cdot Q_1^2$ (Fig. 2). When the productivity changes in the range $Q_1 = 185 \pm 15$ tons, the specific consumption $\omega 1$ changes from 2.2 to 2.55 kWh/ton (13.7%).

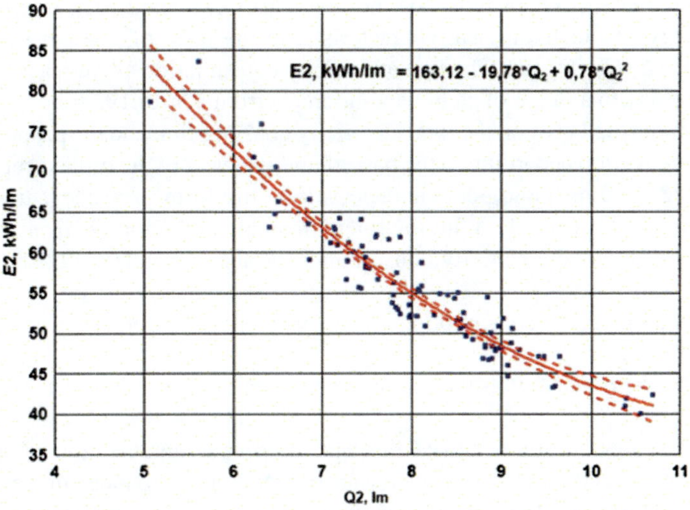

Fig. 1 Energy and technological characteristics $\omega_2 = f(Q_2)$

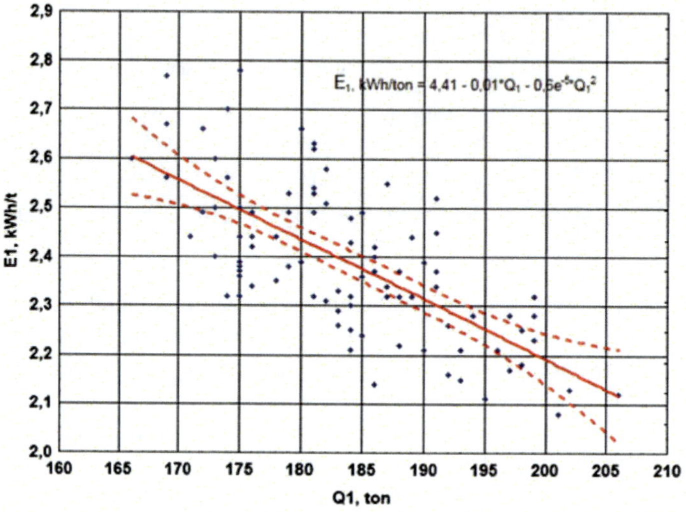

Fig. 2 Energy and technological characteristics $\omega_1 = f(Q_1)$

The specific energy consumption decreases with the growth of the volume of extracted rock. The efficiency of tunneling works depends on compliance with the requirements of the technical conditions: the technology of tunneling, the physical characteristics of the rock, compliance with the parameters of the cross-section of the mine, and the tunneling rate.

The relationship between the total electricity consumption and the productivity of the tunneling section in terms of the volume of extracted rock is represented by the energy and technological characteristic in the form of a linear trend $W = 415.72 + 0.12 \cdot Q_1$ (Fig. 3). The total electricity consumption W varies from 423 to 455 kW (7%) with a range of stable operation Q_1 from 177 to 192 tons.

The relationship between the total electricity consumption and the productivity of the tunneling section according to the passed linear meters of the mine is represented by the energy and technological characteristic in the form of a linear trend $W = 442.44 - 0.52 \cdot Q_2$ (Fig. 4). With the range of stable operation Q_2 from 7.3 to 8.9 linear meters, the total electricity consumption W also varies from 423 to 455 kW (7%).

The increased energy consumption is also due to the violation of the requirements of the technical conditions: the technology of tunneling, the physical characteristics of the rock, compliance with the parameters of the cross-section of the mine, the tunneling rate.

Energy savings can be achieved through the proper organization of tunneling operations, advanced training of drivers and crew members, reduced downtime, and constant monitoring of the operating modes of technological installations.

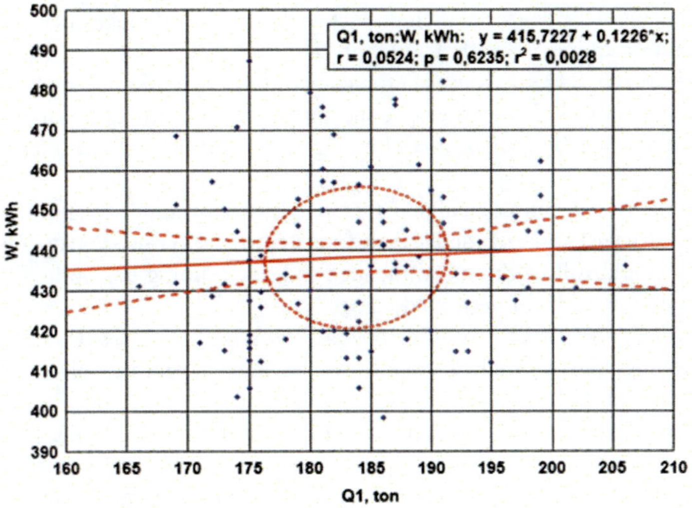

Fig. 3 Energy and technological characteristics $W = f(Q_1)$

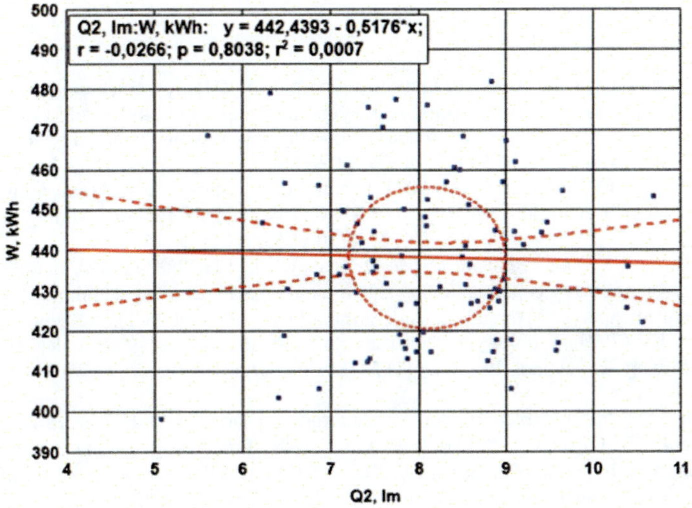

Fig. 4 Energy and technological characteristics $W = f(Q_2)$

4 Conclusion

The conducted analysis allowed drawing the following conclusions and giving some recommendations.

1. The conditions of conducting mining works are defined by a combination of a set of interacting factors (geological, technological, and organizational); the

assessment of the degree of their impact on the energy efficiency of technological processes requires a detailed and thorough study.
2. As the criteria for assessing the efficiency of mining works, the authors propose using the indices of the energy consumption level, efficiency, and quality of mine tunneling by shift teams allowing for the unbiased assessment of their operation.
3. The indices of technological and specific power consumption at mining works change in a vast range of values; therefore, to provide a consistent operation, shift teams should follow the recommended indices defining the optimal tunneling rates and the limits of permitted values.
4. When designing the energy efficiency indices, it is necessary to take into account the mining and geological conditions, technical specifications of tunneling systems, and their operation modes.
5. The developed models have been tested at real mining facilities. The methodological support was successfully used in the educational programs for the training of mining engineers (electrical engineering specialization) [13–16].

References

1. Liu X, Zhang Y, Zhang K (2020) Optimization control of energy consumption in tunneling system of earth pressure balance shield tunneling machine. Eng Lett 28(2):551–558
2. Zakharova A (2017) Bottom-up approach to modeling power use in a coal mine. Gornyi Zhurnal (2):79–82. https://doi.org/10.17580/gzh.2017.02.15
3. Yakonovskaya TB, Zhigulskaya AI (2021) Features of evaluating the economic security of peat industry enterprises in the Tver Region of Russia (the industry review). Mining Sci Technol (Russia) 6(1):5–15. https://doi.org/10.17073/2500-0632-2021-1-5-15
4. Zhang Q, Qu C, Kang Y, Huang G, Cai Z, Zhao Y, Zhao H, Su P (2012) Identification and optimization of energy consumption by shield tunnel machines using a combined mechanical and regression analysis. Tunnelling Underground Space Technol 28(1):350–354
5. Potapova EV (2021) Typology of metro structures for the tasks of geotechnical risk classification. Mining Sci Technol (Russia). 6(1):52–60. https://doi.org/10.17073/2500-0632-2021-1-52-60
6. Busygin AM (2020) Cabled Feeder for Underground Drilling Machines. In: Radionov A, Kravchenko O, Guzeev V, Rozhdestvenskiy Y (eds) Proceedings of the 5th International Conference on Industrial Engineering (ICIE 2019). ICIE 2019. Lecture Notes in Mechanical Engineering. Springer, Cham. https://doi.org/10.1007/978-3-030-22041-9_27
7. Kulikova EY, Shornikov II (2020) Method of estimation of pressure forces from power plant in microtunneling. In: Lecture notes in mechanical engineering, pp 783–789. https://doi.org/10.1007/978-3-030-22041-9_84
8. Moldashi DN (2021) Methods and technical solutions for keeping the path of a geotechnological borehole. Gornye nauki i tekhnologii = Mining Sci Technol (Russia) 6(1):42–51. https://doi.org/10.17073/2500-0632-2021-1-42-51
9. Petrov V Sadridinov A, Pichuev A (2020) Analysis and modeling of power consumption modes of tunnelling complexes in coal mines. In: E3S Web of Conferences, vol 174. https://doi.org/10.1051/e3sconf/202017401006
10. Semenov AS, Kuznetsov NM (2014) An analysis of the results of monitoring the quality of electric power in an underground mine. Meas Techn 57(4):417–420. https://doi.org/10.1007/s11018-014-0470-8

11. Kuznetsov NM, Minin VA, Selivanov VN: (2020) Kola power network development for the sake of the mining industry in the Murmansk region. Gornyi Zhurnal (9):96–10. https://doi.org/10.17580/gzh.2020.09.14
12. Shkrabets F (2017) Electric supply of underground consumers of deep energy-intensive mines. Gornye nauki i tekhnologii = Mining Sci Technol (Russia) (3):25–42. https://doi.org/10.17073/2500-0632-2017-3-25-42
13. Shpiganovich AN, Shpiganovich AA, Zatsepina V, Zatsepin E (2017) State of the issue of the power supply system's reliability. Gornye nauki i tekhnologii = Mining Sci Technol (Russia) (3):47–79. https://doi.org/10.17073/2500-0632-2017-3-47-73
14. Draganescu F, Gheorghe M, Doicin CV (2003) Models of machine tool efficiency and specific consumed energy. J Mater Process Technol 141(1):9–15
15. Petrov VL (2016) Federal Training and Guideline Association on Applied Geology, Mining, Oil and Gas Production and Geodesy – A new stage of government, academic community and industry cooperation. Gornyi Zhurnal (9):115–119. https://doi.org/10.17580/gzh.2016.09.23
16. Petrov VL (2017) Training of mineral dressing engineers at Russian Universities. Tsvetnye Metally (7):14–19. https://doi.org/10.17580/tsm.2017.07.02
17. Puchkov LA, Petrov VL (2017) The system of higher mining education in Russia. Eurasian Mining (2):57–60. https://doi.org/10.17580/em.2017.02.14
18. Klimov IY (2020) Analysis of soft skills-based approach effectiveness in an advanced training program for a mining company. Gornye nauki i tekhnologii = Mining Sci Technol (Russia) 5(1):56–68. https://doi.org/10.17073/2500-0632-2020-1-56-68

Techniques Employed in Renewable Energy Sources Fed Smart Grid—A Comparative Study

M. Nivedha and S. Titus

Abstract The present circumstances of the era involve a vast increment in the need of power requirement due to increased technological developments, population and urbanization. The way in which power is obtained and transferred from the generating source to the consumers plays a major role, and this is done efficiently with the aid of smart grids that helps in regulating the supply from source to the grid. In this paper, a comparative study is made regarding the various techniques employed in the power flow management of Renewable Energy Systems (RES) facilitated by smart grid. For efficient power transfer smart meters and for power calculations multiple VSC's are employed. With the aid of AIT, power governing is increased and by using HESS, the energy along with power density is improved. But certain issues occur in transferring the data amidst the machine. These issues are rectified by the use of IoT, which helps in safe transfer of information during data transfer. This transfer is made efficient with the aid of algorithms like Deep learning, Fuzzy, Neurofuzzy, etc. and hybrid optimization techniques are employed for attaining high efficiency and good accuracy of the system.

Keyword Renewable energy systems (RES) · VSC's (voltage source converters) · AIT (adaptive intelligence technique) · HESS (hybrid energy storage system)

Abbreviations

IoT	Internet of Things
HESS	Hybrid energy storage system
ESS	Electric storage system
DR	Demand response
HEPF	Holomorphic embedded power flow

M. Nivedha
Arasu Engineering College, Kumbakonam, India

S. Titus (✉)
K.Ramakrishnan College of Engineering, Trichy, India

A. Kumar et al. (eds.), *Renewable Energy Towards Smart Grid*, Lecture Notes in Electrical Engineering 823, https://doi.org/10.1007/978-981-16-7472-3_10

ACOPF	AC optimal power flow
VSC	Voltage source converter
DSO	Distribution system operators
GPU	Graphic processing unit
HAN	Home area network
CAMPS	Control and power management system
MMC	Modular multilevel converter
EV	Electric Vehicle
SOC	State of charge
HEMS	Home energy management systems
BLR	Bayesian linear regression
HMS	Hybrid multi-surrogate
GWO	Grey wolf optimization
EM	Energy management
PDN	Power distribution networks
C&CG	Column and constraint
DDUS	Adaptive data-driven uncertainty sets
PCC	Point of common coupling
DER	Distributed energy resources
SCD	Smart controller device
DSO	Distribution substation operators
RL	Reinforcement learning

1 Introduction

There is a diverse necessity and growth in the demand side power management due to increased power consumption by the residential, industrial and commercial buildings. Nowadays, in order to achieve power the sources, namely, solar, wind turbines, fuel cells and storage batteries are fixed at homes, buildings, industries, etc. which modifies the structure of the residential and commercial areas. When there occurs a peak electricity requirement, energy shortage occurs and in order to compensate such requirement, smart meters give a two way on time communication through smart grid based on the necessity of the consumer by accessing their requirement details. With the aid of fast two-way communication, the smart meters and smart grids are equipped which forms the modern power grid capable of controlling the energy consumption that further increments the efficiency related to grid through load management.

RE (Renewable energy) generators are considered to be the normal distribution system as there is drastic change in traditional electric generation to different technologies [1, 2]. An electrical grid capable of delivering energy in a regulated way in an efficient manner starting from generating nodes to active consumers is called as smart grid and the demand response gives responsive, also interactive customers which helps in decreasing the capital cost and the plant investment [3]. Also the data flow amidst the machines can be done with the aid of IoT which helps in availing the power consumption of consumers in the cloud, also in customizing the electronic devices in real time. Such technique helps in consumption of reduced energy at the time of data transfer and also data processing [4].

2 Related Works

In [1], to maintain the energy storage system a variable threshold technique for charging and discharging is represented which helps to attain self-consumption and also to reduce the large amount of power flowing in return to the grid. Here, an AIT method is used to increase the power governing efficiency and also a HESS is employed to get increased energy and also power density that helps in analyzing the defined technique with calculated RES output data. With this technique good load smoothening is attained. While the ESS integration process helps in smoothening the total power transferred to the grid with the RES penetration, it also tracks the profile accepted on a certain basis [5]. But in [2], a supportive distributed regulation technique is implemented for the purpose of power flow features in contradiction to a nano-grid, maintaining the voltage stability and the features give a specified ID to every single individual power load.

And this technique reduces voltage fluctuations but a feedback control is necessary for suppressing the power supply oscillations and also to handle the noise and uncertain data. Considering the real-time updates, the HEPF is very efficient and for that a four-bus scheme is employed in order to define holomorphic system that hinge on the complex analysis and it also introduces a power flow facility on a communication network.

With the holomorphic function properties, the equations related to different types of bus are holomorphically implanted with the aid of a complex variable. It is represented as

$$V_{slack}(s) = 1 + \left(V_p^{spec} - 1\right)s \quad p \in slack \tag{2.1}$$

$$\sum_{q=1}^{N} Y_{pq}^{series} V_q(s) = \frac{sP_p - jQ_p(s)}{V_p^*(s^*)} - S \sum_{\substack{q=1 \\ q \neq p}}^{N} Y_{pq}^{shunt} V_p(S) \quad p \in PVbus \tag{2.2}$$

$$V_p(s) \times V_p^*(s^*) = 1 + \left(\left|V_p^{spec}\right|^2 - 1\right)s \quad For\ p \in PVbus \tag{2.3}$$

$$\sum_{q=1}^{N} Y_{pq}^{series} V_q(s) = \frac{s S_p^*}{V_p^*(s^*)} - S \sum_{\substack{q=1 \\ q \neq p}}^{N} Y_{pq}^{shunt} V_p(S) \; p \in PQbus \tag{2.4}$$

If the voltage solution attained by it exists it finds the solution, but if the solution does not exist it results in voltage collapse [6]. If the solution like equilibrium strategy has to be found for either generator or micro-grid, a fully split technique is defined which updates the decisions considering the voltage angles at buses that is attained by the PMU.

With this technique there is no necessity to transfer data amidst the generators and the micro-grids but the communication distance amidst the grid increases which makes it not to be economical [7], and the double-layer framework in smart grid is represented in Fig. 1. In case of hybrid microgrid system, which involves both dc and ac subgrids, it is necessary to employ novel control techniques for managing and controlling the power. Every single microgrid involves numerous DG units and local loads. In order to enhance the reliability of the system and to provide bidirectional power flow, a VSI (Voltage source inverter) is interconnected between the two entities. When a centralized controller is employed to equalize the loading condition between ac and dc sub grids through multiple VSC's, it increments the count of sensors and the transmitted signals to the central processing unit. However, it is capable of operating in both sub grids irrespective of the control mode of DG units but it depends on power calculations and also works only when both the sub grid are under regulation [8].

An independent smart grid with the aid of bi-level optimization is presented which is a self-directed smart grid that functions at its own networks from lower level and directs its decisions toward upper level relying on the conic AC power flow designs. To calculate the bi-level optimization program, a reformulation decomposition technique is formed and customized by which the solution accuracy is enhanced but the

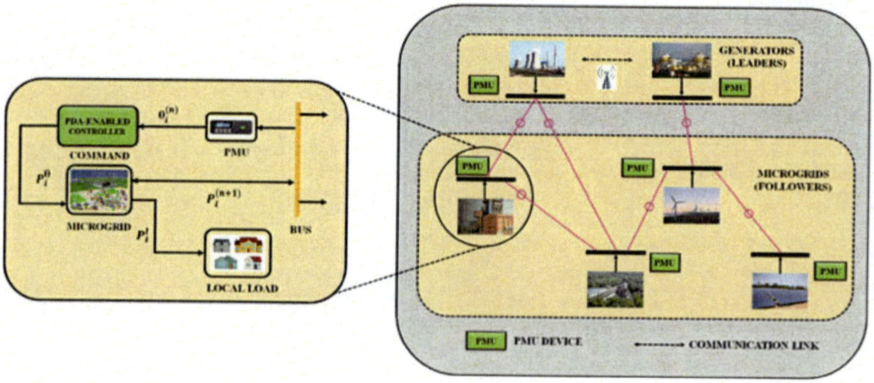

Fig. 1 Two-layer framework of smart grid

storage devices employed are dangerous for obtaining the original optimal outputs of smart grids [9]. A novel technique related to DSO distribution grid management underneath spatial temporal uncertainty gives a protective technique which will be costly during the day ahead stage and is less costly during the operating day. Such technique enhances the reliability of the distribution systems but the level of the strength relies on the formation of uncertainty [10]. Also to bring a reliable energy flow with the presence of fluctuations, real-time processing is employed and by using the GPUs along with the assessed data, and the simulation of energy dispatch scenarios is performed without complexity, which helps in reducing the load energy. It further enhances the reliability of the IGS, while the data of number of buildings relies on the connecting cables, number of possible steps and also the remoteness amidst them [11] whose framework related to smart grid is depicted in Fig. 2.

To regulate the power flow in the converter including all units effectively and to understand the power balance amidst hybrid micro-grid system and the grid, the CAPMS is employed. It gives continuous power supply even at the time of solar shut down or during power fluctuations at the time of unstable irradiance. Such technique does not employ any extra converters, also reduces the control costs [12]. The HEMS management system provides efficiency boosting by reducing the consumption of power, reduced energy cost and load profiles. Here, ANN along with Fuzzy logic and Adaptive neural fuzzy interface are preferred. The Heuristic scheduling algorithm relying on BPSO is relatively ineffective, which makes it unsuitable for applications [13]. In order to restrict the prices that differs from time to time, an Artificial Neural Network (ANN) is employed which helps the home energy management system [14]. For regulating the circulating current, an MMC-based EV fleet in combination with grid is used and it depends on the virtual SOC modelling; it also offers charging

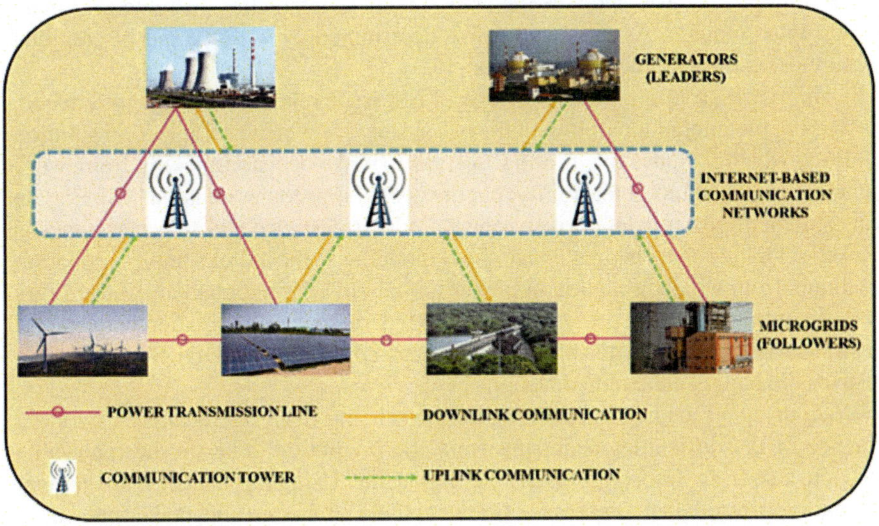

Fig. 2 Framework of smart grid

and discharging power differential regulation, also power management amidst EV's and unity by sustaining the power quality of MMC interface in the grid. The defined MMC relied EVIS is equivalent to a double level converter, whose current, voltage and power relationship in d-q frame is given by

$$\begin{bmatrix} v_{sd} - v_d \\ v_{sq} - v_q \end{bmatrix} = \begin{bmatrix} \left(L + \frac{L_a}{2}\right)\frac{d}{dt} & -\omega\left(L + \frac{L_a}{2}\right) \\ \omega\left(L + \frac{L_a}{2}\right) & \left(L + \frac{L_a}{2}\right)\frac{d}{dt} \end{bmatrix} \begin{bmatrix} i_d \\ i_q \end{bmatrix} \tag{2.5}$$

$$\begin{bmatrix} p \\ q \end{bmatrix} = \begin{bmatrix} v_{sd} & v_{sq} \\ -v_{sq} & v_{sd} \end{bmatrix} \begin{bmatrix} i_d \\ i_q \end{bmatrix} \tag{2.6}$$

Based on the request of EV owners, the power control strategies will be efficient but if there occurs variance in the virtual SOC's of three phases, the circulating current induced will be larger [15] and one more technique helps in regulating the charging and discharging function using solar output by transferring the information amidst home energy management and grid energy management system. Here, with the aid of HEMS EV charges along with discharges which relies on the obtained plan and the real-time regulating data, the operational cost of the residential use is reduced and it also decrements the negative effects created as a result of prediction errors found in power profiles while the HEMS must do the operation depending on the causes produced as a result of prediction error [16]. For central power system management that is not detectable for the purpose of optical co-ordination of vast devices at large scale and in order to aid in difficult grid constraints, a hierarchical signal processing framework is presented. It allows effective and sustainable association of the vast set of power components in the power grid, taking the limits of the physical network into account. The responding capability of this method is very efficient. It also enriches the flexibility of the grid and aids the components in interacting with the grid in an uninterrupted manner. As the nodes make decisions only with the aid of confined measurements there occur inefficiencies [17].

In some cases, a strong tri-level design of EM issue related to PDN is introduced, considering the interaction of it with the gas system for which multiple loop solution is formed which includes two C and CG so as to make decision-making framework that improves the solution feasibility; but the cost of it becomes an issue [18]. For the management of power a data-driven robust model is designed whose performance is checked with the robust budget constrained model with the aid of sample evaluation technique, from which reduction in power loss about 15% is obtained. And it gives good performance with respect to average and risk metrics in association with budget constrained designs. But when high violations are considered the DDUS relied model outstrips the budget constrained designs [19].

When using a receding horizon regulation scheme EMS operates even with the presence of the difficulties occurring from the prediction done on the renewable loads and also from the designs used by components. Using this energy management system, the control action necessary for the active and the reactive power present in the micro-grids is represented, which helps in minimizing the emission of carbon dioxide; it also decreases the cost. But as the micro-turbines are not employed at its maximum power, the emission does not decrement unless the economic cost increases [20].

To provide better control and communication schemes for distributed systems involving generators and loads, a distributed regulation strategy is introduced that helps in exchanging the information amidst the neighbors, and also attains global optimal power outputs. When employing an even triggered condition, it gives details regarding when to sample and transmit information amidst neighbors. By this method it is confirmed that it reduces the frequency related to communication amidst generators and loads having same control performance beneath continuous data communications. But at the time of practical implementation, power transmission loss and capacity limitations must be considered, also when communication networks are used, there occurs packet dropouts, and also transmission delays [21]. While the HAN–IoT helps in transferring the data in an efficient way, it is necessary to test the designed model in an open environment in order to check the linking gateway and coverage boundary [4]. The framework of HAN–IoT is depicted in Fig. 3

So as to compute and accelerate the power system control, linearization of power system model is introduced that involves both the forward and inverse regression model, also BLR relied algorithms are modelled in order to focus on data collinearity

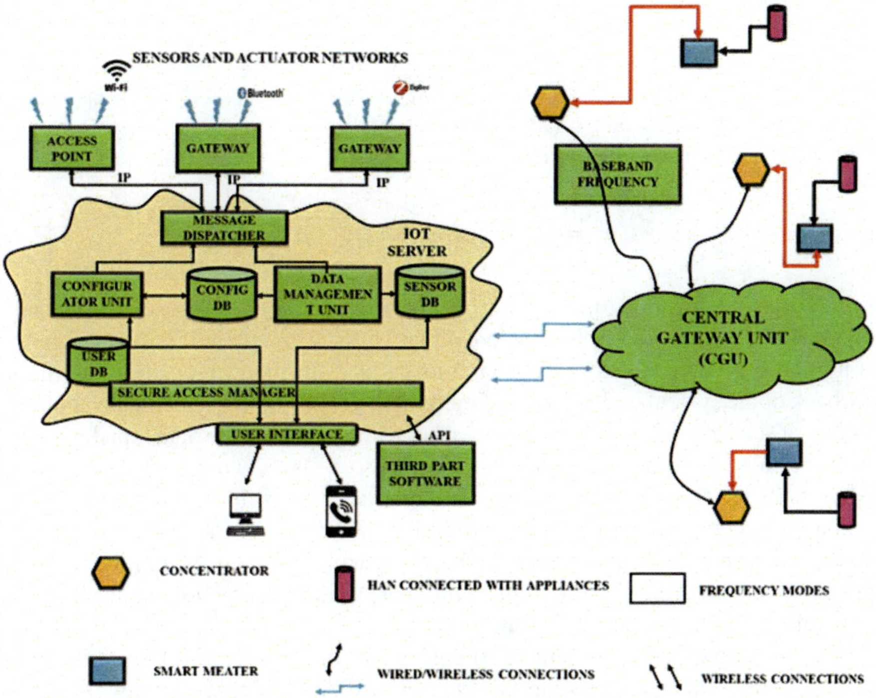

Fig. 3 HAN–IoT framework

and also to prevent over fitting. Such method gives greater calculation accuracy. But it is a tough task that the technique changes frequently with the system change [22].

Another technique that helps in enhancing the accuracy is the use of Monte Carlo technique that also helps in evaluating the dangers caused by high- or low-frequency events. But a single surrogate model is not capable to withhold uncertainties in random branch outages, therefore a HMS technique relying on PCE having low probability enhances the calculational efficiency. As there occurs some distractions in the input parameter models, the QoI obeys certain pdf $q(z)$ and using this, the probability related to target events are expressed as

$$p_f = pr\left(Z \in \Omega_f\right) = \int_{\Omega_f} q(z)dz = \int X\Omega_f(z)q(z)dz \tag{2.7}$$

Here χ represents the characteristic function and it fulfills

$$\chi\Omega_f(z) = \begin{cases} 1 if z \in \Omega_f \\ 0 if z \notin \Omega_f \end{cases} \tag{2.8}$$

Also Ω_f refers to the target domain and is given by

$$\Omega_f \triangleq \{Z : g(z) < 0\} \tag{2.9}$$

Where $g(z)$ represents the limit state function called performance function, which makes the target domain for risk assessment. In the domain, $g(z) < 0$ represents the domain having risk and g $(z) \geq 0$ represents the safe domain. Therefore, the probability related to failure is given by

$$p_f = \int X\{g(z) < 0\}(z)q(z)dz \tag{2.10}$$

From this, the formulation related to risk assessment in PPF analysis is attained. From the results obtained, it is concluded that HMS technique effectively captures the low probability dangers present in association with the rare events. The main drawback of the system is at the time of increment in scale of the system, the attained speedup of such technique is decreased to a large level [23] and the data transfer using IoT in the smart grids are represented in Fig. 4.

The uncertainty quantification of wind source is also done by a Traditional method (Parametric) and Direct PI construction method (Non-Parametric). In addition, stochastic models, Fuzzy logic models along with robust optimization are used. This traditional method will have high complexity in implementation, hence it is not effectively implemented in many sorts [24].

While a middleware is employed to collect information from different devices all through the grid by employing a horizon relied ACOPF related to micro-grids with distributed renewable generation and also storage. It also offers the information to the ACOPF applications that necessitate the forecast on demand found with the

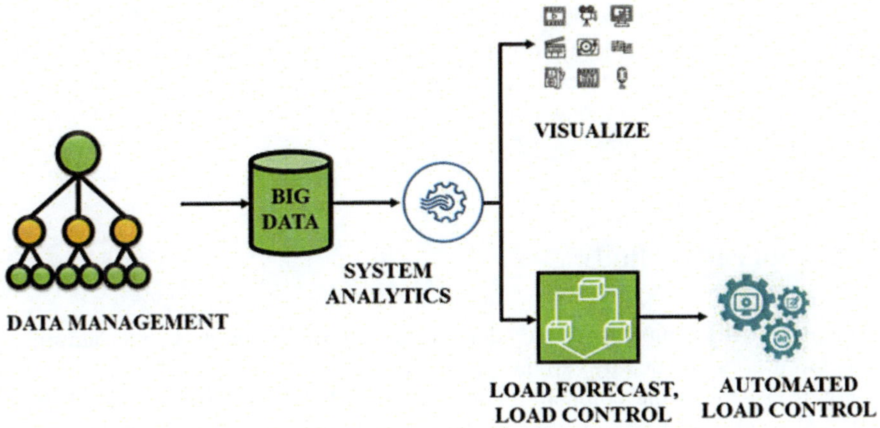

Fig. 4 Transfer of data to micro-grids using IoT

aid of resource allocation and random forest. It also offers global optimal solutions in many instants and it decreases the working cost by 6.54% but the designing of solutions must be done based on the capacity of devices employed in smart grids [6].

If the customers are given the option of choosing their own necessity, there will be a great decrement in the cost, and also increase in the comfort zone of the consumption of electricity which is done with the aid of bio inspired computing relied scheduling algorithms. Also a comparative study is made amidst different algorithms that aim at scheduling the loads and reduction of electricity bills. GWO gives better comfort zone but there is a little delay at the time of scheduling [25].

The users are classified with the aid of SCD in an expanded category and then sending such information to the grid regulation system through DSO. The employed algorithm classifies the load and also the phase in order to get information regarding the specific load and also the phase and by this the voltage regulation and also the power factor of the system is enhanced [26]. In order to apply DR algorithm on various loads, it uses smart direct load control and load shedding. It decreases the grid operation cost and also the power outages and regulates the load in real time. But at certain situations it will not be capable of dispensing the load changes [3].

Also the power in grid tied micro-grids is controlled by reinforcement learning algorithms which does not necessitate any accurate model that belonged to optimization environment so as to attain an optimal solution. The RL method helps in reducing the various issues associated with the grid tied micro-grids, also the problems in RL technique are overcome by combining the RL with the supervised learning which is used online due to their increased efficiency and also speed. But still the improved deep RL method faces issues like data inefficiency and instability, also the limiting factor includes complex balancing and also tuning of algorithmic hyper parameters [19]. By increasing EENS linearly with increase in proportion of load with RC is represented here which further increases the system efficiency. For this process, NN-K-SVD algorithm along with Sequential Monte Carlo (SMC) simulation is preferred.

The EENS generated can be more optimistic than the index. Because of various type of loads, the response to the outrage is different [27] and when ANFIS method is used it provides better efficiency. This provides an optimization of membership function. However, these models are studied only through theoretical and none of the researchers made it practical [28].

3 Comparative Study

Table 1 shows the comparative study done on the renewable energy sources fed smart grid empowerment aided by certain specific techniques.

Figure 5 depicts the comparative efficiency of hybrid multi-surrogate, Fuzzy, ANN, Neurofuzzy and reinforcement learning algorithm in which reinforcement algorithm is considered to be more efficient.

Table 2 shown below clearly explains the applications of renewable energy systems in India.

4 Conclusion

Employing renewable energy sources, the power flow in smart grid is regulated and for this, many methods are used. These techniques aid in improvising the power transfer from the source to the consumers. When adaptive intelligence technique is used, it produces better load smoothening but causes high computational burden. If bi-level optimization technique is used, it brings good accuracy but the storage devices employed are unsafe to derive the outputs. The three-level hierarchical control technique for micro-grid effectively helps in controlling the active and reactive power delivery; meanwhile, at certain situations it is not possible to inject the primary power to the grid. Data-driven technique with forward and reverse regression model aids in enhancing the calculational efficiency; if there occurs any change in the system, the parameters of this technique changes. When HAN–IoT is preferred, it provides energy aware routing but special examination has to be done for the system design. These techniques when employed in smart grid power flow management create inaccuracy, reduced efficiency and inappropriate data transfer. Therefore, a novel optimization technique, namely, Chaotic Ant Lion optimization and Random Forest Algorithm (CALORFA) is introduced and it helps in processing of data transfer, leading to increased flexibility of the network. With the CALORFA technique efficient optimal solution is attained and it reduces the calculational burden of the data.

Table 1 Comparative study

Sl. no.	Author/year of publication	Methodology	Functions	Advantages	Disadvantages
1	Jia et al. [1]	Adaptive Intelligence technique is used	It manages energy storage system and increments the power management efficiency	Attain good load smoothening	Computational burden is 14.6 ms
2	Giorgio et al. [5]	Model predictive control technique for ESS in HV/MV	Controlling and also to smoothen the total power transferred to load	Enables day-ahead power tracking capability	Absence of controlled ESS results in coverage of RES fluctuations
3	Capizzi et al. [29]	Cloud computing aided GPU architecture with soft computing techniques	Enables reliable energy flow and possible to dispatch all possible scenarios	Fast and distributed computation	The data depends on linking power cables and number of possible setups
4	Manimuthu et al. [4]	Advanced cost-effective HAN–IOT is used	Provides energy aware routing	Provides less energy for the purpose of data transfer and data processing	Necessity to examine the designed model
5	Delfino et al. [20]	Energy management platform relying on the receding horizon technique	Helps in optimal regulation of active and reactive power in micro grids	Reduces cost and the carbon dioxide emission	The emission does not decrement unless the economic cost increases
6	Liu et al. [22]	Data-driven approach having forward and reverse regression model	For accelerating and making the calculations easy for power system control	Increased calculation accuracy	Technique changes frequently with the change in system
7	Haghighat et al. [9]	Bi-level optimization technique	Captures the interaction of smart grids and ISO and reformulation decomposition technique is used	Enhances solution accuracy	Storage devices are dangerous to derive the output

(continued)

Table 1 (continued)

Sl. no.	Author/year of publication	Methodology	Functions	Advantages	Disadvantages
8	Brandao et al. [30]	Three-layer hierarchical control method for micro-grids	To control the active and reactive power delivered	Do not necessitate prior knowledge of parameters	At certain situations the primary power cannot be injected in the grid
9	Arwa et al. [31]	Reinforcement learning technique is used	To manage power in grid tied micro-grids	Do not need an accurate model for optimization environment	Data inefficiency and instability
10	Xu et al. [23]	Monte Carlo technique for high fidelity power system along with multi-surrogate method is used	Performs risk assessment in the PPF evaluation	Overcome incapability of the system and also enhances the efficiency	As time increases, the speeding up of technique reduces

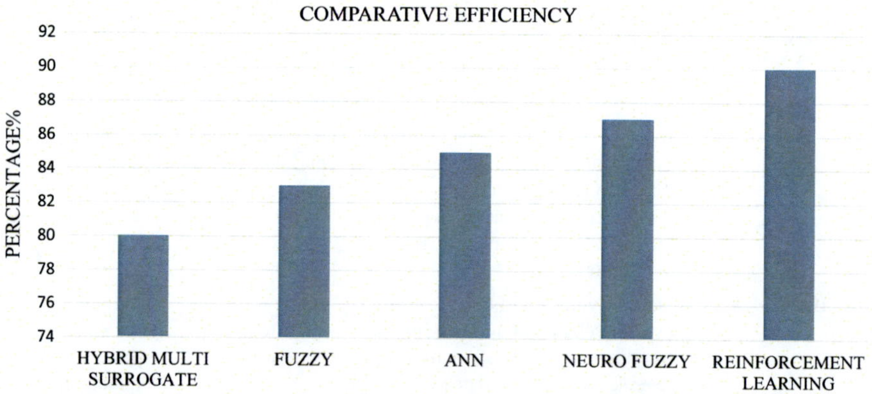

Fig. 5 Comparative Efficiency

Table 2 Applications of RES in India

Sl. no.	State	Type of energy	Volume (MW)	Applications
1	Rajasthan	Solar	2250	Street light
2	Tamil Nadu	Wind	7269.50	Used in factories or mills
3	Uttarakhand	Hydro electric	2400	Used in irrigation, industrial use
4	Andhra Pradesh	Biomass	9806	Used in the production of heat and electricity
5	Himachal Pradesh	Geo-thermal	10000	Employed in cooking, bathing, space heating, and in electrical power generation

References

1. Jia K, Chen Y, Bi T, Lin Y, Thomas D, Sumner M (2017) Historical-data-based energy management in a microgrid with a hybrid energy storage system. IEEE Trans Ind Informat 13(5):2597–2605
2. Javaid S, Kurose Y, Kato T, Matsuyama T (2017) Cooperative distributed control implementation of the power flow coloring over a nano-grid with fluctuating power loads. IEEE Trans Smart Grid 8(1):342–352
3. Mortaji H, Hock Ow S, Moghavvemi M, Almurib HAF (2017): Load shedding and smart-direct load control using internet of things in smart grid demand response management. IEEE Trans Ind Appl 53(6):5155–5163
4. Manimuthu A, Ramesh R (2018) Privacy and data security for grid-connected home area network using Internet of Things. IET Netw 7(6):445–452
5. Di Giorgio A, Liberati F, Lanna A, Pietrabissa A, Priscoli FD (2017) Model predictive control of energy storage systems for power tracking and shaving in distribution grids. IEEE Trans Sustain Energy 8(2):496–504
6. Radhakrishnan KK, Moirangthem J, Panda SK, Amaratunga G (2018) GIS integrated automation of a near real-time power-flow service for electrical grids. IEEE Trans Ind Appl 54(6):5661–5670
7. Chen J, Zhu Q (2018) A Stackelberg game approach for two-level distributed energy management in smart grids. IEEE Trans Smart Grid 9(6):6554–6565
8. Radwan AAA, Abdel-Rady I. Mohamed Y (2017) Networked control and power management of AC/DC hybrid microgrids. IEEE Syst J 11(3):1662–1673
9. Haghighat H, Karimianfard H, Zeng B (2020) Integrating energy management of autonomous smart grids in electricity market operation. IEEE Trans Smart Grid 11(5):4044–4055
10. Maffei A, Srinivasan S, Castillejo P, Martínez JF, Iannelli L, Bjerkan E, Glielmo L (2018) A semantic-middleware-supported receding horizon optimal power flow in energy grids. IEEE Trans Ind Informat 14(1):35–46
11. Yi Z, Dong W, Etemadi AH (2018) A unified control and power management scheme for PV-battery-based hybrid microgrids for both grid-connected and islanded modes. IEEE Trans Smart Grid 9(6):5975–5985
12. Soares T, Bessa RJ, Pinson P, Morais H (2018) Active distribution grid management based on robust ac optimal power flow. IEEE Trans Smart Grid 9(6):6229–6241
13. Shareef H, Ahmed MS, Mohamed A, Al Hassan E (2018) Review on home energy management system considering demand responses, smart technologies, and intelligent controllers. IEEE Access 6:24498–24509

14. Sigamani T, Ponraj RK, Ravindran V (2020) Modified single phase matrix converter with z-source for renewable energy systems. In: Third international conference on smart systems and inventive technology (ICSSIT 2020) DVD Part number: CFP20P17-DVD; ISBN: 978-1-7281-5820-4, pp. 599–605, August 2020
15. Mao M, DingY, Chang L, Hatziargyriou ND, Chen Q, Tao T, Li Y (2019) Multi-objective power management for EV fleet with MMC-based integration into smart grid. IEEE Trans Smart Grid 10(2):1428–1439
16. Kikusato H, Mori K, Yoshizawa S, Fujimoto Y, Asano H, Hayashi Y, Kawashima A, Inagaki S, Suzuki T (2019) Electric vehicle charge–discharge management for utilization of photovoltaic by coordination between home and grid energy management systems. IEEE Trans Smart Grid 10(3):3186–3197
17. Srikantha P, Kundur D (2019) Hierarchical signal processing for tractable power flow management in electric grid networks. IEEE Trans Signal Inf Process Over Netw 5(1): 86–99
18. Sayed AR, Wang C, Zhao J, Bi T (2020) Distribution-level robust energy management of power systems considering bidirectional interactions with gas systems. IEEE Trans Smart Grid 11(3):2092–2105
19. Mancilla-David F, Angulo A, Street A (2020) Power management in active distribution systems penetrated by photovoltaic inverters: a data-driven robust approach. IEEE Trans Smart Grid 11(3):2271–2280
20. Delfino F, Ferro G, Robba M, Rossi M (2019) An energy management platform for the optimal control of active and reactive powers in sustainable microgrids. IEEE Trans Ind Appl 55(6):7146–7156
21. Ding L, Wang LY, Yin GY, Zheng WX, Han Q-L (2019) Distributed energy management for smart grids with an event-triggered communication scheme. IEEE Trans Control Syst Technol 27(5):1950–1961
22. Liu Y, Zhang N, Wang Y, Yang J, Kang C (2019) Data-driven power flow linearization: a regression approach. IEEE Trans Smart Grid 10(3):2569–2580
23. Xu Y, Korkali M, Mili L, Chen X, Min L (2020) Risk assessment of rare events in probabilistic power flow via hybrid multi-surrogate method. IEEE Trans Smart Grid 11(2):1593–1603
24. Quan H, Khosravi A, Yang D, Srinivasan D (2020) A survey of computational intelligence techniques for wind power uncertainty quantification in smart grids. IEEE Trans Neural Netw Learn Syst 31(11):4582–4599
25. Amjad Z, Shah MA, Maple C, Khattak HA, Ameer Z, Asghar MN, Mussadiq S (2020) Towards energy efficient smart grids using bio-inspired scheduling techniques. IEEE Access 8:158947–158960
26. Md. Morshed Alam, Md. Shahjalal, Md. Mainul Islam, Moh. Khalid Hasan, Md. Faisal Ahmed, Jang YM (2020) Power flow management with demand response profiles based on user-defined area, load, and phase classification. IEEE Access 8:218813–218827
27. Li G, Huang Y, Bie Z (2018) Reliability evaluation of smart distribution systems considering load rebound characteristics. IEEE Trans Sustain Energy 9(4):1713–1721
28. Ahmed M, Vahidnia A, Datta M, Meegahapola L (2020) An adaptive power oscillation damping controller for a hybrid AC/DC microgrid. IEEE Access 8:69482–69495
29. Capizzi G, Lo Sciuto G, Napoli C, Tramontana E (2018) Advanced and adaptive dispatch for smart grids by means of predictive models. IEEE Trans Smart Grid 9(6):6684–6691
30. Brandao DI, Ferreira WM, Alonso AMS, Tedeschi E, Marafão FP. Optimal multiobjective control of low-voltage AC microgrids: power flow regulation and compensation of reactive power and unbalance. IEEE Trans Smart Grid 11(2):1239–1252
31. Arwa EO, Folly KA (2020) Reinforcement learning techniques for optimal power control in grid-connected microgrids: a comprehensive review. IEEE Access 8:208992–209007
32. Collotta M, Pau G (2017) An innovative approach for forecasting of energy requirements to improve a smart home management system based on BLE. IEEE Trans Green Commun Netw 1(1):112–120

Numerical Simulation of Time Fractional Advection-Diffusion-Reaction Equation using Exponential B-Splines

A. S. V. Ravi Kanth and Neetu Garg

Abstract In this paper, the numerical simulation of the time-fractional advection-diffusion-reaction equation (TFADR) is presented. The discretization procedure consists of Crank-Nicolson approach in time and the exponential B-spline in space. The stability and convergence analyses are also investigated. Numerical experiments are presented to illustrate the efficiency of the proposed method. A comparison with the existing methods indicates the superiority of the proposed scheme.

Keywords Caputo derivative · Exponential B-splines · Time fractional advection-diffusion-reaction equation

1 Introduction

Consider the TFADR equation of the form ($0 < \theta \leq 1$):

$$\frac{\partial^\theta u}{\partial t^\theta} = \beta(\xi, t)\tfrac{\partial^2 u}{\partial \xi^2} + \gamma(\xi, t)\tfrac{\partial u}{\partial \xi} + \eta(\xi, t)u(\xi, t) + \mathfrak{f}(\xi, t),\ \xi_a \leq \xi \leq \xi_b,\ t > 0, \tag{1}$$

with initial and boundary conditions

$$u|_{t=0} = \psi_0(\xi),\ u|_{\xi=\xi_a} = \psi_1(t),\ u|_{\xi=\xi_b} = \psi_2(t), \tag{2}$$

where $\beta(\xi, t), \gamma(\xi, t), \eta(\xi, t), \mathfrak{f}(\xi, t), \psi_0(\xi), \psi_1(t)$, and $\psi_2(t)$ are sufficiently smooth functions. The time derivative is taken in the Caputo sense [1]. The TFADR is a powerful tool in the modeling of various physical phenomena, such as anomalous diffusion, fluid flow, heat transfer processes, and option pricing [2, 3]. The well-posedness of the TFADR was studied in [4]. In [5], authors utilized implicit meshless spectral method for solving TFADR. The homotopy perturbation method is proposed to solve TFADR in [6]. A meshless method is used to solve time-fractional advection-diffusion

A. S. V. Ravi Kanth (✉) · N. Garg
Department of Mathematics, National Institute of Technology, Kurukshetra 136119, India

A. Kumar et al. (eds.), *Renewable Energy Towards Smart Grid*, Lecture Notes in Electrical Engineering 823, https://doi.org/10.1007/978-981-16-7472-3_11

equation [3]. This paper studies the TFADR equation using Crank-Nicolson approach and the exponential B-spline method. Exponential B–spline method is a very efficient numerical approach consisting of twice continuously differentiable basis functions. It has benefits over the finite difference method to obtain the solutions even between mesh points and over finite element method to get rid of quadrature calculation. This method is a generalization of cubic B–spline method. For applications of the exponential B-spline method, one may refer [7–11].

2 Numerical Procedure

Now, we present the numerical procedure consisting Crank-Nicolson method in time and the exponential B-splines in space. Let the partition of $[\xi_a, \xi_b] \times [0, T]$ be

$$\xi_a + m\Delta\xi, 0 \le m \le M, \; n\Delta t, 0 \le n \le N, \; M, N > 0,$$

with the space and time step sizes $\Delta\xi = \frac{\xi_b - \xi_a}{M}$ and $\Delta t = \frac{T}{N}$, respectively. Discretization of Caputo derivative $\frac{\partial^\theta U}{\partial t^\theta}$ at $(\xi, t_{n+\frac{1}{2}})$ is given as [13]:

$$\frac{\partial^\theta U(\xi, t_{n+\frac{1}{2}})}{\partial t^\theta} = \bar{\varrho}\left(U^{n+1}(\xi) - U^n(\xi)\right) + \varsigma_1 U^n(\xi) - \varsigma_n U^0(\xi) + \sum_{r=1}^{n-1} (\varsigma_{n-r+1} - \varsigma_{n-r}) U^r(\xi) + R_\theta^{n+\frac{1}{2}}, \; 0 < \theta < 1, \tag{3}$$

where

$$U^{n+1}(\xi) = U(\xi, t_{n+1}), \; \bar{\varrho} = \frac{2^{\theta-1}\Delta t^{-\theta}}{\Gamma(2-\theta)},$$

$$\varsigma_q = \frac{1}{\Delta t^\theta \Gamma(2-\theta)}\left(\left(q + \frac{1}{2}\right)^{1-\theta} - \left(q - \frac{1}{2}\right)^{1-\theta}\right), (q = 1, ..., n),$$

and $R_\theta^{n+\frac{1}{2}}$ is the truncation error given as

$$|R_\theta^{n+\frac{1}{2}}| \le C\Delta t^{2-\theta}. \tag{4}$$

Discretizing (1) at $(\xi, t_{n+\frac{1}{2}})$ and utilizing (3), we have

$$
\begin{aligned}
&-\beta^{n+\frac{1}{2}}\frac{\partial^2 U^{n+1}(\xi)}{\partial \xi^2}-\gamma^{n+\frac{1}{2}}\frac{\partial U^{n+1}(\xi)}{\partial \xi}+(2\bar{\varrho}-\eta^{n+\frac{1}{2}})U^{n+1}(\xi)\\
&=\beta^{n+\frac{1}{2}}\frac{\partial^2 U^{n}(\xi)}{\partial \xi^2}+\gamma^{n+\frac{1}{2}}\frac{\partial U^{n}(\xi)}{\partial \xi}+(2\bar{\varrho}+\eta^{n+\frac{1}{2}})U^{n}+2\mathfrak{f}^{n+\frac{1}{2}}\\
&+2\left(\varsigma_n U^0(\xi)-\varsigma_1 U^n(\xi)+\sum_{r=1}^{n-1}(\varsigma_{n-r}-\varsigma_{n-r+1})U^r(\xi)\right)+R_\theta^{n+\frac{1}{2}},
\end{aligned}
\tag{5}
$$

where $\beta^{n+\frac{1}{2}}=\beta(\xi,t_{n+\frac{1}{2}}), \gamma^{n+\frac{1}{2}}=\gamma(\xi,t_{n+\frac{1}{2}}), \quad \eta^{n+\frac{1}{2}}=\eta(\xi,t_{n+\frac{1}{2}})$ and $\mathfrak{f}^{n+\frac{1}{2}}=\mathfrak{f}(\xi,t_{n+\frac{1}{2}})$ for $n=0,1,...,N-1$. Next, the exponential B-splines are utilized to discretize the resulting spatial equation (5). The exponential B-spline functions $EB_l(\xi)$ $(-1\le l\le M+1)$ form a basis, thus we can represent numerical solution U as [12]:

$$
U(\xi)=\sum_{l=-1}^{M+1}\phi_l(t)EB_l(\xi), \tag{6}
$$

where $\phi_l(t)$ are unknowns. The values of $U(\xi)$, $\frac{\partial U}{\partial \xi}$ and $\frac{\partial^2 U}{\partial \xi^2}$ at ξ_m's $(0\le m\le M)$ are as follows:

$$
\begin{aligned}
&U^{n+1}(\xi_m)=\mu_1\phi_{m-1}^{n+1}+\phi_m^{n+1}+\mu_1\phi_{m+1}^{n+1},\quad \frac{\partial U^{n+1}(\xi_m)}{\partial \xi}=\mu_2\phi_{m-1}^{n+1}-\mu_2\phi_{m+1}^{n+1},\text{ and}\\
&\frac{\partial^2 U^{n+1}(\xi_m)}{\partial \xi^2}=\mu_3(\phi_{m-1}^{n+1}-2\phi_m^{n+1}+\phi_{m+1}^{n+1}),
\end{aligned}
\tag{7}
$$

where $\mu_1=\frac{\tilde{s}-p\Delta\xi}{2(p\Delta\xi c-\tilde{s})}$, $\mu_2=\frac{p(1-c)}{2(p\Delta\xi c-\tilde{s})}$, $\mu_3=\frac{p^2\tilde{s}}{2(p\Delta\xi c-\tilde{s})}$, $\tilde{s}=\sinh(p\Delta\xi)$, $\tilde{c}=\cosh(p\Delta\xi)$ and $p>0$ is the tension parameter. Using (7) in (5) at the points ξ_m's, we have

$$
\Upsilon_1|_m^n\,\phi_{m-1}^{n+1}+\Upsilon_2|_m^n\,\phi_m^{n+1}+\Upsilon_3|_m^n\,\phi_{m+1}^{n+1}=\lambda_m^n, \tag{8}
$$

where

$$
\begin{aligned}
\lambda_m^n&=\Upsilon_4|_m^n\,\phi_{m-1}^n+\Upsilon_5|_m^n\,\phi_m^n+\Upsilon_6|_m^n\,\phi_{m+1}^n+2\Big(\varsigma_n Q_m^0-\varsigma_1 Q_m^n-\sum_{r=1}^{n-1}(\varsigma_{n-r+1}\\
&\quad-\varsigma_{n-r})Q_m^r\Big)+2\mathfrak{f}_m^{n+\frac{1}{2}},\\
\Upsilon_1|_m^n&=(2\bar{\varrho}-\eta_m^{n+\frac{1}{2}})\mu_1-\mu_2\gamma_m^{n+\frac{1}{2}}-\mu_3\beta_m^{n+\frac{1}{2}},\ \Upsilon_2|_m^n=2\bar{\varrho}-\eta_m^{n+\frac{1}{2}}+2\mu_3\beta_m^{n+\frac{1}{2}},\\
\Upsilon_3|_m^n&=(2\bar{\varrho}-\eta_m^{n+\frac{1}{2}})\mu_1+\mu_2\gamma_m^{n+\frac{1}{2}}-\mu_3\beta_m^{n+\frac{1}{2}},\\
\Upsilon_4|_m^n&=(2\bar{\varrho}+\eta_m^{n+\frac{1}{2}})\mu_1+\mu_2\gamma_m^{n+\frac{1}{2}}+\mu_3\beta_m^{n+\frac{1}{2}},
\end{aligned}
$$

$$\Upsilon_5|_m^n = 2\bar{\varrho} + \eta_m^{n+\frac{1}{2}} - 2\mu_3\beta_m^{n+\frac{1}{2}},\ \Upsilon_6|_m^n = (2\bar{\varrho} + \eta_m^{n+\frac{1}{2}})\mu_1 - \mu_2\gamma_m^{n+\frac{1}{2}} + \mu_3\beta_m^{n+\frac{1}{2}},$$
$$Q_m^l = \mu_1\phi_{m-1}^l + \phi_m^l + \mu_1\phi_{m+1}^l,\ 0 \le l \le n.$$

The discretized boundary conditions are

$$\begin{aligned}\mu_1\phi_{-1}^{n+1} + \phi_0^{n+1} + \mu_1\phi_1^{n+1} &= \psi_1^{n+1},\\ \mu_1\phi_{M-1}^{n+1} + \phi_M^{n+1} + \mu_1\phi_{M+1}^{n+1} &= \psi_2^{n+1}.\end{aligned} \tag{9}$$

Eliminating ϕ_{-1}^{n+1} and ϕ_{M+1}^{n+1} from (9) and combining with (8), we obtain

$$\mathbb{A}\Phi^{n+1} = \mathbb{F}^n, \tag{10}$$

where

$$\mathbb{A} = \begin{pmatrix} \Upsilon_2|_0^n - \frac{\Upsilon_1|_0^n}{\mu_1} & \Upsilon_3|_0^n - \Upsilon_1|_0^n & 0 & & \\ \Upsilon_1|_1^n & \Upsilon_2|_1^n & \Upsilon_3|_1^n & & \\ & \cdots & \cdots & \cdots & \\ & & \cdots & \cdots & \cdots \\ & & \Upsilon_1|_{M-1}^n & \Upsilon_2|_{M-1}^n & \Upsilon_3|_{M-1}^n \\ & & & \Upsilon_1|_M^n - \Upsilon_3|_M^n & \Upsilon_2|_M^n - \frac{\Upsilon_3|_M^n}{\mu_1} \end{pmatrix},$$

$$\Phi^{n+1} = (\phi_0^{n+1}, \phi_1^{n+1}, ..., \phi_{M-1}^{n+1}, \phi_M^{n+1})',$$

$$\mathbb{F}^n = \left(\lambda_0^n - \frac{\Upsilon_1|_0^n}{\mu_1}\psi_1^{n+1}, \lambda_1^n, ..., \lambda_{M-1}^n, \lambda_M^n - \frac{\Upsilon_3|_M^n}{\mu_1}\psi_2^{n+1}\right)'.$$

3 Stability and Convergence Analysis

Now, we first discuss the stability analysis, then we analyze the convergence of the proposed method.

Lemma 1 ([13]) *The coefficients ς_n hold:*

1. $\varsigma_1 \ge \varsigma_2 \ge ... \ge 0$,
2. $\sum_{r=1}^{n-1}(\varsigma_{n-r} - \varsigma_{n-r+1}) + \varsigma_n = \varsigma_1$.

Theorem 1 *The numerical scheme* (8) *with* (9) *is unconditionally stable.*

Proof The perturbation $\varepsilon_m^n = \phi_m^n - \tilde{\phi}_m^n$, where $\tilde{\phi}_m^n$ is the perturbed solution of (8), satisfying

$$\Upsilon_1|_m^n\, \varepsilon_{m-1}^{n+1} + \Upsilon_2|_m^n\, \varepsilon_m^{n+1} + \Upsilon_3|_m^n\, \varepsilon_{m+1}^{n+1} = \Upsilon_4|_m^n\, \varepsilon_{m-1}^{n} + \Upsilon_5|_m^n\, \varepsilon_m^{n} + \Upsilon_6|_m^n\, \varepsilon_{m+1}^{n} + 2\left(\varsigma_n \bar{Q}_m^0 - \varsigma_1 \bar{Q}_m^n - \sum_{r=1}^{n-1}(\varsigma_{n-r+1} - \varsigma_{n-r})\bar{Q}_m^r\right), \tag{11}$$

where $\bar{Q}_m^l = \mu_1 \varepsilon_{m-1}^l + \varepsilon_m^l + \mu_1 \varepsilon_{m+1}^l,\ l = 0, 1, ..., n$. Substituting $\varepsilon_m^n = \Im^n e^{jm\nu}$ in (11) and simplifying, where $j = \sqrt{-1}$, $\nu = \sigma \Delta\xi$ and σ is mode number and $\Im$ is the amplification factor, we have

$$\Lambda_3 \Im^{n+1} = \Lambda_1 \Im^n + 2\Lambda_2 \left[\varsigma_n \Im^0 - \varsigma_1 \Im^n - \sum_{r=1}^{n-1}(\varsigma_{n-r+1} - \varsigma_{n-r})\Im^r\right], \tag{12}$$

where $\Lambda_1 = \left[\Upsilon_4 e^{-j\nu} + \Upsilon_5 + \Upsilon_6 e^{j\nu})\right]$, $\Lambda_2 = 1 + \mu_1(e^{-j\nu} + e^{j\nu})$, and $\Lambda_3 = \left[\Upsilon_1 e^{-j\nu} + \Upsilon_2 + \Upsilon_3 e^{j\nu}\right]$. Simplification of (12) yields

$$|\Im^{n+1}| \le \left|\left(\frac{\Lambda_1 - 2\varsigma_1 \Lambda_2}{\Lambda_3}\right)\Im^n + 2\left(\varsigma_n \Im^0 - \sum_{r=1}^{n-1}(\varsigma_{n-r+1} - \varsigma_{n-r})\,\Im^r\right)\frac{\Lambda_2}{\Lambda_3}\right|. \tag{13}$$

We prove $\left|\Im^{n+1}\right| \le \left|\Im^0\right|$ by mathematical induction. For $n = 0$ in (13), we have

$$|\Im^1| = |\frac{(2\bar{\varrho} + \eta_m^{n+\frac{1}{2}})\Lambda_2 - 4\beta_m^{n+\frac{1}{2}}\mu_3 \sin^2(\frac{\nu}{2}) - 2j\gamma_m^{n+\frac{1}{2}}\mu_2 \sin(\nu)}{(2\bar{\varrho} - \eta_m^{n+\frac{1}{2}})\Lambda_2 + 4\beta_m^{n+\frac{1}{2}}\mu_3 \sin^2(\frac{\nu}{2}) + 2j\gamma_m^{n+\frac{1}{2}}\mu_2 \sin(\nu)}\Im^0| \le \left|\Im^0\right|. \tag{14}$$

Let us assume

$$\left|\Im^l\right| \le \left|\Im^0\right|,\ l = 1, ..., n. \tag{15}$$

We show that (15) holds for time $l = n + 1$. Using Lemma 1 in (13), we have

$$\left|\Im^{n+1}\right| \le \left(\frac{\left|(2\bar{\varrho} + \eta_m^{n+\frac{1}{2}} - 2\varsigma_1)\Lambda_2 - 4\beta_m^{n+\frac{1}{2}}\mu_3 \sin^2\left(\frac{\nu}{2}\right)\right| + 2\Lambda_2\varsigma_1}{\left|(2\bar{\varrho} - \eta_m^{n+\frac{1}{2}})\Lambda_2 + 4\beta_m^{n+\frac{1}{2}}\mu_3 \sin^2\left(\frac{\nu}{2}\right)\right|}\right)\left|\Im^0\right|. \tag{16}$$

If $(2\bar{\varrho} + \eta_m^{n+\frac{1}{2}} - 2\varsigma_1)\Lambda_2 - 4\beta_m^{n+\frac{1}{2}}\mu_3 \sin^2\left(\frac{\nu}{2}\right) > 0$ in (16), then

$$\left|\Im^{n+1}\right| \le \left(\frac{(2\bar{\varrho} + \eta_m^{n+\frac{1}{2}})\Lambda_2 - 4\beta_m^{n+\frac{1}{2}}\mu_3 \sin^2\left(\frac{\nu}{2}\right)}{(2\bar{\varrho} - \eta_m^{n+\frac{1}{2}})\Lambda_2 + 4\beta_m^{n+\frac{1}{2}}\mu_3 \sin^2\left(\frac{\nu}{2}\right)}\right)\left|\Im^0\right| \le \left|\Im^0\right|.$$

If $(2\bar{\varrho} + \eta_m^{n+\frac{1}{2}} - 2\varsigma_1)\Lambda_2 - 4\beta_m^{n+\frac{1}{2}}\mu_3 \sin^2\left(\frac{\nu}{2}\right) \le 0$ in (16), then

$$\left|\Im^{n+1}\right| \leq \left(\frac{(4\varsigma_1 - 2\bar{\varrho} - \eta_m^{n+\frac{1}{2}})\Lambda_2 + 4\beta_m^{n+\frac{1}{2}}\mu_3 \sin^2\left(\frac{\nu}{2}\right)}{\left|(2\bar{\varrho} - \eta_m^{n+\frac{1}{2}})\Lambda_2 + 4\beta_m^{n+\frac{1}{2}}\mu_3 \sin^2\left(\frac{\nu}{2}\right)\right|} \right) \left|\Im^0\right|. \tag{17}$$

From (17), we have

$$\begin{aligned} &\left|\Im^{n+1}\right| \leq \left|\Im^0\right| \Leftrightarrow \varsigma_1 \leq \bar{\varrho} \\ &\Leftrightarrow \frac{\Delta t^{-\theta}}{\Gamma(2-\theta)} \left(\left(\frac{3}{2}\right)^{1-\theta} - \left(\frac{1}{2}\right)^{1-\theta} \right) \leq \frac{\Delta t^{-\theta}}{2^{1-\theta}\Gamma(2-\theta)} \quad \Leftrightarrow \frac{3}{2} \leq 3^\theta. \end{aligned} \tag{18}$$

Hence $\left|\Im^{n+1}\right| \leq \left|\Im^0\right|$ for $n \geq 0$. Therefore, the proposed scheme is unconditionally stable.

Lemma 2 *The exponential B-splines $EB_{-1}(\xi), EB_0(\xi), ..., EB_{M+1}$ satisfy [7]*

$$\sum_{m=-1}^{M+1} |EB_m(\xi)| \leq \frac{5}{2}, \ \xi_a \leq \xi \leq \xi_b.$$

Theorem 2 *Let $U(\xi, t)$ be unique exponential B-spline interpolation to the exact solution $u(\xi, t)$ of (1)–(2). If $u(\xi, t) \in C^4[\xi_a, \xi_b]$, then $\forall\, t \geq 0$ (see [8])*

$$\left\| \partial_\xi^j (u - U) \right\|_\infty \leq \chi_j \Delta\xi^{4-j}, \ 0 \leq j \leq 2,$$

where χ_j are constants.

Theorem 3 *Let $U(\xi, t)$ be the exponential B-spline approximate to the exact solution $u(\xi, t)$ of (1)–(2). If $\mathfrak{f} \in C^2[\xi_a, \xi_b]$ and $u(\xi, t) \in C^4[\xi_a, \xi_b]$, then*

$$\|(U - u)(\xi, t)\|_\infty \leq \Theta \Delta\xi^2,$$

for a constant Θ.

Proof Let $U(\xi, t) = \sum_{m=-1}^{M+1} \phi_m(t) EB_m(\xi)$ be the exponential B-spline approximation and $\hat{U}(\xi, t) = \sum_{m=-1}^{M+1} \hat{\phi}_m(t) EB_m(\xi)$ be the unique exponential B-spline interpolant to $u(\xi, t)$. Consider (10) for U and $\hat{U}$ $(0 \leq n \leq N - 1)$, then subtracting $\mathbb{A}\hat{\Phi}^{n+1} = \hat{\mathbb{F}}^n$ from $\mathbb{A}\Phi^{n+1} = \mathbb{F}^n$ gives

$$\mathbb{A}\left(\Phi^{n+1} - \hat{\Phi}^{n+1}\right) = \mathbb{F}^n - \hat{\mathbb{F}}^n, \ 0 \leq n \leq N - 1, \tag{19}$$

where $\hat{\phi}^{n+1} = \left(\hat{\phi}_0^n, \hat{\phi}_1^n, ..., \hat{\phi}_M^n\right)'$, $\hat{\mathbb{F}}^n = \left(\hat{\lambda}_0^n - \frac{\Upsilon_1}{\mu_1}\psi_1^{n+1}, \hat{\lambda}_1^n, ..., \hat{\lambda}_M^n - \frac{\Upsilon_3}{\mu_1}\psi_2^{n+1}\right)'$, and $\mathbb{F}^n - \hat{\mathbb{F}}^n = \left(\lambda_0^n - \hat{\lambda}_0^n, \lambda_1^n - \hat{\lambda}_1^n, ..., \lambda_M^n - \hat{\lambda}_M^n\right)'$. Since $\mathbb{A}$ is strictly diagonally dominant, therefore (19) implies

$$\left\|\Phi^{n+1}-\hat{\Phi}^{n+1}\right\|_{\infty} \leq \|\mathbb{A}^{-1}\|_{\infty}\left\|\mathbb{F}^{n}-\hat{\mathbb{F}}^{n}\right\|_{\infty}. \tag{20}$$

To evaluate $\left\|\mathbb{F}^{n}-\hat{\mathbb{F}}^{n}\right\|_{\infty}$, we proceed as follows

$$\begin{aligned}\left|\lambda_m^n-\hat{\lambda}_m^n\right| \leq & \left|\beta^{n+\frac{1}{2}}\left(\frac{\partial^2 U^n(\xi_m)}{\partial\xi^2}-\frac{\partial^2\hat{U}^n(\xi_m)}{\partial\xi^2}\right)\right|+\left|\gamma^{n+\frac{1}{2}}\left(\frac{\partial U^n(\xi_m)}{\partial\xi}\right.\right.\\ & \left.\left.-\frac{\partial\hat{U}^n(\xi_m)}{\partial\xi}\right)\right|+2\sum_{r=1}^{n-1}|\varsigma_{n-r}-\varsigma_{n-r+1}|\left|(U^r(\xi_m)-\hat{U}^r(\xi_m))\right|\\ & +\left|2\bar{\varrho}+\eta^{n+\frac{1}{2}}-2\varsigma_1\right|\left|(U^n(\xi_m)-\hat{U}^n(\xi_m))\right|+2\varsigma_n\left|(U^0(\xi_m)-\hat{U}^0(\xi_m))\right|. \end{aligned} \tag{21}$$

Utilizing ℓ_∞ norm and Theorem 2 in (21) gives

$$\left\|\mathbb{F}^{n}-\hat{\mathbb{F}}^{n}\right\|_{\infty} \leq \Pi h^2, \tag{22}$$

where $\Pi = \|\beta\|_\infty \chi_2 + \|\gamma\|_\infty \chi_1 \Delta\xi + \left(|2\bar{\varrho}-2\varsigma_1| + \|\eta\|_\infty + \varsigma_n + \sum_{r=1}^{n-1}\left|\varsigma_{n-r}\right.\right.$ $\left.\left.-\varsigma_{n-r+1}\right|\right)\chi_0\Delta\xi^2$.

The row sums $\mathfrak{r}_l$ $(0 \leq l \leq M)$ of the matrix $\mathbb{A}$ are

$$\begin{aligned}\mathfrak{r}_0 &= \Upsilon_2+\Upsilon_3-\Upsilon_1\left(1+\tfrac{1}{\mu_1}\right)=(2+\tfrac{1}{\mu_1})(\mu_3\beta_0^{n+\frac{1}{2}}+\mu_2\gamma_0^{n+\frac{1}{2}}),\\ \mathfrak{r}_i &= \Upsilon_1+\Upsilon_2+\Upsilon_3=(1+2\mu_1)(2\bar{\varrho}-\eta_i^{n+\frac{1}{2}}),\ 1\leq i\leq M-1,\\ \mathfrak{r}_M &= \Upsilon_1+\Upsilon_2-\Upsilon_3\left(1+\tfrac{1}{\mu_1}\right)=(2+\tfrac{1}{\mu_1})(\mu_3\beta_M^{n+\frac{1}{2}}-\mu_2\gamma_M^{n+\frac{1}{2}}).\end{aligned}$$

The matrix theory yields

$$\|\mathbb{A}^{-1}\|_\infty \leq \frac{1}{\mathfrak{r}} \leq \frac{1}{|\mathfrak{r}|}, \tag{23}$$

where $\mathfrak{r} = min\{\mathfrak{r}_0, \mathfrak{r}_1, ..., \mathfrak{r}_M\}$. Substituting the equations (22) and (23) into (20) leads to

$$\left\|\Phi^{n+1}-\hat{\Phi}^{n+1}\right\|_{\infty} \leq \frac{\Pi\Delta\xi^2}{|\mathfrak{r}|}. \tag{24}$$

Now, using (24) and Lemma 2, we have

$$\left\|(U-\hat{U})\right\|_\infty = \left\|\sum_{m=-1}^{M+1}(\phi_m-\hat{\phi}_m)EB_m(\xi)\right\|_\infty \leq \frac{5\Pi\Delta\xi^2}{2|\mathfrak{r}|}. \tag{25}$$

Application of the theorem 2 gives

$$\left\|(u-\hat{U})(\xi,t)\right\|_{\infty} \leq \chi_0 \Delta\xi^4. \tag{26}$$

Thus combining equations (25) and (26), we have

$$\|(u-U)(\xi,t)\|_{\infty} \leq \Theta\Delta\xi^2,$$

where $\Theta = \chi_0\Delta\xi^2 + \frac{5\Pi}{2|\tau|}$. Hence, proved.

4 Numerical Examples

Now, we perform the numerical experiments to indicate the effectiveness of the proposed method.

Example 1 Consider

$$\frac{\partial^\theta u}{\partial t^\theta} = \frac{\partial^2 u}{\partial \xi^2} - \frac{\partial u}{\partial \xi} + \frac{2t^{2-\theta}}{\Gamma(3-\theta)} + 2\xi - 2, \; 0 \leq \xi \leq 1,$$

where $u(\xi,t) = \xi^2 + t^2$ is the exact solution.

Example 2 Consider

$$\frac{\partial^\theta u}{\partial t^\theta} = 10\frac{\partial^2 u}{\partial \xi^2} - 4\frac{\partial u}{\partial \xi} + \frac{\Gamma(\theta+4)}{6}t^3\cos(\xi) + t^{\theta+3}(10\cos(\xi) - 4\sin(\xi)), \; 0 \leq \xi \leq 2\pi,$$

where $u(\xi,t) = t^{\theta+3}\cos\xi$ is the exact solution. The initial and boundary conditions are taken from the exact solution.

5 Discussions and Conclusion

The ℓ_2 errors in space and time of Example 1 are computed in Tables 1 and 2. A comparison is presented in Tables 3 and 4, and it is observed that our results are superior than the results in [3]. Tables 5 and 6 record ℓ_2 errors in space and time of Example 2. One can observe that the proposed scheme possesses convergence rate $O(\Delta\xi^2 + \Delta t^{2-\theta})$, which is consistent with theoretical analysis. Figures 1 and 2 exhibit the numerical and exact solution profiles and a good accordance is evident. In a nutshell, we constructed an efficient numerical method for the TFADR equation.

Table 1 ℓ_2 errors in space with $p = 2$ and $\Delta t = \Delta\xi^2$ of Example 1

$\Delta\xi$	$\theta = 0.25$		$\theta = 0.5$		$\theta = 0.75$	
1/8	8.5656e–04	–	8.5886e–04	–	8.5037e–04	–
1/16	2.1565e–04	1.9898	2.1803e–04	1.9779	2.1717e–04	1.9692
1/32	5.4031e–05	1.9969	5.4892e–05	1.9899	5.5032e–05	1.9805
1/64	1.3518e–05	1.9990	1.3769e–05	1.9951	1.3884e–05	1.9868

Table 2 ℓ_2 errors in time with $p = 1$ and $M = 2000$ of Example 1

N	$\theta = 0.5$		$\theta = 0.7$		$\theta = 0.9$	
160	5.5948e–06	–	1.4415e–05	–	1.9571e–05	–
320	1.9927e–06	1.4893	5.8686e–06	1.2965	9.1357e–06	1.0991
640	7.0646e–07	1.4961	2.3853e–06	1.2988	4.2619e–06	1.1000
1280	2.4850e–07	1.5073	9.6759e–07	1.3017	1.9867e–06	1.1011

Table 3 Error comparisons with $p = 0.1$, $\Delta t = 0.01$, $\theta = 0.2$, and $M = 10$ of Example 1

t	Present method			Method in [3]		
	ℓ_∞	ℓ_2	RMS	ℓ_∞	ℓ_2	RMS
0.2	2.9250e–07	7.8147e–07	2.4712e–07	5.0049e–06	1.1651e–05	3.5131e-06
0.4	2.8288e–07	7.4387e–07	2.3523e–07	5.3655e–06	1.2486e–05	3.7645e-06
0.6	3.5280e–07	8.8150e–07	2.7876e–07	5.5582e–06	1.2931e–05	3.8988e-06
0.8	4.1029e–07	9.9720e–07	3.1534e–07	5.6872e–06	1.3229e–05	3.9887e-06
1	4.5434e–07	1.0862e–06	3.4348e–07	5.7830e–06	1.3450e–05	4.0555e-06

Table 4 Error comparisons with $p = 1$, $\Delta t = 0.01$, $\theta = 0.5$, and $M = 50$ of Example 1

t	Present method		Method in [14]	
	ℓ_∞	ℓ_2	ℓ_∞	ℓ_2
0.1	5.4247e–06	4.0091e–06	6.0860e–02	2.6130e–01
0.5	7.2080e–06	5.2689e–06	2.9580e–02	1.2770e–01
1	7.6463e–06	5.5841e–06	2.1140e–02	9.1340e–02
1.5	7.8424e–06	5.7251e–06	1.7320e–02	7.4850e–02
2	7.9597e–06	5.8094e–06	1.5030e–02	6.4940e–02

Table 5 ℓ_2 errors in space with $p = 1$ and $\Delta t = \Delta\xi^2$ of Example 2

Δx	$\theta = 0.25$		$\theta = 0.5$		$\theta = 0.75$	
$\pi/10$	1.5635e–02	–	1.5423e–02	–	1.4499e–02	–
$\pi/20$	4.1398e–03	1.9171	3.9840e–03	1.9529	3.9589e–03	1.8727
$\pi/40$	9.8169e–04	2.0762	9.1248e–04	2.1263	8.9701e–04	2.1419
$\pi/80$	2.4409e–04	2.0079	2.2206e–04	2.0388	2.1288e–04	2.0751

Table 6 ℓ_2 errors in time with $p = 0.01$ and $M = 5000$ of Example 2

N	$\theta = 0.5$		$\theta = 0.7$		$\theta = 0.9$	
100	1.0308e–04	–	2.5679e–04	–	3.1111e–04	–
200	3.7679e–05	1.4520	1.0686e–04	1.2648	1.5111e–04	1.0418
400	1.3743e–05	1.4551	4.4114e–05	1.2764	7.2021e–05	1.0691
800	5.0829e–06	1.4350	1.8186e–05	1.2784	3.4044e–05	1.0810

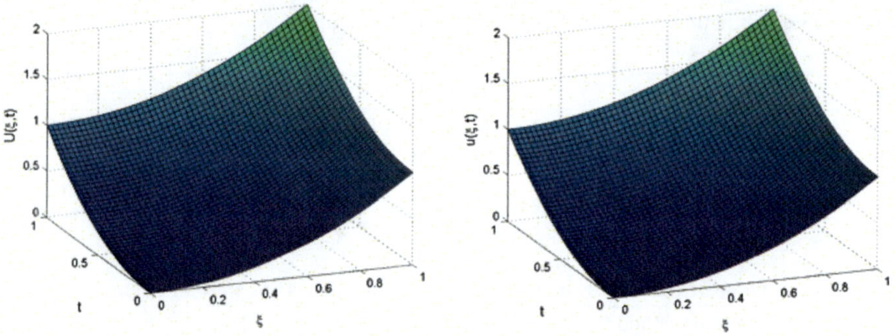

(a) (b)

Fig. 1 **a** Numerical **b** Exact solution with $p = 1$, $N = M = 50$, and $\theta = 0.5$ of Example 1

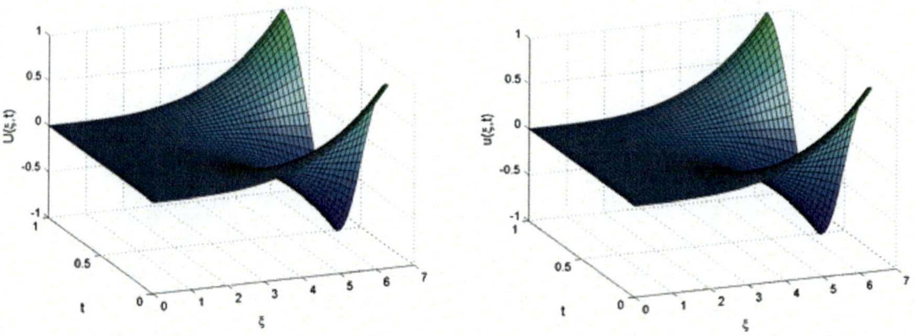

(a) (b)

Fig. 2 **a** Numerical **b** Exact solution with $p = 1$, $N = M = 50$, and $\theta = 0.5$ of Example 2

We discretized the problem using Crank-Nicolson approach in time and exponential B-spline in space. The proposed scheme is unconditional stable with convergence rate $O(\Delta\xi^2 + \Delta t^{2-\theta})$. Numerical experiments exhibit the accuracy and superiority of the proposed scheme. A comparison with existing method confirms the efficiency of exponential B-splines for solving fractional differential equations.

Acknowledgements Authors are grateful to the anonymous reviewers for their insightful comments leading to the improved manuscript. The second author is thankful to the University Grants Commission of India for support under Senior Research Fellowship scheme.

References

1. Kilbas AA, Srivastva HM, Trujillo JJ (2006) Theory and applications of fractional differential equations. Elsevier, North-Holland
2. Hundsdorfer W, Verwer J (2003) Numerical solution of time-dependent advection-diffusion reaction equations. Springer, Heidelberg
3. Mardani A, Hooshmandasl MR, Heydari MH, Cattani C (2018) A meshless method for solving the time fractional advection-diffusion equation with variable coefficients. Comput Math Appl 75:122–133. https://doi.org/10.1016/j.camwa.2017.08.038
4. McLean W, Mustapha K, Ali R, Knio O (2019) Well-posedness of time-fractional advection-diffusion-reaction equations. Fract Calc Appl Anal 22:918–944. https://doi.org/10.1515/fca-2019-0050
5. Haq S, Hussain M, Ghafoor A (2020) A computational study of variable coefficients fractional advection-diffusion-reaction equations via implicit meshless spectral algorithm. Eng Comput 36:1243–1263. https://doi.org/10.1007/s00366-019-00760-x
6. Pandey P, Kumar S, Aguilar JFG (2022) Numerical solution of the time fractional reaction-advection-diffusion equation in porous media. J Appl Comput Mech 8:84–96 (2021). https://doi.org/10.22055/JACM.2019.30946.1796
7. Chandra SRS, Kumar M (2008) Exponential B-spline collocation method for self-adjoint singularly perturbed boundary value problems. Appl Numer Math 58:1572–1581. https://doi.org/10.1016/j.apnum.2007.09.008
8. Mohammadi R (2013) Exponential B-spline solution of convection-diffusion equations. Appl Math 4:933–944. https://doi.org/10.4236/am.2013.46129
9. Ravi Kanth ASV, Garg N (2019) An implicit numerical scheme for a class of multiterm time-fractional diffusion equation. Eur Phys J Plus 134:312. https://doi.org/10.1140/epjp/i2019-12696-8
10. Ravi Kanth ASV, Garg N (2019) A numerical approach for a class of time-fractional reaction-diffusion equation through exponential B-spline method. Comput Appl Math 39:37. https://doi.org/10.1007/s40314-019-1009-z
11. Ravi Kanth ASV, Garg N (2020) An unconditionally stable algorithm for multi-term time fractional advection-diffusion equation with variable coefficients and convergence analysis. Numer Meth Part D E. 1–18 (2020). https://doi.org/10.1002/num.22629
12. McCartin BJ (1991) Theory of exponential splines. J Approx Theory 66:1–23. https://doi.org/10.1016/0021-9045(91)90050-K
13. Karatay I, Kale N, Bayramoglu SR (2013) A new difference scheme for time fractional heat equations based on the Crank-Nicolson method. Frac Calc Appl Anal 16:892–910. https://doi.org/10.2478/s13540-013-0055-2
14. Uddin M, Haq S (2011) RBFs approximation method for time fractional partial differential equations. Commun Nonlinear Sci Numer Simulat 16:4208–4214. https://doi.org/10.1016/j.cnsns.2011.03.021

Combined Economic Emission Dispatch of Thermal and Solar Photo Voltaic Generation Systems by Particle Swarm Optimization

Rajanish Kumar Kaushal and Tilak Thakur

Abstract Nowadays, evolution of renewable energy sources has become a techno-economical viable option for exploitation by electrical utilities as well. Moreover, widespread use of renewable energy sources is also encouraging policy planners and utilities to enhance contribution of green energy for ensuring sustainable economic growth as well in long run. It does not produce any pollution. Solar energy is the most important non-conventional energy source to generate power. In this paper, particle swarm optimization (PSO) is successfully applied to achieve the best or nearest optimum solutions for combined economic emission dispatch (CEED) issue for a scenario involving three conventional thermal power plants and ten solar photovoltaic (PV) plants. Before implementing the presented method, it is validated with several standard optimization examples to ascertain efficacy in solving key issues. In order to highlight effectiveness of the strategy, case study targeting given combination of power generation mix with various practical constraints have been enumerated. The Pareto optimal solution, showing the trade-off relationship among the conflicting objectives, is successfully captured in the multi-objective optimization case resulting in obtaining cost-effective and emission solution in considered situation. To show the benefits of solar PV, outcomes are compared with and without solar PV plants. Results are also compared with teaching learning-based optimization (TLBO) for thermal-solar scheduling.

Keywords Economic · Emission · CEED · PV · TLBO · PSO

1 Introduction

In order to ensure reliable, efficient and economic operation of a power system network having multi-dimensional mix of variety of power generation plants, namely, hydro power plants, thermal power and solar power plants, the scheduling of power generation plays a pivotal role to meet requisite load demand. Thermal power plants provide bulk of the maximum energy demand in the energy system by judicious

R. K. Kaushal (✉) · T. Thakur
EED, Punjab Engineering College, Chandigarh, India

A. Kumar et al. (eds.), *Renewable Energy Towards Smart Grid*, Lecture Notes in Electrical Engineering 823, https://doi.org/10.1007/978-981-16-7472-3_12

generation of electricity considering the techno-economics and environmental issues. Sometimes, it becomes more prudent to plan partial load demand of energy being met from other available energy sources such as hydroelectric energy and most importantly non-conventional power sources as solar or wind, etc. to reduce the fuel prize and fossil fuel emission generated from the thermal units. Sometimes scheduling of hydrothermal power is best option when partial demand is supplied by both thermal and hydro power plant. In present scenario, renewable energy sources are very important and essential energy source. One of the important sources of non-conventional energy is solar energy, which comes directly from the sun's energy and can be used for generating electricity; utilize passive heating systems, and operating light energy systems to share bulk of energy demand in operating different forms of load applications. Solar power can be produced either via sun-based steam operated power plants or through photovoltaic (PV) systems, which convert the sunlight to electricity directly through solar cells. It is known that operational costs of the conventional fuel-based thermal power plant are very high in comparison with the operational costs of the hydro plants and solar power plants. Solar power plants have more initial implementation cost than thermal power plants but have almost no running costs. Sometimes, the combined operation of the thermal and solar power plants in an integrated electrical power network is greatly solicited which may happen to be extra reasonably priced techno-economically by circumventing several undesirable constraints and environmental pollution issues. The short-term scheduling of thermal-solar is extremely important from a monetary and ecological point of view. The main purpose of the scheduling of thermal-solar scheduling is to find optimum electric power generation for a specific demand from thermal and solar plants to reduce the fuel charge and pollution value of thermal generations by meeting both linear and non-linear thermal, solar and power system network constraints. It may be for a short-term retro and long-term retro. Short-term period may be for one hour, more than one hour or a full day. Long-term scheduling may be for one week, one month, more than one month or for full one year. Because of no fuel cost allied with the solar plants, the scheduling of thermal-solar becomes a very difficult optimization problem. In thermal-solar scheduling, our main intention is to diminish the emission and overall cost of production, to minimize the fuel costs of thermal plants by meeting limitations for a scheduled period.

In [1] a fuzzy satisfying optimization technique was presented to explain the STHTS issue. In order to verify the presented technique, a test model with 4-hydro plants and 3-thermal plants was employed with emission as a second objective. There were no comparison of cost and emission with other methods and no comparison of either convergence characteristics or computational time. In [2] power system optimization techniques are discussed and implemented with suitable cases. In [3] study was conducted to evaluate the practical possibility and financial sustainability of a solar power plant to meet the power load of the area, Rajasthan—considering both on-spot and off-spot options. In [4] a generalized and effective model was discussed to the emission generated during generation in the presence of non-conventional energy. The drops in the extent of several contaminants released from the power plants at diverse levels of non-conventional power penetration rates were identified and addressed. In [5] PSO as an optimization tool to solve practical CEED problem with conventional thermal power plants and PV plants was presented. A range of practical limitations was considered of solar and thermal including ramp rate limits. In [6] a solar-PV-diesel-battery hybrid model for the optimal power dispatch was presented. The proposed prototype minimizes the cost of firewood and battery garb and determines the power pour, taking into account the availability of photovoltaic power, charging status of the battery bank and the demand. In [7] a combination of PSO, N-R method and binary integrated technique to solve practical CEED problem with conventional thermal power plants and solar PV power plants was presented. In [8] a two-step process for assessing the storage importance of energy storage in renewable energy systems was presented. In the primary step, the authors express the stochastic unit commitment process to the forecasting of improbability and energy storage for wind power. In another step, the results obtained from the plant commitment were used to run a supple energy storage schedule from ED where the horizon was limited. In [9] a two-stage operational planning process of a virtual power plant (VPP) for short-term operation was presented. In [10] the authors proposed a stochastic programming technique to unravel a multi-retro OPF problem with non-conventional power was presented. In [11] a power management algorithm to solve the power dispatch problem with conventional thermal power plants and solar PV plants was presented. To show the feasibility a solar PV plant, with 8.5 MW/2.25 MWh batteries and a supper capacitors bank, was modelled using MATLAB software environment. In [12] ant lion optimization process (ALOA) for deciding the finest location and sizing of non-conventional energy sources in the distributed network was presented. In [13] a hybrid of both flower pollination algorithm (FPA) and FPA with binary to unravel practical CEED problem with conventional thermal power plants and solar PV plants was presented. In [14] the authors proposed a hybrid of DE and ACO as an optimization tool to solve power dispatch problem with conventional thermal power plants and PV plants. In [15] whale optimization algorithm with unique behavior of humpback whales for deciding the optimal location and sizing of non-conventional energy sources in the distributed network was presented. In [16] a novel whale optimization process is used to solve economic dispatch, CEED, and CEED with renewable energy problems.

1.1 Thermal and Solar Power Plants Characteristic

For combined economic emission ship of three thermal and ten solar photovoltaic power plants are considered. The data of three power plants are taken from [1]; per unit rates and power rating of ten solar PV plants is tabulated in Table 1 and global solar radiation, power demand and temperature details are tabulated in Table 2. Transmission loss matrix is taken from [2]. The solar radiation and temperature detail are taken from pyrometer at Punjab engineering college Chandigarh on July 18, 2020.

Table 1 Per unit rates and power rating of ten solar PV plants

Plant	Unit rate ($/kWh)	Prated (MW)
1	0.25	35
2	0.25	35
3	0.26	40
4	0.27	40
5	0.28	50
6	0.28	50
7	0.285	50
8	0.29	50
9	0.29	50
10	0.29	50

Table 2 Power demand, solar radiation and temperature on July 18, 2020

Time (h)	Power demand (MW)	Global solar radiation (W/m^2)	Temperature (°C)
1:00	750	0	32
2:00	780	0	30
3:00	700	0	29
4:00	650	0	29
5:00	670	10.4	28
6:00	800	120	29
7:00	850	260.3	30
8:00	870	560.3	32
9:00	880	540.5	33
10:00	890	783.7	35
11:00	900	1087	36
12:00	950	1132.4	37
13:00	960	1021.3	38
14:00	930	876.2	38
15:00	910	753.3	38
16:00	960	678	39
17:00	950	396.4	39
18:00	920	220.3	38
19:00	900	40.5	36
20:00	850	0	35
21:00	810	0	35
22:00	860	0	33
23:00	850	0	33
0:00	800	0	33

2 Optimization Technique

To solve the CEED three thermal and ten solar PV power plants are considered. PSO [7] and TLBO [16] are used to solve this multi-objective optimization problem to minimize charge and emission simultaneously. The binary PSO is used to solve conditional decision variables' problems, i.e. real values 0 or 1.

All particles are initialized randomly 0 or 1 as (1)

$$P_i = [P_{i1}, P_{i2}, \ldots\ldots\ldots\ldots\ldots\ldots P_{iN}] \tag{1}$$

$$\forall P_{i1}, P_{i2}, \ldots\ldots\ldots\ldots\ldots\ldots P_{iN} \notin \{0, 1\}$$

BPSO procedures are similar to real-valued PSO as explained detail in [7] except for some minor changes as follows. With probability 0.5, a binary value either zero (0) or one (1) is assigned to each particle in each step of the PSO optimization as (2).

$$f(P) = \begin{cases} 1, if\ r\ and > 0.5 \\ 0, otherwise \end{cases} \tag{2}$$

To scale the velocities between 0 and 1 by using sigmoid function is shown in (3).

$$sigmoid(V_{id}^{j+1}) = \frac{1}{1 + e^{-V_{id}^{j+1}}} \tag{3}$$

3 Problem Description

A multi-objective combined economic emission optimization problem is formed as a thermal-solar planning issue to reduce operating cost and emission generated from thermal power plants. Economic dispatch with valve point effect is shown in (4).

3.1 *Economic and Emission Dispatch*

$$F_T = \sum_{t=1}^{T} \sum_{j=1}^{S} a_i + b_i P_{i,t} + c_i P_{i,t}^2 + \left| e_i * \sin(f_i * (P_i^{low} - P_{i,t})) \right| \tag{4}$$

where

N Number of total units in thermal plant, T: Total time.

Pi^{low}, $P_{i,t}$	Minimum and generated power by the unit *i*th in scheduled period t
ai, bi, ci, ei, fi	*i*th unit fuel charge constants.

Total emission from thermal power plant is shown in (5).

$$E_T = \sum_{t=1}^{T}\sum_{i=1}^{N}(\alpha_i + \beta_i P_{i,t} + \gamma_{i,t}^2 + \eta_i \exp(\delta_i P_{i,t})) \tag{5}$$

ηi, δi, αi, βi, γi are thermal emission constants of *i*th thermal unit.

3.2 Solar Power

The produced power from a solar PV plant can be stated as (6)

$$P_s = P_{rated}\left\{1 + (T_{ref} - T_a) \times \alpha \frac{S_i}{1000}\right\} \tag{6}$$

Prated	Rated solar power, Tref temperature as reference
Ta	Ambient temperature
α	Coefficient of temperature and Si: Solar radiation that incident on earth.

The solar energy cost is shown as follows

$$\text{Solar cost} = \sum_{j=1}^{S} PU\text{Cost}_j \times P_{sj} \times U_{sj} \tag{7}$$

Here PUCost_j is the per unit price of *j*th solar power plant. U_{sj} signifies status of *j*th solar PV which is either 1 (ON) or 0 (OFF).

Linear and Non-Linear Constraints

(i) Power balance limit

$$\sum_{i=1}^{N} P_{i,t} + \sum_{j=1}^{S} P_{S,t} - P_{D,t} - P_{L,t} = 0 \tag{8}$$

$P_{L,t}$ and $P_{D,t}$ are transmission loss and mandate in given schedule.

(ii) Generation boundaries

$$P_i^{low} \leq P_{i,t} \leq P_i^{upper} \tag{9}$$

(iii) Constraint

$$\sum_{j=1}^{S} P_{sj} \times U_{sj} = 0.3 \times P_{D,t} \tag{10}$$

4 Outcomes and Discussion

For the PSO and TLBO population size is 50 and total iteration is 500 for each hour. The MATLAB encoding was written using the Intel (R), i7-8565U, MATLAB R2016a and 8 GB RAM. Table 3 represents the thermal and solar generation, cost and emission at 12:00 h of thermal-solar scheduling for 950 MW demand by PSO and Table 4 represents the thermal and solar generation, cost and emission at 3:00 h of thermal-solar scheduling for 910 MW demand by PSO. Table 5 represents total fuel cost, total loss and total emission for only thermal units scheduling and total

Table 3 Outcomes at 12:00 h of thermal-solar scheduling for 950 MW demand by PSO

Thermal power	P1 (MW)	174.99
	P2 (MW)	294.85
	P3 (MW)	294.53
Solar generation	US1, US2………, US10	1, 01, 0, 1, 1, 0, 0, 0, 0
	Solar power generation	193.2
Cost	Fuel cost ($/h)	2564.2
	Solar cost ($/h)	52,255
	Total loss (MW)	11.25
Others	Emission (kg/h)	1762.2

Table 4 Outcomes at 15:00 h of thermal-solar scheduling for 910 MW demand by PSO

Thermal power	P1 (MW)	175
	P2 (MW)	295.17
	P3 (MW)	266.66
Solar generation	US1, US2, …, US10	1, 1, 0, 1, 1, 1, 0, 0, 0, 1
	Solar power generation	184.22
Cost	Fuel cost ($/h)	2529.1
	Solar cost ($/h)	49,161
	Total loss (MW)	10.64
Others	Emission (kg/h)	1554.6

Table 5 Findings of thermal-solar scheduling

Outputs	PSO (without solar PV)	TLBO (with solar PV)	PSO (with solar PV)
Total fuel cost ($/h)	68,922	62,338	61,915
Total solar cost ($/h)	–	5.8196e + 05	5.8196e + 05
Total emission (kg/h)	68,359	44,931	44,575
Total cost ($/h)	3.1834e + 05	8.0123e + 05	8.0096e + 05
Total loss (MW)	327.48	271.56	274.56

fuel cost, total solar cost, total transmission loss and emission for thermal units and solar photovoltaic power plants for 24 h scheduling by PSO and TLBO. Figure 1 represents the total cost for hour 2, hour 3, hour 5, hour 10, hour 13, hour 15, hour 19, hour 22, by PSO; and Fig. 2 represents the total cost for hour 2, hour 3, hour 5, hour 10, hour 13, hour 15, hour 19, hour 22, by TLBO. The total loss obtained by TLBO shown in Table 5 is less as compared to PSO because the transmission loss depends upon the B-matrix or loss matrix and power generated by each thermal power plants in each interval. The outcomes obtained in terms of total fuel cost, total solar cost,

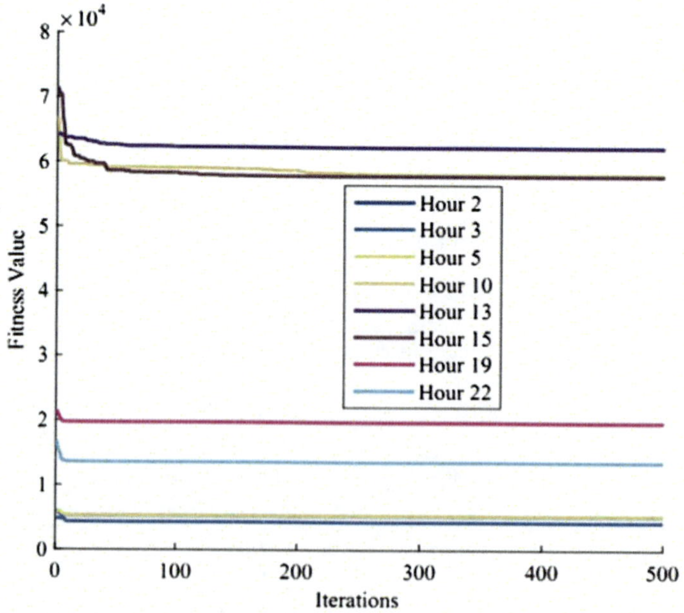

Fig. 1 Total cost in different hours for full solar radiation by PSO

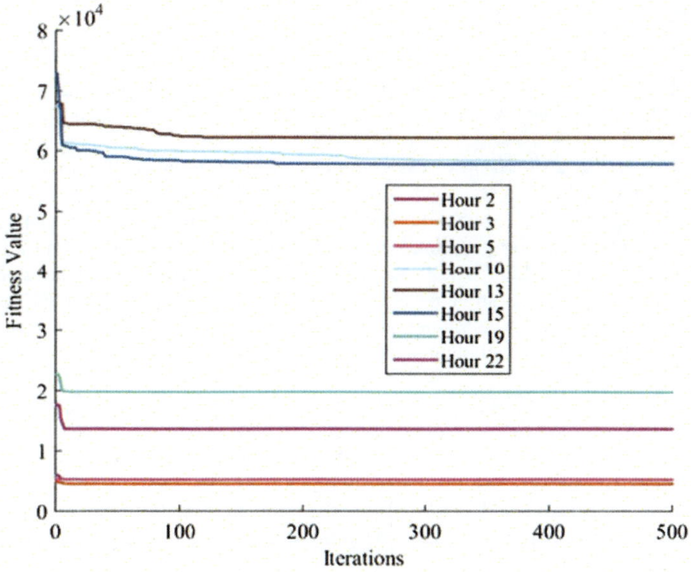

Fig. 2 Total cost in different hours for full solar radiation by TLBO

total emission and total cost (Total cost is sum of total fuel cost, total solar cost and total emission cost) by PSO and TLBO for three thermal and ten solar photovoltaic power plants for 24 h load power demands as given in Table 1 show that for small power system PSO is efficient than the TLBO optimization technique.

5 Conclusion

This paper flourished in executing a PSO for explaining the CEED with actual fuel and environmental contamination costs of thermal units. Outcomes express that the total fuel cost, total emission and total loss via transmission are going to be decreased and total cost is going to be increased. The offered method can, therefore, be widespread well to address the thermal-solar planning on bulky scales. The problems discussed in this research can be further extended in the light of additional constraints. These may include minimum up and down times for generating units, spinning reserve, etc. Grid incorporating non-conventional energy such as wind, geothermal, biomass and tidal energy can be considered in the optimization problems discussed in this research. The problems of multi-objective power system optimization discussed in this research can be reconsidered using other multi-objective approaches to the solution.

References

1. Basu M (2004) An interactive fuzzy satisfying method based on evolutionary programming technique for multiobjective short-term hydrothermal scheduling. Electr Power Syst Res 69(5):277–285
2. Kothari DP, Dhillon JS (2004) Power system optimization. Prentice-Hall of India Pvt. Ltd., New Delhi, India, pp 572–573
3. Chandel M, Agrawal GD, Mathur S, Mathur A (2014) Techno-economic analysis of solar photovoltaic power plant for garment zone of Jaipur city. Case Stud Therm Eng 2:1–7
4. Pazheri FR, Othman F, Malik NH, Al-Arainy AA (2014) Reduction in pollutants emission by increase in renewable penetration: a case study. Res J Appl Sci Eng Technol 7(19):437–4142
5. Khan NA, Ahmed Awan AB, Mahmood A, Sidhu G (2015) Combined emission economic dispatch of power system including solar photo voltaic generation. Energy Convers Manage 92:82–91
6. Tazvinga H, Zhu B, Xia X (2015) Optimal power flow management for distributed energy resources with batteries. Energy Convers Manage 102:104–110
7. Khan NA, Sidhu GAS, Gao F (2016) Optimizing combined emission economic dispatch for solar integrated power systems. IEEE Access 4:3340–3348
8. Uçkun N, Li C, Constantinescu EM, Birge JR, Hedman KW, Botterud A (2016) Flexible operation of batteries in power system scheduling with renewable energy. IEEE Trans Sustain Energy 7(2):685–696
9. Luo F, Dong ZY, Meng K, Qiu J, Yang J, Wong KP (2016) Short-term operational planning framework for virtual power plants with high renewable penetrations. IET Renew Power Gener 10(5):623–633
10. Bukhsh WA, Zhang C, Pinson P (2016) An integrated multiperiod OPF model with demand response and renewable generation uncertainty. IEEE Trans Smart Grid 7(3):1495–1503
11. Wang G, Ciobotaru M, Agelidis VG (2016) Power management for improved dispatch of utility-scale PV plants. IEEE Trans Power Syst 31(3):2297–2306
12. Ali ES, Abd Elazim SM, Abdelaziz AY (2016) Ant lion optimization algorithm for renewable distributed generations. Energy 116(1):445–458
13. Shilaja C, Ravi K (2017) Optimization of emission/economic dispatch using Euclidean Affine flower pollination algorithm (eFPA) and binary FPA (BFPA) in solar photo voltaic generation. Renew Energy 107:550–556
14. Rahmat NA, Aziz NFA, Mansor MH, Musirin I (2017) Optimizing economic load dispatch with renewable energy sources via differential evolution immunized ant colony optimization technique. Int J Adva Sci Eng Inf Technol 7(6):2012–2017
15. Reddy DP, Reddy V, Manohar G (2018) Optimal renewable resources placement in distribution networks by combined power loss index and whale optimization algorithms. J Electr Syst Inf Technol 5(2):175–191
16. Dey B, Roy SK, Bhattacharyya B (2019) Solving multi-objective economic emission dispatch of a renewable integrated micro grid using latest bio-inspired algorithms. Engi Sci Technol Int J 22(1):55–66

Impact of Capacitor Banks on the Nodal Prices of Meshed Distribution System

Karimulla Polisetti, Atma Ram Gupta, and Ashwani Kumar

Abstract The distribution system is the final link between bulk power system and consumer end. The capacitor plays an important role for reactive power management and power factor correction in the distribution systems. The capacitors also have an impact on the nodal prices of both the real and reactive power. In this paper, an optimal power flow model is proposed to determine the nodal prices by minimizing the system operation cost at the power supply point. The method uses marginal loss coefficients (MLCs) to obtain these prices. The results are obtained with ZIP, residential, industrial and commercial (RIC) and seasonal load variations such as winter, summer and spring seasons. The results are obtained for an IEEE 33 bus mesh distribution system using optimization software GAMS 23.4.

Keywords Distribution system · Load flow · Capacitor allocation · Nodal price

1 Introduction

The major objectives of the distribution systems are to provide reliable supply to the consumers, quality power, services to the rural and urban consumers with supply point, and the transparent billing to the consumers. In addition, the good distribution system shall be properly planned and completely automated in the smart grid domain. The smart distribution networks are the need of the present era and shall be operated with higher efficiency for best asset utilization. This is the duty of the key element called as distribution network operator (DNO) which is to maintain and operate the network with minimum power losses and better voltage profile maintaining the highest possible security. The voltage profile and power factor is needed to be maintained in the network for lower losses and better voltage regulation. In this context, capacitor placement plays an important role for maintaining the higher power factor and better voltage profile. Since the competitive structure has been

K. Polisetti · A. R. Gupta · A. Kumar (✉)
Department of Electrical Engineering, NIT Kurukshetra, Kurukshetra, Haryana, India
e-mail: ashwani.k.sharma@nitkkr.ac.in

A. Kumar et al. (eds.), *Renewable Energy Towards Smart Grid*, Lecture Notes in Electrical Engineering 823, https://doi.org/10.1007/978-981-16-7472-3_13

adopted in the power sector for better operation and technical innovations, there is need to address the issues of better market management where energy is sold at the competitive price. This requires a transparent pricing structure to be adopted for the distribution systems.

Capacitor allocation method has been used by authors in [1, 2]. Authors in the reference [3, 4] present a method to minimize the reactive component of branch current by determining the nodes where the capacitor can be located with optimal size using the optimal power flow. Authors proposed for a fixed and switched capacitor placement considering the substation load tap changer setting [5]. The authors consider the capacitor size as discrete variable and used dynamic programming for obtaining the capacitor size in [6]. Capacitor location and sizes can be expressed as continuous variables [7] and a non-linear programming based method was proposed in [8] and considering a voltage regulator problem along with capacitor placement was also proposed using decoupled solution methodology. The method using mixed integer programming for capacitor allocation was proposed in [9, 10].

The techniques of nodal pricing determination used in the transmission system can also be applied to the distribution systems [11]. Authors proposed the nodal pricing in distribution systems [12]. For obtaining the nodal prices in the distribution systems, we need to determine first the marginal loss coefficients (MLCs) [13]. The MLCs are the coefficients that indicate the marginal or incremental deviation in total active power loss due to the changes in active and reactive power injections at a particular node of the system [14].

The effect of solar and wind generation on distribution system prices were analyzed in [15, 16]. The dynamic tariff concept based on the distribution locational marginal prices to solve the congestion problems was introduced in [16]. The nodal price behavior was obtained in [17] considering the ZIP and RIC loads using the load flow based formulation.

In this paper, an optimal power flow based approach is proposed to determine the nodal prices in the distribution system minimizing the system operation cost of substation point called as power supply point (PSP). The obtained results are compared with ZIP, residential, industrial and commercial (RIC) and seasonal load variations of winter, summer and spring which are also considered in the analysis. The results are obtained for an IEEE 33 bus mesh distribution system using optimization software GAMS 23.4 [18]. The comparative analysis has been presented for all types of loads.

2 Proposed Optimal Power Flow Model with Capacitor Banks for Nodal Price Determination

An Optimal power flow (OPF) model is presented in this paper to determine the prices at each nodes. The equivalent fuel cost function of the substation along with the cost of the capacitor banks is minimized and subjected to the constraints. The substation cost function is represented by a quadratic cost function along with the capacitor cost as

Minimize

$$\sum_{i} a_i P_{Gi}^2 + b_i P_{Gi} + c_i + (\mathrm{k} * Q_{ci}) \tag{1}$$

Subject to

$$P_{Gi} - P_{Di} - V_i \sum_{j=1}^{n} V_j \left(G_{ij} cos\theta_{ij} + B_{ij} sin\theta_{ij}\right), i \in S_B \tag{2}$$

$$Q_{Ri} + Q_{ci} - Q_{Di} + V_i \sum_{j=1}^{n} V_j \left(G_{ij} sin\theta_{ij} - B_{ij} cos\theta_{ij}\right), i \in S_B \tag{3}$$

$$P_{Gi}^{min} \le P_{GI} \le P_{Gi}^{max}, i \in S_G \tag{4}$$

$$Q_{Ri}^{min} \le Q_{Ri} \le Q_{Ri}^{max}, i \in S_R \tag{5}$$

$$V_i^{min} \le V_i \le V_i^{max}, i \in S_B \tag{6}$$

$$|P_l| = \left|P_{ij}\right| = \left|V_i V_j \left(G_{ij} cos\theta_{ij} + B_{ij} sin\theta_{ij}\right) - V_i^2 G_{ij}\right| \le P_l^{max}, l \in S_L \tag{7}$$

where $S_{B,} S_{G,} S_R$ and S_l are the set of nodes, generators, reactive power sources and lines, respectively. The voltages, generated real power and reactive power, real and reactive power demand are represented by V_i, P_{Gi}, Q_{Gi}, P_{Di} and Q_{Di}. θ_i is the angle at node i, where as G_{ij} and B_{ij} are the real and imaginary parts of the admittance of line connected between i and j nodes, P_l^{max} is the maximum power flow limit in line l. Q_{ci} is the reactive power supplied by the capacitor and the constant k is 13.24 \$/MVar [19]. The values of cost coefficients assumed are as $\mathrm{a_i} = 0.01$, $\mathrm{b_i} = 40$, $\mathrm{c_i} = 9$.

An optimization problem is modeled and solved using GAMS software with CONOPT solver which solve non-linear quadratic programming problems. The Lagrangian coefficients of the real and reactive power flow equations obtained from the solution give the active and reactive power nodal price at each bus. The realistic ZIP load model, which is a combination of residential, commercial and industrial load has been considered in the analysis. The analysis also considered the seasonal load taking their variation.

2.1 Realistic ZIP Load Model as RIC Load

The load is assumed to be a combination of residential, commercial and industrial load (RIC) [19] which varies with respect to time and is expressed by the following relations:

$$P_i(t) = P_{io}(t)\left[\alpha(t)\left(\frac{V_i(t)}{V_o}\right)^{npr} + \beta(t)\left(\frac{V_i(t)}{V_o}\right)^{npc} + \gamma(t)\left(\frac{V_i(t)}{V_o}\right)^{npi}\right] \quad (8)$$

$$Q_i(t) = Q_{io}(t)\left[\alpha(t)\left(\frac{V_i(t)}{V_o}\right)^{nqr} + \beta(t)\left(\frac{V_i(t)}{V_o}\right)^{nqc} + \gamma(t)\left(\frac{V_i(t)}{V_o}\right)^{nqi}\right] \quad (9)$$

where $\alpha(t)$, $\beta(t)$ and $\gamma(t)$ are the fractions of residential, commercial and industrial load for time t at each bus of the system; $P_i(t)$, $Q_i(t)$ and $V_i(t)$ are the active power, reactive power and voltage at bus i for time t, respectively. V_o is the nominal voltage of the system (1.00 p.u.). The active and reactive power exponents and the fraction of residential, industrial and commercial loads for each hour used are given in Table 1.

3 Numerical Analysis: Results and Discussions for IEEE 33-Bus System for Mesh Network

The nodal prices are obtained for an IEEE 33 bus mesh distribution system. The data taken as with base power of 100 MVA and the base voltage is 12.66 kV. The connected active and reactive power load are 3.72 MW and 2.30 MVAR [19]. Nodal prices at each buses are obtained considering the seasonal loads and RIC load. Subsequent sections discuss the results for a distribution system.

Table 1 RIC Loads exponential coefficient values

Load type	Residential		Commercial		Industrial	
	npr	nqr	npc	nqc	npi	nqi
Coefficients	0.72	2.96	1.25	3.50	0.18	6.00

3.1 Hourly Variation of Nodal Prices for Mesh System Considering Seasonal Load, ZIP Load and RIC Load

The load is assumed to follow the 24 h load profile of winter, summer and spring seasons given in IEEE-RTS. The results are plotted for hour10 at which the peak of the load occurs. Figures 1 and 2 show active and reactive nodal prices for all loads in 10th hour, respectively. In general, 18th bus is more sensitive so price at this bus was shown for all hours.

It is seen from Figs. 1 and 2, the nodal prices are higher for both the real and reactive power for ZIP load compared to seasonal loads. Figures 3, 4 and 5 show the active and reactive nodal prices for all loads at 18th bus and at power supply point (PSP), respectively. It is found that at the PSO, the nodal price for active power is same and the price varies at all other nodes for all RIC and seasonal load variations. It is evident from these figures that nodal prices are higher for spring season among all seasons due to its high load profile, and for ZIP load also nodal prices are higher.

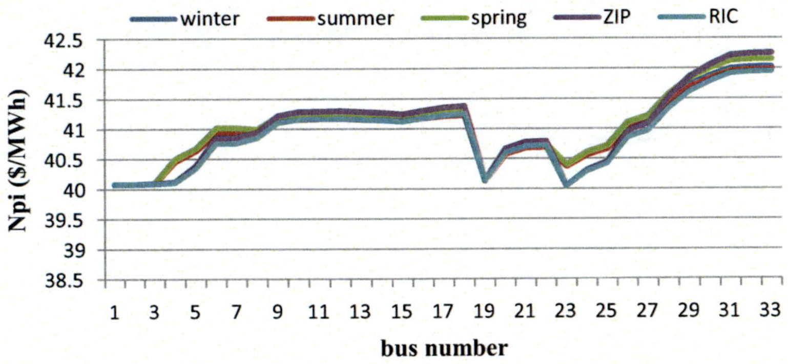

Fig. 1 Nodal prices of active power for 10th hour at all nodes

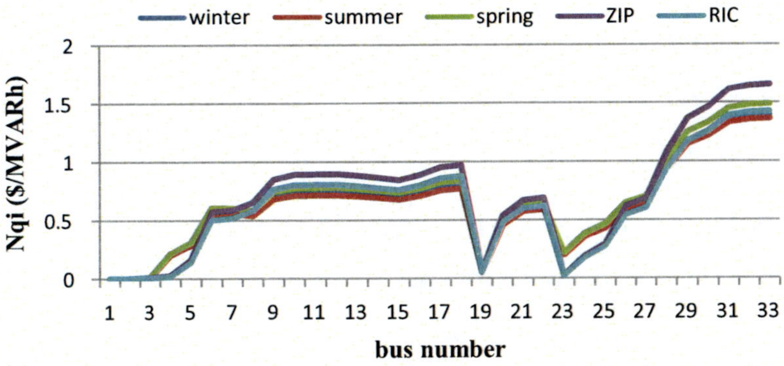

Fig. 2 Nodal price of reactive for 10th hour at all nodes

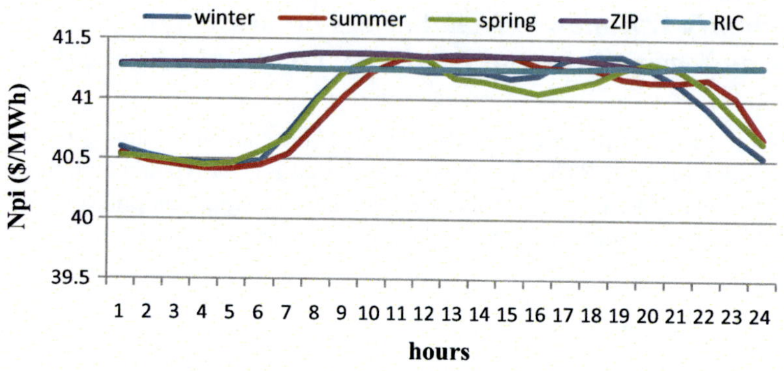

Fig. 3 Nodal prices of active power at 18th bus for 24 h load variation

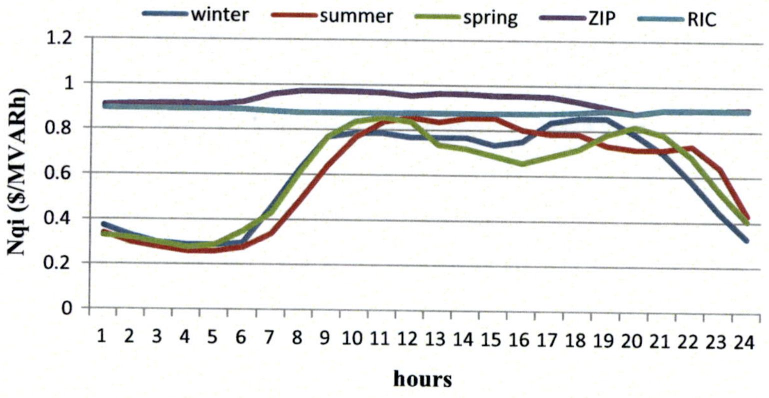

Fig. 4 Nodal prices of reactive nodal prices at 18th bus for 24 h load variation

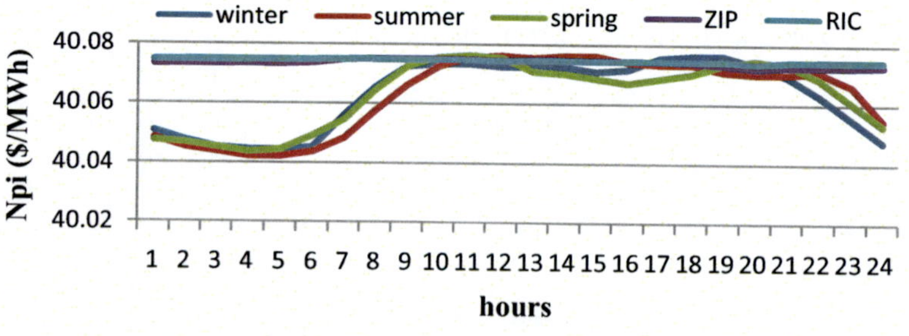

Fig. 5 Nodal prices of Active power at Power supply point (node1)

At 33rd bus, nodal prices are higher at 10th hour for all types of loads as at the last node the losses are higher. At the 18th bus, ZIP load has higher nodal prices among all loads. At PSP, price is higher for ZIP and RIC loads for most of the hours.

3.2 Optimal Capacitor Placement Based on Combined Power Loss Sensitivity

Optimal location of the capacitor banks was determined by determining the combined power loss sensitivity (CPLS) index. This method to obtain CPLS is as follows:

3.2.1 Combined Power Loss Sensitivity Index [19]

In this method, the loss sensitivity factors are determined to obtain the best possible location of capacitor banks. The real power loss (PLS) and reactive power loss (QLS) sensitivity can be calculated using (10) and (11)

$$\frac{\partial Ploss}{\partial Q_2} = \frac{2 * \mathrm{Q}_{2*\mathrm{R[j]}}}{\mathrm{V}_2^2} \tag{10}$$

$$\frac{\partial Qloss}{\partial Q_2} = \frac{2 * \mathrm{Q}_{2*\mathrm{X[j]}}}{\mathrm{V}_2^2} \tag{11}$$

The combination of these two sensitivity factors are obtained for net loss sensitivity as given by (12)

$$\frac{\partial Sloss}{\partial Q_2} = \frac{\partial ploss}{\partial Q_2} + j\frac{\partial Qloss}{\partial Q_2} \tag{12}$$

In a similar way, the loss sensitivity corresponding to real power can be obtained and the combined loss sensitivity with respect to real power can be represented as

$$\frac{\partial Sloss}{\partial P_2} = \frac{\partial ploss}{\partial P_2} + j\frac{\partial Qloss}{\partial P_2} \tag{13}$$

Based on the loss sensitivities, the loss sensitivity matrix can be formulated as

$$\text{Loss sensitivity matrix} = \begin{vmatrix} \frac{\partial ploss}{\partial P_2} & \frac{\partial Qloss}{\partial P_2} \\ \frac{\partial ploss}{\partial Q2} & \frac{\partial Qloss}{\partial Q_2} \end{vmatrix} \tag{14}$$

Based on the CPLS calculation, the best location of the capacitor banks is obtained at 30th bus.

4 Results and Discussions for IEEE 33-Bus System with Capacitor Placement

A fixed capacitor of 100 kVAr is placed at 30th bus in a mesh distribution system. The capacitor is placed at bus 30th based on the sensitivity analysis. Table 2 shows the active and reactive power nodal prices along with the charge of marginal losses per year for both radial and mesh network considering ZIP load with the placement of capacitor. For the mesh distribution system, the prices are slightly lower due to lower power loss in the system which is due to better voltage profile.

Figures 6, 7 and 8 show active and reactive nodal prices for all loads at 10th hour with capacitor for mesh system, respectively. The results were shown for 10th hr due to presence of peak load. In general, 18th bus is more sensitive so price at this bus are shown for all hours. Figures 9, 10, 11, 12, 13 and 14 show the active and reactive nodal prices for all loads at 18th bus and at power supply point (PSP) with placement of capacitor for mesh system, respectively. It is evident from these figures that nodal prices are higher for spring season among all seasons in radial system due to its high load profile, and for mesh system ZIP load is having higher nodal prices. At 33rd bus reactive nodal prices are higher in 10th hour for all types of loads at the end bus due to higher power loss. At 18th bus, ZIP load has higher nodal prices among all loads. At PSP, price is higher for ZIP and RIC loads for most of the hours.

The 24 h active and reactive power nodal prices for each bus of winter season with capacitor are shown in Figs. 15 and 16, respectively. Figures 17 and 18 show the 24 h active and reactive power nodal prices at each bus of summer season with capacitor. Figures 19 and 20 show the 24 h active and reactive power nodal prices for every bus with capacitor considering ZIP load. It is seen from these figures that nodal prices vary according to the load variation in respective seasons.

5 Conclusions

In this paper, an optimal approach was presented for nodal price determination with capacitor placement. Based on the analysis, the following observations were made

- Nodal prices are higher for spring season.
- At the peak load nodal prices are higher due to more loss at peak loads.
- At 33rd bus nodal prices are higher at 10th hour for all types of loads because this is end bus where the losses are higher.
- At 18th bus, ZIP load has higher nodal prices among all loads.
- At PSP, price is higher for ZIP and RIC loads for most of the hours.

Table 2 Nodal prices and charge of marginal losses with ZIP load

Bus no.	N_{Pi} ($/MWh)		N_{Qi} ($/MVARh)		Charge of marginal losses for load ($/year)	
	Radial	Mesh	Radial	Mesh	Radial	Mesh
1	40.0761	40.0755	0	0	0	0
2	40.0798	40.0781	0.0019	0.0013	4.17	2.9
3	40.0991	40.0882	0.0118	0.0081	22.24	12.83
4	40.3422	40.1104	0.1399	0.0203	377.71	50.9
5	40.629	40.3364	0.2857	0.1351	365.69	172.64
6	41.1531	40.7403	0.7486	0.4892	697.19	435.11
7	41.1744	40.7467	0.8062	0.5026	2630.38	1616.2
8	41.6514	40.8301	1.1514	0.5626	3768.49	1814.96
9	41.8799	41.0706	1.3129	0.7368	1178.06	652.11
10	42.0947	41.1239	1.4649	0.773	1317.64	686.47
11	42.1376	41.1312	1.4779	0.7754	1201.04	619.94
12	42.2119	41.1379	1.5026	0.7776	1583.26	796.78
13	42.4266	41.1244	1.6724	0.767	1748.13	786.42
14	42.4761	41.1114	1.737	0.75	3740.15	1614.55
15	42.4904	41.0915	1.7494	0.7323	1422.17	598.14
16	42.5418	41.1437	1.7868	0.7704	1608.98	696.42
17	42.5855	41.1871	1.8453	0.8285	1642.2	729.42
18	42.6074	41.2094	1.8623	0.8458	2648.2	1190.28
19	40.093	40.1292	0.0146	0.0509	18.41	60.13
20	40.188	40.5834	0.1005	0.4655	123.42	563.55
21	40.2026	40.6841	0.1172	0.5787	140.79	682.62
22	40.2139	40.6994	0.1321	0.5972	154.91	701.14
23	40.1798	40.0471	0.0679	0.0204	111.47	31.38
24	40.3769	40.281	0.2243	0.1653	1499.42	1045.56
25	40.4777	40.4007	0.3018	0.2573	2006.38	1647.33
26	41.235	40.8301	0.7759	0.5278	779.04	512.22
27	41.3364	40.946	0.8277	0.587	843.68	586.06
28	41.6589	41.3125	1.1156	0.9141	1027.37	810.33
29	41.8988	41.585	1.3231	1.1499	2727.33	2291.93
30	42.0446	41.7513	1.3953	1.2323	10,782.52	9412.88
31	42.1623	41.8848	1.5116	1.3641	3668.17	3213.87
32	42.1847	41.9101	1.537	1.3928	5225.33	4595.02
33	42.1896	41.9156	1.5445	1.4012	1652	1458.13

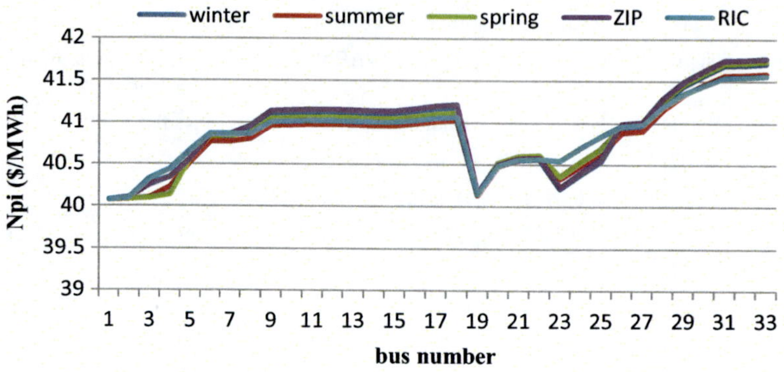

Fig. 6 Active nodal prices for hour 10 at all nodes with capacitor for mesh system

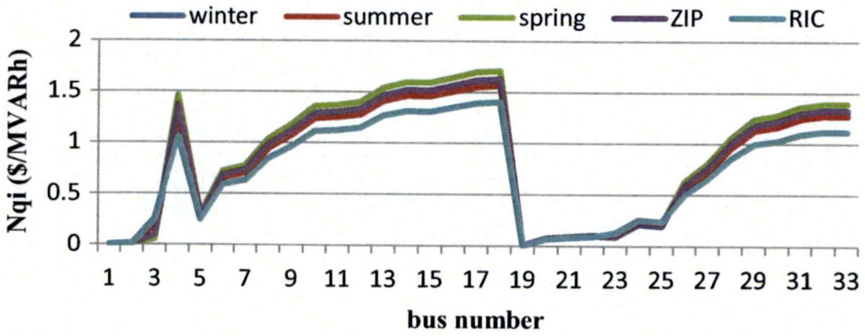

Fig. 7 Reactive nodal prices for hour 10 at all nodes with capacitor for radial system

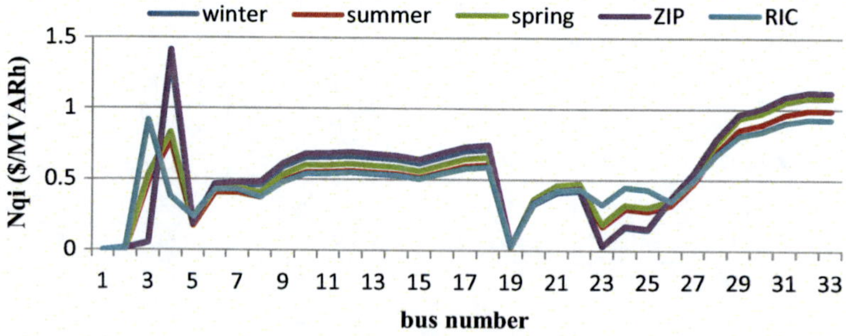

Fig. 8 Reactive nodal prices for hour 10 at all nodes with capacitor for mesh system

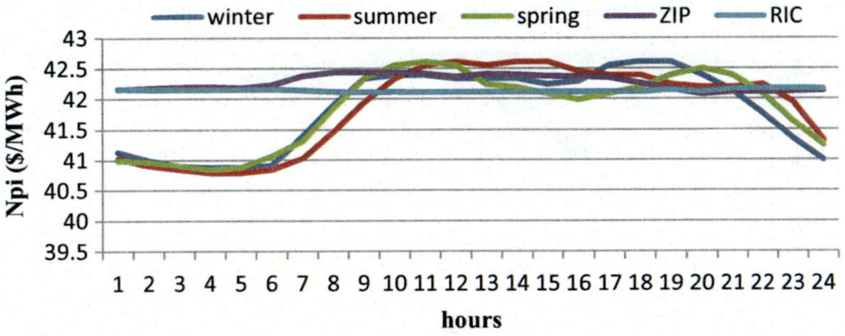

Fig. 9 Active nodal prices at 18th bus for 24 h with capacitor for radial system

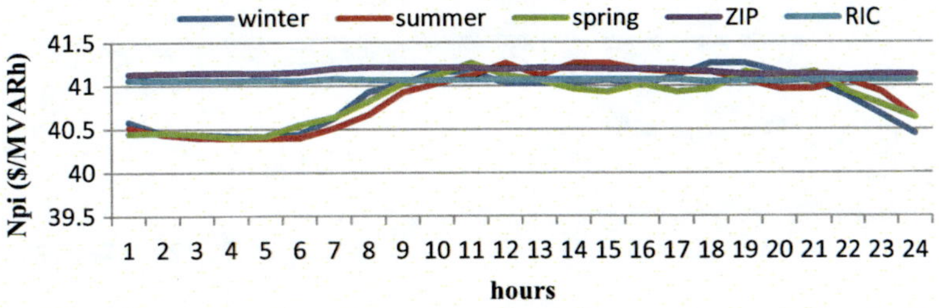

Fig. 10 Active nodal prices at 18th bus for 24 h with capacitor for mesh system

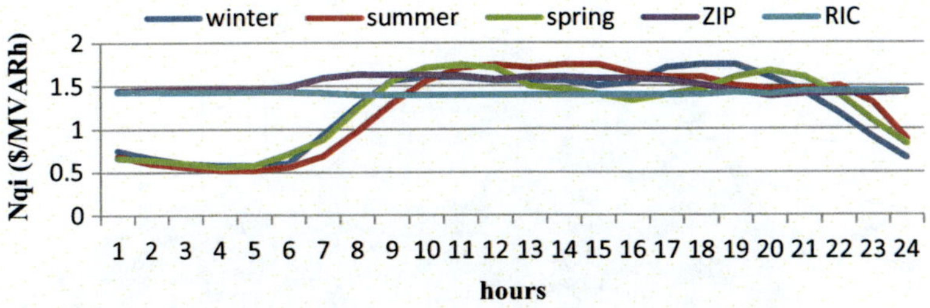

Fig. 11 Reactive nodal prices at 18th bus for 24 h with capacitor for radial system

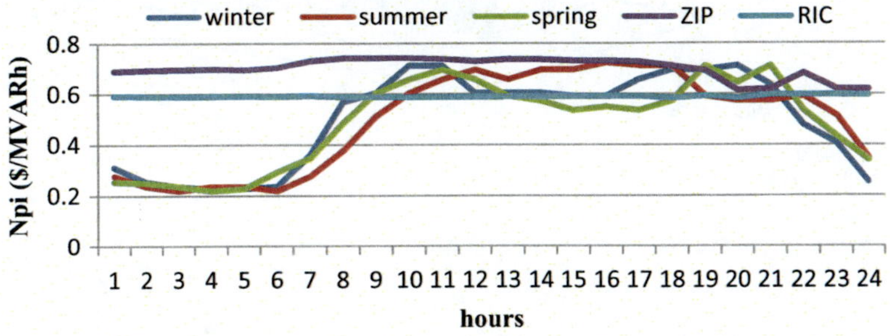

Fig. 12 Reactive nodal prices at 18th bus for 24 h with capacitor for mesh system

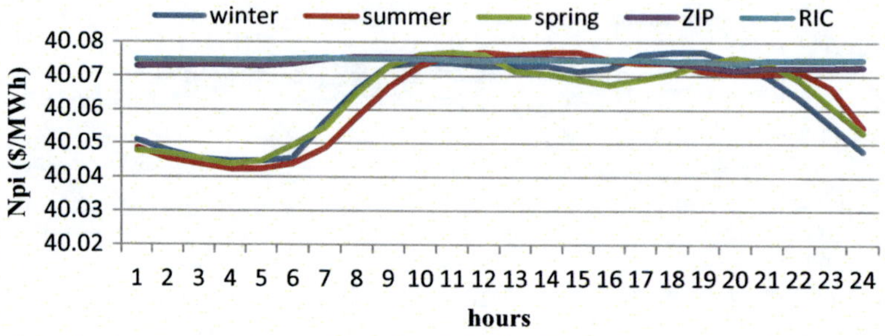

Fig. 13 Active power price at Power supply point (node1) with capacitor for radial system

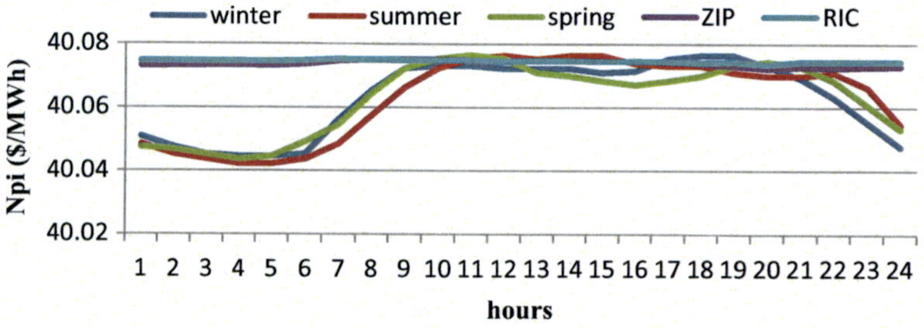

Fig. 14 Active power price at Power supply point (node1) with capacitor for mesh system

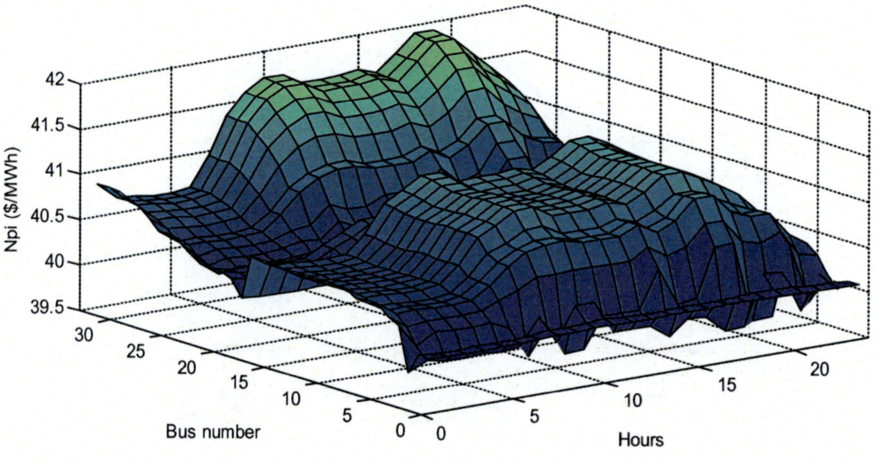

Fig. 15 Hourly active power nodal prices for winter season with capacitor

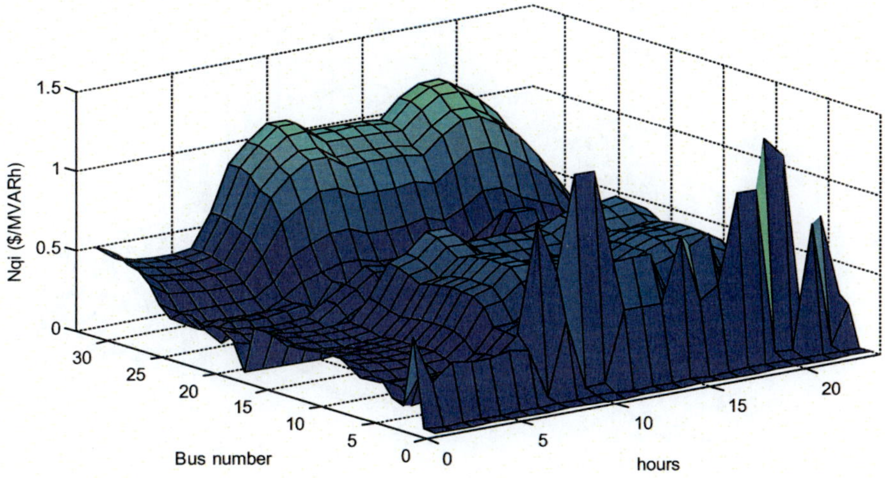

Fig. 16 Hourly reactive power nodal prices for winter season with capacitor

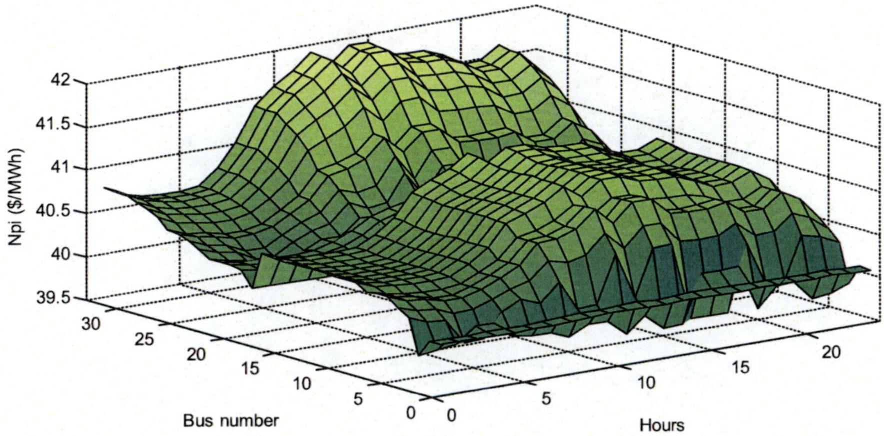

Fig. 17 Hourly active power nodal prices for summer season with capacitor

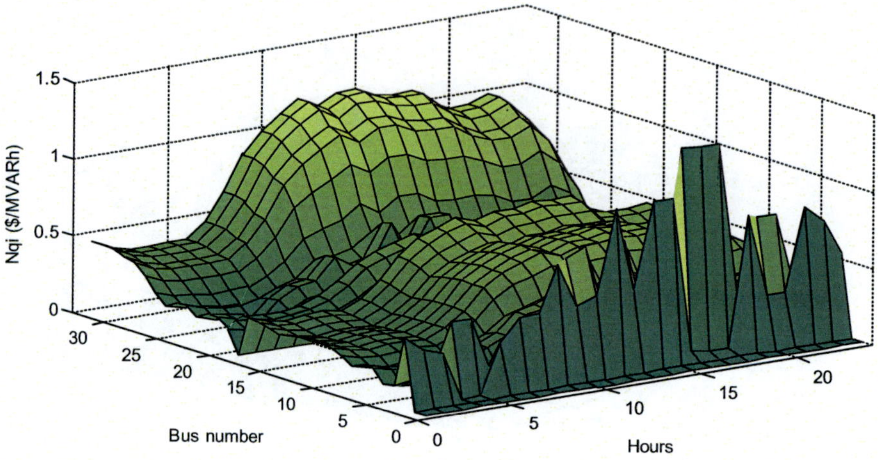

Fig. 18 Hourly reactive power nodal prices for summer season with capacitor

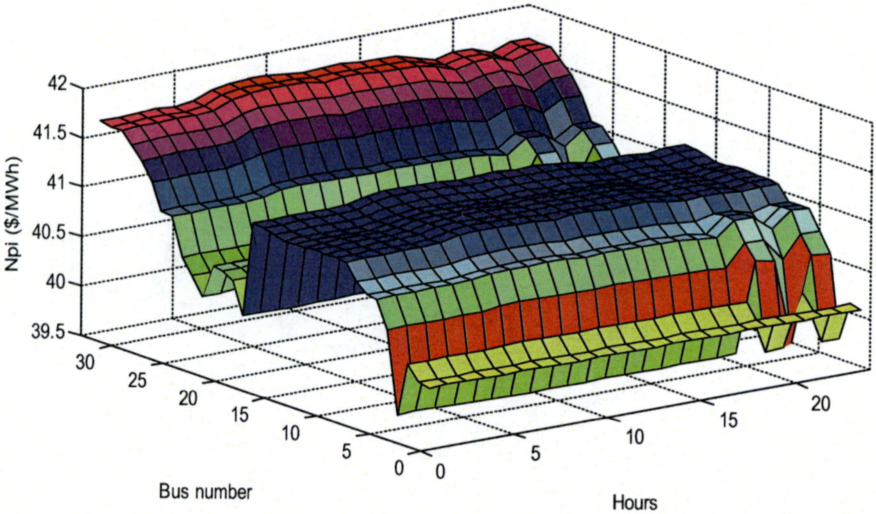

Fig. 19 Hourly active power nodal prices for ZIP load with capacitor

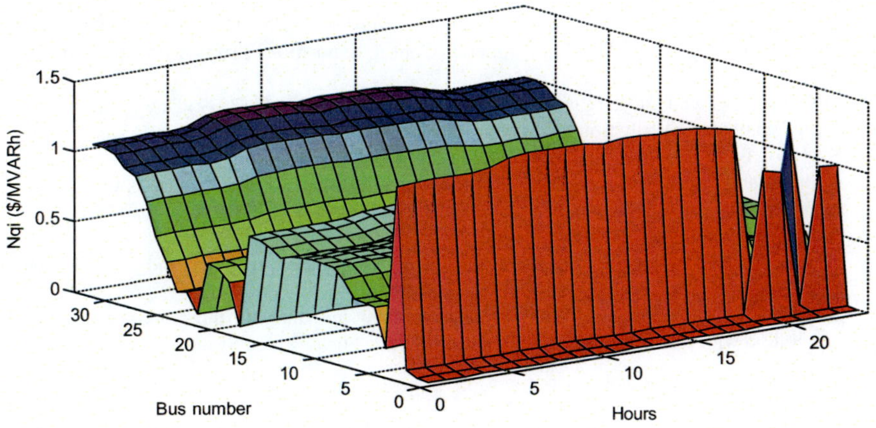

Fig. 20 Hourly reactive power nodal prices for ZIP load with capacitor

References

1. Ramachandra Murthy KVS, Ramalingaraju M, Govinda Rao G, Narasimha Rao K (2010) Comparison of loss sensitivity factor and index vector methods in determining optimal capacitor locations in agricultural distribution. In: Proceedings of 16th national power systems conference, pp 26–30
2. Subrahmanyam JBV, Radhakrishnan C (2010) A simple method for optimal capacitor placement in unbalanced radial distribution systems. Electr Power Comp Syst 38:1269–1284
3. Haque MH (Sept. 1999) Capacitor placement in radial distribution systems for loss reduction. IEEE Proc Gen 146(5):501–505
4. Ahmed R (2013) A. Electr Power Syst Res 95:168–174
5. Duran H (1968) Optimum number, location and size of shunt capacitor in radial distribution feeders: a dynamic programming approach. IEEE Trans Power Syst 87:1769–1774
6. Grainger JJ, Lee SH (1981) optimum size and location of shunt capacitor for reduction of losses on distribution feeders. IEEE Trans Power Syst 3:1105–1118
7. Grainger JJ, Civanlar S (1985) Volt/Var control on distribution system with lateral branches using shunt capacitors as voltage regulators—part I, II and III. IEEE Trans Power Syst 11:3278–3297
8. Baran ME, Wu FF (1989) Optimal sizing of capacitors placed on a radial distribution system. IEEE Trans Power Deliv 1:735–743
9. Baran ME, Wu FF (1989) Optimal capacitor placement on radial distribution system. IEEE Trans Power Deliv 1:725–734
10. Ghayeni M, Ghazi R (2011) Transmission network cost allocation with nodal pricing approach based on ramsay pricing concept. Gen Trans Distr IET 5:384–392
11. Shaloudegi K, Madinehi N, Hosseinian SH, Abyaneh HA (2012) A novel policy for locational marginal price calculation in distribution systems based on loss reduction allocation using game theory. IEEE Trans Power Syst 27(2):811–820
12. Sotkiewicz PM, Vignolo JM (May 2006) Nodal pricing for distribution networks: efficient pricing for efficiency enhancing DG. IEEE Trans Power Syst 21(2):1013–1014
13. Zhao Q, Wang P, Ding Y, Goel K (2010) Impacts of solar power penetration on nodal prices and nodal reliability. IPEC Conf Proc 1134–1139
14. Mutale J, Strbac G, Curcic S, Jenkins N (Jan 2000) Allocation of losses in distribution systems with embedded generation. Gen Trans Distr IEE Procs- 147(1):7–14

15. Sotkiewicz PM, Vignolo JM (2012) The value of intermittent wind DG under nodal prices and amp—mile tariffs. Trans Dist Latin America Conf & Exposition (T&D—LA): 6th IEEE/PES, pp 1–7
16. O'Connell N, Wu Q, Ostergaard J, Nielsen AH, Cha ST, Ding Y (2012) Day ahead tariff for the allevation of distribution grid congestion from electric vehicles. Electr Power Syst Res 92:106–114
17. Sooraj Narayan K, Kumar A (2015) Distribution system nodal pricing analysis with realistic ZIP load and variable wind power source. IEEE India Int Conf (INDICON—2015)
18. GAMS software GAMS Development Corp. USA
19. Kumar A (2004) A zonal congestion management approach using real and reactive power rescheduling. IEEE Trans Power Syst 19(1):554–562
20. Murthy VVSN, Kumar A (2014) Mesh distribution system analysis in presence of distributed generation with time varying load model. Int J Electr Power Energ Syst 62:836–854

Energy Scheduling of Residential Household Appliances with Wind Energy Source and Energy Storage Device

Neelam Jaiswal and Sandeep Kakran

Abstract Home energy management system (HEMS) is rapidly gaining popularity around the world as small-scale renewable power and energy storage has become more viable. This paper represents an optimal scheduling algorithm for residential household appliances in a smart HEMS along with the use of wind energy source as a small renewable energy source and battery as an energy storage device. The optimal scheduling proposes a control strategy for the operation of an air-conditioning system, which is a thermostatically controlled appliance. This also manages the operation of wind energy input and charging or discharging of the battery such that total energy consumption and total cost of the energy consumption of the smart home is minimized. The demand response technique that has been used for the scheduling of energy usage is real-time pricing. Ultimately, mixed-integer linear programming has been used to solve the problem by using CPLEX solver in the GAMS software. Further, effects analyzed from the proposed algorithm on the daily energy usage and electricity bill of the smart home are displayed.

Keywords Home energy management system · Demand response · Wind energy · Air-conditioning system · Real-time pricing

1 Introduction

With the drone of the smart grid era, people are living in a world where electricity prices fluctuate over the course of a day. Therefore, it has become more difficult to track energy consumption, and it leads to diminishing comfort of the user. A home energy management system (HEMS) has been introduced as a solution to this problem. A smart HEMS integrates simultaneously with all the home appliances and allows the user to monitor and control the operation of the appliances and thus controls the energy usage and respective household expenses. One of the main features of a smart grid is that it incorporates maximum volatile renewable energy into the system. Though non-renewable energy sources such as fossil fuels are a rich source of energy

N. Jaiswal (✉) · S. Kakran
Department of Electrical Engineering, National Institute of Technology, Kurukshetra, India

A. Kumar et al. (eds.), *Renewable Energy Towards Smart Grid*, Lecture Notes in Electrical Engineering 823, https://doi.org/10.1007/978-981-16-7472-3_14

as they are highly combustible, but they cause harmful effects to the environment, and they contribute conspicuously to global warming. Also, due to the exhaustion of fossil fuels, the costs of energy produced by them tend to increase. Hence, non-renewable sources have been shifted to renewable energy sources (RES). These sources include many features such as cheaper per unit energy, cheaper transport, better control, better efficiency, and no harmful environmental effects. The most popular renewable sources are solar, wind, hydro, and biomass, out of which solar and wind energy are major sources of energy [1].

Demand side management (DSM) is a great strategy developed by the electricity market. Demand response (DR) is a subset of the overall DSM program, which aims to low voltage networks where end-user customers play an important role in managing the load profile of their households. DR is a tariff introduced to encourage the changes in energy usage by end users, respective to changes in electricity price over time. In this scheme, end users are incentivized so that they can follow the shape of the consumption pattern as well as the time of consumption pattern, which can be matched with the user's renewable generation available. DR programs help the end users to reduce the consumption of energy taken from the main grid as well as the electricity bills.

Many DR programs have been introduced depending upon incentive base and price base. Of these systems, the most preferred tariffs used in the energy industry are time of use (TOU) and real-time pricing (RTP). Though TOU is the basic price-based program and easiest to implement, RTP is considered the most direct and efficient technique which is suitable for a competitive electricity market, and it should be the focus of policymakers [2].

A HEMS is developed by proposing an algorithm for two types of loads: thermostatically controllable and non-thermostatically controllable loads under the use of RTP tariff and small photovoltaic (PV) system in [3]. Zhao et al. [4] proposed an Energy Management System (EMS) strategy considering PV, electric vehicle (EV), and energy storage device (ESD). Mathematical modeling of an air-conditioning system between temperature and power has been proposed to maintain the consumer saving cost of the air-conditioner (AC) in [5]. Also, the advantages of DR programs and disadvantages of non-DR have been explained in this study.

A QoE-aware smart HEMS framework was proposed by Pilloni et al., which relies on the understanding of the irritation experienced by users when appliance operations are altered with regard to the preferences of the ideal user [6]. Therefore, the actions and frustration of the user are documented at the deployment stage to allocate one of the ranges of profiles per appliance. Two separate algorithms then manipulate the assigned profile. The former is aimed at planning managed loads using a greedy strategy based on user profile preferences and electricity prices. The other one reallocates the operations of appliances whenever RES has made a surplus of energy available. In this paper, home appliances are categorized as unregulated appliances, operated by switching and thermostatically controlled. In [7], a stochastic difference equation-based load model of an AC has been utilized to analyze the outcomes of load variables and direct load control behavior.

Another attempt has also been done to predict load behavior during cold load pickup in [8]. A scheme for calculating the aggregate demand of ACs using duty factors is described in this report. It defines the effect of the change in model parameters on energy consumption and the peak duty factor. The effect of reducing supply voltage is also discussed using this model. An energy management method by using renewable energy source and an ESD has been presented in [9] employing the TOU scheme.

In [10], Sandeep Kakran has designed a common energy schedule for energy scheduling of residential end users having smart appliances for multiple users. Mathematical modeling of three types of appliances—interruptible, uninterruptible, and must-run appliances has been done under the scheme of TOU and RTP tariff. In this study, a wind energy source has been used to lower the load of energy demand on the grids and also a delay factor in order to maintain the compensation level of the consumers by decreasing the cost of waiting. Monika Arora suggested an optimizing technique for smart home appliances fitted with a PV panel and an ESD in [11].

2 System Overview

In this study, the issue of optimal scheduling of the residential appliances in the existence of wind energy source and ESD is viewed. The optimal scheduling algorithm proposes a control strategy for the operation of an AC system, which is a thermostatically controlled appliance. This also manages the operation of wind power and charging or discharging of the battery in such a way that total energy consumption and total cost of the consumption of energy of the smart household is minimized. RTP scheme has been used for the scheduling of energy usage. Ultimately, mixed-integer linear programming (MILP) has been used to solve the problem by using CPLEX solver in the GAMS software. This cost-based scheduling method utilizes the ability of the smart grid of delivering additional power to the power grid.

In this study, residential household appliances can be categorized into two types: AC load and non-AC load. Nowadays, one of the key factors that triggered the peak load is AC. For consumers, power providers, and the whole society, this peak load adds more costs. Regular power usage profile of non-AC appliances such as fridge, laptop, TV, iron box, dryer, washing machine, dishwasher, lighting, and fan, etc. attained in [12] have been utilized. ESD has been considered with fundamental characteristics of the rated power of charging/discharging, and maximum/minimum limits for energy storage are also well-chosen. The energy that is to be generated via wind turbine is to be utilized to charge ESD. Finally, results obtained from the above proposed mechanism have been presented to analyze their effect on the daily energy usage and electricity bill of the smart home.

The paper is disciplined in 5 sections. Section 1 constitutes the introduction of HEMS, DR, and distributed energy resources. Research done in this field has also been reviewed in this section. Section 2 illustrates the model of AC and ESD. Section 3 displays the execution, along with the proposed scheduling mechanism, of the storage algorithm for wind energy sources and ESD. Section 4 displays the variation of all the parameters with respect to time and also presents the results obtained from the mechanism. Section 5 comprises the conclusion and after that are the references.

3 System Model

In our study, a smart home has been considered to have AC load, non-AC load, a small wind turbine, and an ESD. The battery has been used as ESD in this study. On the priority basis, the appliance draws power from the wind source. In the absence of wind power, the power required by the load is drawn from the power grid. The battery is charged either from a wind source or from a grid throughout a cost-effective duration. Additional energy available during any time duration is given back to the power grid for getting benefits of energy cost savings.

The system model comprises two models, which are described below:

3.1 Air-Conditioning Model

The model of AC to be used in this study is illustrated in [7]. According to this model, the inside temperature of the house having an AC is given by

$$T(i) = aT(i-1) + (1-a)\big(T_{out}(i) - T_g S_{ac}(i)\big) \tag{1}$$

where

$T(i)$	Inside temperature of the house during time duration i;
a	Constant and $a = e^{-(u/\tau)}$;
u	Time interval duration;
τ	Time constant of the house and $\tau = RC$;
R	Thermal Resistance of the house;
C	Thermal capacity of the house;
$T_{out}(i)$	Ambient temperature during time duration i;
$S_{ac}(i)$	Binary variable denoting the status of AC during time duration i;
T_g	Temperature gain and $T_g = PR$;
P	Power rating of AC.

In order to maintain a comfortable inside temperature of the house, temperature constraints must be defined in the algorithm.

3.2 Energy Storage Device Model

Batteries are the most suitable device for the purpose of residential energy storage. ESD dynamics can be explained as below:

$$ES(i) = ES(i-1) + u(P_w(i) + P_{esd}S_{ch}(i) - P_{esd}S_{dch}(i)) \tag{2}$$

where

$ES(i)$	ESD storage level during time duration i;
u	Time interval duration;
$P_w(i)$	Wind power supplied to ESD during time duration i;
P_{esd}	Rated power of charging/discharging of ESD.
$S_{ch}(i)$	Binary variable to represent the charging status of ESD from the grid during time duration i;
$S_{dch}(i)$	Binary variable to represent the discharging status of ESD from the grid during time duration i;

It is supposed that the charging/discharging power of the battery is the same, and the battery can be either charged or discharged at a time. ESDs play an important role in DR schemes though they are helpful in decreasing peak load required and utilizing renewable energy.

4 Optimization Technique

The objective is to minimize the total cost of the energy consumed by the household appliances from the power grid after utilizing the RES. Therefore, the objective function is given:

$$\min T_{cost} = \sum_{i=1}^{N} E_{Grid}(i)\gamma(i) \tag{3}$$

where

i	Time interval;
N	Total number of time intervals;
T_{cost}	Total electricity cost over the time period;
$E_{Grid}(i)$	Power consumed by the power grid during time duration i;
$\gamma(i)$	Real-time price of electricity during time duration i

The above optimization problem is subjected to some equality and inequality constraints, which are as follows:

4.1 Equality Constraints

- *Thermal Dynamics of Air-conditioner*:-

  The thermal dynamics of AC are specified by Eq. (1).
- *Thermostat Setting of Air-conditioner*:-

  Modeling for the AC thermostat setting is given as follows:

$$T_{set} = Constant \tag{4}$$

where

T_{set} Desired inside temperature over the entire timespan

- *Settings of Temperature Limits*:-

  Minimum and maximum limits of the inside temperature of the house from the point of view of the user's comfort level have been given by

$$T_{min}(i) = T_{set}(i) - T_d \tag{5}$$

$$T_{max}(i) = T_{set}(i) + T_d \tag{6}$$

where

$T_{min}(i)$ Upper level for inside room temperature during time duration i;
$T_{min}(i)$ Lower level for inside room temperature during time duration i;
T_d Acceptable Temperature band around the desired inside temperature.

- *ESD Storage Dynamics*:-

  ESD storage dynamics are specified by Eq. (2).
- *Constraint for ESD Status*:-

  A battery can be either charged or discharged during any time duration i, i.e.,

$$S_{ch}(i) + S_{dch}(i) = 1 \tag{7}$$

- *Power Balance Equation*:-

  Total energy supplied by the grid to the household can be indicated as follows:

$$E_{Grid}(i) = u(P_{ac}S_{ac}(i) + P_{nonac}(i) + P_{esd}S_{ch}(i) - P_{esd}S_{dch}(i)) \tag{8}$$

where

P_{ac} Rated power of AC;
$P_{nonac}(i)$ Power consumed by non-AC appliances during time duration i

4.2 Inequality Constraints

- *State Constraints of AC Thermostat*:-

  In the above optimization problem, the state of operation of the AC thermostat has been specified by $S_{ac}(i)$ which is a binary variable, according to which,

$$S_{ac}(i) = \begin{cases} 0, & if\ AC\ is\ OFF; \\ 1, & if\ AC\ is\ ON \end{cases} \tag{9}$$

- *Inside Temperature Constraints*:-

  The inside temperature of the house must satisfy the following constraints:

$$T_{min}(i) \leq T(i) \tag{10}$$

$$T(i) \leq T_{max}(i) \tag{11}$$

- *Constraints for ESD status*:-

  The charging status of ESD from the grid can be represented by a binary variable $S_{ch}(i)$ and discharging status of ESD can be represented by a binary variable $S_{dch}(i)$.
- *Constraints for ESD Level*:-

  The energy storage level of ESD must **satisfy the following constraints:**

$$ES_{min} \leq ES(i) \leq ES_{max} \tag{12}$$

where

ES_{min} Minimum limit of ESD energy storage;
ES_{max} Maximum limit of ESD energy storage.

5 Results

This section displays the solution to the problem of scheduling of household devices as a consequence of the implementation of the methodologies explained in Sect. 3. A combination of AC dynamics and ESD dynamics is employed to attain the outcomes of the scheduling that determines the AC operation and ESD operation. To solve the above optimization problem, the MILP optimization method is used. Here, we have considered that surplus generation from a wind power source will be introduced into the utility grid. The rate at which excessive power is fed back to the utility grid at that time is equal to the RTP of electricity. The results indicate that a greater reduction in the cost of energy is achieved by the MILP. GAMS and MATLAB applications have been used in the implementation of the method.

The smart home is fitted with an AC of 1.5 and a 5 kW wind turbine along with the ESD. The input data needed for the execution of the program are ambient temperature, electricity RTP, produced wind power, and power demand for non-AC appliances are displayed in Figs. 1, 2, 3 and 4, respectively. Data of RTP is obtained from the Indian Energy Exchange and it is seen in units Rs per kWh [13]. The software runs for a day, that is, 24 h. Every day is split into 96-time durations, each 15 min long.

In this case, the desired inside temperature of AC is considered as 22 °C. The acceptable temperature band is considered as 2 °C, according to which minimum and maximum temperature limits are 20 °C and 24 °C, respectively. The initial inside temperature of the house is taken as 25 °C [11]. 'a' is a constant which gives the value of system inertia. Its value is considered as 0.96. The thermal resistance of the house is 20 °C/kW, the thermal capacity of the house is 1.25 kWh/°C, and the temperature gain is 30 °C [7].

AC thermostat is turned ON if the inside house temperature obtained during the previous time duration violates temperature limits. On the basis of this, variation in inside room temperature with respect to time is shown in Fig. 5. ESD will be charged

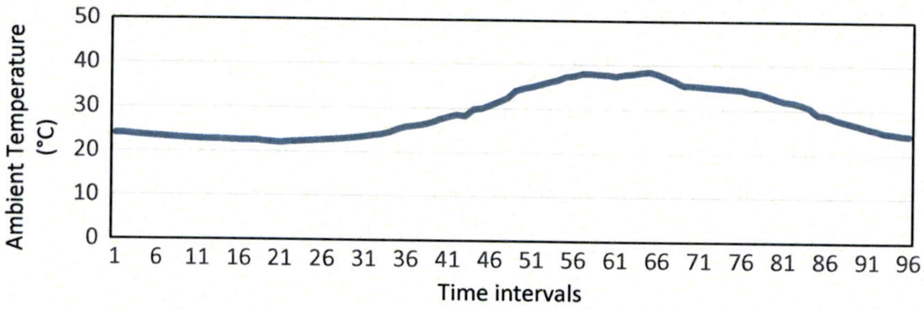

Fig. 1 Variation in ambient temperature with time

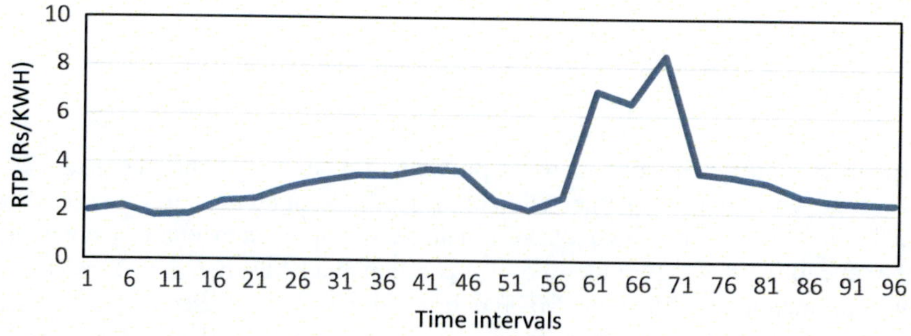

Fig. 2 Variation in RTP with time

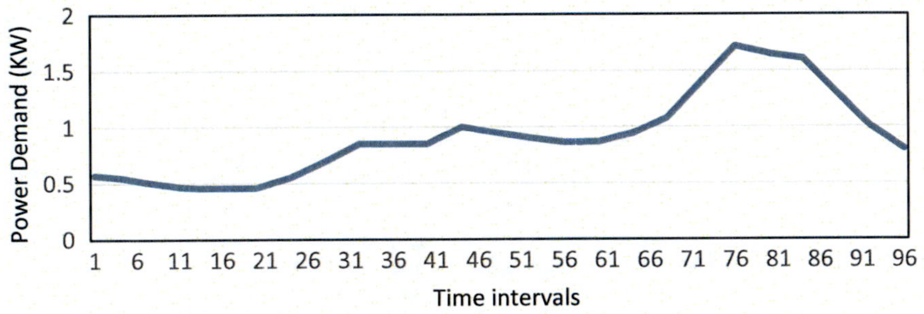

Fig. 3 Power generated from the wind source

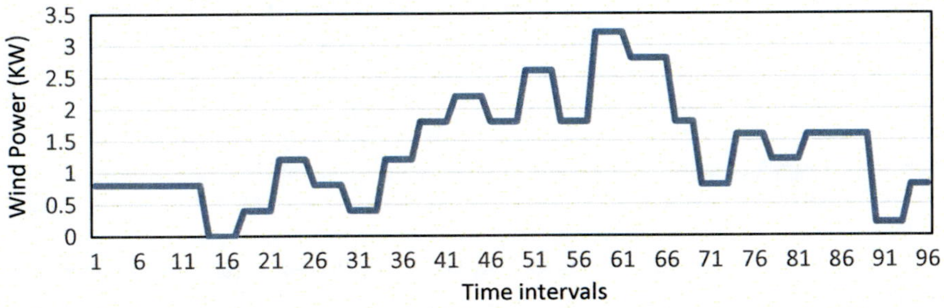

Fig. 4 Power consumed by non-AC appliances

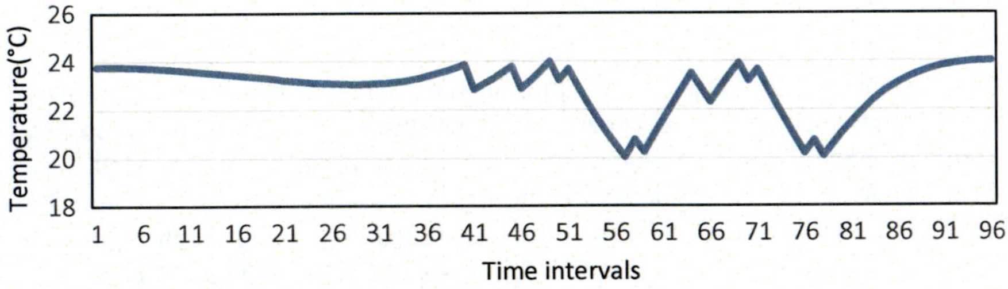

Fig. 5 Variation in inside temperature with time

or discharged under the storage level constraints. It will be charged from the wind source on priority. In this case, if the wind energy is not available during any time duration, ESD will be charged from the grid. The wind turbine will generate power when the speed of wind exceeds its cut-in speed [10].

The total cost of energy used by household appliances is given by the above scheduling algorithm. The variation of total energy consumed from the grid and electricity cost in each time duration are plotted in Figs. 6 and 7, respectively. The optimal solution of the scheduling mechanism gives the electricity cost of Rs. 31.151.

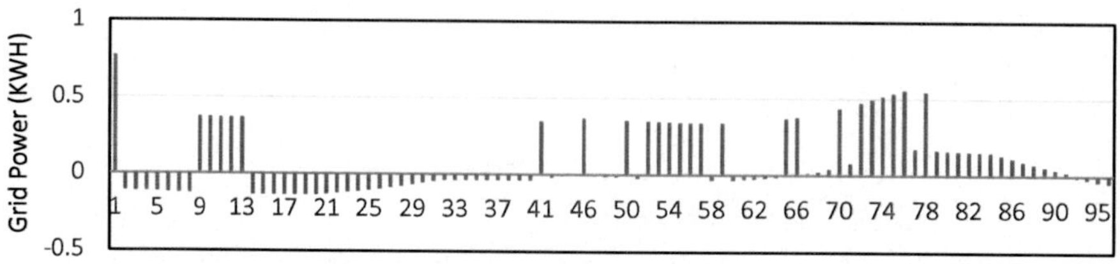

Fig. 6 Variation of grid energy consumed by the household with time

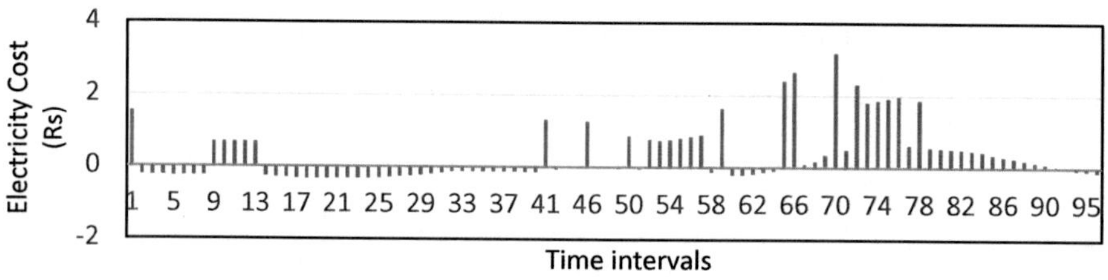

Fig. 7 Variation of electricity cost with time

While energy demand from the grid decreases to a lower value, the total cost of electricity also reduces.

6 Conclusion

In the existence of an ESD and a wind turbine, this paper has presented an optimizing method for controlling the total energy usage of residential appliances in the smart household under RTP. In this work, the simpler paradigm of AC and ESD, with their mathematical model, followed by the AC thermostat control strategy has been described in detail. Then, a MILP-based energy scheduling model for the appliances has been proposed and solved by the CPLEX solver of GAMS software. The obtained results have been presented in figures. The desired temperature inside the room has been successfully achieved by scheduling the AC under RTP. Furthermore, the cost of electricity consumption also reduces due to the use of available wind energy. This cost further reduces to a lower value on selling the extra wind power back to the grid. Results have clearly shown that the MILP scheduling methodology presents a good saving in energy bills under RTP-based DR programs.

References

1. Kåberger T (2018) Progress of renewable electricity replacing fossil fuels. Glob Energy Interconnect 1(1):48–52. ISSN 2096-5117. http://www.sciencedirect.com/science/article/pii/S2096511718300069
2. Albadi MH, El-Saadany EF (2007) Demand response in electricity markets: an overview. In: 2007 IEEE Power engineering society general meeting, Tampa, FL, 2007, pp 1–5. https://doi.org/10.1109/PES.2007.385728
3. Paterakis NG, Medeiros MF, Catalão JPS, Siaraka A, Bakirtzis AG, Erdinc O (2015) Optimal daily operation of a smart-household under dynamic pricing considering thermostatically and non-thermostatically controllable appliances. In: 2015 IEEE 5th International conference on power engineering, energy and electrical drives (POWERENG), Riga, 2015, pp 389–393. https://doi.org/10.1109/PowerEng.2015.7266348
4. Zhao J, Kucuksari S, Mazhari E, Son YJ (2013) Integrated analysis of high-penetration PV and PHEV with energy storage and demand response. Appl Energy 112:35–51
5. Zhu N, Bai X, Meng J (2011) Benefits analysis of all parties participating in demand response. Power Energy Eng Conf (APPEEC), pp 1–4
6. Pilloni V, Floris A, Meloni A, Atzori L (May 2018) Smart home energy management including renewable sources: a QoE-driven approach. IEEE Trans Smart Grid 9(3):2006–2018. https://doi.org/10.1109/TSG.2016.2605182
7. Uncap C, Caglar R (1998) The effects of load parameter dispersion and direct load control action on aggregated load. In: Proceedings of international conference on power system technology (POWERCON), vol 1, pp 280–284, August 1998
8. Pahwa A, Brice CW (1985) Modeling and system identification of residential air conditioning load. In: IEEE Dansactzons on power apparatus and systems, vol PAS-104, pp 1418–1425, June 1985
9. Boynuegri AR, Yagcitekin B, Baysal M, Karakas A, Uzunoglu M (2013) Energy management algorithm for smart home with renewable energy resources. In: 4th International conference on power engineering, energy and electrical drives, pp 1753–1758, May 2013
10. Kakran S, Chanana S (2017) An energy scheduling method for multiple users of residential community connected to the grid and wind energy source. Build Serv Eng Res Technol 39:014362441773453. https://doi.org/10.1177/0143624417734536
11. Arora M, Chanana S (2014) Residential demand response from PV panel and energy storage device. In: 2014 IEEE 6th India international conference on power electronics (IICPE), Kurukshetra, 2014, pp 1–6. https://doi.org/10.1109/IICPE.2014.7115731
12. IEA energy conservation in buildings and community systems. http://www.ebcs.org/docs/Annex_42_Domestic_Energy_Profiles.pdf
13. Indian Energy Exchange. http://www.iexindia.com/Reports/AreaPrice.aspx

Real Power Loss Reduction by Hybridization of Augmented Particle Swarm Optimization with Improved Crow Search Algorithm

Lenin Kanagasabai

Abstract In this work Hybridization of augmented particle swarm optimization (APS) algorithm with improved crow search (ICS) algorithm is applied to solve the power loss lessening problem. In hybridized algorithm (HAPSICS) both augmented particle swarm optimization (APS) algorithm and improved crow search (ICS) algorithm will be executed simultaneously. In specific number of iterations in the procedure few individuals are chosen from both the algorithms APS and ICS by selection approach, and then it will be exchanged sequentially. Then a local search operator has been utilized to augment eminence of the solution. Legitimacy of the HAPSICS algorithm has been corroborated in IEEE 30 Bus system (with and devoid of voltage permanence index). Projected approach meritoriously abridged the power loss.

Keywords Optimal reactive power · Transmission loss · Augmented particle swarm · Improved crow search

1 Introduction

Power loss minimization is a major problem in the Electrical power transmission. Many methods [1–6] are utilized to solve the problem. In the swarm-based algorithms balancing the exploration and exploitation [7–13] will play a principal role in order to reach the preeminent solution [14–20]. Hybridization of augmented particle swarm optimization (APS) algorithm with improved crow search (ICS) algorithm is applied to solve the problem in this paper and both algorithms are executed simultaneously. Particle swarm optimization is augmented by amalgamating the properties of PSO and GA. This procedure perks up the prominence of the solution. Crow search algorithm (CSA) imitates the behavior of crows. Dynamic attentiveness probability (DAP) is used for augmentation of candidate solution. Hybridization improves the balance between exploration and exploitation. Proposed HAPSICS algorithm is substantiated in IEEE 30 bus system (with and deprived of L-index).

L. Kanagasabai (✉)
Prasad V. Potluri Siddhartha Institute of Technology, Kanuru, Vijayawada, Andhra Pradesh 520007, India

A. Kumar et al. (eds.), *Renewable Energy Towards Smart Grid*, Lecture Notes in Electrical Engineering 823, https://doi.org/10.1007/978-981-16-7472-3_15

2 Problem Formulation

Power loss minimization is defined by

$$MinO\tilde{B}F(\overline{r},\overline{u}) \tag{1}$$

Subject to

$$L(\overline{r},\overline{u}) = 0 \tag{2}$$

$$M(\overline{r},\overline{u}) = 0 \tag{3}$$

$$r = \left[VLG_1, .., VLG_{Ng}; QC_1, .., QC_{Nc}; T_1, .., T_{N_T}\right] \tag{4}$$

$$u = \left[PG_{slack}; VL_1, .., VL_{N_{Load}}; QG_1, .., QG_{Ng}; SL_1, .., SL_{N_T}\right] \tag{5}$$

$$F_1 = P_{Minimize} = Minimize\left[\sum_{m}^{NTL} G_m\left[V_i^2 + V_j^2 - 2 * V_i V_j cos\emptyset_{ij}\right]\right] \tag{6}$$

$$F_2 = Minimize\left[\sum_{i=1}^{N_{LB}} \left|V_{Lk} - V_{Lk}^{desired}\right|^2 + \sum_{i=1}^{Ng} \left|Q_{GK} - Q_{KG}^{Lim}\right|^2\right] \tag{7}$$

$$F_3 = MinimizeL_{MaxImum} \tag{8}$$

$$L_{Maximum} = Maximum\left[L_j\right]; j = 1; N_{LB} \tag{9}$$

And

$$\begin{cases} L_j = 1 - \sum_{i=1}^{NPV} F_{ji}\dfrac{V_i}{V_j} \\ F_{ji} = -[Y_1]^1[Y_2] \end{cases} \tag{10}$$

$$L_{Maximum} = Maximum\left[1 - [Y_1]^{-1}[Y_2] \times \frac{V_i}{V_j}\right] \tag{11}$$

$$0 = PG_i - PD_i - V_i \sum_{j \in N_B} V_j\left[G_{ij}cos\left[\emptyset_i - \emptyset_j\right] + B_{ij}sin\left[\emptyset_i - \emptyset_j\right]\right] \tag{12}$$

$$0 = QG_i - QD_i - V_i \sum_{j \in N_B} V_j\left[G_{ij}sin\left[\emptyset_i - \emptyset_j\right] + B_{ij}cos\left[\emptyset_i - \emptyset_j\right]\right] \tag{13}$$

$$\mathrm{P}_{\mathrm{gslack}}^{\mathrm{minimum}} \leq \mathrm{P}_{\mathrm{gslack}} \leq \mathrm{P}_{\mathrm{gslack}}^{\mathrm{maximum}} \tag{14}$$

$$\mathrm{Q}_{\mathrm{gi}}^{\mathrm{minimum}} \leq \mathrm{Q}_{\mathrm{gi}} \leq \mathrm{Q}_{\mathrm{gi}}^{\mathrm{maximum}}, \mathrm{i} \in \mathrm{N}_{\mathrm{g}} \tag{15}$$

$$\mathrm{VL}_{\mathrm{i}}^{\mathrm{minimum}} \leq \mathrm{VL}_{\mathrm{i}} \leq \mathrm{VL}_{\mathrm{i}}^{\mathrm{maximum}}, \mathrm{i} \in \mathrm{NL} \tag{16}$$

$$\mathrm{T}_{\mathrm{i}}^{\mathrm{minimum}} \leq \mathrm{T}_{\mathrm{i}} \leq \mathrm{T}_{\mathrm{i}}^{\mathrm{maximum}}, \mathrm{i} \in \mathrm{N}_{\mathrm{T}} \tag{17}$$

$$\mathrm{Q}_{\mathrm{c}}^{\mathrm{minimum}} \leq \mathrm{Q}_{\mathrm{c}} \leq \mathrm{Q}_{\mathrm{C}}^{\mathrm{maximum}}, \mathrm{i} \in \mathrm{N}_{\mathrm{C}} \tag{18}$$

$$|SL_i| \leq S_{L_i}^{maximum}, \mathrm{i} \in \mathrm{N}_{\mathrm{TL}} \tag{19}$$

$$\mathrm{VG}_{\mathrm{i}}^{\mathrm{minimum}} \leq \mathrm{VG}_{\mathrm{i}} \leq \mathrm{VG}_{\mathrm{i}}^{\mathrm{maximum}}, \mathrm{i} \in \mathrm{N}_{\mathrm{g}} \tag{20}$$

$$\begin{aligned} MOF = F_1 + r_1 F_2 + u F_3 = F_1 + \Bigg[& \sum_{i=1}^{NL} x_v \left[VL_i - VL_i^{min}\right]^2 \\ & + \sum_{i=1}^{NL} x_v \left[VL_i - VL_i^{min}\right]^2 \Bigg] + r \end{aligned} \tag{21}$$

$$VL_i^{minimum} = \begin{cases} VL_i^{max}, VL_i > VL_i^{max} \\ VL_i^{min}, VL_i < VL_i^{min} \end{cases} \tag{22}$$

$$QG_i^{minimum} = \begin{cases} QG_i^{max}, QG_i > QG_i^{max} \\ QG_i^{min}, QG_i < QG_i^{min} \end{cases} \tag{23}$$

3 Augmented Particle Swarm Optimization Algorithm

Movement and velocity of the swarm in PSO is defined by

$$v_i^{t+1} = wv_i^t + c_1 \times rand \times \left(pbest_i - x_i^t\right) + c_2 \times rand \times (gbest - x_i^t) \tag{24}$$

$$x_i^{t+1} = x_i^t + v_i^{t+1} \tag{25}$$

Properties of PSO and GA combined to form the augmented particle swarm optimization (APS) and it perks up the eminence of the solution [21, 22].

$$C_1 r_1 + C_2 r_2 > 0 \tag{26}$$

$$\frac{C_1 r_1 + C_2 r_2}{2} - \omega < 0.98 \tag{27}$$

$$\omega < 1 \tag{28}$$

Knowing that $r_1, r_2 \in [0, 1]$, then

$$0 < C_1 + C_2 < 3.96 \tag{29}$$

$$\frac{c_1 + c_2}{2} - 1 < \omega < 1 \tag{30}$$

$$\omega^{t+1} = K_W \omega^t \tag{31}$$

$$P_{mi} = 0.49 \times \left[\frac{F_{maximum} - F_i}{F_{maximum} - F_{average}} \right] if\, F_i \geq F_{average} \tag{32}$$

$$P_{mi} = \left[\frac{F_{average} - F_i}{F_{maximum} - F_{average}} \right] if\, F_i < F_{average} \tag{33}$$

$$x_i^{(1,t+1)} = x_i^{(1,t)} + (r_i - 0.49)\Delta_i \tag{34}$$

$$\Delta_i = 0.5 \times (\max(x_i) - \min(x_i)) \tag{35}$$

$$\Delta_i = (0.025 \sim 0.075) \times ave(x_i) \tag{36}$$

Procedure of augmented particle swarm optimization algorithm given as below

a. Initialization of population
b. p_{best} which possess best value is considered as g_{best}
c. Particle's position and movement computed by

$$v_i^{t+1} = wv_i^t + c_1 \times rand \times \left(pbest_i - x_i^t\right) + c_2 \times rand \times (gbest - x_i^t)$$

$$x_i^{t+1} = x_i^t + v_i^{t+1}$$

d. Is max. no of iter reached otherwise round to step c till convergence.
e. Create novel population size particles.
f. Gen = Gen + 1.
g. Most excellent solution is output.

4 Improved Crow Search Algorithm

Crow search algorithm (CSA) is modeled on the behavior of crows [23]. Location $X_{i,k}$of the crow "i" in a definite iteration k is defined as

$$X_{i.k} = \left[x_{i,k}^{1}, x_{i,k}^{2}, .., x_{i,k}^{n}\right]; i = 1, 2, .., N; k = 1, 2, .., \text{Maximum iteration} \quad (37)$$

$M_{i,k}$ location to cover food until the present iteration given by

$$M_{i,k} = \left[m_{i,k}^{1}, m_{i,k}^{2}, .., m_{i,k}^{n}\right] \quad (38)$$

$$X_{i,k+1} = \begin{cases} X_{i,k} + r_i \cdot fi_{i,k} \cdot (M_{i,k} - X_{i,k}) r_i > \text{attentiveness probabilty (AP)} \\ random \quad otherwise \end{cases} \quad (39)$$

$$M_{i,k+1} = \begin{cases} F\left(X_{i,k+1}\right) & F\left(X_{i,k+1}\right) < F\left(M_{i,k}\right) \\ M_{i,k} & otherwise \end{cases} \quad (40)$$

$$DAP_{i,k} = 0.90 \cdot \frac{F\left(X_{i,k}\right)}{\omega V} + 0.10 \quad (41)$$

$$Z_i = \frac{\boldsymbol{a}}{|\boldsymbol{b}|^{1/\beta}} \quad (42)$$

$$a \sim N(0, \sigma_a^2); \; b \sim N(0, \sigma_b^2) \quad (43)$$

$$\sigma_a = \left\{ \frac{\Gamma(1+\beta)\sin(\pi\beta/2)}{\Gamma[(1+\beta)/2]\beta^{2(\beta-1)/2}} \right\} \quad (44)$$

$$\sigma_b = 1 \quad (45)$$

Factor L is computed by,

$$L = 0.01 - Z_j \in \left(X_{i,k} - X^{best}\right) \quad (46)$$

New-Fangled position $X_{i,k+1}$ is given by,

$$X_{i,k+1} = X_{i,k} + L \quad (47)$$

Procedure of Improved crow search algorithm is given below:

a. Begin
b. Primary parameters are fixed
c. Capriciously engender the Crows position
d. Memory has been encumbered with reference to preliminary position

e. For every position fitness position has been computed
f. Dynamic attentiveness probability computation by; $DAP_{i,k} = 0.90 \cdot \frac{F(X_{i,k})}{\omega V} + 0.10$
g. Stimulate capricious value r_i for every crow i
h. $r_i > DAP$; If "yes" then compute novel position by $X_{i,k+1} = \begin{cases} X_{i,k} + r_i \cdot fi_{i,k} \cdot (M_{i,k} - X_{i,k}) r_i > \text{attentiveness probabilty (AP)} \\ random \ \ otherwise \end{cases}$
i. If "No" compute novel position by $X_{i,k+1} = X_{i,k} + L$
j. Modernize the memory when improvement in fitness value
k. Is send condition met? if yes stop the iteration otherwise go to step f.

5 Hybridization of Augmented Particle Swarm Optimization Algorithm with Improved Crow Search Algorithm

In the projected hybridized algorithm (HAPSICS) both augmented particle swarm optimization (APS) algorithm and improved crow search (ICS) algorithm will be executed simultaneously.

After the positions are modernized then the center particle (CEPE) is added to the population by

$$x^{t+1}_{cepe,j} = \frac{\sum_{i=1}^{N-1} x^{t}_{i,j}}{N-1}, \ forj = 1, 2, .., d \tag{48}$$

$$V_{maximum} = \left(1 - \left(\frac{iteration}{iteration_{maximum}}\right)^{h}\right) \times V_{maximum0} \tag{49}$$

$$V_{maximum0} = \alpha \times (x_{maximum} - x_{minimum}) \tag{50}$$

$$Probability_i = \frac{Fitness_i}{\sum_{i=1}^{N} Fitness_i} \tag{51}$$

$$x^{t+1}_{i,j} = \begin{cases} x^{t}_{i,j}; iff\left(x^{t}_{i,j}\right) < f\left(x^{t+1}_{i,j}\right) \\ x^{t+1}_{i,j} : otherwsie \end{cases} \tag{52}$$

Procedure for Hybridization of Augmented Particle Swarm Optimization Algorithm with Improved Crow Search Algorithm

a. Step a. Start
b. Step b. With reference to APS and ICS agents are initialized
c. Step c. Then the center of particle and crow is obtained by

$$x_{cepe,j}^{t+1} = \frac{\sum_{i=1}^{N-1} x_{i,j}^{t}}{N-1}, \; forj = 1, 2, .., d$$

d. Step d. Apply the APS procedure

i. Initialization of population
ii. Particle's position and movement computed by

$$v_i^{t+1} = wv_i^t + c_1 \times random \times \left(pbest_i - x_i^t\right) + c_2 \times rand \times \left(gbest - x_i^t\right)$$

$$x_i^{t+1} = x_i^t + v_i^{t+1}$$

iii. Is max no of iter reached or else loop to step c
iv. Create novel population size particles.
v. Gen = Gen + 1, then step c is carried out.
vi. Output the preeminent solution

e Step e. Apply the ICS procedure

i. Begin
ii. Primary parameters are fixed
iii. Dynamic attentiveness probability computation by; $DAP_{i,k} = 0.90 \cdot \frac{F(X_{i,k})}{\omega V} + 0.10$
iv. Stimulate capricious value r_i for every crow i
v. $r_i > DAP$; If "yes" then compute novel position by $X_{i,k+1} = \begin{cases} X_{i,k} + r_i \cdot fi_{i,k} \cdot (M_{i,k} - X_{i,k}) r_i > \text{attentiveness probabilty (AP)} \\ random \quad otherwise \end{cases}$
vi. If "No" compute novel position by $X_{i,k+1} = X_{i,k} + L$
vii. IS end condition is met; if yes stop the iteration otherwise go to step f

f. Step f. Local operator is applied

$$x_{i,j}^{t+1} = \begin{cases} x_{i,j}^t; iff\left(x_{i,j}^t\right) < f\left(x_{i,j}^{t+1}\right) \\ x_{i,j}^{t+1} : otherwsie \end{cases}$$

g. Step g. When specific iteration number reached implement the roulette-wheel approach

$$Probability_i = \frac{Fitness_i}{\sum_{i=1}^{N} Fitness_i}$$

h. Step h. When end criterion is not met then go to step "c"
i. Step i. Output the superlative solution

6 Simulation Results

Projected Hybridization of augmented particle swarm optimization (APS) algorithm with improved crow search (ICS) algorithm (HAPSICS) has been substantiated in IEEE 30 bus system [24] with and devoid of stability. Tables 1, 2 and 3 show the appraisal of values. Figures 1 and 2 show the graphical appraisal. Voltage deviation and stability index values of HAPSICS are 0.0859 and 0.1002.

Table 1 Assessment of whole power loss

Method	Real power loss (MW)
Standard PSO-TS [7]	4.5213
Standard TS [7]	4.6862
Standard PSO [7]	4.6862
ALO [8]	4.5900
QO-TLBO [9]	4.5594
TLBO [9]	4.5629
Standard GA [10]	4.9408
Standard PSO [10]	4.9239
HAS [10]	4.9059
Standard FS [11]	4.5777
Improved FS [11]	4.5142
Standard FS [13]	4.5275
APS	4.5015
Improved CS	4.5014
HAPSICS	4.5012

Table 2 Comparison of loss with reference to IEEE—0 system

Parameter	Active power loss in MW	Percentage of reduction in power Loss
Base case [19]	17.5500	0.0000
M-PSO[19]	16.0700	8.40000
Basic -PSO [18]	16.2500	7.4000
EP [16]	16.3800	6.60000
S -GA [17]	16.0900	8.30000
PSO [20]	17.5246	0.14472
DEPSO [20]	17.52	0.17094
JAYA [20]	17.536	0.07977
APS	14.12	19.54
ICS	14.10	19.65
HAPSICS	14.09	19.71

Table 3 Convergence characteristics

IEEE 30 bus system	Real power loss in MW (with L-index)	Real power loss in MW (without L-index)	Time (s) (with L-index)	Time (s) (without L-index)	Number of iterations (with L-index)	Number of iterations (without L-index)
APS	4.5015	14.12	18.49	14.37	24	19
ICS	4.5014	14.10	18.51	14.42	27	21
HAPSICS	4.5012	14.09	18.52	14.54	29	23

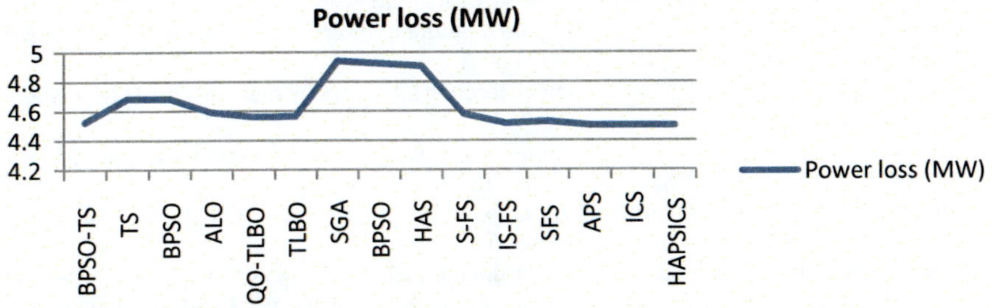

Fig. 1 Power loss assessment

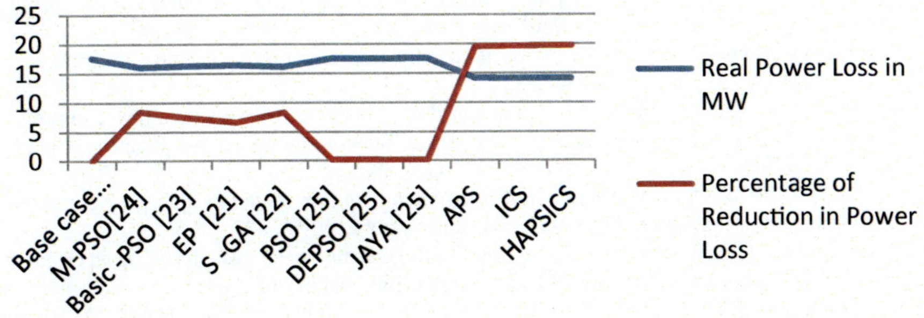

Fig. 2 Appraisal of real power loss

7 Conclusion

In this work Hybridization of augmented particle swarm optimization (APS) algorithm with improved crow search (ICS) algorithm has been condensed the power loss meritoriously. Centre particle (CEPE) is added to the population, roulette-wheel approach, local search operator has been applied in the projected algorithm to improve the quality of the solution. In the improved crow search (ICS) algorithm attentiveness probability (AP) is swapped by a dynamic attentiveness probability (DAP) for enrichment and it is accustomed by the dominance of candidate solution. Cogency of Proposed HAPSICS algorithm is substantiated in IEEE 30 bus test system (with and devoid of L-index). Active Power loss minimization has been accomplished.

References

1. Lee K (1984) Fuel-cost minimisation for both real and reactive-power dispatches. Proc Gener Transm Distr Conf 131(3):85–93
2. Deeb N (1998) An efficient technique for reactive power dispatch using a revised linear programming approach. Electr Power Syst Res 15(2):121–134
3. Bjelogrlic M (1990) Application of Newton's optimal power flow in voltage/reactive power control. IEEE Trans Power Syst 5(4):1447–1454
4. Granville S (1994) Optimal reactive dispatch through interior point methods. IEEE Trans Power Syst 9(1):136–146
5. Grudinin N (1998) Reactive power optimization using successive quadratic programming method. IEEE Trans Power Syst 13(4):1219–1225
6. Khan I (2016) Distributed control algorithm for optimal reactive power control in power grids. Int J Electr Power Energy Syst 83(1):505–513
7. Sahli Z (2014) Hybrid PSO-tabu search for the optimal reactive power dispatch problem. In: Proceedings of the IECON 2014–40th annual conference of the IEEE industrial electronics society, vol 40, pp 3536–3542. Dallas, TX, USA
8. Mouassa S (2017) Ant lion optimizer for solving optimal reactive power dispatch problem in power systems. Eng Sci Technol Int J 20(3):885–895
9. Mandal B (2013) Optimal reactive power dispatch using quasi-oppositional teaching learning based optimization. Int J Electr Power Energy Syst 53(1):123–134
10. Khazali H (2011) Optimal reactive power dispatch based on harmony search algorithm. Int J Electr Power Energy Syst 33(3):684–692
11. Tran H (2019) Finding optimal reactive power dispatch solutions by using a novel improved stochastic fractal search optimization algorithm. Telecommun Comput Electron Control 17(5):2517–2526
12. Polprasert J (2016) Optimal reactive power dispatch using improved pseudo-gradient search particle swarm optimization. Electr Power Comp Syst 44(5):518–532
13. Thanh L (2020) Optimal reactive power flow for large-scale power systems using an effective metaheuristic algorithm. Hindawi J Electr Comput Eng 20(1):1–11
14. Mukherjee A (2015) Solution of optimal reactive power dispatch by Chaotic Krill Herd algorithm. IET Gener Transm Distrib 9(15):2351–2362
15. Li J (2020) Optimal reactive power dispatch of permanent magnet synchronous generator-based wind farm considering levelised production cost minimisation. Renew Energy 1459(1):1–12
16. Dai C, Chen W, Zhu Y, Zhang X (2009) Seeker optimization algorithm for optimal reactive power dispatch. IEEE T. Power Syst. 24(3):1218–1231
17. Subbaraj P, Rajnarayan PN (2009) Optimal reactive power dispatch using self-adaptive real coded Genetic algorithm. Electr Power Syst Res 79(2):374–438
18. Pandya S, Roy R (2015) Particle swarm optimization based optimal reactive power dispatch. In: Proceeding of the IEEE international conference on electrical, computer and communication technologies (ICECCT), pp 1–5
19. Hussain AN, Abdullah AA, Neda OM (2018) Modified particle swarm optimization for solution of reactive power dispatch. Res J Appl Sci Eng Technol 15(8):316–327. https://doi.org/10.19026/rjaset.15.5917
20. Vishnu, Mini, Sunil (2020) An improved solution for reactive power dispatch problem using diversity-enhanced particle swarm optimization. Energies 13:2862, 2–21. https://doi.org/10.3390/en13112862

21. Leke Z (2018) Concepts, methods, and performances of particle swarm optimization, backpropagation, and neural networks. Appl Comput Intell Soft Comput 7(1):1–7
22. Türkylmaz A (2015) A hybrid algorithm for total tardiness minimisation in flexible job shop: genetic algorithm with parallel VNS execution. Int J Prod Res 53(6):1832–1848
23. Rajput S (2016) Paras Optimization of benchmark functions and practical problems using Crow Search Algorithm. In: Proceedings of the 2016 fifth international conference on eco-friendly computing and communication systems (ICECCS), Bhopal, India, 8–9 December 2016, pp 73–78
24. Illinois Center for a Smarter Electric Grid (ICSEG). https://icseg.iti.illinois.edu/ieee-30-bus system/. Accessed 25 Feb 2019

Comparative Analysis of Peak Limiting Strategies in the Home Energy Management System

Vikas Deep Juyal and Sandeep Kakran

Abstract The paper considers a residential consumer living in Ontario, who is a customer of Hydro One, an electricity transmission and distribution service provider. The customer is the owner of a smart home with smart appliances who can participate in any demand response (DR) program by any communication medium. The customer has a solar photovoltaic set up on the roof, a battery energy storage system, and an electric vehicle with some thermostatically and non-thermostatically controlled appliances. The price tariffs of the country are rapidly changing due to the economic attack of Covid-19. These changing tariffs are considered as scenarios, and the effect of the tariffs on the customer's electricity bill is analyzed. The scenarios are formulated as a simple mixed integer linear programming problem. All the events are optimized by different DR programs using the CPLEX solver of GAMS software for minimization of the cost. This paper also investigates the peak to average ratio of the power demand in each scenario. Different strategies for controlling the power transfer from the grid are employed as DR programs, and the results are analyzed in detail.

Keywords Home Energy Management System · Demand Response · Time of Use · Power Import Limit

Abbreviations

P_a^h	Energy consumption by appliance 'a' during timeslot 'h'
$b_{h,a}$	Binary variable to decide the state of type-1 appliance 'a'
$E_{n,h,a}$	Energy consumed by appliance 'a' during 'h' timeslot for nth schedule
T_h^{HW}	The temperature of hot water
T_h^a	Air temperature (°C)
ΔT	Time interval
r_h	Hot water usage

V. D. Juyal (✉) · S. Kakran
Department of Electrical Engineering, National Institute of Technology, Kurukshetra, India

A. Kumar et al. (eds.), *Renewable Energy Towards Smart Grid*, Lecture Notes in Electrical Engineering 823, https://doi.org/10.1007/978-981-16-7472-3_16

$T^{HWmx.}$	The maximum range of hot water (°C)
q	The capacity of the EWH (2 kW)
d_h^{EWH}, d_h^{AC}	Binary variables to decide the state of EWH and AC
R, C	Thermal resistance and capacitance of EWH
T_h^r	Room temperature (°C)
M_a	Mass of air in kg
C_a	The thermal capacity of air
$R_{eq.}$	Equivalent thermal resistance
$Stpt_h$	Setpoint for required room temperature
V_h^u, V_h^l	Upper and lower limit for deviation of indoor temperature from setpoint
$L_{mx}(h)$	Limit for power import during timeslot 'h'

1 Introduction

In this era of modernization, energy plays an important role in human lifestyle, various services, and nearly all types of infrastructures; hence, efficient use of energy is a must to manage the energy demand and supply [1]. In the last few decades, the traditional power grid is changing toward an intelligent, reliable, and automated smart grid [2]. Demand side management and advanced metering infrastructure comprising 'information & communication technology (ICT) and smart meters' are the key elements of a modern or smart grid [3]. The participation of the consumers is possible using various types of demand response (DR) programs [4].

It has been stated [5] that buildings have a consumption of around 40% of the overall energy consumption, which will increase in upcoming years [6]. Home energy management system (HEMS) schedules the home appliances, which depends on the various constraints used in the DR program and various external data inputs like environmental parameters and updated grid prices, etc.

Paper [7] discussed the various concepts of the smart grid, including various programs of DR in detail. For the optimization models of DR schemes, it is said that objective function is generally multi-objective for large systems, and heuristic approaches have been proved very useful to get optimal and fast solutions.

The paper [3] proposed a HEMS, including solar PV as a renewable energy resource (RER) and battery as an energy storage system to reduce cost and peak to average ratio (PAR). Single and multi-objective optimization was done using particle swarm optimization (PSO) and binary PSO. The paper [8] used model predictive control for energy management in an isolated microgrid by employing a DR program. Various research papers are based on the heuristic and metaheuristic approaches for the energy management system, which provides a fast solution [9–12]. The heuristic techniques are based on a high-level procedure to find the suboptimal solution, which may be global, but the results can vary in different runtime, and they can give irrelevant output [1].

Real-time pricing (RTP), flat pricing, and time of use (ToU) are the most popular price-based DR. In a flat pricing scheme, the prices remain the same for the whole day, and the only way to reduce the electricity consumption bill is to reduce the electricity usage, although the prices may vary seasonally [13]. The ToU tariff divides the whole day into three or four time zones according to the average energy transfer in that time zone [14]. The time zones may be classified as off-peak, mid-peak, and peak and the prices will be higher in the zones having higher demand. The RTP scheme works on the hourly varying price according to the real-time wholesale price of electricity [15].

Many countries follow the tiered electricity tariffs, especially two-tiered or three-tiered, but some countries have implemented ToU and RTP tariffs also, and Ontario is one of them. Hydro One, a publicly traded company on the Toronto Stock Exchange, is an electricity transmission and distribution service provider, which deals with the ToU tariff. The tariff is decided each year on 1 Nov for the summer season and winter season. However, the tariff is generally fixed for the year, but due to Covid-19, the whole world is facing economic problems, and that's why every service provider is also affected. Keeping these facts under consideration, a special revised flat electricity tariff has been implemented with a lower price from 01-01-2021 up to 28-01-2021. These events are considered as different scenarios in this study.

This paper includes the study of different seasonal changes in electricity tariff and implementation with different power import limiting strategies to reduce the PAR and the cost of electricity consumption. A residential customer is considered to have non-thermostatically and thermostatically controlled appliances, electric vehicle (EV), battery energy storage system (BESS), and solar rooftop photovoltaic panels in the household. Electric water heater (EWH) and air-conditioner (AC) are taken as thermal loads used in the household. Different scenarios have been considered in the paper according to the change in electricity tariff applied. Various constraints are used for the proper management of the appliances. The scheduling of household appliances and sources is formulated as a mixed-integer linear programming (MILP), and the objective function is solved using CPLEX solver of GAMS software.

The remaining paper is organized in the following sections: The mathematical modelling of the appliances and sources are discussed in Sect. 2. All the equations and constraints used in the modelling are also described. The case study and results are represented and discussed in Sect. 3. The conclusion and future extension of the study are discussed in the Sect. 4.

2 Methodology

In this study, it is assumed that a home energy management unit (HEMU) is employed at the household of the consumer for the scheduling of non-thermostatically controlled appliances, EWH, AC, EV, BESS, and PV. An hourly divided time horizon of a day is considered for the scheduling. The first schedule is considered from 12 to 1 AM and similarly, the other twenty-three schedules are also considered in the sequence i.e., H = {1,2.........24}.

2.1 Non-Thermostatically Controlled Appliances

The scheduling of these appliances is done by the HEMU and the energy consumption vector P_a can be written as

$$P_a \triangleq [P_a^1, P_a^2 \ldots P_a^{24}] \tag{1}$$

Total energy consumption by any appliance 'a' can be written as

$$\sum_{h \in H} P_a^h = E_a \, \forall a \tag{2}$$

The included appliances of this category are divided according to their operational characteristics:

- Category-1: Interruptible appliances have been put into this sub-category who consume B_a^{max} energy during ON condition and B_a^{min} during OFF condition.

$$P_a^h = b_{h,a} \cdot B_a^{max} + \left(1 - b_{h,a}\right) B_a^{min} \, \forall h \tag{3}$$

The total ON duration (S_a) for these appliances can be represented by

$$\sum_{h \in H} (b_{h,a} * K_{a,h}) = S_a \tag{4}$$

$K_{a,h}$ represents the ON/OFF state of the appliance in the total time horizon.

- Category-2: Uninterruptible appliances have been put in this sub-category. A binary variable 'd' is defined for the modelling of these appliances such that

$$\sum_{n \in N} d_{n,a} = 1 \tag{5}$$

where the value of 'n' is 1,2,3,4....... N and N are the total possible schedules for the appliance 'a'. Now we can present the total energy consumption of these appliances by

$$P_a^h = \sum_{n \in N} d_{n,a} * E_{n,h,a}) \, \forall h \tag{6}$$

Thus, it can be said that the total consumption of energy by the non-thermostatically controlled appliances will be

$$P_h^{appl.} = \sum_{a \in A} P_a^h \, \forall h \tag{7}$$

2.2 Thermostatically Controlled Appliances

These are the appliances that affect the comfort zone of the consumer hence the preference of consumers must be considered in the scheduling of these types of appliances. In this study, EWH and AC are included as thermostatically controlled loads.

- *EWH mathematical model*: The mathematical model of the EWH can be represented by the following equations.

$$T_{h+1}^{HW} = T_h^a + q.\,R.d_h^{EWH} - (T_h^a - T_h^{HW}).e^{-\frac{\Delta T}{R.C}} \, \forall h, r_h = 0 \tag{8}$$

$$T_{h+1}^{HW} = \frac{T_h^{HW}.(U - r_h) + T_h^{CW}.u_h}{V} \, \forall h, \; r_h > 0 \tag{9}$$

$$T_h^{HW} \le T^{HWmx.} \, \forall h \tag{10}$$

$$T_h^{HW} \ge T^{HWmn.} \, \forall h \tag{11}$$

$$P_h^{EWH} = q \cdot d_h^{EWH} \, \forall h \tag{12}$$

The maximum and minimum limit for the temperature of the hot water tank is shown by (10) and (11), whereas (12) shows the total power consumption by EWH.

- *AC mathematical model:* A linearized model of an AC can be represented as

$$\begin{aligned} T_h^r = T_{h-1}^r + \left(\frac{\Delta T}{M_a.C_a.R_{eq.}}\right) \cdot (T_{h-1}^a - T_{h-1}^r) \\ -m_{h-1}^{AC}\left(\frac{\beta.P^{AC}.\Delta T}{0.000277.M_a.C_a}\right) \forall h > 1 \end{aligned} \tag{13}$$

$$T_h^r \le Stpt_h + V_h^u \, \forall h \tag{14}$$

$$T_h^r \leq Stpt_h - V_h^l \,\forall h \tag{15}$$

$$P_h^{AC} = P^{AC} \cdot d_h^{AC} \,\forall h \tag{16}$$

Mathematical equations for the modelling of PV, BESS, and EV are taken from [16], and these equations are taken as constraints for the scheduling and operation of the devices.

2.3 Power Import Limit

In this paper, each time slot is provided with different power import limits from the grid, and therefore it can be written that

$$P_h^{gr.} \leq L_{mx}(h) \,\forall h \tag{17}$$

2.4 Objective Function

This study is focused on the minimization of electricity consumption cost of the household by proper scheduling of the appliances and hence a simple objective function can be presented as

$$Cost = \sum_{h \in H} \left(P_h^{gr.} . \lambda_h^{buy} \right) \tag{18}$$

The power balance equation can be written as

$$\begin{aligned} P_h^{gr.} + P_h^{PV.hm} + P_h^{BE.hm} + P_h^{EV.hm} = P_h^{appl.} + P_h^{BE.chr} \\ + P_h^{EV.chr} + P_h^{AC} + P_h^{EWH} \,\forall h \end{aligned} \tag{19}$$

where $P_h^{PV.hm}$, $P_h^{BE.hm}$, $P_h^{EV.hm}$ are the power utilized by home by PV, BESS, and EV, respectively, and $P_h^{BE.chr}$, $P_h^{EV.chr}$ are the power used for the charging of BESS and EV. λ_h^{buy} is the rate of electricity tariff during 'h' timeslot.

Proposed solution: The Eqs. (1) to (17) along with the mathematical equations taken from [16] are included as the constraints for the problem, which are solved as a MILP problem by CPLEX solver of GAMS.

3 Case Study and Results

A residential household customer having smart appliances is in this study. The data for non-thermostatically controlled loads are taken from [17]. The solar power generated by the panels is shown in Fig. 1.

The thermal parameters of 2 kW, 50-gallon capacity EWH are taken from [18]. The customer uses hot water at 6 PM and 6 AM with the preferred range of 35 °C to 50 °C and 30 °C to 45 °C, respectively. The flow of water is considered to be 2.5 gallons/min. For the operation of a 2 kW AC, the setpoint of the thermostat is 25 °C ± 2 °C, C_a is 1.01 kJ/Kg °C, $R_{eq.} = 3.1965 \times 10^{-3}$ h. °C/KJ. The consumer wants the operation of the AC from 10 AM to 5 PM. The initial state of energy of 1 kWh BESS is assumed to be 0.5 kWh with an efficiency of 0.95 and a charging/discharging rate of 0.2 kWh. It is also assumed that the energy produced by PV and unutilized energy to EV and BESS is used in the household itself.

The real events that occurred in Ontario's electricity tariffs are considered in different scenarios. Fixed power import limits (FPIL) and variable power import limits (VPIL) are applied in these scenarios, and the effect of these limits on the overall cost of electricity consumption and PAR is also considered in the analysis. The data is taken from [19]

- *Scenario-1*: Scheduling of the household appliances with a flat rate tariff of 12.8 cents/kWh during Covid-19 till 31/10/2020 with an FPIL of 12 kW in each time slot.
- *Scenario-2*: Scheduling of the household appliances with a flat rate tariff of 12.8 cents/kWh during Covid-19 till 31/10/2020 with VPIL of 5 kW to 12 kW in each time slot.
- *Scenario-3*: Scheduling of the household appliances with a ToU tariff set for winters onwards from 01/11/2020 with an FPIL of 12 kW in each time slot.
- *Scenario-4*: Scheduling of the household appliances with a ToU tariff set for winters onwards from 01/11/2020 with VPIL of 5 kW to 12 kW in each time slot.
- *Scenario-5*: Scheduling of the household appliances with a ToU tariff set for summers onwards from 01/11/2020 with an FPIL of 12 kW in each time slot.

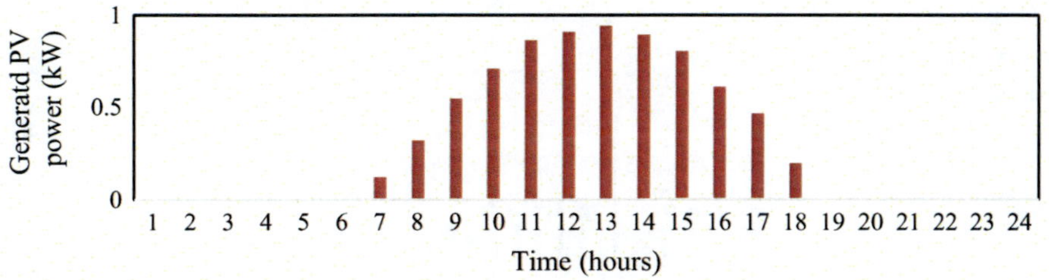

Fig. 1 Generated solar PV power

- *Scenario-6:* Scheduling of the household appliances with a ToU tariff set for summers onwards from 01/11/2020 with VPIL of 5 kW to 12 kW in each time slot.
- *Scenario-7*: Scheduling of the household appliances with a special revised flat-rate tariff of 8.5 cents/kWh from 01/01/2021 till 28/01/2021 with an FPIL of 12 kW in each time slot.
- *Scenario-8*: Scheduling of the household appliances with a special revised flat-rate tariff of 8.5 cents/kWh from 01/01/2021 till 28/01/2021 with VPIL of 5 kW to 12 kW in each time slot.

The scenario-1 represents the duration of Covid-19 before 01-Nov-2021 during which the Ontario electricity board has applied a flat rate tariff of 12.8 cents/kWh during all hours of the day. The scheduling of appliances of the customer is done by a DR program by applying a fixed power import limit of 12 kW in each slot and the resulted scheduling is shown in Fig. 2. The non-thermostatically controlled appliances and others are scheduled according to the customer's preferences. The peak demand is found to be 9.5 kW with a PAR of 3.5144 and an overall cost of 651.5931 cents (Fig. 3).

In scenario-2, to limit the peak demand and to improve the PAR, variable limits are the constraints on the power import from the grid. These limits affect the scheduling of the appliances as shown in Fig. 4. It can be seen in Fig. 4 that the peak is reduced from 9.5 kW to 5.5 kW, which is a significant amount for the reduction of stress in

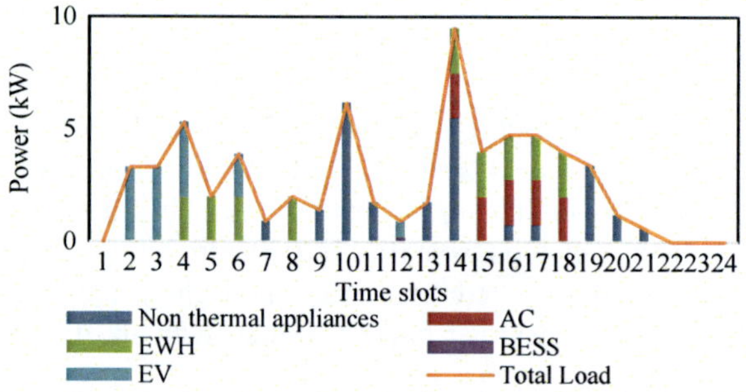

Fig. 2 Scheduling of household appliances with a flat rate tariff DR with FPIL strategy

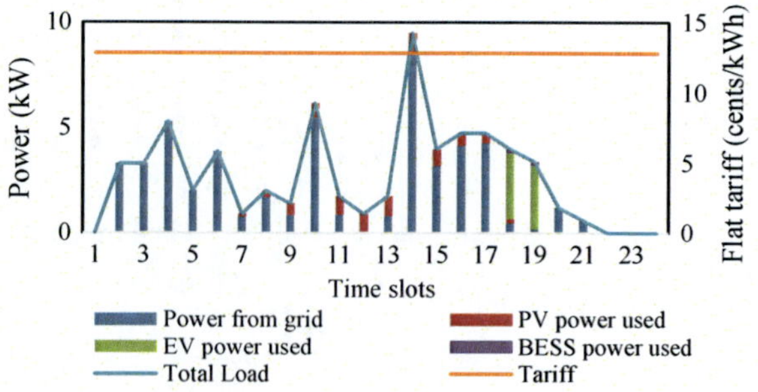

Fig. 3 The commitment of various power sources toward the customer's demand for scenario-1

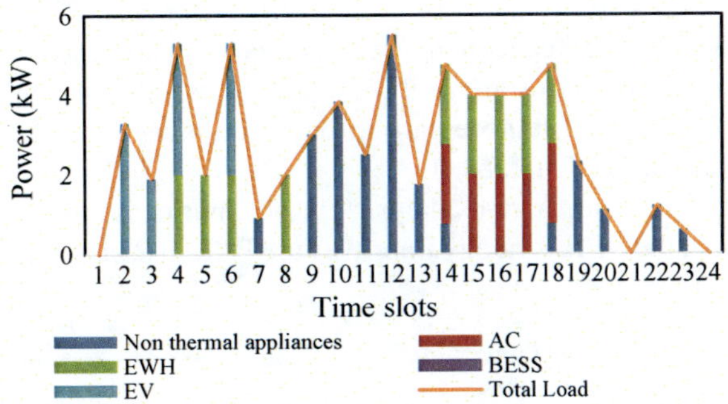

Fig. 4 Scheduling of household appliances with a flat rate tariff DR with VPIL strategy

the system network. Some of the peak load is shifted toward off-peak time slots. The cost for energy consumption, in this case, is 642.2688 cents and the PAR is 2.063565, which is also far decreased in comparison with the FPIL strategy in the last scenario (Fig. 5).

Scenario-3 and 4 are simulated by considering the ToU tariff applied for winters in Ontario. The scheduling after implementation of DR programs with FPIL and VPIL with the imposed ToU tariff are represented in Figs. 6 and 8, respectively. In these scenarios, there is a peak reduction from 9.25 kW to 6.49 kW and an improvement

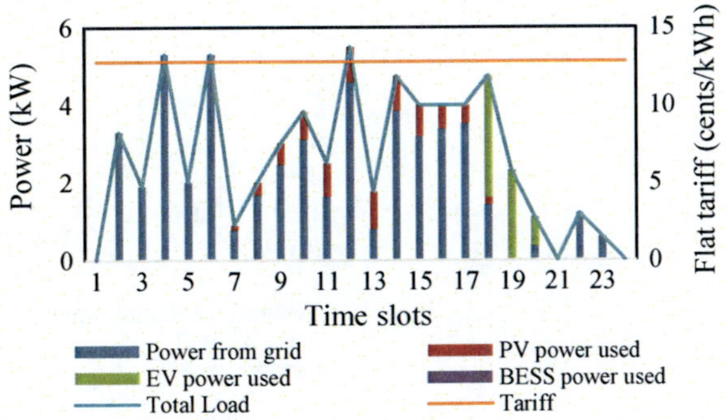

Fig. 5 The commitment of various power sources toward the customer's demand for scenario-2

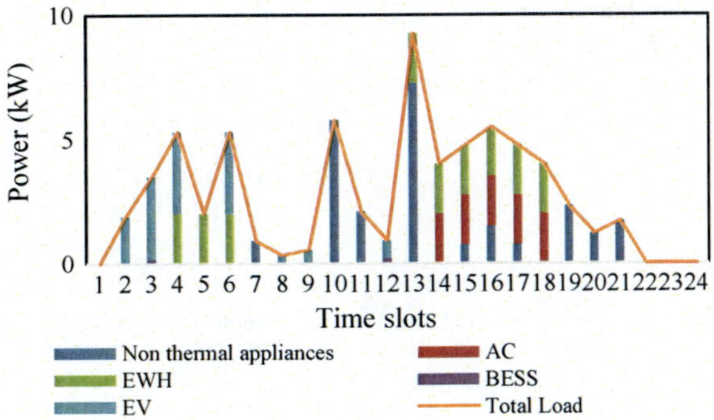

Fig. 6 Scheduling of household appliances with a flat rate tariff DR with FPIL strategy for winters

of PAR from 3.368114 to 2.410616. The cost of energy consumption is found to be 725.5355 cents and 708.9366 cents in scenario-3 and 4, respectively. Similarly, all the scenarios are analyzed, and the results are represented in the Figs. 9, 10, 11, 12 and 13. The necessary information is mentioned in the graphs, and the rest are summarized in Table 1.

The comparative analysis of all scenarios is represented in Table 1. It can be seen that the VPIL strategy improves the PAR in each scenario, which is beneficial for the

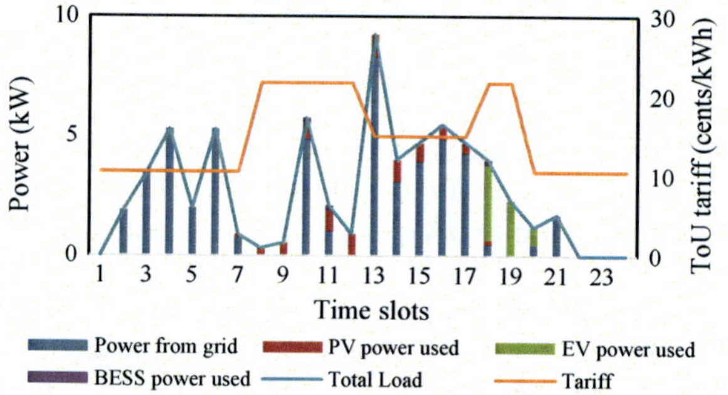

Fig. 7 The commitment of various power sources toward the customer's demand for scenario-3

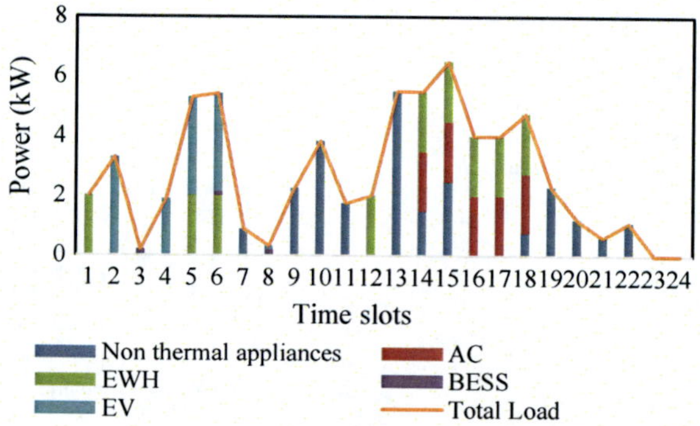

Fig. 8 Scheduling of household appliances with a flat rate tariff DR with VPIL strategy in winters

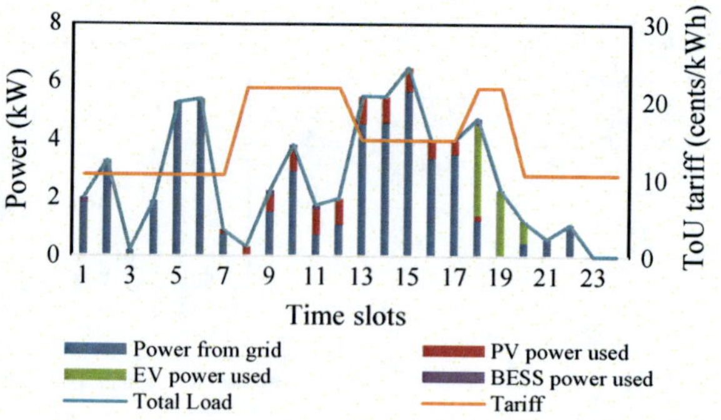

Fig. 9 The commitment of various power sources toward the customer's demand for scenario-4

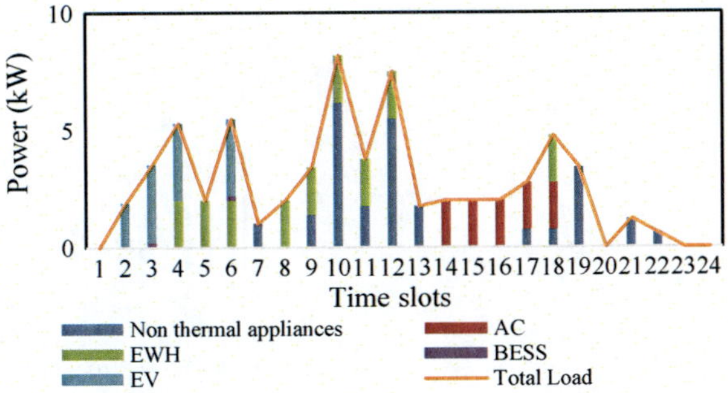

Fig. 10 Scheduling of household appliances with a ToU tariff DR with FPIL strategy for summers

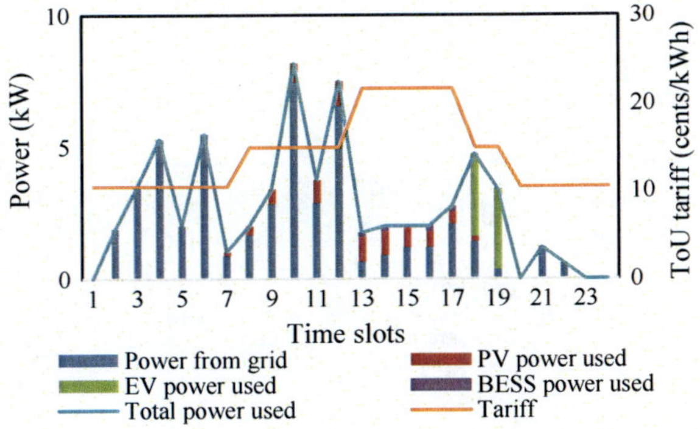

Fig. 11 The commitment of various power sources toward the customer's demand for scenario-5

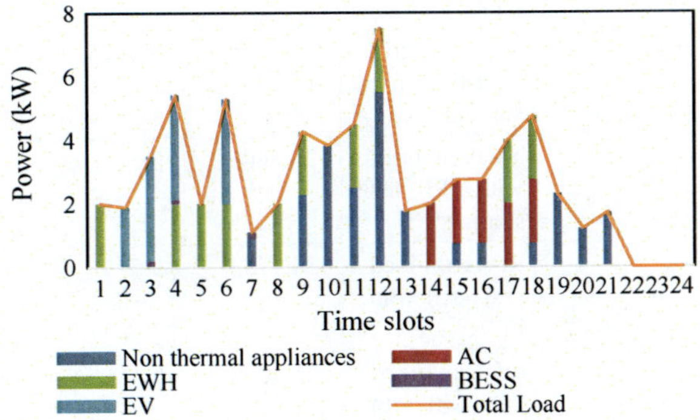

Fig. 12 Scheduling of household appliances with a ToU tariff DR with VPIL strategy for summers

service provider. The objective function is formulated for minimization of cost of energy consumption, and it can be analyzed that VPIL strategy reduces the cost in all scenarios except for summers due to reversed prices by the utility during mid-peak and peak demand hours.

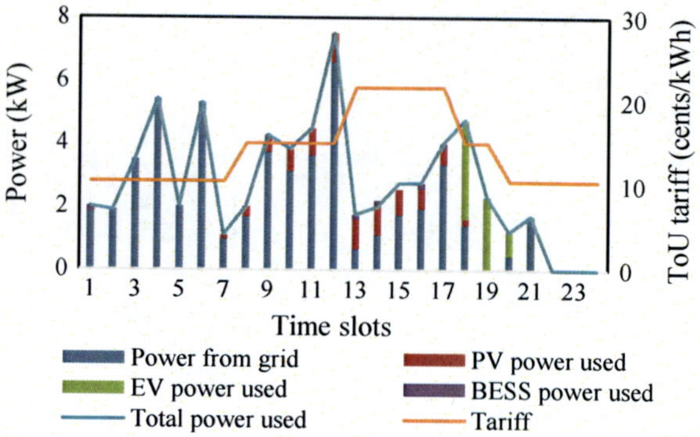

Fig. 13 The commitment of various power sources toward the customer's demand for scenario-6

Table 1 Comparative analysis of presented scenarios

Scenario No	Description	Cost (cents)	Peak demand (kW)	Average demand (kW)	PAR
1	Fixed rate tariff with FPIL	651.5931	9.5	2.703164	3.5144
2	Fixed rate tariff with VPIL	642.2688	5.5	2.665291	2.063565
3	ToU rate tariff with FPIL for winters	725.5355	9.25	2.746344	3.368114
4	ToU rate tariff with VPIL for winters	708.9366	6.49	2.692257	2.410616
5	ToU rate tariff with FPIL for summers	700.0976	8.1725	2.68722	3.041247
6	ToU rate tariff with VPIL for summers	735.8406	7.5	2.770554	2.70704
7	Revised Fixed rate tariff with FPIL	439.1752	9.5	2.742433	3.464077
8	Revised Fixed rate tariff with VPIL	426.5067	5.5	2.665291	2.063565

4 Conclusion

It can be concluded from all scenarios in Table 1 that by imposing limits on the power drawn from the grid, PAR can be improved. However, these limits must be modified with the change in the tariff for the optimized results. A single set of limits may give a different cost for different scenarios. In this study, the same set of limits are used in each VPIL scenario. The applied strategy gives improved results in comparison with FPIL, except in the summers when the ToU tariff was changed and limits constraints were kept the same as in other VPIL scenarios. In the summers, PAR was improved but the overall cost is more than that of FPIL strategy. Reversed pricing by the utility during mid-peak and peak demand hours in summers and allotment of load in peak demand slot are the cause of this increment of cost in scenario-6. Therefore, it can be observed that the limiting strategy should be adaptive which must change according to the tariff. A single strategy would not give satisfactory results for all tariffs. The implemented strategy resulted in a significant reduction in PAR; that's why it could be a perfect choice for the service providers.

In this study, various limiting strategies have been implemented with different tariffs. The appliances and sources have been scheduled according to the preference of the customer, and thus, the comfort of the customer has also been kept in mind. All the scenarios are simulated by CPLEX using GAMS software, and results are shown in the graphs. The CPLEX solver uses branch and cut algorithm to solve the linear programming problem. For the future scope of the study, an adaptive power import limiting strategy may be designed. Energy selling to the grid has not been considered in this paper, which can be included in the extension of the paper.

References

1. Leitao J, Gil P, Ribeiro B, Cardoso A (2020) A survey on home energy management. IEEE Access 8:5699–5722
2. Kakran S, Chanana S (2018) Smart operations of smart grids integrated with distributed generation : a review. Renew Sustain Energy Rev 81 August 2017: 524–535
3. Dinh HUYT, Member S, Yun J, Kim DMIN, Lee K, Kim D (2020) A home energy management system with renewable energy and energy storage utilizing main grid and electricity selling. IEEE Access 8:49436–49450
4. Albadi MH, El-Saadany EF (2008) A summary of demand response in electricity markets. Electr Power Syst Res 78(11):1989–1996
5. Chen P, Xu J, Gu F, Schmidt, Li W (2018) Measures to improve energy demand flexibility in buildings for demand response (DR): a review. Energy Build 177. Elsevier Ltd, pp 125–139. Oct 15
6. Yahia Z, Pradhan A (2018) Optimal load scheduling of household appliances considering consumer preferences: an experimental analysis. Energy 163:15–26
7. Vardakas JS, Zorba N, Verikoukis CV, Member S (2015) A survey on demand response programs in smart grids : pricing methods and optimization algorithms. IEEE Commun Surv Tutor 17(1):152–178
8. Solanki BV, Raghurajan A, Bhattacharya K, Canizares CA (2017) Including smart loads for optimal demand response in integrated energy management systems for isolated microgrids. IEEE Trans Smart Grid 8(4):1739–1748
9. Kakran S, Chanana S (2019) Energy scheduling of residential appliances by a pigeon-inspired algorithm under a load shaping demand response program. Int J Electr Eng Inf 11(1):18–34
10. Bouakkaz A, Haddad S, Martín-García JA, Gil-Mena AJ, Jiménez-Castañeda R (2019) Optimal scheduling of household appliances in off-grid hybrid energy system using PSO algorithm for energy saving. Int J Renew Energy Res 9(1):427–436
11. Yousefi M, Hajizadeh A, Soltani MN (2019) A Comparison study on stochastic modeling methods for home energy management systems. IEEE Trans Ind Inf 15(8):4799–4808

12. Molla T, Khan B, Moges B, Alhelou HH, Zamani R, Siano P (2019) Integrated energy optimization of smart home appliances with cost-effective energy management system. CSEE J Power Energy Syst 5(2):249–258
13. Doostizadeh M, Ghasemi H (2012) A day-ahead electricity pricing model based on smart metering and demand-side management. Energy 46(1):221–230
14. Shareef H, Ahmed MS, Mohamed A, Al Hassan E (2018) Review on home energy management system considering demand responses, smart technologies, and intelligent controllers. IEEE Access 6: 24498–24509
15. Samadi P, Mohsenian-Rad A-H, Schober R, Wong VWS, Jatskevich J (2010) Optimal real-time pricing algorithm based on utility maximization for smart grid. First IEEE Int Conf Smart Grid Commun 2010:415–420
16. Kakran S, Chanana S (2019) Operation scheduling of household load, EV and BESS using real time pricing, incentive based dr and peak power limiting strategy. Int J Emerg Electr Power Syst 20(1):1–13
17. Kakran S, Chanana S (2018) An energy scheduling method for multiple users of residential community connected to the grid and wind energy source. Build Serv Eng Res Technol 39(3):295–309
18. Du P, Lu N (2011) Appliance commitment for household load scheduling. IEEE Trans Smart Grid 2(2):411–419
19. Ontario electricity board prices. https://www.hydroone.com/rates-and-billing/rates-and-charges/electricity-pricing-and-costs. Accessed 04 Jan 2021

Generation Scheduling Considering Emissions in Cost-Based Unit Commitment Problem

Vineet Kumar, R. Naresh, Veena Sharma, and V. Kumar

Abstract Electricity plays an important role for economic and financial development of any country. As population increases, electricity demand also increases rapidly, and generation facilities are under tremendous pressure to generate more and more power with minimum operating cost. The proper scheduling, planning and maintenance of generation companies is one of the main task in electricity market to improve the lowest operating cost of generation. As steam generating units emit greenhouse gases which lead to emissions, they are responsible for global climatic and environmental changes. Thus, an attempt is made here for the proper scheduling of generators along with emissions in conventional environment. In this work, IEEE 39 bus system is considered (10 thermal generating units at 24-time horizon) as it is having quadratic cost characteristics and simulations are carried out in General Algebraic Modeling System by applying Branch and Reduce Optimization Navigator solver. The effectiveness of proposed solver has been successfully examined in terms of total operating cost and emissions generated by thermal units while comparing the results obtained with other MINLP solvers.

Keywords General Algebraic Modeling System (GAMS) · Cost-based Unit Commitment (CBUC) · Branch and Reduce Optimization Navigator (BARON) · Environmental Emission (EM)

V. Kumar (✉) · R. Naresh · V. Sharma · V. Kumar
National Institute of Technology Hamirpur, Hamirpur, Himachal Pradesh, India
e-mail: vineet@nith.ac.in

R. Naresh
e-mail: rnaresh@nith.ac.in

V. Sharma
e-mail: veena@nith.ac.in

V. Kumar
e-mail: vineet.kumar@nith.ac.in

A. Kumar et al. (eds.), *Renewable Energy Towards Smart Grid*, Lecture Notes in Electrical Engineering 823, https://doi.org/10.1007/978-981-16-7472-3_17

1 Introduction

In the era of industrialization, electricity plays a vital role in day-to-day life. The total power production from thermal generators is around 70% of the total world's power production, across the globe. Thermoelectricity accounted for 80% of the total volume of electricity globally in 2006 [1]. The main issue facing the thermal energy sector is the environmental pollution generated by the extensive espousal of thermal generators. The thermal generation cause consists of transforming energy stored in fossil fuels into electricity by burning/ heating up of fuel and it mainly follows three steps. During the first phase, in a boiler furnace, oil and coal both are burned to heat up the H_2O into extreme temperature and at extreme pressure steam. At the same time, as a by-product, emissions such as NO_x, SO_x and CO_2 are generated. The subsequent high-pressure vapor is whistled into a stream generator in the second stage, and the steam contraction initiates the stream generator to rotate. During the final step, mechanical energy is provided by the revolution of the steam turbine, which activates the generator to generate electrical energy. A large amount of fuel is used in the thermal generation process and a huge number of pollutants are released [2]. An instance, it takes roughly 335 g of coal to produce 1 units of energy, which emits roughly 870 g of electric power that leads to huge environmental pollution in power industry [3]. The power industry's exposures to environmental emissions raise concerns about the safety of the atmosphere and the methods of mitigating emissions from power stations, either through design or operational strategies. Thus, it is mandatory to reduce the emission by switching the fuels if possible that leads in minimum emission and having the objective to minimize the emission for all the generating units. Many papers considered emissions, but in the field of economic dispatch (ED), only optimal dispatch of each thermal unit is carried out, but on/off status of thermal generating unit considering unit commitment constraints have not been considered [4–6].

Many classical and metaheuristic approaches have been employed on cost-based unit commitment problem. The classical techniques are, namely, priority method, brute force enumeration method, dynamic programming, Lagrange relaxation method, branch and bound (B&B) method [1], etc. Likewise, many meta-heuristic optimization techniques are also incorporated by many researchers in the field of UC like genetic algorithm [7], evolutionary programming [8], differential evolution [9], memetic algorithm [10], shuffled frog leaping algorithm [11], quasi oppositional teaching learning-based optimization [12], particle swarm optimization [13], grey wolf optimization [14], teaching learning-based optimization [15], whale optimization algorithm [16], and many more [17–19]. The drawback of classical approaches is that these approaches are applicable to small-sized problems and computational time taken by these techniques are very extensive [1]. Although, meta-heuristic optimization techniques solve large-scaled test problems easily but these approaches do not guarantee the global optimal solutions [20]. So, keeping these points in view many researchers also applied hybrid algorithms on CBUC problems like dynamic programming linked with artificial neural networks, hybrid artificial bee colony and many more. The detailed overview on classical, metaheuristic and hybrid approaches are carried out in [6, 21, 22, 26]. From the literature review, it is determined that a lot of work has been carried out in UC problem.

The goal of the paper is to find optimum unit commitment solution by using BARON to minimize operating and start-up costs while satisfying the diverse constraints. The main advantage of this software is that researchers have to identify the nature of problem and according to that there is choice in the selection of the appropriate solver. Remaining paper is structured as, Sect. 2 represents problem formulation in GAMS environment, and detailed solution technique is provided in Sect. 3. Case study and its discussion is carried in Sect. 4 which is followed by conclusion in Sect. 5.

2 Problem Formulation

The complete single objective function corresponding to the total generation cost along with emission can be approached to have a quadratic cost curve. Exactly, complete objective function is denoted as

$$\text{Total objective function } (F) = \text{Min } (TGC) + \text{Min } (E_{\text{cost}}) \tag{1}$$

Here, *TGC* is total generating cost and E_{cost} is total cost of emission.

$$TGC = \text{Min}\left(\sum_{t=1}^{T}\sum_{i=1}^{N}[F_i(P_i^t)U_i^t + U_i^t(1 - U_i^{t-1})SUC_{i,t} + U_i^{t-1}(1 - U_i^t)SDC_{i,t}]\right) \tag{2}$$

$$\text{Emission}(EM) = \text{Min}\left(\sum_{t=1}^{T}\sum_{i=1}^{N}E(P_i^t)U_i^t\right) \tag{3}$$

$$\text{Total cost of emission, } E_{\text{cost}} = T_{\text{carbon}} * EM \tag{4}$$

Here, T_{carbon} is assumed to be 20 Rs. /ton [25].

In above equations, generator and emission characteristics are considered as quadratic, and $SDC_{i,t}$ is considered in this work as zero, whereas $SUC_{i,t}$ is defined as

$$SUC_{i,t} = \left\{\begin{array}{l} HSC_i, \text{ if } T_{i,\text{down}} \le T_{i,\text{off}}^t \le T_{i,\text{down}} + T_{i,\text{cold}} \\ CSC_i, \text{ if } T_{i,\text{off}}^t > T_{i,\text{down}} + T_{i,\text{cold}} \end{array}\right\} \tag{5}$$

In this paper, thermal characteristics is given as

$$F_i(P_i^t) = a_i + b_i P_i^t + c_i (P_i^t)^2 \tag{6}$$

and emission curve is represented as

$$E_i(P_i^t) = d_i + e_i P_i^t + f_i (P_i^t)^2 \tag{7}$$

Nomenclature:			
Sets			
I	Index for thermal generating units	T	Index for time periods
Variables			
a_i, b_i, c_i	Thermal cost coefficient of ith unit	$P^{\min}$	Lower generation capacity in (MW)
EM	Emission produced/generated	$P^{\max}$	Upper generation capacity in (MW)
d_i, e_i, f_i	Emission coefficient of thermal generators	U	On/off status
N	Total fuel units	T_{off}	Continuously off time in (hours)
$F_i\ (P_i{}^t)$	Production/generation cost in (Rs. /h)	T_{cold}	Cold start-up time limit in (hours)
CSC	Cold start-up cost in (Rs. /h)	$P_i{}^t$	Active power production in (MW)
HSC	Hot start-up cost (Rs. /h	T	Denotes total time horizons in h
$P_{\mathrm{load}}{}^t$	Power demand at tth time (MW)	T_{down}	Minimum-down time
T_{up}	Minimum-up time in (hours)	T_{on}	Continuously on time

Constraints imposed on CBUC problem considering emissions are as follows:

2.1 Power Balance Constraint

For CBUC problem, the power demand of respective interval must be equal to sum/addition of power produced/generated by each unit [23].

$$\sum_{i=1}^{N} P_{i,t} U_i^t = P_{\mathrm{load}}^t \tag{8}$$

2.2 Generator Limits

The generator limits always be in the range of upper and lower bounds.

$$P_i^{\min} U_i^t \le P_i^t \le P_i^{\max} U_i^t \tag{9}$$

2.3 Minimum up/down Constraints

These constraints indicate that a device would not be switched off instantly when in operation, and that when turned off, it would not be switched on without a lag in minimum time interval. As a result, the generating schedule/plan is essential in compliance with the time limits and generating/production units to fulfil the time-based constraint [6].

$$(T_{\text{on}}(i, t-1) - T_{\text{up}}(i))\,(U(i, t-1) - U(i, t)) \ge 0 \tag{10}$$

$$(T_{\text{off}}(i, t-1) - T_{\text{down}}(i))\,(U(i, t) - U(i, t-1)) \ge 0 \tag{11}$$

where T_{on} (i, t) represents continuous on time of ith unit at t time, T_{up} (i) and $T_{\text{down}}(i)$ denotes minimum up/down times, and T_{off} (i, t) is off time of unit i at t time, respectively.

2.4 Spinning Reserve Constraints

It is described as a pre-determined portion or amount of predicted peak demand and that must be accessible throughout the planning horizon and is referred as [24]:

$$\sum_{i=1}^{N} P_{i,t}^{\max} U_i^t \ge P_{\text{load}}^t + SR^t \tag{12}$$

where SR^t denotes the spinning reserve of test system during time t.

2.5 Emission Constraint

The sum of emission produced by each thermal generator must be equal to overall emission level of the system [3].

$$\sum_{t=1}^{T}\sum_{i=1}^{N} E(P_i^t)U_i^t = EM_t \tag{13}$$

3 Solution Methodology

The UC problem formulated in Sect. 2 is a highly nonlinear and mixed integer optimization problem and it is modeled in GAMS software which is a very useful approach for solving the optimization problems and also creates an accurate model. GAMS model includes sets, input data, variables, equations, type of model and, at last, display/print of output is the basic structure that followed in GAMS [6]. In order to obtain best solution of nonlinear algebraic programs (NLPs) and mixed integer nonlinear programs (MINLP), a computational system BARON solver can be employed. The complete layout of proposed BARON solver in CBUC problem along with constraints is presented in Fig. 1. The pseudocode of above problem is shown in Fig. 2.

4 Results and Discussions

The effectiveness of proposed solver is carried out on IEEE 39 bus system having 10 thermal units considering emissions. The overall single line figure of IEEE 39 bus case having 10 generator buses is presented in Fig. 5, which can be referred from Appendix. The unit commitment and emission of thermal generation data is considered from [6] and [3]. The main aim of choosing BARON solver in this case system is that BARON solver produces better quality solutions in UC problem [6]. The simulation program file is tested in GAMS 25.1 version and performed on personal PC having specifications Intel core i5, 4 GHz processor. BARON method (GAMS) is applied to the 10 thermal units in absence of ramp rate limits considering the quadratic cost characteristics with 10% spinning reserve hourly load. Optimum UC operation for IEEE 39 bus test system comprising 10 thermal units over 24-h time scheduling horizon is presented in Table 1, wherein the zero entries indicate the off status of units and total value of generating cost is the addition of the production cost and start-up costs, given in last two columns.

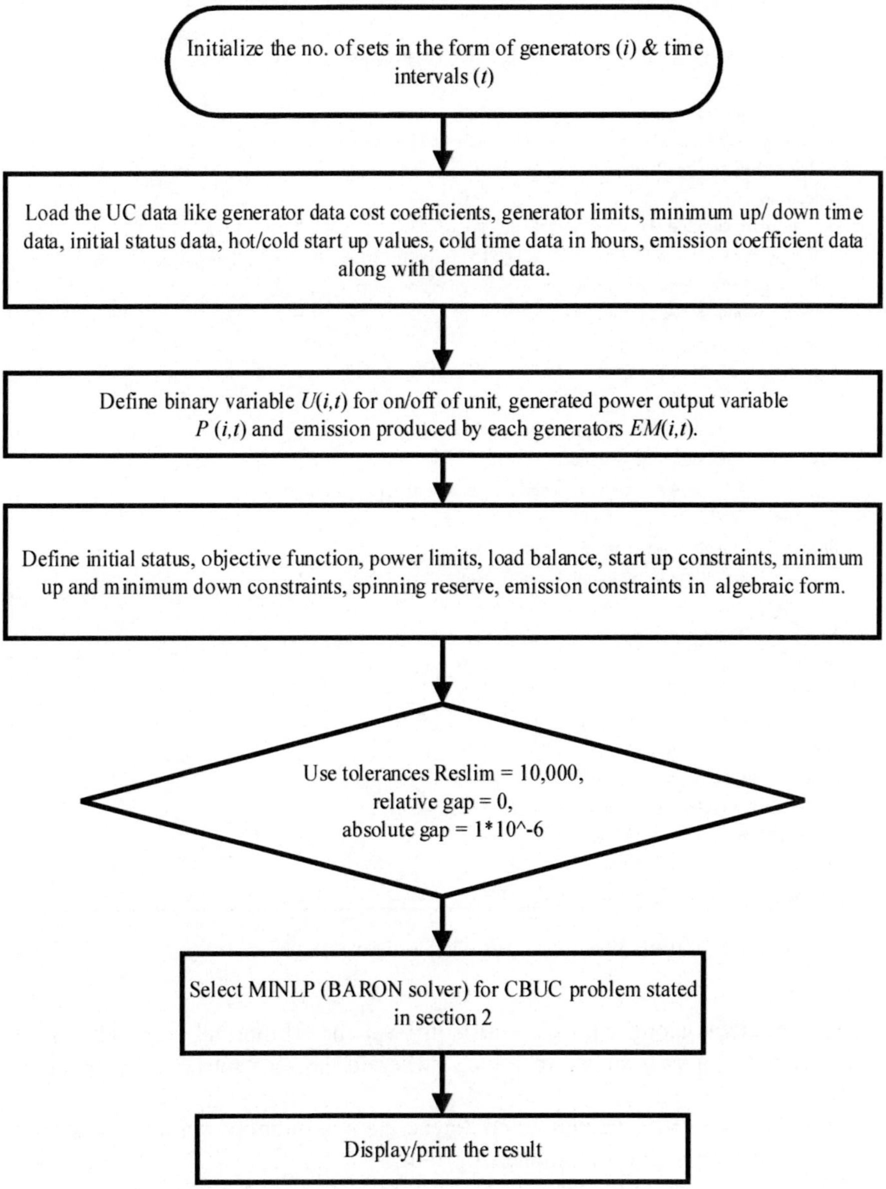

Fig. 1 Flowchart depicting layout of CBUC problem in BARON solver

```
Set
Sets for thermal generators i /1*10/, t time /1*24/; (10 generators are considered in the system)
Data
table for generation data (i, *); (Load the generation data)
table for power demand data (t, *); (Load the power demand data)
table for emission coefficients (i, *); (Load the emission coefficients data)
Variables
define binary variable U (t, i); (For ON and OFF of units)
define positive variable like P (t, i), F (t, i), SUC (t, i), EM (t, i), Ecost (t, i); (For power generation, total
cost, start-up cost, emission produced and total emission cost)
objective minimize the operating cost;
outline initial status for all thermal generators
Equations
F          Total fuel cost
Ecost      Total emission cost produced
EM         Total emission generated
P          Total power generation within lower and upper power limits
Objective  Total cost function as given in eq 1
SUC        Equation for start-up constraint
Tup        Equation for minimum up constraint
Tdown      Equation for minimum down constraint
SR         Spinning reserve
Model
model pf /all/;
option optca=0; (For absolute gap)
option optcr=0; (For relative gap)
option minlp=baron; (Solver used)
option Reslim=10000; (Maximum time limit in seconds)
solve pf us minlp min objective;
Display
display objective.l, U.l, P.l, EM.l, Ecost.l, F.l,
```

Fig. 2 Pseudocode of IEEE 39 bus test system in UC problem

The unit status along with the emission level for 10 thermal units determined using BARON solver is shown in Table 2, wherein the zero entries indicate the off status of units.

Figure 3a shows that the unit 1 and 2 generate maximum power and are utilized throughout the day. Energy attained by unit 1, 2, 3, 4, 5, 6, 7, 8, 9 and 10 is 7697.5, 7513.5, 2340, 2470, 3564, 1200, 1105, 605, 330 and 275 MWh, respectively. Figure 3b shows that the unit 1 and 2 generate maximum emission level and this is obvious because these thermal units are utilized throughout the day and these units are of higher rating having high efficiency. The emission level produced obtained for unit 1, 2, 3, 4, 5, 6, 7, 8, 9 and 10 is 6101.6, 5808.6, 1137.6, 1200.8, 1331, 376, 434.2, 281.6, 151.8 and 141.5 tons, respectively.

Figure 4a shows the total power generation and emission cost for 10 thermal units, over 24-h time horizon. From this, it is evident that the power demand of

Table 1 Optimum power generation of 10 thermal units considering BARON solver

T	Generating units' power in (MW)										Total load (MW)	Total operation cost (Rs. /h)	SUC (Rs. /h)
	1	2	3	4	5	6	7	8	9	10			
1	353.8	346.2	0	0	0	0	0	0	0	0	700	13,770.7	0
2	378.8	371.2	0	0	0	0	0	0	0	0	750	14,621.3	0
3	347.8	340.2	0	0	162	0	0	0	0	0	850	17,312.5	900
4	397.7	390.3	0	0	162	0	0	0	0	0	950	19,014.3	0
5	252.9	245.1	130	130	162	80	0	0	0	0	1000	22,038.9	1280
6	260.4	252.6	130	130	162	80	85	0	0	0	1100	25,136.4	260
7	257.9	250.1	130	130	162	80	85	55	0	0	1150	27,149.8	30
8	310.3	302.7	130	130	162	80	85	0	0	0	1200	26,831.2	0
9	332.8	325.2	130	130	162	80	85	55	0	0	1300	29,693.3	30
10	327.8	320.2	130	130	162	80	85	55	55	55	1400	33,898.7	60
11	352.8	345.2	130	130	162	80	85	55	55	55	1450	34,748.3	0
12	377.8	370.2	130	130	162	80	85	55	55	55	1500	35,598.9	0
13	327.8	320.2	130	130	162	80	85	55	55	55	1400	33,898.7	0
14	305.4	297.6	130	130	162	80	85	55	55	0	1300	30,931.2	0
15	282.9	275.1	130	130	162	80	85	55	0	0	1200	27,996.6	0
16	317.8	310.2	130	130	162	0	0	0	0	0	1050	22,045.8	0
17	292.9	285.1	130	130	162	0	0	0	0	0	1000	21,197.6	0
18	342.9	335.1	130	130	162	0	0	0	0	0	1100	23,733.1	0
19	282.9	275.1	130	130	162	80	85	55	0	0	1200	27,996.6	460
20	327.8	320.2	130	130	162	80	85	55	55	55	1400	33,898.7	60

(continued)

Table 1 (continued)

T	Generating units' power in (MW)										Total load (MW)	Total operation cost (Rs. /h)	SUC (Rs. /h)
	1	2	3	4	5	6	7	8	9	10			
21	332.8	325.2	130	130	162	80	85	55	0	0	1300	29,693.3	0
22	302.9	295.1	130	130	162	80	0	0	0	0	1100	23,733.1	0
23	307.8	300.2	0	130	162	0	0	0	0	0	900	18,814.6	0
24	322.8	315.2	0	0	162	0	0	0	0	0	800	16,463.1	0
Total value of generation cost (Rs.) = 613,296.9											27,100	610,216.9	3080
Total objective function value (Rs.) = (613,296.9 + 338,802.4 = 952,099.3)													

Table 2 Optimum level of emissions produced for 10 thermal units considering BARON solver

T	Generating units' emissions in (Tons)										Emission (Tons/h)	Emission cost (Rs. / h)
	1	2	3	4	5	6	7	8	9	10		
1	314.4	299.7	0	0	0	0	0	0	0	0	614.1	12,281.2
2	365.3	349.6	0	0	0	0	0	0	0	0	714.9	14,298.7
3	302.7	288.3	0	0	60.5	0	0	0	0	0	651.5	13,030.3
4	406.7	390.2	0	0	60.5	0	0	0	0	0	857.4	17,146.5
5	148.1	137.8	63.2	63.2	60.5	23.5	0	0	0	0	496.3	9925.2
6	158.3	147.7	63.2	63.2	60.5	23.5	33.4	0	0	0	549.8	10,993.5
7	154.8	144.3	63.2	63.2	60.5	23.5	33.4	25.6	0	0	568.5	11,370.8
8	235	222.2	63.2	63.2	60.5	23.5	33.4	0	0	0	701	14,017.7
9	274.6	260.8	63.2	63.2	60.5	23.5	33.4	25.6	0	0	804.8	16,094.2
10	265.5	251.9	63.2	63.2	60.5	23.5	33.4	25.6	25.3	28.3	840.4	16,808.1
11	312.4	297.8	63.2	63.2	60.5	23.5	33.4	25.6	25.3	28.3	933.2	18,663.4
12	363.2	347.5	63.2	63.2	60.5	23.5	33.4	25.6	25.3	28.3	1033.7	20,674.7
13	265.5	251.9	63.2	63.2	60.5	23.5	33.4	25.6	25.3	28.3	840.4	16,808.1
14	226.6	214	63.2	63.2	60.5	23.5	33.4	25.6	25.3	0	735.3	14,705.9
15	190.9	179.2	63.2	63.2	60.5	23.5	33.4	25.6	0	0	639.5	12,789.3
16	247.8	234.7	63.2	63.2	60.5	0	0	0	0	0	669.4	13,386.4
17	206.4	194.3	63.2	63.2	60.5	0	0	0	0	0	587.6	11,749.5
18	222.5	233.5	63.2	63.2	60.5	0	0	0	0	0	642.9	12,855.7
19	190.9	179.2	63.2	63.2	60.5	23.5	33.4	25.6	0	0	639.5	12,789.3
20	265.5	251.9	63.2	63.2	60.5	23.5	33.4	25.6	25.3	28.3	840.4	16,808.1

(continued)

Table 2 (continued)

T	Generating units' emissions in (Tons)										Emission (Tons/h)	Emission cost (Rs. / h)
	1	2	3	4	5	6	7	8	9	10		
21	274.6	260.8	63.2	63.2	60.5	23.5	33.4	25.6	0	0	804.8	16,094.2
22	222.5	210	63.2	63.2	60.5	23.5	0	0	0	0	642.9	12,855.7
23	230.8	218.1	0	63.2	60.5	0	0	0	0	0	572.6	11,449.8
24	256.6	243.2	0	0	60.5	0	0	0	0	0	560.3	11,206.2
Total value of emission cost (Rs.) = 338,802.4												

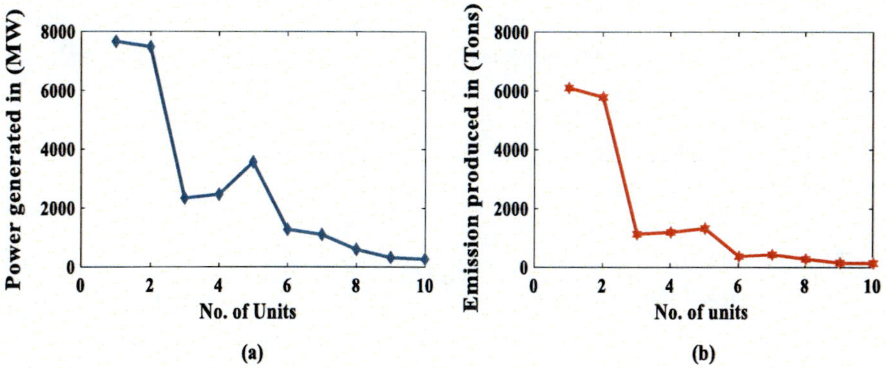

Fig. 3 Optimum power generation, emission produced by each unit **a** Total power generated by each unit **b** Total emissions produced by each unit

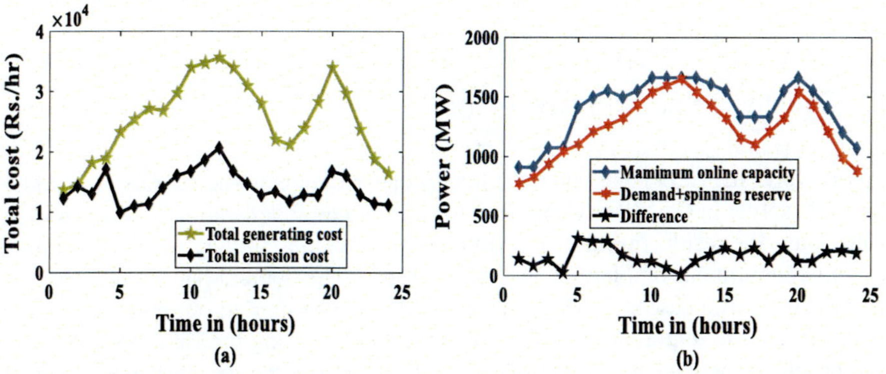

Fig. 4 Total hourly power generation and emission produced and comparison between available maximum online capacity of power units and also sum of power demand with spinning reserve of 10 thermal units **a** hourly power generation and emission produced **b** maximum online capacity of generating units and sum of power demand with spinning reserve

1500 MW during 12th hour presents the highest operating and emission cost, while demand of 700 MW during 1st hour presents the lowest operating cost. Therefore, it can be concluded that higher the load demand higher is the operating cost and vice-versa. From Fig. 4b, it is clearly justified that total online capacity of units thoroughly follows the power demand curve including spinning reserve except hours 15–17 where minimum up/down constraint of some generating units are binding. Difference between the MOC and demand considering spinning reserve is depicted in Fig. 4b. Observing unit commitment results in Tables 1 and 2, and it is clear that all the diverse constraints like minimum up and minimum down times, generator limits, spinning reserve, etc. are satisfied. The same system has also been solved now using ANTIGONE, KNITRO, COUENNE, LINDO, DICOPT and BARON. The

Table 3 Assessment of results on 10 thermal units with other MINLP solvers available in GAMS

Solvers/ Techniques	Cost Value (Rs.)	Emission value (Rs.)	Total objective value (Rs.)	Execution time (sec)
ANTIGONE	613,316.4	338,889.2	952,205.6	1536
KNITRO	613,299.7	338,811.5	952,111.2	1683
COUENNE	613,302.6	338,843.1	952,145.7	1490
LINDO	613,307.1	338,879.8	952,186.9	1584
DICOPT	613,312.3	338,833.7	952,146.0	1385
BARON	**613,296.9**	**338,802.4**	**952,099.3**	**1240**

comparison of results has been made in terms of total operating cost and execution CPU time for various solvers available in GAMS, as mentioned in Table 3.

It is evident that BARON solver provides better results in comparison with other MINLP solvers in terms of total operating cost and the execution CPU time taken.

5 Conclusion and Future Scope

In this work, IEEE 39 bus system having 10 thermal units is successfully solved by BARON solver in GAMS environment considering total generation cost along with emissions. The results reveal that BARON solver satisfies all the diverse constraints of unit commitment problem as well as emission. The authors have considered 10 thermal units as a case study, but this work can also be carried out for large-scaled test system. It is important to note here that when applying the BARON solver on UC problem, the execution time taken by the system is higher, this is because of MINLP nature of problem. Therefore, to handle this it is better to linearize all the nonlinear equations into linear one and then apply MIP solver like CPLEX, by doing this execution time taken for UC problem will be less as compared to BARON solver and this can be used as one important aspect in future research work.

Acknowledgements This work has been funded by the Council of Scientific and Industrial Research (CSIR), New Delhi, India under project grant 22(0815)/19/EMR-II of the Human Resources Development Group, sanctioned to second author.

Appendix

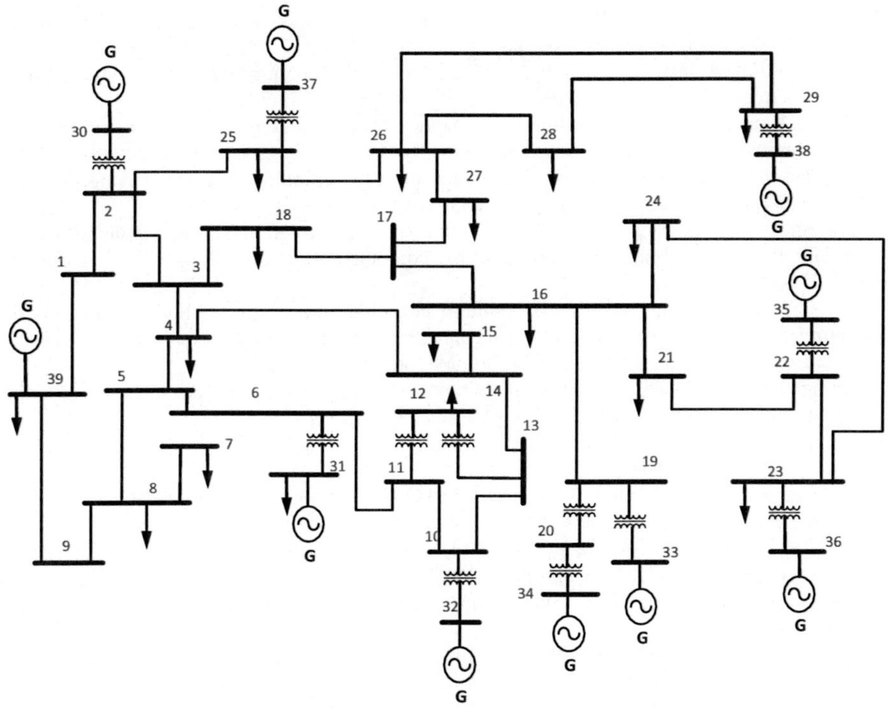

Fig. 5 Overall single line figure of IEEE 39 bus test case having 10 thermal generator buses

References

1. Allen JW, Bruce F (2007) Wollenberg: Power generation operation and control. 2nd edn. India: Wiley 131–155
2. Tang L, Che P (2013) Generator scheduling under a CO_2 emission reduction policy in the deregulated environment. IEEE Trans Eng Manage 60(2):387–397
3. Senthivadivu A, Gayathri K, Asokan K (2018) Exchange market algorithm-based profit-based unit commitment GENCOs considering environmental emissions. Int J Appl Eng Res 13(21):14997–15010
4. Hamdi Abdi (2020) Profit-based unit commitment problem: A review of models, methods, challenges, and future directions. Renew Sustain Energy Rev
5. Padhy NP (2003) Unit commitment problem under deregulated environment-a review. In: Proceeding of IEEE power engineering society general meeting, pp 1088–1094
6. Kumar V, Naresh R (2020) Application of BARON solver for solution of cost-based unit commitment. Int J Electr Eng Inf 12(4):807–827
7. Damousis IG, Bakirtiz AG, Dokopolus PS (2004) A solution to the unit commitment problem using integer coded genetic algorithm. IEEE Trans, Power Syst 19:1165–1172
8. Othman MNC, Rahman TKA, Mokhlis H, Aman MM (2015) Solving unit commitment problem using multi-agent evolutionary programming incorporating priority list. Arab J Sci Eng 40: 3247–3261
9. Jatinder Singh Dhaliwal (2018) Jaspreet Singh Dhillon: modified binary differential evolution algorithm to solve unit commitment problem. Electric Power Compon Syst 46(8):900–918

10. Zhu Y, Liu X, Deng R, Zhai Y (2020) Memetic algorithm for solving monthly unit commitment problem considering uncertain wind power. J Control, Autom Electr Syst 31:511–520
11. Ebrahimi J, Hosseinian SH, Gharehpetian GB (2011) Unit commitment problem solution using shuffled frog leaping algorithm. IEEE Trans Power Syst 26:573–581
12. Provas, Kumar Roy (2014) Ranadhir Sarkar: Solution of unit commitment problem using quasi-oppositional teaching learning-based algorithm. Int J Electr Power Energy Syst 60:96–106
13. Khorasani J (2012) A new heuristic approach for unit commitment problem using particle swarm optimization. Arab J Sci Eng 37:1033–1042
14. Lokesh Kumar, Panwara Srikanth Reddy, Ashu Verma B, Panigrahi K, Rajesh Kumar (2018) Binary grey wolf optimizer for large scale unit commitment problem. Swarm Evolut Comput 38: 251–266
15. Manikshetti MK, Kalage AA (2017) Optimum unit commitment using teaching learning-based optimization Algorithm. In: International conference on current trends in computer, electrical, electronics and communication (CTCEEC), Mysore, pp 267–272
16. Kumar V, Kumar D (2020) Binary whale optimization algorithm and its application to unit commitment problem. Neural Comput App 32:2095–2123
17. Pourjamal Y, Ravadanegh SN (2013) HSA based solution to the UC problem. Int J Electr Power Energy Syst 46:211–220
18. Saravanan B, Vasudevan ER, Kothari DP (2014) Unit commitment problem solution using invasive weed optimization algorithm. Int J Electr Power Energy Syst 55:21–28
19. Yuan X, Ji B, Zhang S, Tian H, Hou Y (2014) A new approach for unit commitment problem via binary gravitational search algorithm. Appl Soft Comput 22:249–260
20. Vikram, Kumar Kamboj (2012) Unit commitment in power system: a review. Int J Electr Power Eng 6(1):51–57
21. Mallipeddi R, Suganthan PN (2014) Unit commitment-a survey and comparison of conventional and nature inspired algorithms. Int J Bio-Inspired Comput 6(2):71–90
22. Vikram, Kumar Kamboj (2016) A novel hybrid PSO-GWO approach for unit commitment problem. Neural Comput Appl 27(6):1643–1655
23. Kumar V, Naresh R (2019) Amita Singh: solution approach to unit commitment problem using GAMS environment. In: International conference on issues and challenges in intelligent computing techniques (ICICT), IEEE, Ghaziabad
24. Kumar V, Naresh R (2020) Amita Singh: Scheduling of energy on deregulated environment of profit-based unit commitment problem. In: 9th Power India International Conference (PIICON), IEEE, Haryana
25. Partha P, Biswas P, Suganthan N, Gehan A, Amartunga J (2017) Optimal power flow solutions incorporating stochastic wind and solar power. Energy Convers Manage 148: 1194–1207
26. Kumar V, Naresh R, Singh A (2021) Investigation of solution techniques of unit commitment problems: a review. Wind Eng, February (2021)

Unit Commitment Including Wind and Hydro Generators Using DWOA

Ankit Uniyal and Ashwani Kumar

Abstract A proper hourly scheduling of thermal generation units (TGUs) satisfying variable hourly power demand results in minimized operating cost of the power network. This process of commiting and decommiting units and optimizing their schedules becomes a complex process when renewable energy resources (RERs) are included in the system owing to the emergence of large number of constraints. Hence proper selection of an optimization technique plays a significant role in obtaining the most optimized schedules. The present work aims at solving a TGU commitment problem for a transmission system having RERs like wind and pump storage systems considering minimization of operating cost along with emission cost. The proposed method has employed a metaheuristic technique namely Dual-Whale optimization algorithm (DWOA) and is implemented on a standard IEEE-30 bus test system.

Keywords Optimization algorithm · Pump storage · Transmission system · Unit commitment · Wind power

Symbols

d, T	Index for time and total no. of hours.
u, N_u	Index for thermal generation unit (TGU) and total TGUs.
$P_{u,d}, U_{u,d}$	Power generation using TGUs and on/off decision of TGU.
$Fu\text{-}Co, Em\text{-}Co, ST\text{-}Co, SH\text{-}Co$	Fuel, emission, start up, shut down cost of TGU.
a_u, b_u, c_u	Fuel cost coefficients of TGU.

A. Uniyal
National Institute of Technology Uttarakhand, Srinagar, India

A. Kumar (✉)
National Institute of Technology Kurukshetra, Kurukshetra, Haryana, India
e-mail: ashwani.k.sharma@nitkkr.ac.in

A. Kumar et al. (eds.), *Renewable Energy Towards Smart Grid*, Lecture Notes in Electrical Engineering 823, https://doi.org/10.1007/978-981-16-7472-3_18

A_u, B_u, C_u	Emission cost coefficients of TGU.
$\alpha_u, \beta_u, \gamma_u$	Start up cost coefficients of TGU.
H_d	Fuel cost scaling factor for dth time.
M, t_{off}	Time step duration and time for which units are off.
$PS_{p,d}, PS_{g,d}$	Motor and generator power in PS unit for dth time.
η_m, η_g	Efficiency of motoring and generation operation in PS unit.
$PS_{p,d,max}, PS_{p,d,min}$	Maximum and minimum motor power in PS unit.
$PS_{g,d,max}, PS_{g,d,min}$	Maximum and minimum generated power in PS unit.
$PL_d, d1$	Reservoir level at dth time and maximum reserve in %.
PL_{min}, PL_{max}	Minimum and maximum reservoir level.
$P_{w,d}, D_d$	Wind power and total power demand at dth time.
S_R, r	System up spinning reserve and reserve percentage.
$Pmin_u, Pmax_u$	Minimum and maximum TGU rating.
DS_u, US_u	Down and up spinning reserve contribution of TGU.
$DSmax_u, USmax_u$	Maximum down and up spinning reserve contribution of TGU.
$DSmin_u, USmin_u$	Minimum down and up spinning reserve contribution of TGU.
$Ton, Toff$	Minimum up and down time of TGU.
t, X_{rand}	Current iteration number and random position vector in WOA.
Δ, χ	Coefficient vectors.
X, X^*	Current and best position vector.
$r1$	Random number in [0, 1].
$\sigma, \sigma', \sigma''$	Distance between HW and prey in encircling, spiral and exploration modes.
B, L	Shape constant for spiral and random number in 0 to 1.

1 Introduction

The unit commitment (UC) is a sophisticated, significant, and large-scale process in power system which is used to analyze the optimal operation cost in time bound duration which may be weekly or daily considering different operational conditions while optimizing different objectives [1, 2]. In recent years there have been developments in the modeling and planning of RERs as alternatives to fossil-based generations which have aided in curbing of the environmental emissions [1].

The UC in the present times which the inclusion of RERs and energy storage options have become more complex owing to their rigorous operational constraints. Thus, there have been significant developments in the arithmetical and logical programming methods including heuristic optimization techniques to solve such rigorous UC problems. In years to come the complexity and dynamicity of UC problems will certainly rise and thus more efficient methods would need to be developed to cope with such large complexities.

The literature contains a lot of works in optimizing the UC of a power network. Nikzad et al. [1] have proposed a robust UC following GA-PL (TOAT) where emission costs have been minimized including energy storage systems. Verma et al. [2] have proposed a UC solution for wind-integrated hybrid system considering the minimization of operating cost and emission cost. In this work, the effect of pump storage along with wind generation has been taken into account. They have used GAMS software-based MINLP solvers to solve the UC problem. Li et al. [3] developed a model to perform combined UC of hydro and thermal generation units (TGUs), considering the hydrodynamic constraints with the aim of improving overall economy of power system operation. They optimized the watershed part using NL programming and used PL-based DP to perform UCs. Christiansen et al. [4] used a special type of GA to solve a UC problem, which not only consisted of basic GA operators viz crossover and mutation but five other special GA operators. Xiaohui et al. [5] formulated a UC problem based on profit under the deregulated electricity market (PBUC). They used an improved discrete binary PSO and the standard value PSO to solve this PBUC problem. Rajan et al. [6] solved the UC problem (short term) using tabu search considering banking and cooling constraints. Xiao et al. [7] solved the UC problem by dividing it into two parts, the first part solved the unit scheduling problem using ant colony optimization (ACO) and the second part solved economic dispatch problem using particle swarm optimization (PSO). Mori et al. [8] solved wind-hydro thermal coordination based UC using gravitational search algorithm. They dealt with equality constraints using pseudo-code-based algorithm to reduce the time consumption of the process. The wind scenarios are generated here using Monte Carlo simulation. Shukla et al. [9] proposed a multi-level UC for a regional power system. The interactions and coordination between transmission, active distribution, and microgrid systems have been proposed in this method. Ji et al. [10] have integrated wind generation uncertainties into frequency dynamic constraint UC considering the availability of plug-in electric vehicles and reserves.

All above works have proposed the wind and thermal UC problem, however very few have analyzed the effect of pump storage systems on the UC problem. Moreover, the works lack in exploitation of metaheuristic optimization techniques in utmost way. Once metaheuristic optimization parameters are carefully selected and modified as per the selected problem the global optima can be obtained reliably and even better than those obtained using commercial optimization solvers like GAMS that use conventional optimization methods.

Thus, the highlights of the present work are proposing an optimal 24 h UC problem considering the effect of wind energy system and pump storage plant and solving it using dual-whale optimization problem (DWOA) which is an amalgamation of continuous-WOA and binary-WOA in MATLAB®coding platform [13] which is a novel and recently developed technique. The proposed method has been tested on standard IEEE-30 bus system [2]. The results have shown that proposed DWOA method has performed better as compared to those obtained using CONOPT solver of GAMS software. The paper has been organized with problem formulation at the start followed by methodology used, then discussions of the results and finally significant conclusions have been drawn.

2 Problem Formulation

The optimal UC process sets the objective to minimize the total cost of the system. It comprises four parts as given in Eq. (1). The first part is the fuel cost of all GUs, second part represents total emission cost of the GUs, third and fourth part represents the start up and shut down costs of GUs. The optimization problem is subjected to constraints given from Eq. (2) to Eq. (19).

$$OB = min\left(\sum_{d=1}^{D}\sum_{u=1}^{Nu}\left[Fu\text{-}Co(P_{u,d})\cdot U_{u,d} + H_d\cdot Em\text{-}Co(P_{u,d})\right] + \sum_{d=1}^{T}\sum_{u=1}^{Nu}\left[U_{u,d}\cdot(1-U_{u,d})\cdot ST\text{-}Co_{u,d} + \omega_{u,d}\cdot SH\text{-}Co_{u,d}\right]\right) \tag{1}$$

$$Fu\text{-}Co = a_u + b_u\cdot P_{u,d} + c_u\cdot P_{u,d}^2 \tag{2}$$

$$Em\text{-}Co = A_u + B_u\cdot P_{u,d} + C_u\cdot P_{u,d}^2 \tag{3}$$

$$ST\text{-}Co = \alpha_u + \beta_u\cdot\left(1 - e^{\frac{-t_{off,u}}{\gamma_u}}\right) \tag{4}$$

$$PL_{d+1} = PL_d + M\cdot\left(\eta_m\cdot PS_{p,d} - \frac{PS_{g,d}}{\eta_g}\right) \tag{5}$$

$$PS_{g,d,min} \le PS_{g,d} \le min\left(PS_{g,d,max}, \eta_g\cdot\frac{PL_d}{M}\right) \tag{6}$$

$$PS_{p,d,min} \le PS_{p,d} \le PS_{p,d,max} \tag{7}$$

$$PL_{min} \le PL_d \le PL_{max} \tag{8}$$

$$\sum_{d=1}^{Nu}(P_{u,d}) \cdot (U_{u,d} + P_{w,d} + PS_{g,d} - PS_{p,d}) = D_d \tag{9}$$

$$\sum_{d=1}^{Nu}(P_{max,u,d}) \cdot U_{u,d} + P_{w,d} + PS_{g,d} - PS_{p,d}) \geq D_d + S_R + r\% \cdot P_{w,d} \tag{10}$$

$$P_{min,u} \leq P_{u,d} \leq P_{max,u} \tag{11}$$

$$\sum_{d=1}^{Nu}(US_u \cdot U_{u,d}) \geq S_R + r\% \cdot P_{w,d} \tag{12}$$

$$\sum_{d=1}^{Nu}(DS_u \cdot U_{u,d}) \geq r\% \cdot P_{w,d} \tag{13}$$

$$DR_u \geq P_{u,d} - P_{u,d-1} \geq UR_u \tag{14}$$

$$USmax_u = d1\% \cdot Pmax_u, DSmax_u = d1\% \cdot Pmin_u \tag{15}$$

$$US_u = min(USmax_u, (Pmax_u - P_{u,d})) \tag{16}$$

$$DS_u = min(DSmax_u, (P_{u,d} - Pmin_u)) \tag{17}$$

$$|Ton_{u,d-1} - Ton_{u,d}| \cdot |U_{u,d-1} - U_{u,d}| \geq 0 \tag{18}$$

$$|Toff_{u,d-1} - Toff_{u,d}| \cdot |U_{u,d-1} - U_{u,d}| \geq 0 \tag{19}$$

Equations (2)–(4) represent fuel, emission and start up cost of each generation unit u at dth hour. Equation (5) represents the available energy in the reservoir at $d + 1$th interval. Equations (6) and (7) represent the generation limits of the PS unit during its operation as a generator and motor, respectively. Equation (8) represents the storage capacity limit of the reservoir. Equation (9) represents the hourly generation-load balance of the system. Equation (10) represents the reserve constraint in the system. Here $r\%$ of wind power is kept as reserve. Equation (11) represents the power generation limits for TGUs. Equation (12) represents the system up spinning reserve constraint. Equation (13) represents the system down spinning reserve constraint. Equation (14) represents ramping up and down constraints of TGUs respectively. Equation (15) represents maximum down and up reserve contribution of TGUs. Equations (16) and (17) represent actual up and down spinning reserve contribution respectively of TGUs. Equations (18) and (19) represent the minimum up and down time constraints, respectively, for the TGUs.

3 Methodology

The present UC-based optimization problem is solved using Dual-WOA which is amalgamation of Binary-WOA (BWOA) and Continuous-WOA (CWOA). BWOA optimizes the TGUs ON/OFF status as 1 and 0, respectively. The CWOA optimizes the power generations by the TGUs as real numbers.

3.1 DWOA

The WOA, a metaheuristic technique [11], is based on hunting and searching pattern of humpback whales (HWs) for their prey. The HWs forage by creating upward spirals and double loops around their prey and thus get maneuvered toward their prey. The movement of HWs involves encircling target, exploitation phase (bubble-net attack), and exploration phase (search for target). Encircling target: Although the target position is known a priori to HWs, but in algorithm the global optima (GO) is not known, thus WOA always assumes current best as the global optima which is updated after each iteration. This behavior is mathematically represented by Eqs. (20)–(21). Where Δ and χ are given using Eqs. (22)–(23).

$$\sigma = |\chi \cdot X^*(t) - X(t)| \tag{20}$$

$$X(t+1) = |X^*(t) - \Delta \cdot \sigma| \tag{21}$$

$$\Delta = 2 \cdot \delta \cdot r1 - \delta \tag{22}$$

$$\chi = 2 \cdot r1 \tag{23}$$

Exploitation phase: The bubble-net attack is modeled using shrinking encircling method and spiral updating method [11, 12]. In shrinking encircling method [11], value of δ is decreased from 2 to 0 linearly over course of iterations which decreases Δ, and now equation can be used to get updated position of HW. In spiral updating of HW position, a spiral equation is used to mimic helix-shaped movement of HWs which is given in Eq. (24).

$$X(t+1) = \sigma' \cdot e^{BL} \cdot \cos(2\pi L) + X^*(t) \tag{24}$$

$$\sigma' = |X^*(t) - X(t)| \tag{25}$$

HWs follow both shrinking circle and spiral-shaped path simultaneously, so a probability (pr) of 0.5 has been assumed to follow each of the mentioned paths.

Exploration: In this values of $|\Delta| > 1$ to force search agent to keep movement far away from a reference HW, this allows WOA to perform global search efficiently.

$$\sigma'' = |\chi \cdot X_{rand} - X| \tag{26}$$

$$X(t+1) = X_{rand} - \Delta \cdot \sigma'' \tag{27}$$

The WOA proposed by [12] works on continuous real numbers. The binary version of WOA (BWOA) was proposed by Kumar et al. [13] in 2020, where they modeled toggling between 0 and 1 by position updating after each iteration. The value of current bit is changed with a probability that is calculated in form of C_{st} variable which is a sigmoidal function of the distance between the position of prey and HW. The equations for updating the position of HWs, i.e., Equations (21), (24), and (27) after every iteration used in case of CWOA are replaced by Eqs. (28)–(33) for BWOA.

Shrinking encircle movement:

$$C_{st} = \frac{1}{1 + e^{(-10\cdot(\Delta\cdot\sigma - 0.5))}} \tag{28}$$

$$X(t+1) = \begin{cases} Complement(X(t)), & \text{if } rand < C_{st}. \\ X(t), & \text{otherwise.} \end{cases} \tag{29}$$

The modification done in bubble-net behavior of HWs makes use of Eqs. (30) and (31).

$$C'_{st} = \frac{1}{1 + e^{(-10\cdot(\Delta\cdot\sigma' - 0.5))}} \tag{30}$$

$$X(t+1) = \begin{cases} Complement(X(t)), & \text{if } rand < C'_{st}. \\ X(t), & \text{otherwise.} \end{cases} \tag{31}$$

The modification done in exploration phase is done using Eqs. (32) and (33)

$$C_{st}'' = \frac{1}{1 + e^{(-10\cdot(\Delta\cdot\sigma'' - 0.5))}} \tag{32}$$

$$X(t+1) = \begin{cases} Complement(X(t)), & \text{if } rand < C_{st}''. \\ X(t), & \text{otherwise.} \end{cases} \tag{33}$$

The pseudo-code for DWOA is shown in Fig. 1.

```
initialize the population of HWs: each search agent having binary and continuous variables.
calculate the score of each search agent
while (t<maximum iterations), do
  for1 (each search agent), do
    update the values of δ, Δ, χ, L and pr as given in eq.(20)-eq.(23).
    if1 (score < best score)
    update the score as best score.
    update the search agent as best agent.
    end if1
    for2 (each search agent), do
      if2 (pr<0.5)
        if3 (|Δ|>=1)
          select a random switch array (X_rand)
          update the search agent using eq.(27) (CWOA) and eq.(32)-eq.(33) (BWOA)
          else if3 (|Δ|<1)
          update the search agent using eq.(21) (CWOA) and eq.(28)-eq.(29) (BWOA)
        end if3
        else if2 (pr>=0.5)
        update the search agent using eq.(24) (CWOA) and eq.(30)-eq.(31) (BWOA)
      end if2.
    end for2
  end for1
  t=t+1
end while
Store and display the best score and best agent.
```

Fig. 1 Pseudo-code for DWOA

3.2 *Proposed Method*

The proposed DWOA method is shown using flowchart in Fig. 2. Firstly, system, pump storage, wind data are send as input. Then a population of HWs is initialized using Gen(i)hour for power generations of TGUs and Bin(i)units for ON/OFF states of the units. The important parameters related to PSP and wind units are updated and all the constraints are checked. Once the population satisfies all the required constraints the HW population is send to optimization subroutine. Till the stopping criteria are met, HWs position (Gen(i)hour and Bin(i)hour) are optimized. Once it is done, total cost (objective) and unit commitments for this hour (d) are stored. This process is repeated till all 24 h schedules are obtained and finally results are stored and printed if required.

4 Results and Discussions

The analysis of unit commitment considering the fuel and emission cost of thermal generators has been done for IEEE-30 bus system. The value of r, d1 has been assumed as 20% and 10%, respectively. The ramp up and down capabilities of system has

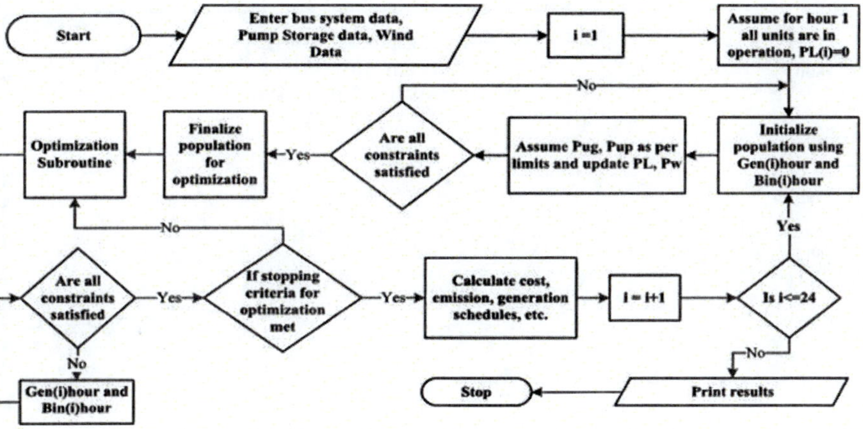

Fig. 2 Flowchart of proposed method

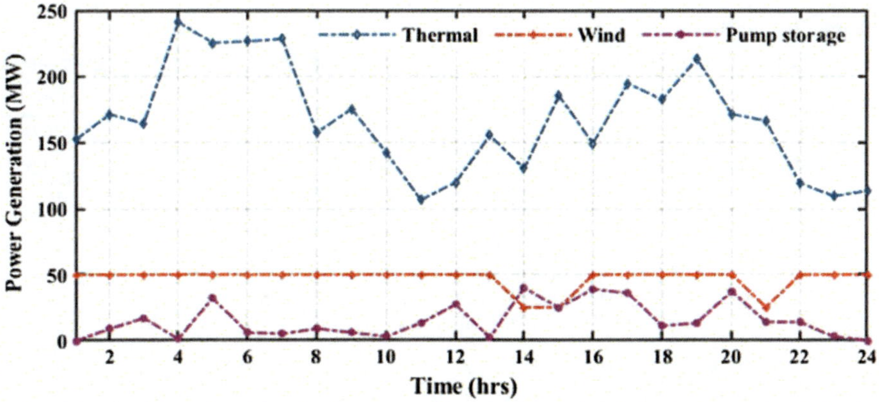

Fig. 3 Power generation profile for 24 h

been kept as 25%. For DWOA, maximum iterations and population size has been kept as 100 and 20, respectively. The TGU schedules obtained for 24 h is given in Table 1. The net emission (lb/hr) and fuel costs ($/hr) obtained using proposed DWOA, i.e., 13663.74 and 24643.49 are obtained less as compared to 15595.27 and 28304.93, respectively, using GAMS [2]. There is better utilization of all six TGUs using proposed method while TGUs 2 and 4 are set to off in all 24 h in GAMS. Figure 3 shows the power generation using TGUs, wind, and pump storage for 24 h using proposed method. Figure 4 shows the hourly fuel and emission costs using DWOA and GAMS. The hourly costs obtained in case of proposed DWOA are less as compared to GAMS. This shows that proposed technique has performed better in searching for global minima of given cost objective (OB).

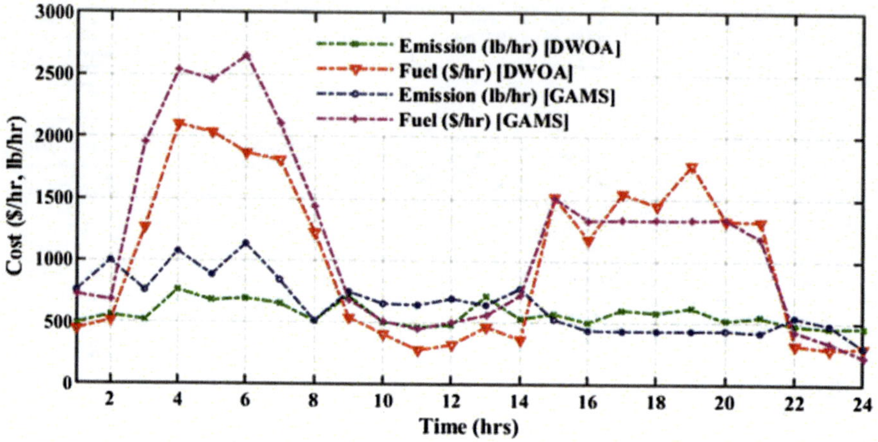

Fig. 4 Fuel and emission costs of thermal units for 24 h

Table 1 TGU schedules for 24 h

hr	TGU schedule (DWOA) [proposed]						TGU schedule (GAMS) [2]					
	G1	G2	G3	G4	G5	G6	G1	G2	G3	G4	G5	G6
1	50.6	50.5	18.3	0	19.8	12.8	72.28	0	32	0	25.67	26.04
2	53.7	59.3	24.5	0	19	14.5	83.44	0	40	0	30	32.55
3	50.1	54.4	30.3	0	29.2	0	68.26	0	50	0	30	40
4	60.5	72	43	0	25.5	40	85.33	0	50	0	30	40
5	55.4	76	43.3	0	28.2	22.1	75.55	0	50	0	30	40
6	57.3	67.9	46.2	0	23	31.6	88.23	0	50	0	30	40
7	53.3	73.1	44.3	0	28.4	28.9	73.22	0	50	0	30	40
8	50.6	52.2	30.1	0	24.1	0	54.91	0	43.98	0	30	40
9	64	66.9	20.7	0	23.4	0	68.64	0	43.36	0	30	40
10	51.1	54.6	23	0	13.3	0	65.98	0	32.58	0	22.5	30
11	51	39	16	18.3	22	27.2	66.71	0	25.6	0	22.19	22.5
12	50.8	47.4	20.8	30.7	28.7	20.7	68.86	0	32	0	23.53	25.6
13	64.5	66.6	23.9	0	0	0	64	0	40	0	24	32
14	54.1	53.7	22.2	0	0	0	69.22	0	50	0	30	40
15	50.7	68.8	40.1	0	0	25.4	51.92	0	50	0	30	40
16	50.8	50.9	26	0	0	21.3	45.91	0	50	0	30	40
17	51.5	75.8	40.6	0	0	26.1	45.65	0	50	0	30	40
18	50.9	79	21.1	0	0	31	45.63	0	50	0	30	40
19	53.3	68.8	31.6	0	25.3	34	45.64	0	50	0	30	40
20	50.7	49.8	30.2	0	24.8	15.5	45.78	0	50	0	30	40

(continued)

Table 1 (continued)

hr	TGU schedule (DWOA) [proposed]						TGU schedule (GAMS) [2]					
	G1	G2	G3	G4	G5	G6	G1	G2	G3	G4	G5	G6
21	51.9	57.5	34.1	0	22.5	0	48	0	37.5	0	28.5	30
22	50.5	49.3	19.2	0	0	0	60	0	28.12	0	21.37	22.5
23	50.2	39.3	19.5	0	0	0	57	0	21.09	0	16.03	16.87
24	50.2	44.3	18.5	0	0	0	45	0	15.82	0	12.03	12.66
Net Em-Co [DWOA]				**Net Fu-Co [DWOA]**			**Net Em-Co** [2]			**Net Fu-Co** [2]		
13663.74 lb/hr				24643.49 $/hr			15595.27 lb/hr			28304.93 $/hr		

5 Conclusion

In present work, UC problem of TGUS with inclusion of wind power and pump storage systems was mathematically modeled and solved using a metaheuristic technique DWOA. The results were compared with those obtained using concept solver of GAMS software which is a commercially popular optimization software. The net objective of emission and operating costs of TGUs were set as the objective. The emission cost was included using a penalty factor. The proposed DWOA approach provided more optimized TGU schedules with lesser costs as compared to GAMS-based approach. Thus, it could be said that if metaheuristic techniques are properly tuned they could be utilized in more accurate way than the commercial mathematical solvers like GAMS which use conventional optimization methods. In future, the proposed technique will be used for smart grid environment having multi-energy storage units and electric vehicles.

References

1. Nikzad HR, Abdi H (2020) A robust unit commitment based on GA-PL strategy by applying TOAT and considering emission costs and energy storage systems. Electr Power Syst Res 180:106154
2. Verma YP, Kumar A (2011) Economic-emission unit commitment solution for wind integrated hybrid system. Int J Energy Sect Manag (IJESM) 5:287–303
3. Li C, Hsu E, Svoboda AJ, Tseng CL, Johnson RB (1997) Hydro unit commitment in hydro-thermal optimization. IEEE Trans Power Syst 12(2):764–769
4. Christiansen JC, Dortolina CA, Bermudez JF (2000) An approach to solve the unit commitment problem using genetic algorithm. In: PES summer meeting, vol 1, pp 261–266
5. Xiaohui Y, Yanbin Y, Cheng W, Xiaopan Z (2005) An improved PSO approach for profit-based unit commitment in electricity market. In: IEEE/PES transmission & distribution conference & exhibition, pp 1–4
6. Rajan CCA (2006) An evolutionary programming based tabu search method for unit commitment problem with cooling-banking constraints. In: IEEE power India conference

7. Xiao G, Li S, Wang X, Xiao R (2006) A solution to unit commitment problem by ACO and PSO hybrid algorithm. In: Proceedings of the 6th world congress on intelligent control and automation, vol 2, pp 7475–7479
8. Mori H, Ohkawa K (2008) Application of hybrid meta-heuristic method to unit commitment in power systems. In: Electric power and energy conference
9. Shukla A, Singh SN (2014) Cluster based wind-hydro-thermal unit commitment using GSA algorithm. In: IEEE/PES general meeting conference & exhibition
10. Ji X, Zhang Y, Han X, Ye P, Xu B, Yu Y (2021) Multi-level interactive unit commitment of regional power system. Electr Power Energy Syst 125:106464
11. Mirjalili S, Lewis A (2016) The whale optimization algorithm. Adv Eng Softw 95:51–67
12. Kumar V, Kumar D (2020) Binary whale optimization algorithm and its applications to unit commitment. Neural Comput Appl 32:2095–2123
13. MATLAB R2018a 1994-2018 The math works

Generation and Reserve Scheduling Under Frequency Linked Pricing Regime

Yajvender Pal Verma and Ashwani Kumar Sharma

Abstract Power system operators (SO) generally use reserve to manage the disturbances caused by the outage of power generating units and unscheduled load disconnections. The over-scheduling of Spinning Reserve (SR) increases the overall operating cost of electric utility whereas, under-scheduling results in load shedding in case of contingencies leading to security risk for the power system. This paper investigates the scheduling of electric power generation and SR finding the operating status and production level of all units over the operating horizon, considering the unit characteristics and system restrictions. The Generation Scheduling (GS) and SR have been obtained for a test system of Tamil Nadu state in India considering frequency prediction sensitive approach. This approach prefers the units with lower operating costs over the dearer units on the basis of predicted frequency and thus, reduces the overall system operating cost of the system. Simulations results give efficient schedule for generation and reserve with low computation time.

Keywords Deregulation · Generation scheduling · Spinning reserve · ABT · Optimization · GAMS

1 Introduction

Ancillary services (AS) play a key role in exerting security of power systems in competitive electricity market [1]. SR helps the SO to respond to the load-generation imbalance caused by varying generations of power units from their GS, their sudden outage and due to load forecasting errors. The availability of SR tries to mitigate the economic cost of interruptions due to occasional outages. Traditionally, the reserve in any system is kept equal to or greater than the capacity of the largest generating unit. This practice ensures that the there is no need of any load shedding in case of any generation unit's failure. However, this also does not ensure the secure operation

Y. P. Verma (✉)
UIET Panjab University Chandigarh, Chandigarh 160014, India

A. K. Sharma
National Institute of Technology, Kurukshetra, Haryana 136119, India

A. Kumar et al. (eds.), *Renewable Energy Towards Smart Grid*, Lecture Notes in Electrical Engineering 823, https://doi.org/10.1007/978-981-16-7472-3_19

of the power system in the occurrence of sudden outage of more than one generating units [2]. Although, the overall risk of not meeting the load demand and the probability of an unprotected contingency can be reduced by selecting an appropriate value of SR but, this continuous provision of maximum SR is costly as it requires the commitment of additional generating units all the times, which raises both the operating cost and wastage of resources of electric utility. The SO would therefore set the SR on the basis of risk index to ensure that risk for the system during operation remains within acceptable level.

A number of optimization techniques have been proposed for solving the SR problems which are categorized as probabilistic and deterministic methods. Ref. [3] presented an approach to determine SR considering the probabilistic nature of outage. In this approach, it is required to achieve unit commitment (UC) risk during the GS by means of increasing SR provision with uniform level of risk. The problem associated with this technique is that the SR is not optimized itself, but instead, it simply varies the SR until the desired level of risk is achieved. In [4], a new approach that considered the unit ramp rate constraints to schedule the required SR in the power system is presented through Lagrangian Relaxation (LR) method. Here, the assessment of the reserve is done using probability to meet the specified risk index. The load forecast intermittency and reliability of generating units at each time instant is suggested to evaluate the risk index in [5].

Reserve Management strategy has been discussed in restructured power system considering loads reliability [6].The risk index value is obtained on the basis of trade off between the cost of Expected Energy Not Served (EENS) and the reserve cost To quantify the risk at all the time instant, an approximation of Loss of Load Probability (LOLP) is integrated in UC optimization problem [7]. The desired LOLP level is achieved by estimating the optimal value of reserve required in the system and it can be approximated using an exponential function. Two reliability matrices, ELNS and LOLP have been incorporated to implicitly set the reserve requirements. A factor of penalty is introduced in the objective function to avoid the problem of finding optimal value of ELNS ceiling and avoiding the utilization of ELNS and LOLP ceiling [8]. The two conflicting objectives i.e. the minimization of the reserve power and EENS have been combined for optimization. The reserve cost increases as the cost of interruption decreases. Similarly the more provision of reserve also tends to reduce the cost of EENS. A cost/benefit analysis based approach to compute SR is presented in [9]. The Reserve requirement in presence of renewable and role of storage in providing reserves has been discussed in [10].

In most of the above discussed UC problems, a fixed amount of reserve is kept through constraints in any optimization problem. For each power system, this fixed value of SR is computed to attain a desired level of risk. This technique gives sub-optimal level of reserve as the balance between the cost of SR and the cost of interruption that average consumer serves is not attained. Therefore, it can allocate more SR which is not economically justifiable or less SR which may be insufficient and endanger the power system security. The higher level of SR reduces the risk associated with the outage of generating units but in that case the cost associated with SR is usually greater than the operating cost of the generating units; therefore economically SR is required to be set at low level. As a result, a balance between these two SR levels is required to obtain i.e. optimal SR value/level; so that the power system operates at minimized total system cost by attaining a target level of risk.

This paper proposes a solution to obtain optimized generation schedule and SR level on the basis of frequency prediction using ANN under ABT environment. The cost/benefit analysis has been performed internally for each generating units. The optimal reserve is obtained considering the cost associated with EENS and VOLL.

2 Problem Formulation

2.1 Frequency Linked Pricing Mechanism

The grids in India were operating in unsatisfactory manner prior to the introduction of Availability Based Tariff (ABT) mechanism. Under this system of scheduling or dispatch, both the generator and beneficiary are required to pre-commit to day-ahead schedules. ABT scheme is basically for unscheduled interchange (UI) now known as deviation settlement charges (DSC) of energy. Whatever energy is taken from the central generating units is paid according to three charge components viz: Capacity, energy and DSC charge. DSC charges are imposed at the frequency sensitive rate for UI. The DSC variations with frequency are shown in Fig. 1. Therefore, ABT helps to regulate frequency and thereby ensure grid discipline. The utilities get an opportunity to either sell or buy power based on the prices depending upon frequency at that time [11].

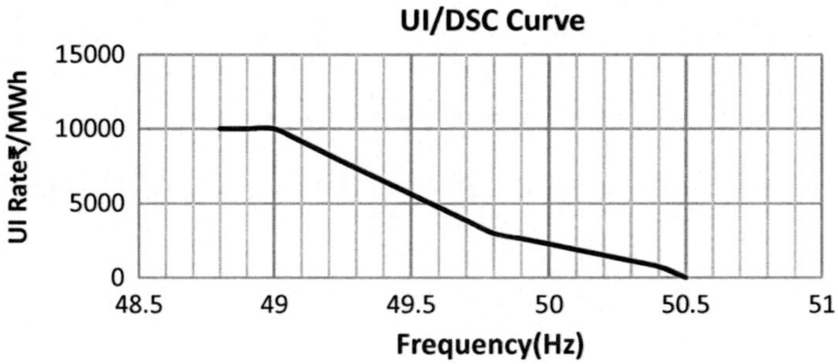

Fig. 1 UI curve

2.2 Objective Function

This paper proposes an approach to obtain optimized spinning reserve for a power system, with an aim to provide the maximum benefits to the state utilities by reducing overall operating cost and the cost of generation outage. To accomplish this aim, the socio-economic cost of outage along with the cost of spinning reserve has been considered in the objective function of UC problem. Therefore, proposed objective function is expressed as follows:

$$Minimize$$

$$\left\{\sum_t \left[\left(\sum_i (BC_i * PG_{it})\right) + CUI_t + CSI_t + RC_t + EC(R_t)_t\right]\right\} \forall n = 1, 2 \ldots N; \forall t 1, 2..24 \quad (1)$$

where,

PG_{it} = Power of ith generating unit at tth hour in MW. The different costs involved in the objective functions are discussed below. For each time interval 't', the total cost of DSC can be obtained as given below:

$$CUI_t = \sum_n PU_{nt} * UI(f_t) \ \forall n = 1, 2, 3 \ldots .N) \quad (2)$$

At each interval t, the cost of scheduled interchange can be obtained by the formula given below:

$$CSI_t = \sum_n PS_{nt} * SI_n \ \forall n = 1, 2, 3 \ldots .N \quad (3)$$

The cost of EENS can be formulated as given by (4) below:

$$EC(R_t)_t = VOLL * EENS_t \quad (4)$$

The cost of spinning reserve cost of ith unit at interval 't' ($RC_{i,t}$) can be expressed as:

$$RC_{i,t} = 1.2 * BC_i * SR_{i,t} \quad (5)$$

where, $SR_{i,t}$= SR obtained from ith generating unit in tth hour in MW.

$PU_{nt,}$ and PS_{nt} are UI and (SI) power of nth central generating unit in tth hour in MW.

SI_n= Scheduled interchange cost of nth central generating unit at time 't'.

The power generated by each generating unit is limited by the generation limits of each units. These are incorporated in the model by using (6) below:

$$PG_{max,i} \geq PG_{it} \geq PG_{min,i} \forall t = 1, 2, 3 \ldots .24 \tag{6}$$

where,

$PG_{min,i}$ = Minimum generation limit on ith generator in MW.
$PG_{max,i}$ = maximum operating capacity of ith generating unit in MW.

Frequency has been computed using ANN algorithm presented in [12] and used to calculate the frequency dependent UI charge ($UI\ (f_t)$) using as explained in [13]. The cost of reserve is more than the running cost of the generating units and decided by the SO. In this approach, it is assumed to be the 1.2 times the operating cost of the particular generating units. The optimization automatically computes the optimal value of SR that minimizes the sum of power system's operating cost, reserve cost and cost of expected energy not served EENS caused due to the outage of generating units or unavailability of sufficient power to ample the forecasted load demand.

2.3 System Constraints

Generation Load balance

The proposed objective function of obtaining SR is subjected to many constraints and the fundamental and the most prominent among them is to maintain balance between generation and the load demand of state utility at each time step as expressed by (7) and (8) below:

$$\sum_t \left(\sum_i PG_{it} + \sum_n P_{nt} + EENS_t \right) = \sum_{\mathrm{t}} \mathrm{PD_t} + \mathrm{P_{Loss}} \tag{7}$$

$$P_{nt} = PU_{nt} + PS_{nt} \qquad \forall\ \mathrm{n} = 1, 2, 3;\ \forall\ \ \mathrm{t} = 1, 2, 3, 4 \ldots 24 \tag{8}$$

where,

P_{Loss} = Total Power Loss in MW.
P_{nt} = The power exchanged in MW by nth central generating unit at time 't'.

Besides the system operating constraints, UI has also associated maximum and minimum constraints. This paper considers the situation of over-drawl of power, the maximum value UI is limited to ten percent of maximum operating capacity of unit owned by central government. In case of under-drawl, the units can deviate maximum of 10 percent negatively from SI and the restrictions are imposed by (9) below:

$$(0.1 * P_n) \geq PU_{nt} \geq -(0.1 * P_n)\ n = 1, 2, 3 \ldots .\mathrm{N};\ \forall \mathrm{t} = 1, 2, \ldots ..24 \tag{9}$$

SR Constraints

The following constraint is used to make sure that the minimum SR is supplied during scheduling at each time interval:

$$R_t - \sum_i SR_{i,t} \leq 0 \quad (10)$$

and

$$\sum_i SR_{i,t} \leq 0.1 * PG_{max,i} \quad (11)$$

where, $SR_{i,t}$ is the reserve contribution of ith unit at 't' hour.

The complete flow chart of the implementation of the proposed approach is given in the flow chart I Fig. 2.

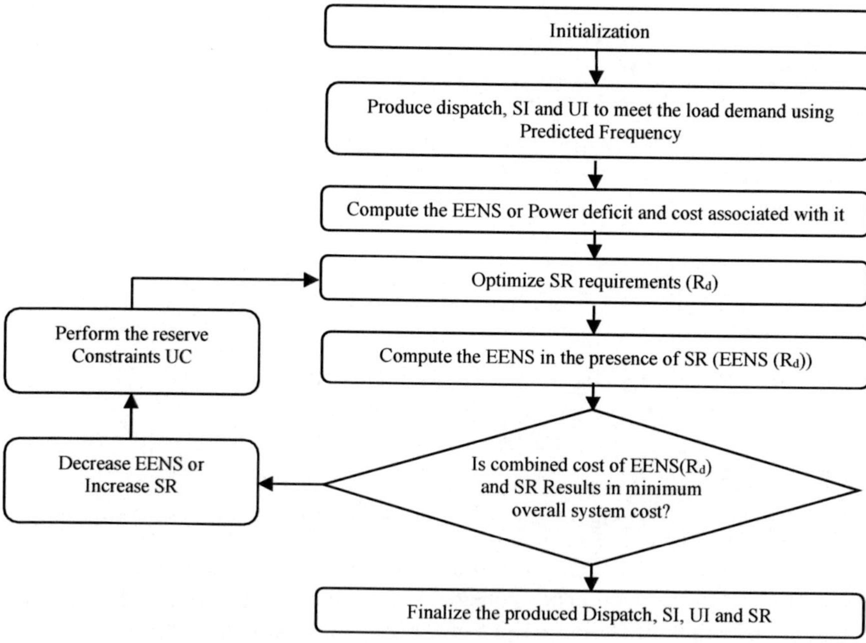

Fig. 2 Flowchart of proposed approach

3 Results and Discussion

The model proposed has been implemented on a test system of a state in India named Tamil Nadu which has 44 units owned by state, centre and Independent Power Producers (IPPs). The data used for the analysis of the proposed approach have been obtained from [14]. In this approach, the predicted frequency has been used as the parameter which greatly influences the selection of committed units, UI and optimal value of reserve so as to obtain optimal operating cost of the system for the state utility. In the proposed approach, out of the total generation capacity of central generating units 10 percent is kept reserve for UI and 10% of the generation capacity of other units is kept for reserve level selection. The scheduled interchange power of the state is kept 7000 MW and 700 MW of power is kept for UI from the central generating units with total generation capacity of the state generators at 10,140 MW.

Out of the total generation capability of state generators some power is reserved (10%), which is to be utilized in the situation of outage. Therefore, when the power demand is less than maximum generation capacity of the state utility (9369 MW), it is easily served by units owned by the state only and thus frequency is maintained close to the nominal value of 50 Hz. In this situation, the cost can be minimized by drawing power from UI pool or SR at the corresponding less prices. Therefore, optimization of generation schedule along with the spinning reserve not only helps in serving the load during outage but also helps in reducing the overall operating cost of the state utility. The system operation has been investigated under following two modes of operations.

(a) UC with SR Provision
(b) UC with SR Provision Considering Outage

3.1 UC with SR Provision

In this mode of operation, it is assumed that all the generating units are functioning accurately or there is no outage. Therefore, the first step is computation of an EENS on the basis of load demand and available capacity (state/IPPs/Central), which in this case would be same as of Power Deficit (PD). The VOLL has been assumed to be Rs. 2900/MWh. Using cost/benefit analysis the EENS and spinning reserve values are computed at each time instant.

Considering single period situation, at hour 4 the forecasted system load is 17600 MW. Figure 3 shows curves of cost of interruptions and system expected cost with varying system SR capacity. From this plot, it can be observed that as the SR level increases the cost of EENS cost decreases and the system operating cost comes out to be Rs. 20470749 where the SR capacity lies in the range 850 MW 900 MW. After this point, the system expected total cost tends to increase with the increase in SR capacity. This can be concluded that it is not appropriate to increase

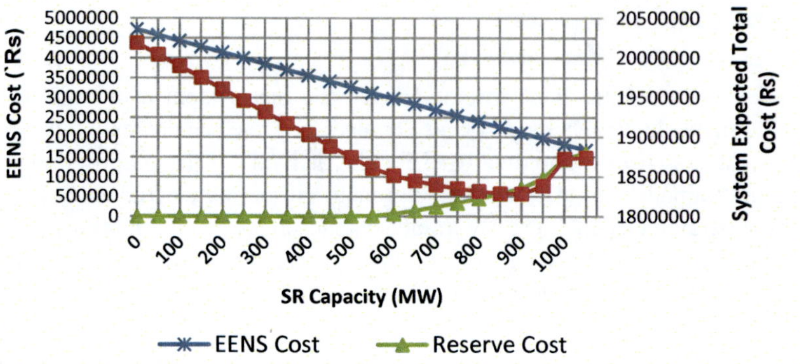

Fig. 3 Effect of Varying SR Capacity on EENS Cost and System Expected Cost

the value of SR beyond a limit with the motive to cut down the cost of the EENS. Therefore, using the proposed approach the optimal SR requirements is 870.1 MW, EENS cost is Rs. 2193560 with EENS of 756.4 MW and the system expected total cost is Rs. 20470749.

The spinning reserve capacities are plotted in Fig. 4 for whole day (24 h). The results show that the reserve capacities have flexible characteristics in terms of their selection procedure which is the cost/benefit analysis, rather than fixed deterministic criteria. This not only results in increased cost but also waste the available resources (energy). Therefore, the proposed method maximizes the social benefit. It can also be observed that the shape of SR curve is almost same to that of load demand, which indicates that all the SR capacities are utilized in the situation of peak load occurrence, and the system reliability can be well guaranteed.

It can be seen in Fig. 5 that at every period of horizon, the proposed UC approach with reserves attains lower EENS in comparison to UC problem without reserve provision.

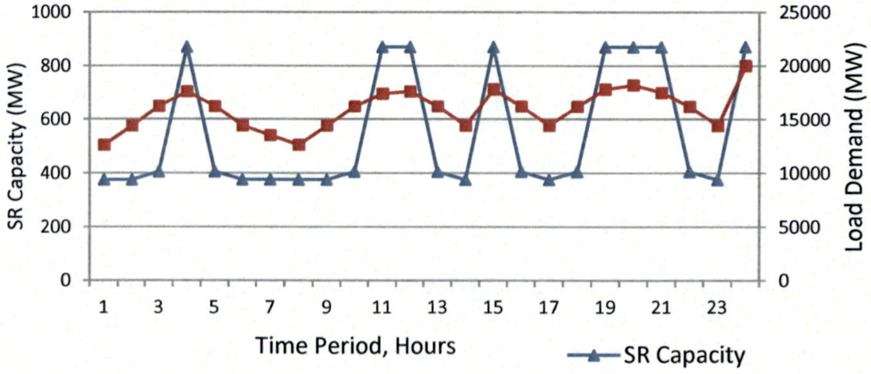

Fig. 4 Computed SR capacity and load demand over the testing horizon

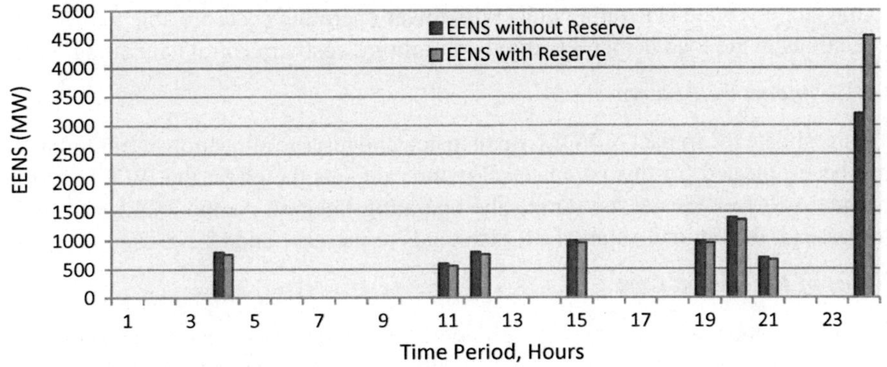

Fig. 5 Comparison Plot of EENS with and without SR provision

Effect of SR Provision on Generation Schedule

Figure 6 contrasts the schedule of committed generating units obtained during the investigation of UC problem, which is solved with and without SR provision and show that optimizing SR has significant effect on UC. However, no effect was observed in the commitment of central generating units even without reserve. This can be due

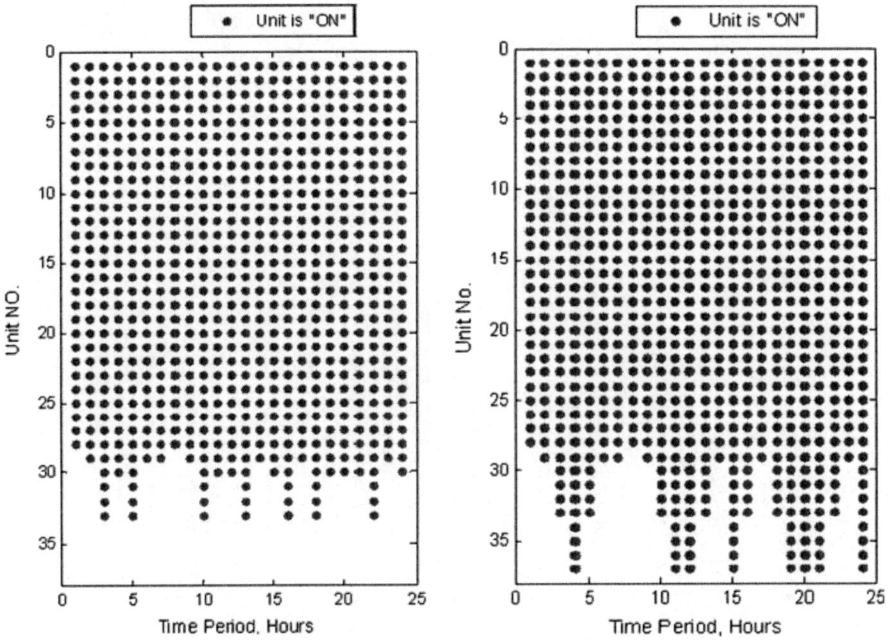

Fig. 6 Scheduling of state generating units with and without reserve

to the fact that state generating units with lower operating costs are able to meet the deviations in the load demand without committing costlier central generating units.

VOLL impact on Reserve

To investigate the impact of VOLL on dispatch and reserve allocation, the simulation has been repeated for this parameter. Figure 7 depicts the effect that VOLL on the optimal value of SR needed during the operating interval. As the VOLL (all in ₹) increases it the optimal value of SR also tends to increase and EENS decreases.

Effect of Reserve on Cost

The cost comparison for the UC problem with and without reserve is shown in Fig. 8. The total system cost per hour is either almost same or lower for UC with optimized SR requirement during the test duration. For lower VOLL, it is obtained

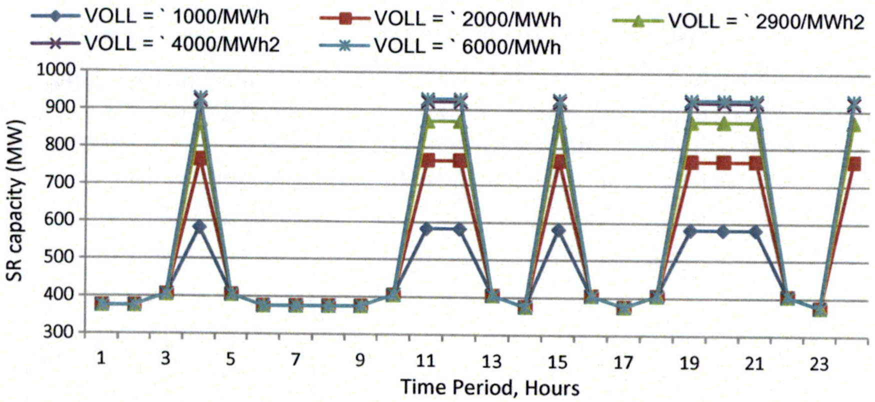

Fig. 7 SR Sensitivity to VOLL in ₹/MWh

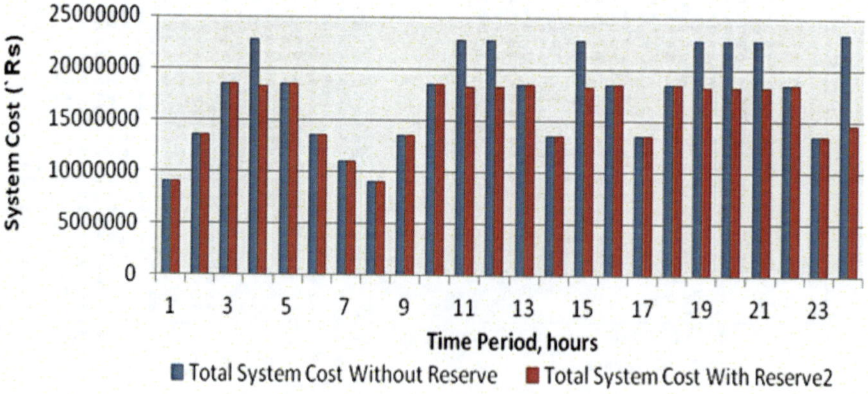

Fig. 8 Comparison of system cost with and without SR

by increasing the EENS cost and reducing SR, on the other hand for Higher VOLL the cost reduction is obtained by decreasing the cost of EENS and raising the SR using cost / benefit technique.

3.2 UC with SR Provision Considering Outage

The proposed model is also investigated to see the effect of the outage of generating units on the overall optimization of the GS and SR over a 24 h scheduling horizon. In this mode of operation, the VOLL is assumed to be ₹0.2900/MWh. Using cost/benefit analysis the EENS and spinning reserve values are computed at each time instant by considering the outage of different number of units.

Effect of outage on Generation Schedule and Reserve

In this case study, it is assumed that all the units are working properly except the unit no. 10 with maximum operating capacity of 240 MW. Under this situation, unit commitment with and without reserve is shown in Fig. 9. In both cases unit no. 10 is out of service due to failure over the 24 h and other expansive units are brought into service to compensate outage of this unit. The results also contrast the generation schedules obtained after solving the UC problem with SR and without SR provision.

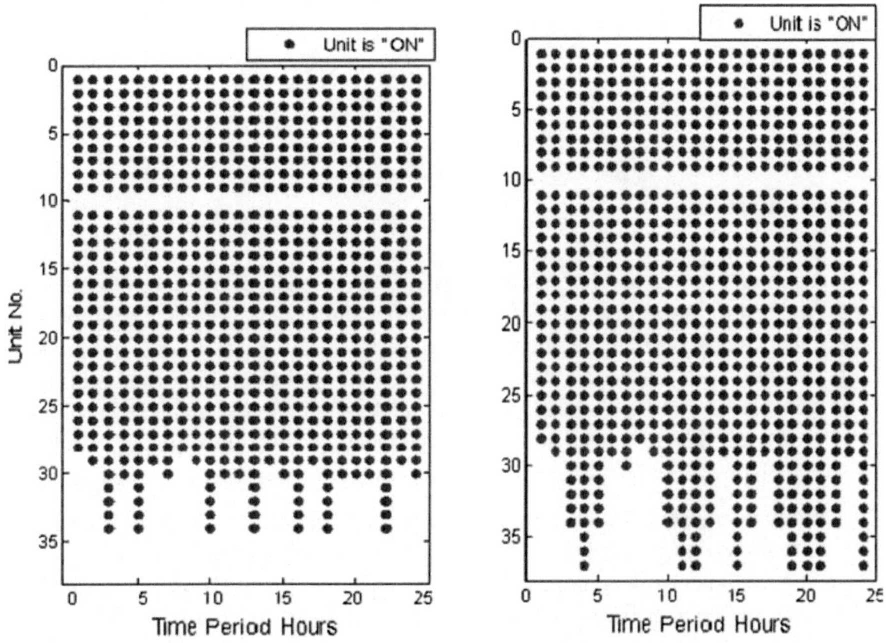

Fig. 9 Unit commitment with and without SR provision with outage

It is clear for this figure that optimizing the SR requirements in UC function can impact the UC and SR provision. In case of UC without the provision of SR the outage of a single generating unit is compensated either by putting on the expensive generating units or by load shedding at the particular time instant. On the other hand in case of the proposed UC with SR provision, this outage is compensated either by utilizing the SR or by bringing on the expansive generating units or by load shedding on the basis of social benefit analysis. Here, the threshold/maximum value of reserve is set at ten percent of the highest generation capacity of the units. Therefore, schedule of reserves can be changed according to the present situation to overcome the outage, after the analysis of benefits it provides.

Effect on EENS

Table 1 illustrates the effect that provision of reserve has on EENS. In the case of single unit failure, it is assumed that unit 10 with operating capacity 240 MW is not working. In case of double units' failure, the units 10 and 15 are assumed out of service with their operating capacities 240 MW and 124 MW respectively. In the situation of three unit failure, the units 10, 15 and 16 are assumed not working.

As stated above, the main motive of proposed approach with reserve provision is to minimize the load shedding and maximizing the benefits with its provision. Therefore, in case of failure of any generating unit the load shedding is increased by an amount approximately equal to the operating capacity of respective unit, if there is no provision of reserves. In the system having provision for reserve, the load shedding is less because of the utilization of power from the reserve poll. The effect of the VOLL on the EENS is also shown in the Table 1. As the VOLL increases the EENS decreases and SR increases with the aim to minimize the operating cost of the system.

4 Conclusions

In this paper, an optimal level of SR is computed on the basis of predicted frequency over all the testing horizons. The SR values are calculated on the basis of trade-off between the reserve cost and the benefits derived from it in terms of cost associated with interruptions or lost load. Therefore, this results in minimizing the combined cost of EENS and SR cost. The full EENS calculation is included in the UC problem formulation itself. The proposed approach has advantages over the other approaches, which are listed as:

- The optimal value of SR is determined implicitly by internally adjusting the EENS due to random outage until the cost of SR do not justify its provision.
- This approach is novel as it does not require cost/benefits analysis externally.
- The proposed approach can be applied to assess reserve level with large number of renewable and storage system in the system.

Table 1 Effect of reserve provision on EENS

	EENS Without Reserve (MW)			EENS With Reserve (MW) at VOLL Rs.2900/MWh		
Time (Hour)	With 1 unit failure (MW)	With 2 units failure (MW)	With 3 units failure (MW)	With single unit failure (MW)	With double units failure (MW)	With three units failure (MW)
1	0	0	0	0	0	0
2	0	0	0	0	0	0
3	0	0	0	0	0	0
4	1010	1136	1244	996.4	1136.4	1256.4
5	0	0	0	0	0	0
6	0	0	0	0	0	0
7	0	0	0	0	0	0
8	0	0	0	0	0	0
9	0	0	0	0	0	0
10	0	0	0	0	0	0
11	810	936	1044	796.4	936.4	1056.4
12	1010	1136	1244	996.4	1136.4	1256.4
13	0	0	0	0	0	0
14	0	0	0	0	0	0
15	1210	1336	1444	1196.4	1336.4	1456.4
16	0	0	0	0	0	0
17	0	0	0	0	0	0
18	0	0	0	0	0	0
19	1210	1336	1444	1196.4	1336.4	1456.4
20	1610	1736	1844	1596.4	1736.4	1856.4
21	910	1036	1144	896.4	1036.4	1156.4
22	0	0	0	0	0	0
23	0	0	0	0	0	0
24	3410	3536	3644	4796.4	4936.4	5056.4

Acknowledgements The authors are thankful to MHRD, Govt. of India vide letter no. 17-11/2015-PN.1 for providing funds to establish a Design and Innovative Centre (DIC) at UIET, PU Chandigarh.

References

1. Introduction of Ancillary Services in Indian electricity market, April, 25 2013. http://www.cercind.gov.in/2013/whatsnew/SP13.pdf
2. Anstine LT, Burke RE, Casey JE, Holgate R (1963) Application of probability methods to the determination of spinning reserve requirements for the Pennsylvania-New Jersey-Maryland Interconnection. IEEE Trans Power Appar Syst 82(68):726–735
3. Wu H, Gooi HB (1999) Optimal scheduling of spinning reserve with ramp constraints. Power Engineering Society 1999 Winter Meeting IEEE 2:785–790
4. Dalabeih DM, Allan RN (1991) Statistical determination of unit commitment risk level. In: Probabilistic methods applied to electric power systems. Third international conference on, pp 127–130, IET
5. Gooi H B, Mendesi DP, Bell KRW, Kirschen DS (1999) Optimal scheduling of spinning reserve. IEEE Trans Power Syst 14(4), November
6. Hashemi SZ, Mohammad H H, Azam Z (2017) Reserve management in restructured power system considering loads reliability. Eur J Eng Res Sci 2(12), December
7. Chattopadhyay D, Baldick R (2002) Unit commitment with probabilistic reserve. In Power Engineering Society Winter Meeting, IEEE 1:280–285
8. Bouffard F, Galiana FD, Conejo AJ (2005) Market-clearing with stochastic security—Part I: formulation. IEEE Trans Power Syst 20(4):1818–1826 Nov
9. Ortega-Vazquez MA, Kirschen DS (2007) Optimizing the spinning reserve requirements using a cost/benefit analysis. IEEE Trans Power Syst 22(1), February
10. Javadi MS, Lotfi M, Gough M, Nezhad AE, Santos S F, Catalão JPS (2019) Optimal spinning reserve allocation in presence of electrical storage and renewable energy sources. In: 2019 IEEE international conference on environment and electrical engineering and 2019 IEEEindustrial and commercial power systems Europe (EEEIC / I&CPS Europe), Genova, Italy, pp 1–6. https://doi.org/10.1109/EEEIC.2019.8783696
11. Verma YP, Kumar A (2013) Economic-emission load dispatch in renewable integrated system under availability based tariff (ABT) environment. Sustain Energy Technol Assess 4:78–88
12. Kaur S, Agrawal S, Verma YP (2013) Power grid frequency prediction considering the stochasticity of wind power using ANN. In: IEEE international conference on computational intelligence and communication networks (*CICN 2013*), 2013, GLA University, Mathura, India
13. Kaur S, Verma YP, Agrawal S (2013) Optimal generation scheduling in power system using frequency prediction through ANN under ABT environment. Frontiers Energy, Springer 7(4):468–478
14. Verma YP (2015) Investigation on operational aspects of power system with integration of renewable energy sources. PhD Thesis

Impact of Wind Generation Participation on Congested Power System

Smriti Singh, Atma Ram Gupta, and Ashwani Kumar

Abstract With the restructured power system operation in the competitive electricity markets there is need to optimally utilize the facility of the transmission system accommodating all transactions. But with load growth and addition of the renewable sources, the network may be congested. Wind Power Impact on Congestion Management is assessed in this article. Main contribution of the paper is to mitigate congestion and considering the effect of addition of the wind power plants in the system. The cost of producing power during congestion is taken as the cost function which is minimized taking into consideration the constraints. Wind power plant location bus is identified based on Bus Sensitivity Factor. Generation Rescheduling is considered to alleviate congestion in the transmission network. Modified IEEE 24 bus RTS system is considered for the result.

Keywords Deregulation · Electricity market · Wind generation · Congestion management · GAMS

1 Introduction

Electricity demand in today's world is increasing day by day. The stable economy is highly dependent on reliable and secure electricity service. The question is how the secure reliable and bulk electricity can be guaranteed. Along with individual consumers, industry professionals and business owners are more sensitive for the security in electricity grid. For instance, if failure in transmission network happens it must be removed instantly, because this failure may result in huge financial loss to the business owners. More-over during last 50 years' electricity is being generated from polluting sources like coal or oil which produces greenhouse gases and lead to global warming. The challenge ahead is thus to add more green and sustainable energy resources to generate electricity. These sustainable resources are hydro, wind, solar and biomass. Though occurrence of Congestion might be a short term phenomenon,

S. Singh · A. R. Gupta · A. Kumar (✉)
Department of Electrical Engineering, NIT Kurukshetra, Kurukshetra, Haryana, India
e-mail: ashwani.k.sharma@nitkkr.ac.in

A. Kumar et al. (eds.), *Renewable Energy Towards Smart Grid*, Lecture Notes in Electrical Engineering 823, https://doi.org/10.1007/978-981-16-7472-3_20

but if not dealt with on time can lead to drastic situations resulted from consecutive tripping of lines, this is what makes it an important issue. In De-regulated electricity markets this problem magnifies. Congestion management (CM) is the responsibility of the entity that operates and controls the interconnected transmission system. This entity is Independent System Operator (ISO).

The congestion can be removed by different methods. They are market based methods and non-market based methods. Further, CM depends on system condition and it may be either static congestion management or dynamic congestion management. The CM patterns have been categorized into three main different ways to tackle the network congestion [1–3]. The various congestion management methods are described and evaluated in [4, 5]. The non-market methods are considered as technical solutions which consist of outage the line, implementation of FACTS devices and operation of transformer tap changer to mitigate congestion [6]. The non-market method, as the name suggests does not involve market mechanism. The other type is the market based method. In order to manage congestion the Independent System Operator (ISO) takes the bids from the GENCOS [7, 8]. Many authors addressed the issues of congestion management (CM), based on the techniques of sensitivity factors, reactive power support, rescheduling of generators, demand and zonal based methods [4, 9]. Stochastic approach has also been used [10]. Authors presented the congestion management approach based on security in [11–13]. Verma and Kumar proposed congestion management with hydro-thermal system in [13]. Price based zonal methods are well proposed in [14, 15]. CM considering voltage stability of power systems was proposed in [16]. Demand elasticity and demand response can be adopted as one of the techniques for CM as demand can play an important role to control congestion [17–21]. Following are the methods for Congestion Management along with the research work done.

In this paper, impact of integration of the Wind Generation to manage Congestion in the system is proposed. Congestion cost is minimized taking the bids offered by the thermal and wind based plants. Wind Plant position bus is identified based on Bus Sensitivity Factor. Generation Rescheduling is considered to alleviate congestion in the transmission network. IEEE 24 bus RTS system (Modified) is considered. The problem formulated for congestion management has been solved in GAMS 23.4 [22].

2 Methods of Congestion Management

There are various possible ways to deal with problem of transmission network congestion; some of them are listed below:

- Generation Rescheduling
- Demand Rescheduling and Load Shedding
- Congested Line Outage
- Rerouting of Line Flows (Series reactive power compensation)

- Use of FACTS Controllers
- Use of various Demand Response Methods [23–25]

Out of the various Congestion Management techniques, which are listed above, line outage of congested line is not a much effective and practical solution, as it might lead to further system failure and other issues. In the real time, the generation rescheduling based Congestion Management is an effective solution to mitigate the Congestion. In this paper, generation scheduling based approach has been implemented, taking into consideration the influence of the Wind Power Generation on Congestion Management.

3 Pool and Hybrid Market Model in Deregulated Electricity Market

Electricity sector deregulation gave birth to three major components i.e. Gencos, Transcos and Discoms. There exists several market structures of which are two main market models are, Pool and Bilateral Contracts. The hybrid market model is where both the pool with bilateral and multilateral transactions exists. The bilateral contract model can be expressed as [13].

$$T \equiv \begin{bmatrix} GG\ GD\ GE \\ DG\ DD\ DE \\ EG\ ED\ EE \end{bmatrix} \tag{1}$$

In the bilateral contract model, mainly the transactions occur between Gencos and Discoms (GD). In Eq. (2) a bilateral contract between the supplier P_g and customer P_d is represented. The sum of row i s the total power produced by the generator i and the sum of column j is the total power consumed at load j.

$$T \equiv GD \equiv \begin{bmatrix} t_{11} & \cdots & t_{1,nd} \\ \vdots & \ddots & \vdots \\ t_{ng,1} & \cdots & t_{ng,nd} \end{bmatrix} \tag{2}$$

The P_g (power generation) and P_d (power demand) are vectors of two dimensional matrices given as:

$$\begin{bmatrix} P_d \\ P_g \end{bmatrix} = \begin{bmatrix} T^T & 0 \\ 0 & T \end{bmatrix} \begin{bmatrix} u_g \\ u_d \end{bmatrix} \tag{3}$$

where, u_g and u_d are matrices with dimension of n_g and n_d respectively and are the column vectors of ones. The AC model line flow can be expressed as below if $u_d = u_g = u$.

$$P_{line} = ACDF\left[P_g - P_d\right] \quad (4)$$

where, the distribution factors are ACDF [1].

Objective function is the minimization of the bilateral transactions deviations as:

$$\text{Minimize}\left\{\sum_i \sum_j b_{ij}\left(GD_{ij} - GD_{ij}^0\right)^2\right\} \quad (5)$$

Subject to: The power balances equations of real and reactive powers (6) and (7) respectively.

$$P_i = P_{gi} + P_{wi} - P_{di} = \sum_j^N V_i V_j Y_{ij} \cos\left(\delta_i - \delta_j - \theta_{ij}\right) \ \forall i = 1.2\ldots..N \quad (6)$$

$$Q_i = Q_{gi} + Q_{wi} - Q_{di} = \sum_j^N V_i V_j Y_{ij} \sin\left(\delta_i - \delta_j - \theta_{ij}\right) \ \forall i = 1.2\ldots N \quad (7)$$

The power balance equations are:

$$P_{db} = \sum_i GD_{ij} \quad (8)$$

$$P_{gb} = \sum_j GD_{ij} \quad (9)$$

$$P_g = P_{gb} + P_{gp} \quad (10)$$

$$P_d = P_{db} + P_{dp} \quad (11)$$

The bilateral power flow equation is given by (12).

$$P_{fb} = ACDF\left(P_{gb} - P_{db}\right) \quad (12)$$

The power flow equation for the power in pool is given by (13).

$$P_{fbp} = ACDF\left(P_{gp} - P_{dp}\right) \quad (13)$$

The equivalent power flow is given by (14).

$$P_f = P_{fb} + P_{fp} \quad (14)$$

Generators generate within its capacity of generation. The Eqs. (15) and (16) gives the real and reactive power generation capacities of thermal and wind generators.

$$P_{gi}^{min} \leq P_{gi} \leq P_{gi}^{max};\ \forall i \in G \tag{15}$$

$$P_{wi}^{min} \leq P_{wi} \leq P_{wi}^{max};\ \forall i \in W \tag{16}$$

The reactive power limits of the generator units are specified by (17) and (18) for thermal and wind generators respectively.

$$Q_{gi}^{min} \leq Q_{gi} \leq Q_{gi}^{max}; \forall i \in G \tag{17}$$

$$Q_{wi}^{min} \leq Q_{wi} \leq Q_{wi}^{max}; \forall i \in W \tag{18}$$

The voltage and angle limits on the lines are incorporated as given by (19) and (20) respectively.

$$V_i^{min} \leq V_i \leq V_i^{max} \tag{19}$$

$$\delta_i^{min} \leq \delta_i \leq \delta_i^{max} \tag{20}$$

The transaction limits between the seller bus-i and buyer bus-j is given by (21)

$$GD_{ij}^{min} \leq GD_{ij} \leq GD_{ij}^{max} \leq \min\left(P_{gi}^{max}, P_{dj}\right) \tag{21}$$

The Apparent Power (MVA) limit for the lines are expressed by

$$|MVA_{ij}| \leq MVA_{ij}^{max} \tag{22}$$

The line voltage variations are taken between 1.05 p.uto 0.95 p.u. The secure bilateral transaction matrix has been utilized in the congestion management model.

4 Congestion Management Considering Wind Generation

The total Congestion Management cost which is to be minimized consists of up cost and down cost of bid components of the thermal and wind generating units.

Minimize:

$$CC = \sum_{i \in g}\left(C_i^{up} \Delta P_{gi}^{up}\right) + \left(C_i^{down} \Delta P_{gi}^{down}\right) + \sum_{i=w}(C_{wi}^{up} \Delta P_{wi}^{up}) + (C_{wi}^{down} \Delta P_{wi}^{down}) \tag{23}$$

Subject to the constraints:
Equality Constraints:

$$\mathrm{P_i} = \sum_{j}^{N} \mathrm{V_i V_j Y_{ij} \cos(\delta_i - \delta_j - \theta_{ij}) \forall i = 1.2 \ldots\ldots\ldots N} \tag{24}$$

$$\mathrm{Q_i} = \sum_{j}^{N} \mathrm{V_i V_j Y_{ij} \sin(\delta_i - \delta_j - \theta_{ij}) \forall i = 1.2 \ldots\ldots\ldots N} \tag{25}$$

The Eqs. (24) and (25) give the real power and reactive power injections at bus i. The net up and down generations in managing congestion is given by (26).

$$\sum_{i}^{N} \Delta P_{gi}^{up} - \sum_{i}^{N} \Delta P_{gi}^{down} + \sum_{i}^{N} \Delta P_{wi}^{up} - \sum_{i}^{N} \Delta P_{wi}^{down} + \sum_{i}^{N} \Delta P_{di}^{up} - \sum_{i}^{N} \Delta P_{di}^{down} = 0; \forall i \in N \tag{26}$$

The generation and demand on bus i after rescheduling is given by (27)–(29).

$$P_{gni} = P_{gi}^{0} + \Delta P_{gi}^{up} - \Delta P_{gi}^{down}; \forall i \in G \tag{27}$$

$$P_{gwi} = P_{wi}^{0} + \Delta P_{wi}^{up} - \Delta P_{wi}^{down}; \forall i \in W \tag{28}$$

$$P_{dni} = P_{di}^{0} + \Delta P_{di}^{up} - \Delta P_{di}^{down}; \forall i \in D \tag{29}$$

The real and reactive power which is injected can thus be given by (30) and (31) respectively.

$$P_i = P_{gni} + P_{gwi} - P_{dni} \tag{30}$$

$$Q_i = Q_{gni} + Q_{gwi} - Q_{dni} \tag{31}$$

Inequality Constraints

The inequality constraints on thermal and wind generating units along with the demand.

$$\mathrm{P_{gi}^{min} \le P_{gi}^{0} + \Delta P_{gi}^{up} - \Delta P_{gi}^{down} \le P_{gi}^{max}; \forall i \in G} \tag{32}$$

$$\mathrm{P_{wi}^{min} \le P_{wi}^{0} + \Delta P_{wi}^{up} - \Delta P_{wi}^{down} \le P_{wi}^{max}; \forall i \in W} \tag{33}$$

$$P_{di}^{min} \leq P_{di}^{0} + \Delta P_{di}^{up} - \Delta P_{di}^{down} \leq P_{di}^{max};\ \forall i \in D \tag{34}$$

The limits of the units for reactive power are given by.

$$Q_{gi}^{min} \leq Q_{gi}^{0} \leq Q_{gi}^{max};\ \forall i \in G \tag{35}$$

$$Q_{wi}^{min} \leq Q_{wi}^{0} \leq Q_{wi}^{max};\ \forall i \in W \tag{36}$$

The limits voltage and angle are given by (37) and (38) respectively.

$$V_{i}^{min} \leq V_{i} \leq V_{i}^{max} \tag{37}$$

$$\delta_{i}^{min} \leq \delta_{i} \leq \delta_{i}^{max} \tag{38}$$

The power flow through the congested transmission lines should be limited to their MVA limit as given by (39)

$$P_{ij}^{2} + Q_{ij}^{2} \leq \left(S_{ij}^{max}\right)^{2} \tag{39}$$

The ramp rate up and down limits for all the generating units are considered in the analysis as:

$$0 \leq \Delta P_{gi}^{up} \leq \Delta P_{gi}^{up,ramp};\ \forall i \in G \tag{40}$$

$$0 \leq \Delta P_{gi}^{down} \leq \Delta P_{gi}^{down,ramp}\ \forall i \in G \tag{41}$$

$$0 \leq \Delta P_{wi}^{up} \leq \Delta P_{wi}^{up,ramp};\ \forall i \in W \tag{42}$$

$$0 \leq \Delta P_{wi}^{down} \leq \Delta P_{wi}^{down,ramp}\ \forall i \in W \tag{43}$$

5 Wind Power Generation

Since the renewable sources of energy are highly intermittent and sporadic in nature. Here we try to optimize the generation cost by placing wind at strategic location in the given power system. The output power from the wind turbine depends on the velocity of the wind at a given time instant. Equation (44) gives the output power from a wind turbine [26]:

$$P(v) = \frac{1}{2}C_p \rho A v^3 \tag{44}$$

where,

C_p: Power coefficient of the turbine

ρ: Area swept by the rotor blades of the turbine (in sq. m)

A: Density of air (1.226 kg/m^3)

v: Air velocity (m/s)

The value of C_p is limited to 0.59 known as Betz limit. This means a maximum of 59% of power present in the air mass can be extracted by any wind turbine.

5.1 *Wind Turbine Power Output Modeling*

The power output from the wind turbine as a function of wind velocity at various instants of time. The typical power output curve for a pitch controlled wind turbine is given in Fig. 1. The power output for a pitch-controlled wind turbine can be expressed generally as [27]:

$$\begin{aligned} &0 \ v\langle v_{ci} \cup v \rangle v_{co} \\ &P_W(v) = P_r\left(\frac{v - v_{ci}}{v_r - v_{ci}}\right) \ v_{ci} \le v < v_r \\ &P_r \ v_r \le v \le v_{co} \end{aligned} \tag{45}$$

where,

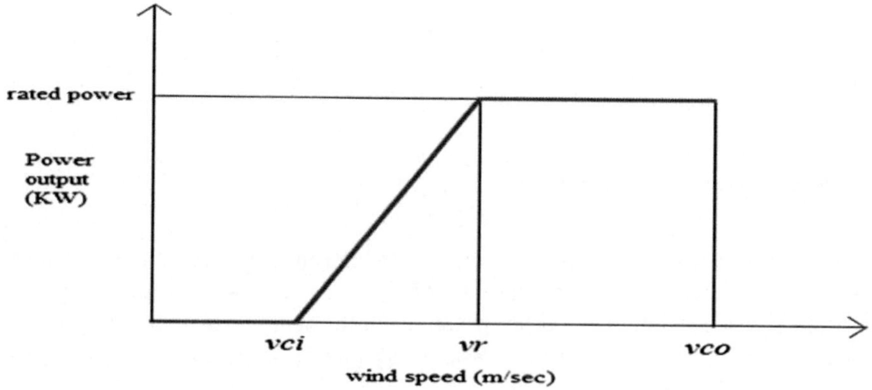

Fig. 1 Power output curve for wind turbine

P_w is the output power from the wind turbine in MW or KW

v is the velocity of wind in m/s

v_{ci}, v_{co} are the cut-in and cut-out velocity of the wind turbine in m/s

v_r is the rated velocity of the wind turbine in m/s

P_r is the rated output power of the wind turbine in MW or KW

Here P_{Wi} is the injected wind power at ith bus in the system.

5.2 Probabilistic Wind Speed Modeling

The power generation through wind depends on wind velocity which is variable with time. This behaviour of wind speed can be mathematically modelled by the Weibull probability distribution function. [28, 29].

$$f(v) = \frac{k}{c}\left(\frac{v}{c}\right)^{k-1} exp\left(\frac{-v}{c}\right)^{k}, 0 \leq v \leq \infty \tag{46}$$

where,

v: wind velocity in m/s

k: the Weibull shape parameter

c: the Weibull scale parameter in m/s.

The Weibull shape (k) and scale (c) parameters are calculated from the historical wind speed data for a given site.

$$k = \left(\frac{\sigma}{\overline{v}}\right)^{-1.086} \tag{47}$$

$$c = \frac{\overline{v}}{\Gamma\left(1 + \frac{1}{k}\right)} \tag{48}$$

where,

$\overline{v}$: the mean wind speed in m/s

σ: the standard deviation of the wind speed in m/s

A historical wind speed data for a period of study is obtained and from this, the mean and variance of wind speed can be determined. From this the value of k and c is obtained and thus a wind speed probabilistic model is obtained from (46). Monte

Carlo sampling (MCS) has been utilized to obtain the deterministic power output of the wind turbine.

5.3 *Wind Speed Samplings and Levels*

The samples of wind speed are obtained using MCS approach. To bring out a complete picture of probabilistic nature of wind 20,000 samples were drawn from Weibull probability distribution function. Further, wind levelling procedure is used in which the wind samples are divided into a number of levels. When a wind sample is chosen for a given level its corresponding wind turbine power output is also segregated into that level and average power for each level is calculated. The probability of occurrence for each level can be calculated using:

$$Prob_i = \frac{samples_i}{total\,samples} \tag{49}$$

where, $Prob_i$ denotes probability of a wind sample lying in ith level, $samples_i$ are the number of samples corresponding to the ith level and $totalsamples$ represent the total number of samples under consideration. The *actual power output* for each level can be calculated as:

$$actual\ power\ output_i = average\ power\ output_i * Prob_i \tag{50}$$

This *actual power output* for different levels is considered for Congestion Management calculation by placing the wind, so as to minimize the cost.

6 Results and Discussions

In this paper, Congestion Management analysis has been done on Modified IEEE-24 Reliability Test System with Thermal and Wind Generating Power Plants. The Congestion cost function is minimized using CONOPT solver in GAMS 23.4 [22, 30]. The modified IEEE 24 Reliability Test System has 24 buses, 17 loads and 38 transmission lines, 5 transformers and 12 generators. At bus 19, the wind generating unit is added. Line 7–8, 11–13 and 14–16 are assumed as the congested lines. Rating of 7–8 is taken as 150 MW instead of 175 MW, 11–13 is taken as 225 MW instead of 500 MW and 14–16 is taken as 226 MW instead of 500 MW.

GENCOs biddings are taken from [31]. Table 1 gives VESTAS-110 wind turbine.

Weibull distribution of wind velocity samples is shown in Fig. 2. Thirty-five levels of Wind velocities are taken and number of samples corresponding to each level is plotted. Further, wind generation in p.u MW is calculated and plotted corresponding

Table 1 Specifications for VESTAS-110 wind turbin

Rated Output power(P_r in MW)	Cut-in velocity (v_{ci} in m/s)	Rated velocity (v_r in m/s)	Cut-out velocity (v_{co} in m/s)	Shape parameter (k)	Scale parameter (c in m/s)
2.00	3	11.5	20	1.75	8.78

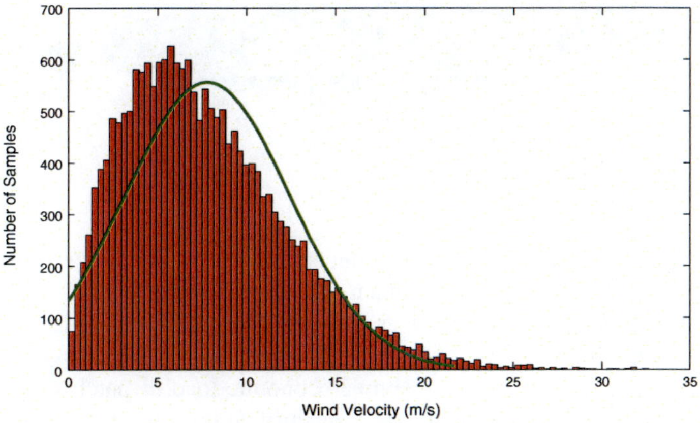

Fig. 2 Histogram for 20,000 Weibull distribution wind velocity sample

to velocity level. It is seen that it follows the same pattern as the probability distribution function. Figure 3 shows the wind active power generation. Assumption: speed level of 25 and number of turbines is 200. With this the power output from wind generation, result is obtained.

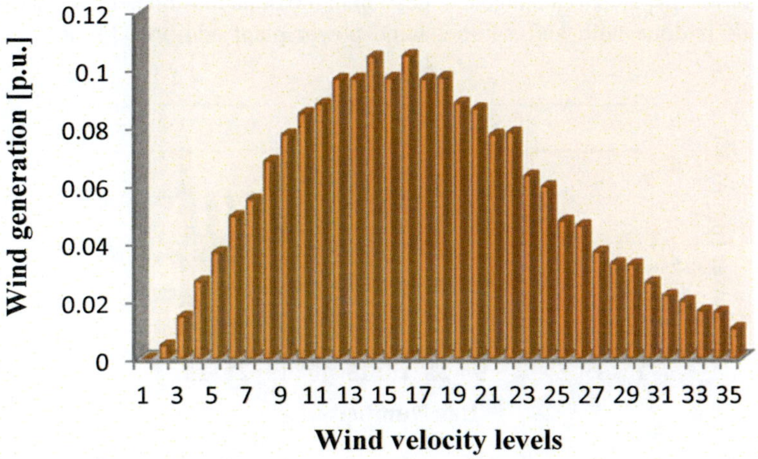

Fig. 3 Active power wind generation corresponding to each wind velocity levels in p.u. MW

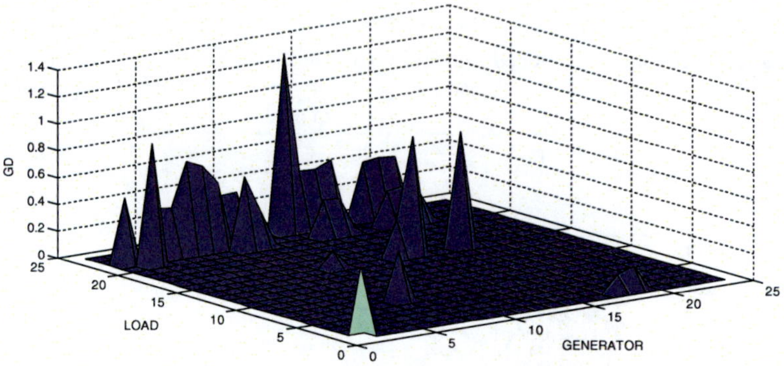

Fig. 4 Secure bilateral transaction between generator and load units

Now, at first payment to the power producers for alleviating congestion is calculated without addition of wind power plant. This is done with the help of two stage process. The Independent System Operator clears the market based on base case generation. The CONOPT solver of GAMS 23.4 using GAMS-MATLAB interfacing is used to minimize the fuel cost. A simple economic load dispatch solution gives the base case generations. Congestion in transmission lines is checked. If it occurs Independent System Operator tries to clear the market by collecting linear bids from the generators. Figure 4 shows the GD matrix for secure bilateral trading considering wind units in addition to thermal unit.

6.1 Generator Rescheduling Without Wind

Generator Re-dispatch to manage Congestion in the transmission network is considered. This is done with and without wind power plant addition. Figure 5 shows

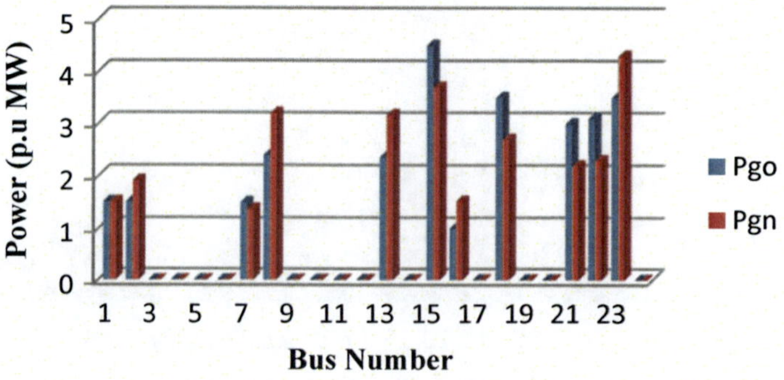

Fig. 5 New generation schedule for 3L congestion without wind

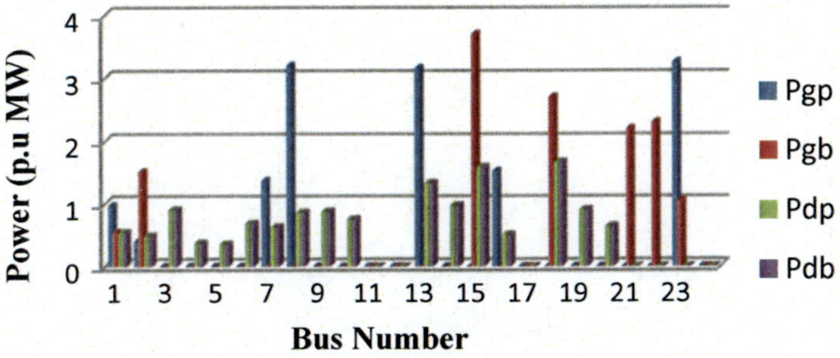

Fig. 6 Pool and bilateral generation and demand without wind

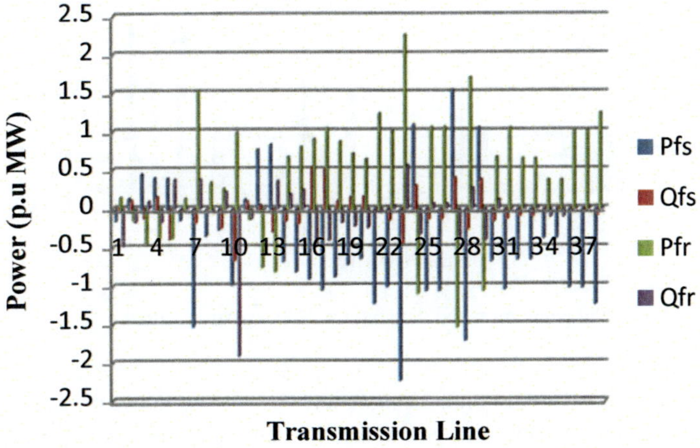

Fig. 7 Sending and Receiving end active and reactive Power flow in the transmission line without Wind

the power generation before and after the congestion is removed. It is seen that thermal generator at bus no. 2 and bus no. 23 contributes more while that at bus no. 7 generation reduces.

The pool and bilateral sharing of generator and the load after secure bilateral transaction is shown in Fig. 6.

Further, Fig. 7 depicts the active and reactive power flow at sending and receiving ends of the transmission lines. There are 38 transmission lines.

6.2 Generation Rescheduling with Wind

Now, similar procedure is followed by adding wind generation at bus no. 19. The wind power at wind speed level of 25 is considered. Total number of turbines taken

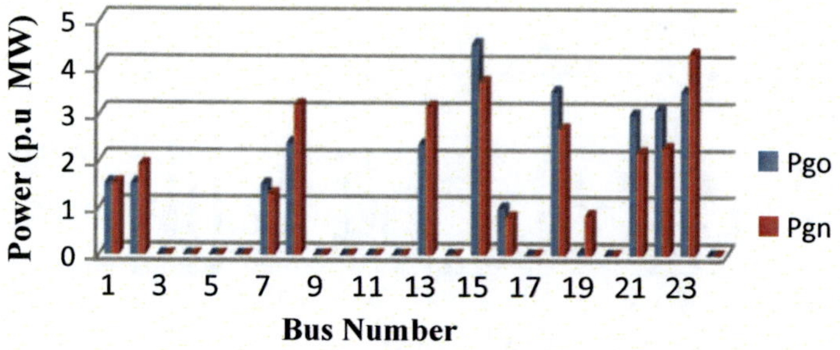

Fig. 8 New generation schedule for schedule for 3L congestion with wind

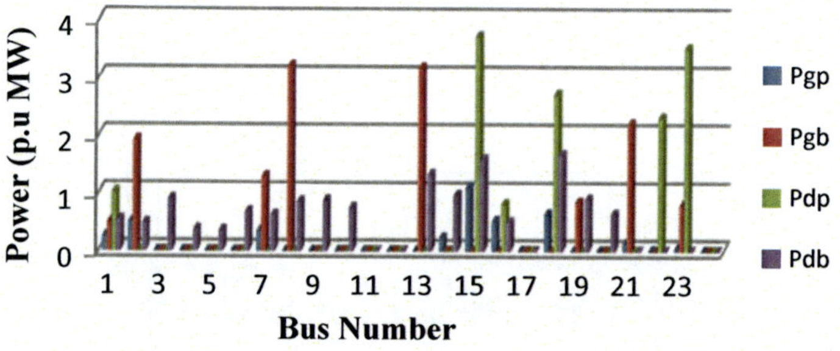

Fig. 9 Pool and bilateral generation and demand for 3L congestion with wind

is 200. With this wind generation Congestion Cost using generation rescheduling is obtained. Figure 8 shows the generation re-dispatches. We see, in addition to earlier rescheduling of thermal plants, here a contribution of wind generator reduces the overall cost paid to the power producers to manage congestion. Wind generator at bus no.19 produces 0.8484 p.u MW thus managing congestion in transmission network.

Figure 9 gives the amount of power generation up and down to remove Congestion. Here up and down generation of Wind is also taken. We see that an up generation of 0.8484 p.u MW is observed.

In the Fig. 9, the pool and bilateral generations and demands are shown. Figure 10 shows the pattern of active power flow and reactive power flow at sending and receiving end in the transmission line with the Wind.

6.3 *Congestion Cost and Power Loss Comparison*

Payment to the power producers for managing Congestion is compared. It is seen that without wind generation congestion cost is high 3521.91$/hr which reduces to 3260.29$/hr when Wind Generator is added to the system. Thus without considering

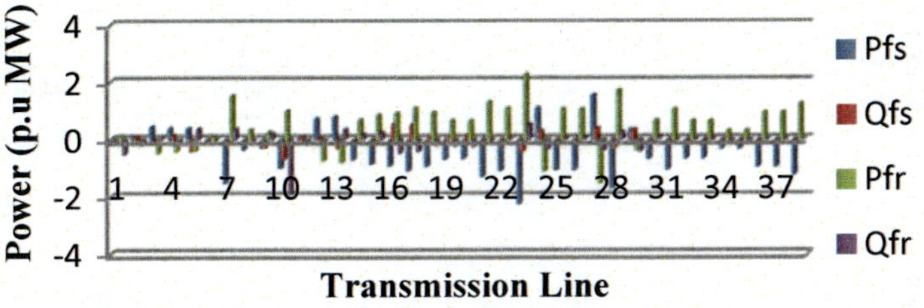

Fig. 10 Active and reactive power flow at sending and receiving end in the transmission line with wind

intermittent nature of wind and a constant wind velocity we observe that addition of wind brings down the Congestion Management cost as shown in Fig. 11.

Similarly, when power loss in transmission network is compared we observe that there is a drastic reduction in the active power loss when Wind Generation is included in the system. Thus, we can say renewable energy resource addition helps in alleviating Congestion at a lower cost, reducing the participation of thermal power plants and the active power loss also. Power loss in transmission network with and without Wind addition is shown in Fig. 12. It is observed that the power loss reduces

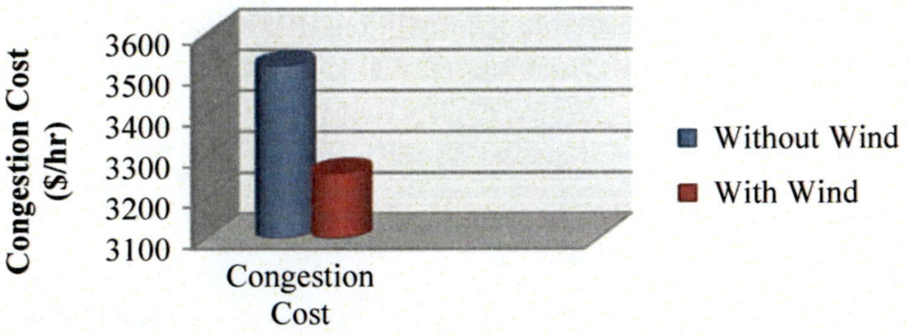

Fig. 11 Cost comparison for 3L congestion with and without wind addition

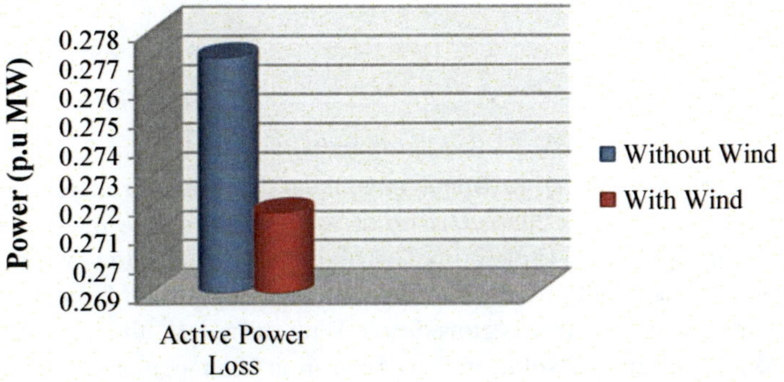

Fig. 12 Power loss in transmission network with and without wind addition

Table 2 Congestion cost and power loss for with and without wind

Analysis	Scenario	
	Without wind	With wind
Congestion cost ($/hr)	3521.91	3260.29
Active power loss (p.u MW)	0.277192	0.271878

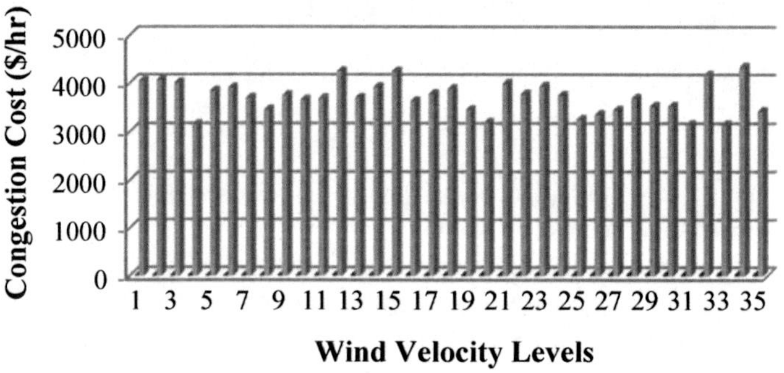

Fig. 13 Congestion cost variations as per wind levels

in the transmission system with wind power addition due to the lower power flow in the lines as wind power is available locally at bus 19. In Table 2. the Congestion Cost and power loss with and without wind are given.

6.4 Congestion Cost at Different Power Levels

Here all the results have been obtained for a wind level of 25. However, if wind speed is varied, power generation through it also varies and hence its contribution to grid changes. Thus, congestion cost for all velocity levels must be analysed. Figure 13 shows variation of Congestion Cost with the variation in velocity levels.

7 Conclusions

Wind penetration on Congestion Management has been analyzed for Multi line congestion case. Intermittent characteristic of wind is taken care of by generating wind velocities using Weibull Probability Distribution Function. However, in order to study particularly, the effect of wind participation on Congestion Management, only a certain range of wind velocity is considered. With the result, it is found that Wind proved to be an economical solution for Congestion Management in a supportive market structure. The cost paid to the power producers to minimize congestion is reduced when Wind Generation is used in addition to the thermal generation. Also since wind carries a part of load that is carried by thermal generators earlier, power loss in the transmission network is less.

References

1. Kumar A, Srivastava SC, Singh SN (2004) A zonal congestion management approach using ac transmission congestion distribution factors. Electr Power Syst Res 72(1):85–93 June
2. Shirmohammadi D, Wollenberg B, Vojdani A, Sandrin P, Pereira M, Rahimi F et al (1998) Transmission dispatch and congestion management in the emerging energy market structures. IEEE Trans Power Syst 13(4):1466–1476
3. Singh SN, David AK (2001) Optimal location of FACTS devices for congestion management. Electric Power Syst Res 58(2):71–79
4. Kumar A, Srivastava SC, Singh SN (2005) Congestion management in competitive power market: a bibliographical survey. Electric Power Syst Res 76(1–3):153–164 July
5. Acharya N, Mithulananthan N (2007) Locating series FACTS devices for congestion management in deregulated electricity markets. Electr Power Syst Res 77(3):352–360
6. Kumar A, Sekhar C (2012) DSM based congestion management in pool electricity markets with FACTS devices. Energy Procedia 14:94–100
7. Yamin HY, Shahidehpour SM (2003) Transmission congestion and voltage profile management coordination in competitive electricity markets. Electr Power Energy Syst 25(10), June
8. Singh AK, Parida SK (2013) Congestion management with distributed generation and its impact on electricity market. Electr Power Energy Syst 48:39–47 Jan
9. Esmaili M, Amjady N, Shayanfar HA (2010) Stochastic congestion management in power markets using efficient scenario approaches. Energy Convers Manage 51(11):2285–2293 March
10. Pillay A, Karthikeyan SP, Kothari DP (2015) Congestion management in power systems-a review. Electr Power Energy Syst 70:83–90 February
11. Conejo AJ, Milano F, Bertrand RG (2006) Congestion management ensuring voltage stability. IEEE Trans Power Syst 21(1):357–364 February
12. Bonab SMM, Rabiee A, Ivatloo BM (2015) Voltage stability constrained multi-objective optimal reactive power dispatch under load and wind power uncertainties: a stochastic approach. Renew Energy 85:598–609 July
13. Verma YP, Sharma AK (2014) Congestion management solution under secure bilateral transactions in hybrid electricity market for hydro-thermal combination, Electr Power Energy Syst 64: 398–407, August
14. Barbulescu, Kilyeni S, Cristian DP, Oprea DJ (2006) Congestion management using open power market environment electricity trading. In: IEEE 45th International UPEC, pp 1–6, September
15. Xiao Y, Wang P, Goel L (2009) Congestion management in hybrid power markets. Elect Power Syst Res 2009(10):1416–1423 May
16. Esmaili M, Shayanfar HA, Amjady N (2009) Congestion management considering voltage stability of power systems. Energy Convers Manage 50(10):2562–2569 July
17. Aalami H, Moghaddam MP, Yousefi G (2010) Demand response modeling considering interruptible/curtailable loads and capacity market programs. Appl Energy 87(1):243–250
18. Bompard E, Carpenato E, Chicco G, Gross G (2000) The role of load demand elasticity in congestion management and pricing. In: Proceedings of IEEE PES, summer meeting, vol 4, pp 2229–34, July
19. Gan D, Bourcier DV (2002) Locational market power screening and congestion management: experience and suggestions. IEEE Trans Power Syst 17(1):180–185
20. Smriti S, Kumar A (2017) Congestion management using demand response program. In: 2017 International conference on power and embedded drive control (ICPEDC), pp 83–88. IEEE

21. Smriti S, Kumar A (2017) Demand response program solution to manage congestion in transmission network considering uncertainty of load. In: 2017 8th international conference on computing, communication and networking technologies (ICCCNT). IEEE
22. GAMS (General Algebraic Modeling System) software package. http://www.gams.com
23. He T, Kolluri S, Mandal S, Galvan F, Rastgoufard P (2004) Identification of weak locations in bulk transmission systems using voltage stability margin index. In: iEEE international conference on probabilistic methods applied to power systems, pp 878–882, September
24. Deb S, Gope S, Kumar Goswami A (2015) Congestion management considering wind energy sources using evolutionary algorithm. Electric Power Compon Syst 43(7): 723–732
25. Vergnol A et al (2009) Real time grid congestion management in presence of high penetration of wind energy. In: 2009 13th European conference on power electronics and applications. IEEE
26. Sheen J, Tsai M, Wu S (2013) A benefit analysis for wind turbine allocation in a power distribution system. Int J Energy Convers Manage 68:305–312 April
27. Chang T, Liu F, Ko H, Cheng S, Sun L, Kuo S (2014) Comparative analysis on power curve models of wind turbine generator in estimating capacity factor. Int J Energy 73:88–95 August
28. Yu Z, Tuzuner A (2009) Fractional weibull wind speed modeling for wind power production estimation. In: IEEE, power & energy society general meeting, pp 1–7
29. Martin D, Zhang W, Chan J, Lindley J (2014) A comparison of Gumbel and Weibull statistical models to estimate wind speed for wind power generation. In: Power Engineering Conference (AUPEC), 2014 Australasian Universities, pp 1–6, 28 September 2014–1 October 2014
30. Gupta A, Yajvender PV, Chauhan A (2020) Wind-Hydro combined bidding approach for congestion management under secured bilateral transactions in hybrid power system. IETE J Res: 1–14
31. Singh SN, Erlich I (2006) Wind power trading options in competitive electricity market. In: IEEE Power Engineering Society General Meeting, June

Opposition-Based Competitive Swarm Optimizer for Optimal Sizing and Siting of DG Units in Radial System

Soumyabrata Das and Amar Kumar Barik

Abstract This paper endeavors the Opposition-based competitive swarm optimizer (OCSO) algorithm for solving the Dispersed Generation (DG) placement problems in the Distribution System (DS), proposing a multi-objective function. DS's different issues in the modern power system might be improved with optimal DG placement for the quality supply to the consumers. The authors have utilized the OCSO algorithm to find the optimal location and sizes of the DG in order to maximize the voltage stability index by minimizing the power losses and voltage deviation. Initially, the single objective problems are solved with proposed methods and then the multi-objective problems. The OCSO algorithm's effectiveness is tested on standard 33 bus and 69 bus Radial Distribution System. Thereafter, the results are compared with the original CSO algorithm and other reported methods to confirm the proposed method's superiority.

Keywords DG placement · Competitive swarm optimizer algorithm · Distribution system · Multi-objective problem

1 Introduction

The demand for electrical power is growing enormously due to the faster explosion in industrialization and population. Consequently, the supply of electrical power generation is also grown massively with various new dimensions. However, Thermal Power Plants (TPP) play the leading role in alleviating the energy demand until today. It becomes difficult for TPP with the growing need to satisfy the power quality of the supply [1]. The Dispersed Generation (DG) plays an essential part in the power distribution system to achieve the broader commercial, technical and environmental importance worldwide towards minimizing the overpower demand. Moreover, it becomes necessary to supply the additional power demand with some eco-friendly

S. Das (✉)
Motilal Nehru National Institute of Technology Allahabad, Uttar Pradesh, Prayagraj, India

A. K. Barik
Gandhi Institute for Technology, Bhubaneswar, Odisha, India

A. Kumar et al. (eds.), *Renewable Energy Towards Smart Grid*, Lecture Notes in Electrical Engineering 823, https://doi.org/10.1007/978-981-16-7472-3_21

sustainable/renewable power-based DG [2]. However, to reduce environmental pollution and waste proliferation, the DG should be designed with waste-to-energy-based sustainable hybrid systems in the form of isolated [1] or interconnected microgrids [2]. DG usually consists of a small decentralized power generation plant, mostly located close to the customer side. DG is regularly distributing the power generated from the small generating units to the Distribution System (DS) [3]. The assistances of DG are depended on the locations, sizes and types of connection [3]. Moreover, if their placements of DG are not proper, then the DG reimbursement may have limited effect in improving the system voltages and reducing the power losses [3]. Therefore, identifying optimal DG positions and sizes in DS is vital and challenging for power system research.

Since DG placement is a real-time complex optimization problem, many researchers have tried to solve the DG placement problem. There are various optimization techniques used to solve different objective functions. Earlier times, analytical methods [4] were utilized to solve the DG settlement problem. Gradually, the Power DS networks became more complicated due to growing load demand and expansion of existing lines. Therefore, the numerical methods [5, 6] are preferred in solving the DG placement problems. Such methods are healthy and recursive, which are having exceptional characteristics of convergence to solve linear optimization problems. However, initial speculation is the key issue with this approach. Such techniques mainly depend on the initial guess, which may be trapped into local optima instead of global ones.

The heuristic method becomes more accessible to solve DG placement problems. It produces a fast and concrete strategy that provides quick and hands-on policies that diminish the extensive search space and lead to an optimal solution. Such approaches have the property to solve the location and size issue simultaneously. Till today several meta-heuristic algorithms are used to solve the DG settlement problems, such as Genetic Algorithm (GA) [7], Particle Swarm Optimization (PSO) [8], artificial bee colony [9], ant colony optimization [10], Teaching Learning-Based Optimization (TLBO) [11], Elephant herding optimization [12], wale optimization algorithm [13].

This article presents an Opposition-based Competitive Swarm Optimizer (OCSO) algorithm to solve the DG settlement problem. The OCSO algorithm's main advantage is its capability to identify a universal best solution in a small interval with fewer free parameters. The proposed methods are experienced on standard 33 bus and 69 bus Radial Distribution System (RDS) networks. The proposed algorithm's performance is matched with some other reported optimization methods to confirm its effectiveness and superiority. The prime focus of this work is:

(a) To include a sustainable DG with Solar photovoltaic (SPV), Biogas, and Biodiesel generators.
(b) To formulate a multi-objective-based cost function considering minimization of power loss and Voltage Deviation (VD) with simultaneous Voltage Stability Index (VSI) maximization.
(c) To test the proposed method on standard 33 bus and 69 bus RDS networks.

(d) To verify the proposed algorithm's effectiveness and superiority by comparing the performances with some other reported optimization techniques.

2 DG Structure

A stochastic DG [14] of 2000 kW including SPV arrays, Biogas generators (5 × 400 kW) and Biodiesel generators (4 × 250 kW) units are considered for real power injection to connected RDS in this work. A schematic configuration of the projected DG unit is demonstrated in Fig. 1. The SPV units are expected to inject real power during the availability of sunlight (in normal day scenario), whereas the biogas generators take its place during unavailability of SPV generation (in bad weather or night time scenario) [1]. However, the biodiesel generators support both of them to meet the desired active power demand injection to RDS.

The bio-generators (biogas/biodiesel) are expected to utilize local bio-degradable wastes for power generation with the principle of waste to power [1]. In this way, the proposed DG will inject eco-friendly renewable energy into the RDS. The concerns for optimal DG settlement are discussed in the following sections.

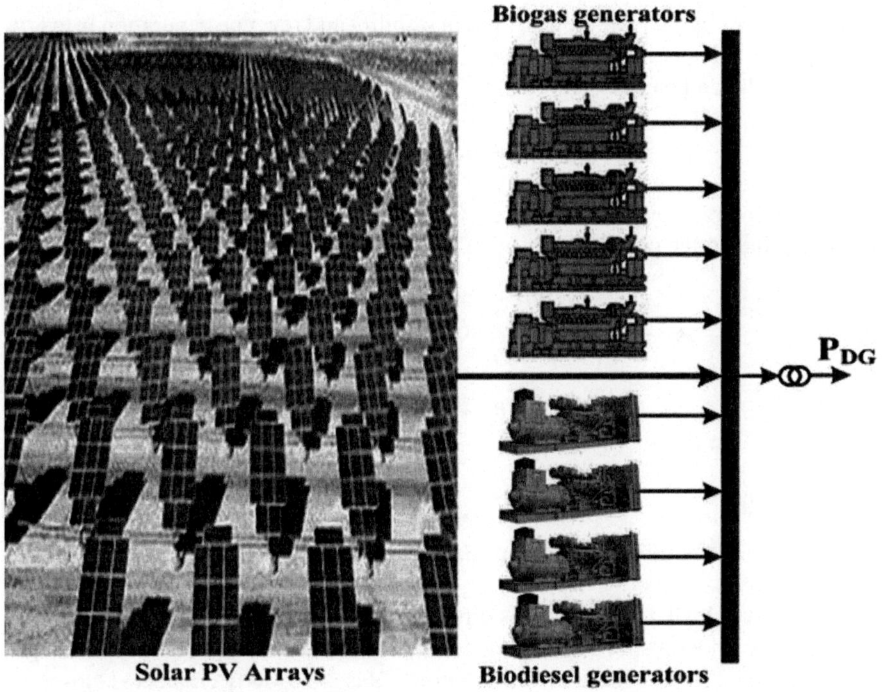

Fig. 1 Schematics of proposed DG

3 Problem Formulation

This work's prime focus is to estimate the best position and size of DG in the RDS with the lowest power loss, voltage deviation, and maximum VSI. The objective function is distributed into three parts [11], and they are solved individually as a single objective function. Thereafter, these three are solved simultaneously as a multi-objective function, considering a perfect circumstance of DG settlement, where the position and size of DGs are chosen freely.

3.1 Objective Functions

The power system researcher's main objective is to maximize net saving by maintaining the power system's smooth operation. The overall aim of this work is subdivided into the following.

3.1.1 Power Loss Minimization

The active power plays a vital role in system stability and revenue generation in every power system. Hence, it is highly essential to reduce the active power loss (P_{LOSS}) of RDS, which can be estimated by (1).

$$OF_1 = Minimize(P_{LOSS}) \tag{1}$$

Hence, the first objective function (OF_1) of DG placement problem towards minimizing the active power losses in RDS is expressed by (2).

$$P_{LOSS} = \sum_{i=1}^{n_b}\sum_{j=1}^{n_b} \frac{r_{ij}}{V_i V_j}\cos(\partial_i - \partial_j)(P_i P_j + Q_i Q_j) + \frac{r_{ij}}{V_i V_j}\sin(\partial_i - \partial_j)(Q_i P_j + P_i Q_j) \tag{2}$$

3.1.2 Voltage Deviation (VD) Minimization

The increment in load demand creates difficulty for system operators to preserve the voltage profile under a specific loading condition. Therefore, the chance of voltage collapse is more. To overcome such a situation, DGs are usually connected closer to the load center. The voltage profile improvement functions can be expressed considering the rated bus voltage (Vrated = 1.0 p.u) [11]. Hence, the second objective function (OF_2) of the DG settlement problem to minimize the voltage deviations in RDS is expressed by (3).

$$OF_2 = \sum_{i=1}^{n_b} (V_i - V_{rated})^2 \tag{3}$$

3.1.3 Voltage Stability Index Maximization

It is crucial to improve the VSI in order to maintain the voltage profile. The VSI of RDS is given by (4).

$$VSI_i = |V_j|^4 - 4[P_i x_{ij} + Q_i r_{ij}]|V_j|^2 - 4[P_i x_{ij} - Q_i r_{ij}]^2 \tag{4}$$

where VSI_i is the VSI of the ith bus. For the stable process of the power system $VSI_i \geq 0$. The bus with the lowest VSI is considered as weakest bus and has more chances of voltage collapse. Therefore, it is necessary to maximize the VSI as given in (5) [34]. Hence, the third objective function (OF_3) of the DG settlement problem is expressed by maximizing the RDS voltage stability index (5).

$$OF_3 = \max imise(VSI) = \min imise(\frac{1}{VSI}) \tag{5}$$

The multi-objective optimization problem is formulated by a linear combination of all these three objective functions together. Thereafter, the overall fitness function could be expressed by (6).

$$FF = \min imise(OF_1 + OF_2 * PC_1 + OF_3 * PC_2) \tag{6}$$

where PC_1 and PC_2 are the penalty coefficient and the value of these coefficients are 0.6 and 0.35, respectively. In (6), all of the single objective functions are normalized by dividing its corresponding base values, making the (6) dimensionless. The above-mentioned fitness function is supposed to have the following equality and inequality constraints.

3.2 *Equality Constraints (EC)*

The present work's EC is described by the real and reactive power balance equation shown in (7) and (8).

$$P_i^g - P_i^d - U_i \sum_{j=1}^{N} U_j Y_{ij} \cos(\phi_i - \phi_j - \theta_{ij}) = 0 \tag{7}$$

$$Q_i^g - Q_i^d - U_i \sum_{j=1}^{N} U_j Y_{ij} \sin(\phi_i - \phi_j - \theta_{ij}) = 0 \tag{8}$$

3.3 *Inequality Constraints*

3.3.1 Voltage Limits

The voltage constraints are considered by stipulating the lower and upper limits of the voltage violation at each RDS network node. This constraint is expressed by using (9) as shown below

$$V_i^{\min} < V_i < V_i^{\max} \tag{9}$$

3.3.2 Thermal Limits

The thermal limit of a line should not surpass the maximum limit of the thermal limits. The expression of thermal limits is shown in (10).

$$S_{ij} \leq S_{ij}^{\max} \tag{10}$$

3.3.3 Real Power Limits

The output power of DG must be restricted to its minimum (100 kW) and maximum limits (1500 kW). The mathematical expression of these facts is expressed in (11).

$$P_{gi}^{\min} < P_{gi} < P_{gi}^{\max} \tag{11}$$

Here, the defilements on node voltages and thermal limits are managed by enforcing the penalties. The overall multi-objective-based fitness function considering penalties on constraint violations could be proposed as (12).

$$FF = Min(OF_1 + OF_2 * PC_1 + OF_3 * PC_2 + \lambda_i |V_i - V_i^{\lim}| + \lambda_{sli} |S_{li} - S_{li}^{\lim}|) \tag{12}$$

4 Overview of OCSO Algorithm

The OCSO algorithm [15, 16] is derivative from the Competitive Swarm Optimizer (CSO) algorithm with an aim to increase the conversion speed. Like CSO, in OCSO also a pairwise competition is performed. But after the competition, the winners are categorized into two subdivisions according to their fitness. Among them, the particles having higher fitness values are transferred to the next iteration; whereas the remaining winner particles are gone through the opposition-based learning phase. On the other hand, the loser particles update their velocity and positions by learning from winner particles. In OCSO, seventy-five percentage of the total swarm is modernized in each iteration and enriches the original CSO's exploration property [17, 18]. The main reason behind OCSO use is its higher exploration capability and fewer control parameters. Moreover, the local and global best concept is also absent in OCSO; therefore, the memory requirement is also less in the OCSO algorithm.

5 Results and Discussions

The OCSO algorithm is utilized here to solve the optimal DG settlement problem. This projected problem's efficacy is verified on a standard 33 bus and 69 bus systems. The proposed method is developed on MATLAB 2014 platform in a Lenovo desktop (4 GB, i7-4770 CPU, 3.40 GHz).

5.1 Standard 33 Bus Test System

The data for the 33 bus RDS is taken from [19]. The real (reactive) power demands of 33 bus RDS is 3.72 MW (2.3 MVAr). The system is worked with a nominal bus voltage of 12.66 kV and 1000 kVA base. A multi-objective and three single objective functions are solved in this work to validate the proposed algorithms' efficacy.

5.1.1 Single Objective

In single objective cases, three different objective functions are solved individually. They are power loss minimization, VD minimization and VSI maximization. The simulation results obtained from OCSO and CSO for finest DG sitting and sizing in power loss minimization are depicted in Table 1. It has been witnessed from Table 1 that the finest value of DG injection from the CSO algorithm for 33 bus RDS are 700, 1016 and 959 kW at bus 14, 24 and 30, respectively. The corresponding power loss for the obtained solution is 72.38 kW. Whereas the power loss obtained from

Table 1 Outcomes for 33 bus RDS using CSO and OCSO

	CSO		OCSO	
Power loss minimization	Location	Size	Location	Size
	14	700	14	812
	24	1016	24	1160
	30	959	30	1018
	Power loss (kW)	72.38	Power loss (kW)	71.68
	VD	0.0181	VD	0.0127
	VSI	0.8635	VSI	0.8778
VD minimization	Location	Size	Location	Size
	14	1104	14	1057
	27	1200	27	1200
	32	1123	31	1182
	Power loss (kW)	108.37	Power loss (kW)	107.54
	VD	0.00093	VD	0.00084
	VSI	0.9316	VSI	0.9318
VSI maximization	Location	Size	Location	Size
	10	1200	13	1200
	31	1198	26	1198
	32	1188	29	1188
	Power loss (kW)	167.02	Power loss (kW)	108.73
	VD	0.0036	VD	0.0014
	VSI	0.9326	VSI	0.9341

the OCSO is 71.68 kW with the help of DG placement at buses 14 (812 kW), 24 (1160 kW), and 30 (1018 kW) with some compromise in VD and VSI.

Similarly, the results for the VD's minimization are listed in Table 1. It has been observed from Table 1 that the VD minimization for the OCSO algorithm is 0.00084 p.u., which is better than the 0.00093 p.u. obtained by the CSO algorithm with a small compromise in power losses. The simulation results for VSI maximization are listed in Table 1, and it has been found that the VSI obtained from the OCSO algorithm is 0.9318 p.u. Whereas the maximum VSI obtained from the CSO algorithm is 0.9316 p.u. with some compromise in power losses. From the results of Table 1, we can conclude that all three objective functions are performed superior with the OCSO algorithm compared to the CSO algorithm.

5.1.2 Multi-Objective

Further, a multi-objective function is considered to validate the proposed OCSO algorithm's efficacy for simultaneous maximization of the VSI, minimization of VD and real power losses. The simulation results obtained from OCSO and CSO are depicted in Table 2 and compared with those of some other state-of-the-art literature. It has been observed from Table 2 that the Final Fitness (FF) obtained by the OCSO algorithm is 0.3978 p.u. which is better than 0.4034 p.u. obtained by the CSO algorithm and other algorithms. The best results correspond to a social factor of 0.6, and the swarm size is 80. The outcomes of the OCSO and CSO algorithm are analyzed and compared with GA [8], PSO [8], GA-PSO [8], TLBO [11], QOTLBO [11]. From comparing the result, it is clear that the OCSO algorithm has outperformed the other methods for minimizing the fitness function. The convergence plot of the fitness function achieved by the OCSO, CSO algorithm is shown in Fig. 2. It has been detected from Fig. 2 that the OCSO converges faster w.r.t the CSO algorithm. Figure 3 presents the system voltage profile with and without the DG placement for 33 bus RDS.

Table 2 Results for simultaneous optimization of 33-bus RDS with different algorithms

Algorithm	GA [8]	PSO [8]	GA-PSO [8]	TLBO [11]	QOTLBO [11]	CSO	OCSO
DG location (Size in kW)	11 (1500) 29 (423) 30 (1071)	8 (1177) 13 (982) 32 (829)	11 (925) 16 (863) 32 (1200)	12 (1183) 28 (1191) 30 (1186)	13 (1083) 26 (1187) 30 (1199)	13 (1081) 24 (1148) 31 (1159)	14 (1022) 25 (1156) 30 (1148)
Loss (p.u)	0.1063	0.1053	0.1034	0.1246	0.1034	0.0792	0.0779
VD (p.u)	0.0407	0.0335	0.0124	0.0011	0.0011	0.0046	0.0061
VSI (p.u)	0.9490	0.9256	0.9508	0.9503	0.9530	0.9183	0.9033
Value of FF	0.4578	0.4493	0.4436	0.4578	0.4376	0.4034	0.3978

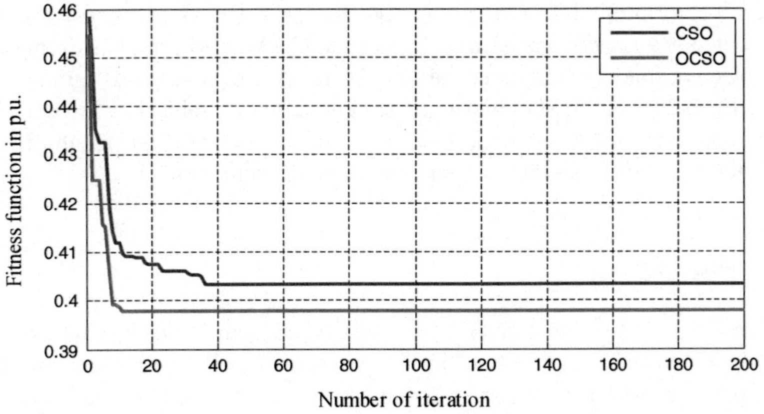

Fig. 2 Convergence plot for 33 bus RDS

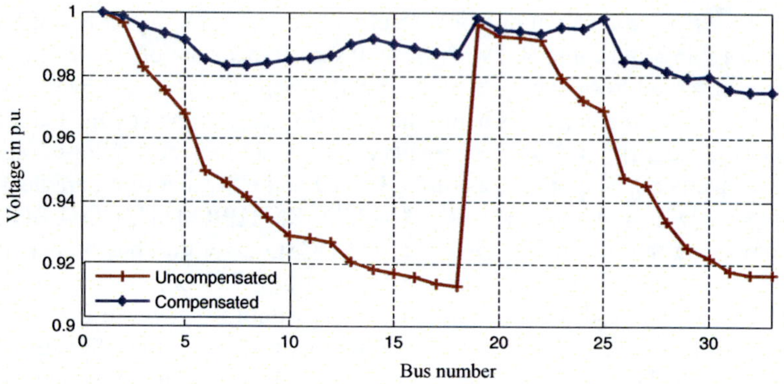

Fig. 3 Voltage profile for 33 bus RDS

5.2 *Standard 69 Bus Test System*

The proposed OCSO and CSO algorithm's performance in the DG placement problem is verified on a vast RDS such as 69 bus RDS. The system data and single line diagram for 69 bus RDS are collected from [15, 17]. The real (reactive) power demands for this system are 3.8 MW (2.69 MVAr), whereas the uncompensated case's losses are 224.7 kW and 102.13 kVAr. The nominal bus voltage for 69 bus RDS is 12.66 kV. Like earlier, both single and multi-objective problems have been solved and compared here.

5.2.1 Single Objective

The OCSO algorithm's effectiveness is also validated in a more extensive system, and the outcomes are compared with that of the CSO algorithm considering three single objective functions as stated earlier. The results for power loss minimization obtained by OCSO and CSO algorithms are compared in Table 3, which confirms the supremacy of OCSO over the CSO algorithm. Similarly, the results for VD minimization and VSI maximization are also examined in Table 3.

5.2.2 Multi-Objective

Again, a multi-objective function is considered to validate the efficacy of the proposed OCSO algorithm for simultaneous maximization of the VSI, minimization of VD and real power losses. Here also, three numbers of DG are installed optimally. The results of the proposed OCSO algorithm are depicted in Table 4, including the results of GA [8], PSO [8], GA-PSO [8], TLBO [11], QOTLBO [11] algorithm along with the original CSO algorithm. Table 4 shows that the best fitness of 0.4119 is obtained

Table 3 Outcomes for 69 bus RDS using CSO and OCSO

	CSO		OCSO	
Power loss minimization	Location	Size	Location	Size
	17	476	17	567
	50	614	50	1182
	51	1157	53	546
	Power loss (kW)	72.20	Power loss (kW)	71.67
	VD	0.00713	VD	0.0060
	VSI	0.9179	VSI	0.9219
VD minimization and VSI maximization	Location	Size	Location	Size
	16	751	16	760
	48	1200	48	1162
	50	1200	51	1250
	Power loss (kW)	84.63	Power loss (kW)	85.17
	VD	0.00044	VD	0.00041
	VSI	0.9769	VSI	0.9769

Table 4 Results for simultaneous optimization of 69 bus RDS with different algorithms

Algorithm	GA [8]	PSO [8]	GA-PSO [8]	TLBO [11]	QOTLBO [11]	CSO	OCSO
DG location (Size in kW)	21 (929) 62 (1075) 64 (985)	17 (992) 61 (1199) 63 (795)	21 (910) 61 (1193) 63 (885)	13 (1013) 61 (990) 62 (1160)	15 (811) 61 (1147) 62 (1002)	18 (728) 50 (1037) 51 (1050)	17 (530) 50 (1086) 51 (986)
Loss(p.u)	0.0890	0.0832	0.0811	0.0821	0.0805	0.0774	0.0748
VD (p.u)	0.0012	0.0049	0.0031	0.0008	0.0007	0.0012	0.0029
VSI (p.u)	0.9705	0.9676	0.9768	0.9745	0.9769	0.9647	0.9583
Value of FF	0.4294	0.4248	0.4248	0.4236	0.4228	0.4157	0.4119

for 80 swarm size and 0.4 social factor. It is clear from Table 4 that the fitness value obtained by the OCSO algorithm is the least among all other algorithms. The convergence plots of fitness function obtained by the OCSO, CSO is displayed in Fig. 4, and the voltage profile of the 69 bus system is presented in Fig. 5.

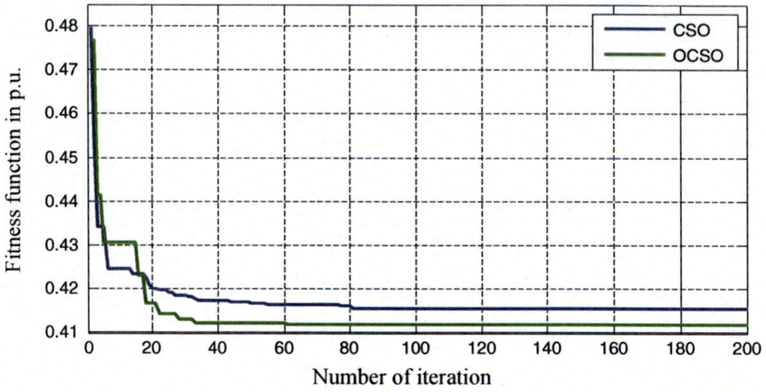

Fig. 4 Convergence plot for 69 bus RDS

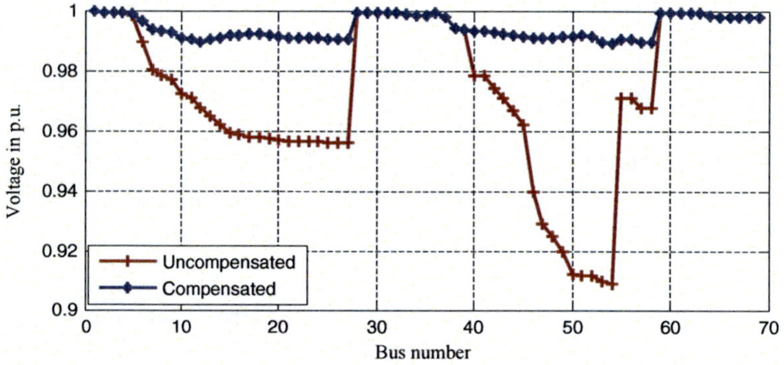

Fig. 5 Voltage profile for 69 bus RDS

6 Conclusion

A multi-objective problem has been formulated to solve the optimal DG location and sizing problem to minimize the real power losses and VDs with the maximization of the VSI. The proposed OCSO algorithm has a better search capacity due to integrating the opposition-based learning strategies. Moreover, the OCSO algorithm has fewer parameters to be adjusted. It is identified from the convergence plot that the OCSO algorithm converges earlier than the basic CSO algorithm. It is interpreted from the results that the OCSO algorithm can estimate the optimal locations and sizes of DG, minimize the power losses, reduce the VDs and maximize the VSI simultaneously. The projected method's efficiency is studied on standard 33 bus and 69 bus RDS for different objective functions. It is noticed from the outcomes that the OCSO algorithm outperforms other methods in solving optimal DG settlement issues for both single objective and multi-objective problems. The work could be further extended with the probabilistic approach.

References

1. Barik AK, Das DC (2018) Expeditious frequency control of solar photovoltaic/biogas/biodiesel generator based isolated renewable microgrid using grasshopper optimisation algorithm. IET Renew Power Gener 12(14):1659–1667
2. Barik AK, Das DC (2019) Proficient load-frequency regulation of demand response supported bio-renewable cogeneration based hybrid Microgrids with quasi-oppositional selfishherd optimisation. IET Gener Trans Distrib 13(13):2889–2898
3. Sultana U et al (2016) A review of optimum DG placement based on minimization of power losses and voltage stability enhancement of distribution system. Renew Sustain Energy Rev 63:363–378
4. Wang C, Nehrir MH (2004) Analytical approaches for optimal placement of distributed generation sources in power systems. IEEE Trans Power Syst 19(4):2068–2076
5. Borghetti A (2012) A mixed-integer linear programming approach for the computation of the minimum-losses radial configuration. IEEE Trans Power Syst 27(3):1264–1273
6. Rueda-Medina AC et al (2013) A mixed-integer linear programming approach for optimal type, size and allocation of distributed generation in radial distribution systems. Elect Power Syst Res 97:133–143
7. Soroudi A, Ehsan M (2011) Application of a modified NSGA method for multi-objective static distributed generation planning. Arab J Sci Eng 36(5):809
8. Moradi MH, Abedini MA (2012) combination of genetic algorithm and particle swarm optimisation for optimal DG location and sizing in distribution systems. Int J Electr Power Energ Syst 34(1):66–74
9. Abu-Mouti FS, El-Hawary ME (2011) Optimal distributed generation allocation and sizing in distribution systems via artificial bee colony algorithm. IEEE Trans Power Deliv 26(4):2090–2101
10. Wang LF, Singh C. Reliability-constrained optimum placement of reclosers and distributed generators in distribution networks using an ant colony system algorithm. IEEE Trans Syst Man Cybernet Part C: Appl Rev 38(6):757–764
11. Sultana S, Roy PK (2014) Multi-objective quasi-oppositional teaching learning based optimisation for optimal location of distributed generator in radial distribution systems. Int J Electr Power Energy Syst 2014(63):534–545
12. Prasad CH, Subbaramaiah K, Sujatha P (2019) Cost–benefit analysis for optimal DG placement in distribution systems by using elephant herding optimization algorithm. Renew Wind Water Solar 6(1):2
13. Yahyazadeh M, Rezaeeye H (2020) Optimal placement and sizing of distributed generation using wale optimization algorithm considering voltage stability and voltage profile improvement, power loss and investment cost reducing. Iran J Sci Technol Trans Electr Eng 44(1):227–236
14. Pereira BR et al (2016) Optimal distributed generation and reactive power allocation in electrical distribution systems. IEEE Trans Sustain Energy 7(3):975–984
15. Das S, Malakar T (2019) An emission constraint capacitor placement and sizing problem in radial distribution systems using modified competitive swarm optimiser approach. Int J Ambient Energy https://doi.org/10.1080/01430750.2019.1587723
16. Das S, Malakar T (2020) Estimating the impact of uncertainty on optimum capacitor placement in wind-integrated radial distribution system. Int Trans Electr Energy Syst 30(8):e12451
17. Das S, Malakar T (2021) Optimal capacitor placement and sizing in distribution system using competitive swarm optimiser algorithm. Int J Adv Intell Paradigms https://doi.org/10.1504/IJAIP.2021.10034544

18. Cheng R, Jin Y (2015) A competitive swarm optimiser for large scale optimisation. IEEE Trans Cybern 5(2):191–204
19. Baran ME, Wu FF (1989) Network reconfiguration in distribution systems for loss reduction and load balancing. IEEE Trans Power Delivery. 4(2):1401–1407

An Optimization Model for Commercial Loads Under Time of Use and Real-Time Pricing Scheme

Manish Sharma and Sandeep Kakran

Abstract Commercial loads share a huge amount (8.24% in India) of the total power consumption of any country. This paper presents an optimization model for commercial loads, especially food courts and commercial kitchens. Food courts and restaurants are service and process industries; that's why scheduling techniques used for these types of loads are different from each other. Scheduling of thermostatically controlled loads is done on an empirical basis considering their ageing factor into account. Then the model utilizes the mixed-integer linear programming technique for minimizing the electricity cost by scheduling the loads satisfying all the related and needed constraints. The CPLEX solver of GAMS software is used to apply the above technique. The proposed model uses two different dynamic pricing schemes (i) Time-of-Use, (ii) Real-Time Pricing. Results reveal that the proposed model has the potential to significantly reduce the prices of electricity consumed by any restaurant or food court when it is subjected to uneven pricing.

Keywords Demand-side management · Demand response · Time of use · Real-time pricing · Commercial kitchen

1 Introduction

Since the past some decenniums due to the high exploitation of natural resources everyone has experienced the increased cost of every product and sub-product of these natural resources and electricity is one of them. The reason for this abusive escalation in the cost of electricity is due to increased prices of burning fuel and capital expenditure for the construction of new generating plants. To avoid these costly peaky units and expansion of units, Demand-Side Management (DSM) can play an effective role. There are various types of consumers from which commercial consumption of electricity shares a considerable amount (8.24% in India) of the total electricity consumption of any country. Load Management (LM) of commercial loads is minimal, and it is due to a lack of knowledge about the demand pattern of

M. Sharma (✉) · S. Kakran
Department of Electrical Engineering, National Institute of Technology, Kurukshetra, India

A. Kumar et al. (eds.), *Renewable Energy Towards Smart Grid*, Lecture Notes in Electrical Engineering 823, https://doi.org/10.1007/978-981-16-7472-3_22

the appliances to be needed. However, Commercial Load Management (CLM) can participate heavily in shaping peaky load curves and saving on electricity bills.

This paper presents the scheduling of commercial loads like food courts, food and beverages (F&B), and commercial kitchens. A food court or a restaurant is mainly having a front end (dining area, bars, etc.) and a back end (kitchen). Both ends have different types of loads, so the scheduling method is also different for both. The restaurant's front end is having mainly uncontrollable loads like lights, fans, etc. and on the other hand, the back end of the restaurant is having a combination of uncontrollable and controllable loads like a deep fryer, steam table, lights, etc. In this paper, during scheduling, uncontrollable loads (fixed loads) are considered as constant values for each hour throughout the day. In controllable loads, switch-controlled appliances are considered the major appliances of scheduling and thermostatically controlled appliances are scheduled on an empirical basis. All the appliances used in the commercial kitchen experience abusive treatment in terms of maintenance and care. Due to this behaviour, appliances start consuming more power than usual. Here, an ageing factor is also considered, which is taken into account annually or at every two or five years.

A wide range of research has been done in this field. Scheduling of industrial load with a case study is done in [1]. Scheduling of residential and commercial load is done in [2]. Electricity used in the commercial kitchen [3] explains very wisely the total energy consumption of any commercial kitchen. The dynamics of cooling and heating load in [4] explains the theoretical method to schedule the thermostatic loads, but they are limited to face some white noise coefficient and some unknown disturbances. This creates a massive difference between the theoretical power consumption and practical power consumption of any cooling or heating appliance. In [5], a stochastic computer model for heating and cooling load is proposed. In [6], a model is proposed to increase the quality of experience for the users who participate in the demand response. A physically based load model for the industry which includes the utilization factor is proposed in [7].

In the proposed model, all the appliances used in a commercial kitchen (Table 1) went through experiments to calculate some experimental data. Then this data is taken as constant for this scheduling. The ageing factor and noise factor of each appliance are also considered in this paper. Food court or restaurant is assumed to be opened from 09:00 to 22:00 of a day. So, most of the appliances used must be scheduled in this period only except smatter appliances like deep-freezer, hot storage, etc. They must be kept in on state for the whole day. The inside temperature of some food storage appliances is kept below/above (as required) the specific temperature throughout the day as per the recommendation of the Health Science Policy Advisor at the FDA's Center for Food Safety and Applied Nutrition. Here it is to notice that the energy consumption of any food court indeed depends upon the number of meals served and the type of menu they serve. A survey of average energy consumption per average meal served is done in [3].

2 System Model

2.1 *Classification of Loads*

Loads of a food court are classified as: (a) uncontrollable loads which are not considered for any load scheduling algorithm, and (b) controllable loads. Further classification of controllable loads is

- Switch-controlled loads,
- Thermostatically controlled loads.

2.2 *System Model*

With the enactment of the smart grid and advanced dynamic pricing schemes, there will be two-way communication between user-to-grid and grid-to-user. Dynamic prices of energy will be conveyed to the users according to the pricing scheme by the utility. It is also to be considered that each smart meter is also equipped with the optimization device or function (Fig. 1).

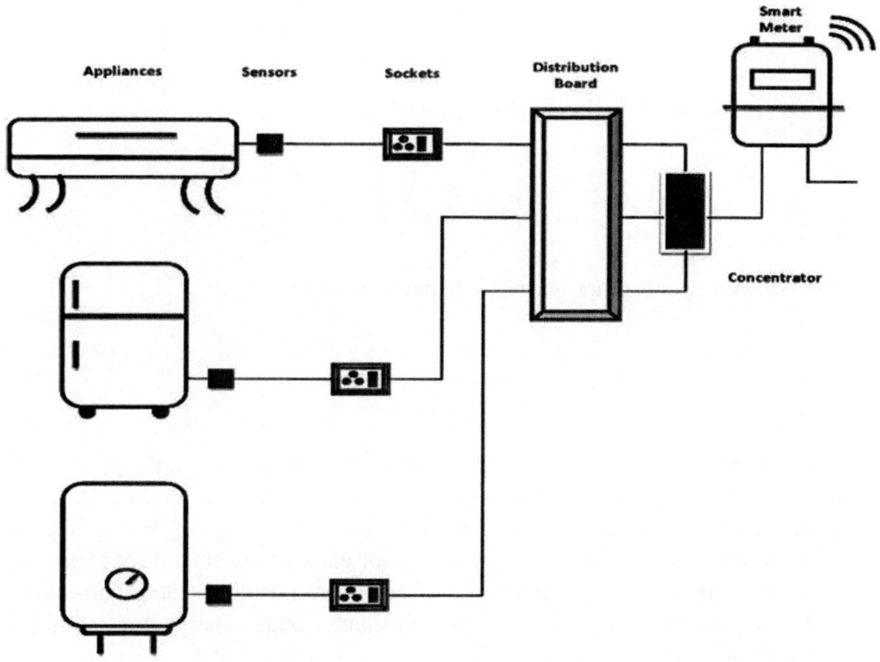

Fig. 1 Connection diagram of thermostatically controlled loads

A set of $I = 12$ appliances is considered, which contains all the loads, i.e. fixed loads, switch controlled loads and thermostatically controlled loads as

$$I = [1, 2, 3 \ldots\ldots\ldots\ldots\ldots, 12] \quad (1)$$

It is taken that the scheduling interval is one day, which is divided into 24 small intervals, each of one-hour duration. Let T be the set of time intervals such that

$$T = [1, 2, 3 \ldots\ldots\ldots\ldots\ldots, 24] \quad (2)$$

Energy consumed by appliance $i \in I$ in tth $\in T$ interval will be

$$E_{i,t} = P_{i,t} * \alpha_{i,t} \quad (3)$$

where

$E_{i,t}$ Energy consumed by appliance i in tth hour.
$P_{i,t}$ Power rating of appliance i in tth hour.
$\alpha_{i,t}$ On/Off status of appliance i in tth hour.

and,

$$E_{i,t} = \begin{cases} P_{i,t}, \alpha_{i,t} = 1 \\ 0, \alpha_{i,t} = 0 \end{cases} \quad (4)$$

So total energy consumption in tth hour will be

$$E_t = \sum_{i=1}^{I} E_{i,t} \quad (5)$$

And total energy consumption in 24 h (one day) will be

$$E_{total} = \sum_{t=1}^{T} E_t \quad (6)$$

- **Uncontrollable Loads**. Uncontrollable loads are those, that can't go through any kind of scheduling algorithm. They are free to be on/off whenever required, e.g. range, display, lights, billing computer, etc. In this model, scheduling of uncontrollable loads is done as a fixed load for each slot of time throughout the day. In this paper, appliance number 1 is considered as an uncontrollable load which is a combination of various uncontrollable loads. Energy consumed by fixed loads will be-

$$F_t = \sum_{t=1}^{T} E_t^{fixed} \tag{7}$$

- **Switch-Controlled Loads**. Switch-controlled loads are those that can go through any kind of scheduling algorithm. Their task must be completed in any of the preferred possible schedules, for example, mixer, grinder, ice crusher, etc. In this paper, appliance number 2 to 4 are considered switch-controlled loads. Energy consumed by switch-controlled loads will be-

$$E_{i,t}^{sc} = \sum_t \sum_i E_{i,t} * \alpha_{i,t} \ \forall\, t \in T, \forall i \in [2, 3, 4] \tag{8}$$

- **Thermostatically Controlled Loads**. These load's power consumption is not fixed as it depends on various factors like thermal resistance, thermal capacity, ambient temperature, set temperature, system inertia, temperature gain, the thermal time constant, etc. [4], but even after considering all these parameters along with white noise coefficient, it is not possible to find the actual energy consumption because it also depends upon various other physical factors like how many persons are using the room, for how much time door/windows get opened, etc. for AC. Similarly, there are some other physical factors for every appliance.

The proposed method to find the energy consumption of these types of loads is an empirical method. Data required for this method is taken by doing experiments with every appliance. Although whenever a new device gets connected to the smart meter for the first time, then it fetches some data from that device. This data is collected only for the first time. This period is called the test period. Now energy consumed by appliance '*i*' in time slot '*t*' will be

$$E_{i,t} = \left[E_i^0 * \left\{ \frac{T_{i,t}^{out} - T_{i,t}^{in}}{T_{i,0}^{out} - T_{i,0}^{in}} \right\} + E_i^{Base} \right] * \alpha_{i,t} \tag{9}$$

where

$E_{i,t}$	Energy consumed by appliance i in $t$he period.
E_i^0	Energy consumed by appliance i in the test period.
$T_{i,t}^{out}$	Outside/Ambient temperature at time t.
$T_{i,t}^{in}$	Inside/Set temperature of appliance i in time slot t.
$T_{i,0}^{out}$	Outside/Ambient temperature in the testing period.
$T_{i,0}^{in}$	Inside/Set temperature of appliance i in the testing period.
E_i^{Base}	Base/idle power consumed by the appliance i.
$\alpha_{i,t}$	ON/OFF status of appliance i in the tth time slot.

In commercial kitchens, maintenance and care of appliances is a big problem as most of the appliances are bulky and it requires a downtime for the appliance. Due to low maintenance and ageing of the appliance, it starts consuming some increased power. To compensate for this excessive energy consumption a factor of ageing is considered. So, after considering the ageing factor and an unknown noise factor to make the equations more realistic for each appliance, Eqs. (8) and (9) becomes

$$E_{i,t}^{sc} = \sum_{t=1}^{24}\sum_{i=1}^{I} E_{i,t}(1+\lambda_i+\gamma_i) * \alpha_{i,t} \tag{10}$$

$$E_{i,t} = \left[E_i^0 * \left\{\frac{T_{i,t}^{out} - T_{i,t}^{in}}{T_{i,0}^{out} - T_{i,0}^{in}}\right\}(1+\lambda_i+\gamma_i) + E_i^{Base}(1+\lambda_i+\gamma_i)\right] * \alpha_{i,t} \tag{11}$$

where

$\lambda{=}_i$ Ageing factor of appliance *i* (to be considered on annual basis).
$\gamma{=}_i$ Noise factor for appliance *i*.

So total energy consumption for thermostatically controlled loads will be

$$E_{i,t}^{tc} = \sum_{t=1}^{24}\sum_{i=1}^{I} E_{i,t} * \alpha_{i,t} \tag{12}$$

Objective function

Min.

$$Cost = \left[\left\{\sum_{t=1}^{24}\sum_{i=1}^{I}\left(E_{i,t} * \alpha_{i,t}\right)\right\} + \sum_{t=1}^{24} E_t^{fixed}\right] * Rs_t \tag{13}$$

where
Rs_t = Electricity price (¢/kwh).

and

$$E_{i,t} = E_{i,t}^{sc} + E_{i,t}^{tc} \tag{14}$$

Subjected to

$$\left[\sum_{t=1}^{T}\left(\sum_{i=1}^{I} E_{i,t} + F_t\right)\right] \le E_t^{max} \tag{15}$$

and

$$T_{min,i}^{in} \le T_{i,t}^{in} \le T_{max,i}^{in} \tag{16}$$

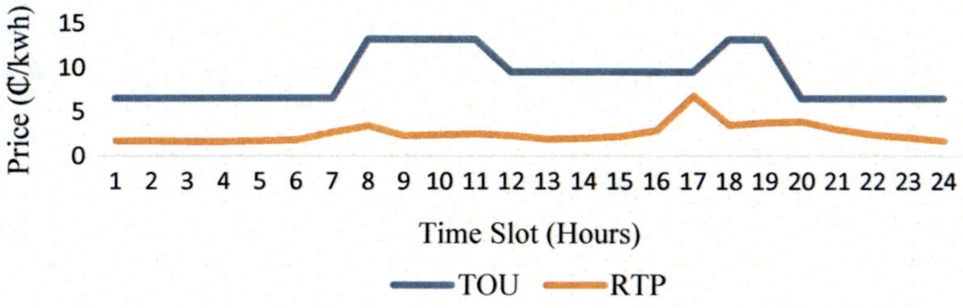

Fig. 2 Pricing schemes used in the scheduling

where $T_{min,i}^{in} and T_{max,i}^{in}$ are the minimum and maximum set temperature limits, respectively for appliance *i*.

2.3 Pricing Schemes

Pricing schemes are provided by the utility to the consumers. In [8], pricing schemes are classified as a) static pricing scheme and b) dynamic pricing scheme. In static one, the price of energy consumption per kWh is constant throughout the time horizon. Contrarily in a dynamic pricing scheme, the price of energy consumption per kWh varies frequently according to the utility.

- **Time-of-Use**. It mainly has time-varying prices. The cost of energy varies in blocks in this scheme. Majorly, it has 2 to 5 blocks of prices depending upon forecasted demand fixed by the utility. This scheme motivates the consumers to run their appliances in the low-price time slot. Figure 2. shows the time-of-use (TOU) prices used in this paper.
- **Real-time pricing**. It is also a type of dynamic pricing scheme, but the difference from TOU is that cost of energy varies for every single slot of time (15 min. or one hour). These prices of electricity are conveyed to the users just before a day or an hour. Figure 2. shows the prices of real-time pricing (RTP) used in this paper.

3 Case Study and Result

This section presents the solution to the above scheduling problem with a case study. The result shows a detailed comparison between the cost of electricity and energy with scheduling and without scheduling. Appliances used in this case study for the commercial kitchen with their type are given in Table 1.

The cost of electricity for each category appliance is calculated using both the pricing schemes. Data required for the execution of this program like ambient temperature and the set temperature is taken from the experiment. Figure 2 shows the prices of both pricing schemes. These prices are taken from [9, 10].

Table 1 Appliances and their data

Sr. No	Appliances	No	Type	Average Wattage(kW)	Operating Time(Hr.)	Energy consumed in test hour (E_0)[1]	$T_0 = (T^{out} - T^{in})$[2]
1	Uncontrolled	1	UC	0.4,4.55	10,14	–	–
2	Mixer	1	SC	2	1	–	–
3	Ice crusher	1	SC	0.4	1	–	–
4	Grinder	1	SC	1.5	1	–	–
5	Air-conditioner	4	TC	–	14	4.8	10
6	Steam table	1	TC	–	14	3.75	−50
7	Hot storage	1	TC	–	24	3	−30
8	Water heater	1	TC	–	2	1.5	−70
9	Heating ventilation	1	TC	–	14	2	4
10	Deep fryer	1	TC	–	14	2	−320
11	Deep freezer	1	TC	–	24	0.29	47
12	Walk-in fridge	1	TC	–	24	1.7	26

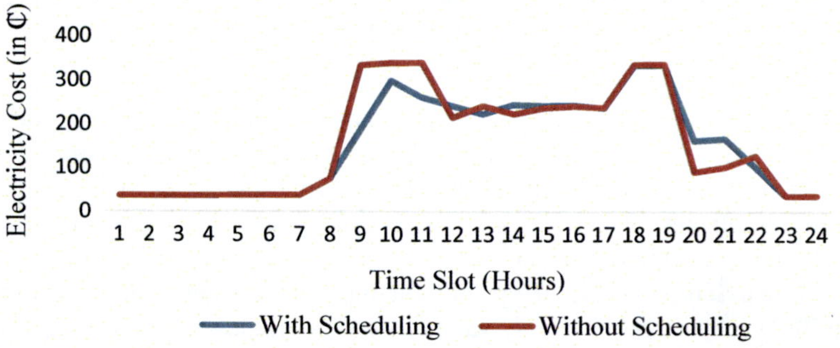

Fig. 3 Electricity consumption cost under TOU pricing scheme with and without scheduling

The difference between electricity consumption cost with scheduling and without scheduling under the TOU pricing scheme is shown in Fig. 3. A total saving of 126.9108 ¢ per day is achieved by scheduling under the TOU pricing scheme. It is to be noted that the number of meals served and the type of menu served in the restaurant directly affect the power consumption by each appliance.

Figure 4 shows the comparison between the hourly electricity consumption cost with scheduling and without scheduling under the RTP scheme and a total saving of 13.3943 ¢ per day is achieved.

[1] Energy consumed in testing hour by each appliance in kWh.

[2] Difference between ambient and set temperature in testing hour in °C.

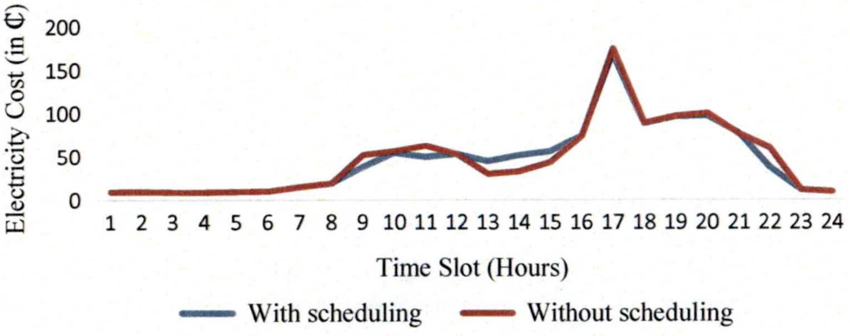

Fig. 4 Electricity consumption cost under RTP scheme with and without scheduling

The hourly cost of electricity is shown in Fig. 5 under both the pricing schemes. Figures 6 and 7 show the hourly energy consumption under both the pricing schemes.

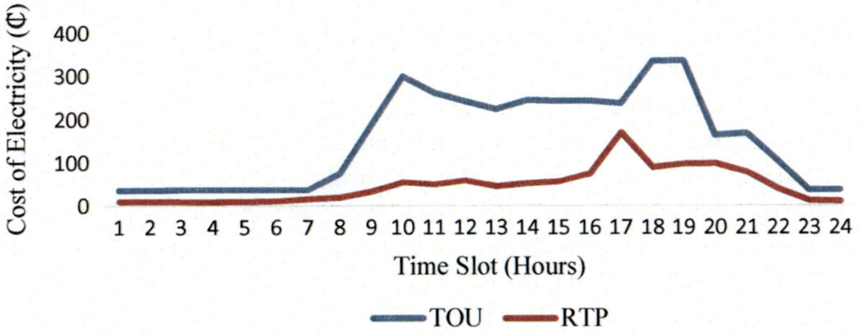

Fig. 5 Hourly electricity consumption cost under both pricing schemes

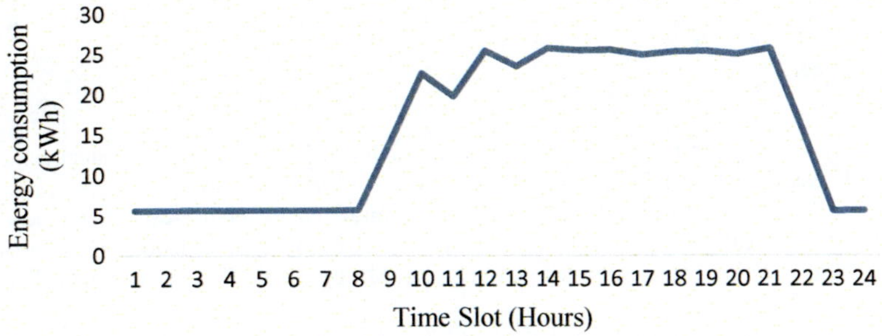

Fig. 6 Hourly energy consumption under TOU pricing scheme

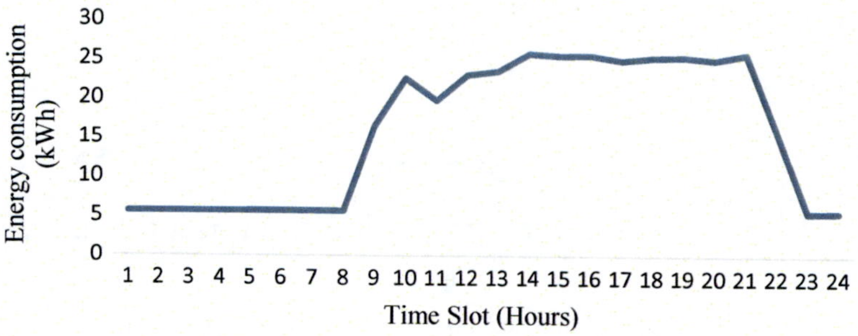

Fig. 7 Hourly energy consumption under RTP scheme

4 Conclusion

This study is based on the optimal scheduling of commercial loads, especially for food courts and commercial kitchens. Modeling of fixed loads, switch-controlled loads and thermostatically controlled loads have been done and explained. The mixed-integer linear programming based scheduling strategy using CPLEX solver in GAMS software has been considered for the solution. The results presented in the next section have shown a detailed comparison between energy consumption costs under TOU and RTP schemes. A comparison between with and without scheduling subjected to each pricing scheme has also been shown in the result. The results have shown that the considered scheduling technique has sufficiently reduced the electricity bills. From graphs, it can be seen that if the number of switch-controlled loads is increased, then saving in the cost can also be increased. In the future, the cost-saving can be increased if this study is carried out on small-time slots and with some other required constraints.

References

1. Ashok S, Banerjee R (2001) An optimization mode for industrial load management. IEEE Trans Power Syst 16(4):879–884. https://doi.org/10.1109/59.962440
2. Alwan HO, Sadeghian H, Wang Z (2018) Decentralized demand side management optimization for residential and commercial load. In: 2018 IEEE International Conference on Electro/Information Technology (EIT), Rochester, MI, pp 0712–0717.https://doi.org/10.1109/EIT.2018.8500213.
3. Mudie S, Essah EA, Grandison A, Felgate R (2016) Electricity use in the commercial kitchen. Int J Low-Carbon Technol 11(1):66–74. https://doi.org/10.1093/ijlct/ctt068
4. Mortensen RE, Haggerty KP (1990) Dynamics of heating and cooling loads: models, simulation, and actual utility data. IEEE Trans Power Syst 5(1):243–249. https://doi.org/10.1109/59.49112
5. Mortensen RE, Haggerty KP (1988) A stochastic computer model for heating and cooling loads. IEEE Trans Power Syst 3(3):1213–1219. https://doi.org/10.1109/59.14584

6. Pilloni V, Floris A, Meloni A, Atzori L (2018) Smart home energy management including renewable sources: A QoE-Driven Approach. IEEE Trans Smart Grid 9(3):2006–2018. https://doi.org/10.1109/TSG.2016.2605182
7. Manichaikul Y, Schweppe FC (1979) Physically based industrial electric load. In: IEEE Transactions on Power Apparatus and Systems, vol. PAS-98, no. 4, pp 1439–1445. https://doi.org/10.1109/TPAS.1979.319346
8. Albadi MH, El-Saadany EF (2007) Demand response in electricity markets: an overview. In: IEEE Power Engineering Society General Meeting. Tampa, FL 2007:1–5. https://doi.org/10.1109/PES.2007.385728
9. Kakran S, Chanana S (2018) Energy scheduling of smart appliances at home under the effect of dynamic pricing schemes and small renewable energy source. Int J Emerg Electr Power Syst 19. https://doi.org/10.1515/ijeeps-2017-0187
10. Real time pricing data powered by Comed an Exelon company https://hourlypricing.comed.com/live-prices/. (Accessed 25 Dec 2020)

Multi-objective Stochastic Volt/VAR Optimization in AC-DC Hybrid Distribution Network Considering Soft Open Point

Vijay Babu Pamshetti, V. V. S. N. Murty, S. P. Singh, and Ashwani Kumar Sharma

Abstract To cope with different types of distributed energy sources (DERs) and AC/DC loads, combined AC and DC distribution network has emerged as a potential solution for the forthcoming distribution network. However, upward integration of DERs imposes several problems such as voltage violations and system power losses. In order to minimize the system power loss and voltage deviations at the same time, a two-stage multi-objective stochastic Volt/VAR optimization for HDN considering soft open points (SOP) is proposed in this paper. In state 1, a coordination optimization model considering legacy Volt/VAR control devices, status of network switches, DERs, voltage source converter (VSC) and SOP has been established; in stage 2, model predictive control (MPC) based rolling optimization control model has been built, to eliminate the forecasted deviations caused by renewable energy sources and loads. Besides, the original non-convex mixed-integer nonlinear programming (MINLP) problem has been transformed into a mixed-integer second-order cone programming (MISOCP) problem for a feasible solution. The proposed scheme has been verified and authenticated on IEEE 33 bus AC-DC DN. The simulation results reveal the reputation of the proposed methodology on the aforementioned problems.

Keywords Energy losses · Voltage deviation · Volt/VAR control

1 Introduction

Due to the eco-friendly concerns, the proliferation of renewable energy distributed generations (DGs) such as solar photovoltaic (PV) generation (i.e. DC-based DGs) and wind generation (i.e. AC-based DGs) has been increasing rapidly in a distribution

V. B. Pamshetti (✉)
Chaitanya Bharathi Institute of Technology, Hyderabad, Telangana 500075, India

V. V. S. N. Murty · A. K. Sharma
National Institute of Technology Kurukshetra, Kurukshetra, Haryana 136119, India

S. P. Singh
Indian Institute of Technology (Banaras Hindu University) Varanasi, Varanasi, Uttar Pradesh 221005, India

A. Kumar et al. (eds.), *Renewable Energy Towards Smart Grid*, Lecture Notes in Electrical Engineering 823, https://doi.org/10.1007/978-981-16-7472-3_23

network (DN). However, the conventional DN is still AC based, it could be beneficial to operate the AC DN with the DC DN cooperatively by employing the voltage source converter (VSC) to form an combined AC and DC distribution network [1]. On the other hand, Volt/VAR control (VVC) operation [2] is well-known scheme for peak shavings as well as voltage violation mitigation. The traditional VVC scheme has been executed by legacy voltage-regulated devices such as capacitor banks (CBs), on-load tap changing transformers (OLTC), and voltage regulators (VRs), these devices could not deal with the sudden voltage changes [3]. However, an advanced voltage control device (i.e. smart inverter (SI) and soft open point (SOP)) can effectively handle the sharp voltage violations owing to their quick action. In order to effective utilization of legacy VVC devices and cutting-edge VVC devices, different multi-stage VVC methodologies have been published in recent times [4].

The potential benefits of combined AC and DC distribution network based on voltage source converter (VSC) have been explored in [1]. Advanced Power Electronic (PE) devices such as SI and SOP [5] are regular in use owing to their quick action nature. In [6, 7], the applications and benefits of SOP have been studied.

The coordinated operation of OLTC, CBs, switches, DER, VSC and SOP for the VVO procedure has not been fully explored. In the present paper, the coordinated operation of legacy and cutting-edge VVC devices has been investigated. Besides, an advanced two-stage coordination VVO scheme has been suggested to decrease voltage deviations and energy losses. The main contributions of the present paper include:

- A horizon rolling optimization model is established to the combined operation of the legacy and cutting-edge VVC devices for well-organized control in AC-DC hybrid DN.
- A two-stage coordination VVO considering DER, VSC and SOP has been suggested.

2 Advanced Devices

2.1 Soft Open Points (SOP)

Normally, neighbouring feeders are linked with normally open points (NOP) in the distribution network. For flexible power control, the SOP device [7] is mounted instead of NOP as shown in Fig. 1a.

2.2 Voltage Source Converter (VSC)

The VSC device has been employed between the AC and DC distribution network [8] for demand transfer and power flow control, which is operated in PQ control

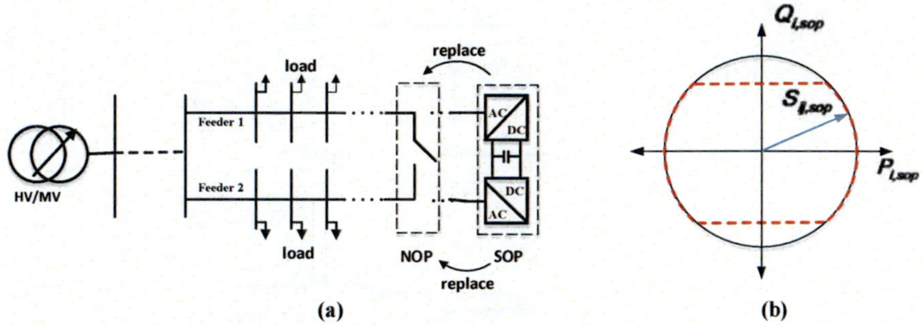

Fig. 1 **a** Sample diagram of soft open point (SOP) **b** P Q capability curve

mode. The operation control constraints of VSC are as follows:

$$\left.\begin{array}{l}\sqrt{\left(P_{t,i}^{VSC}\right)^2+\left(Q_{t,i}^{VSC}\right)^2} \le S_i^{VSC} \\ S_i^{VSC}=2V_{i.DC}I_{i.DC}\end{array}\right\} \quad (1)$$

In (1), $P_{t,i}^{VSC}$, $Q_{t,i}^{VSC}$ represents the active and reactive power injection at VSC. S_i^{VSC} represent the apparent power capacity. $V_{i.DC}$, $I_{i.DC}$ represents the voltage and current at ith DC bus, respectively.

3 Proposed Two-Stage Coordinated Volt/VAR Optimization Organization

A two-stage coordinated VVO policy has been introduced to decrease the energy losses in addition to mitigating the voltage violations in AC-DC HDN. Figure 2 illustrates the proposed scheme, which consists of two stages specifically day-ahead scheduling stage and inner-day dispatch stage. Normally, legacy voltage control devices (e.g. OLTC, CB) could not operate regularly owing to their mechanical switching nature. Therefore, in the proposed day-ahead scheduling stage, legacy voltage control devices are operated. Whereas, advanced voltage control devices (e.g. DER, SOP and VSCs) are operated in both stages owing to their quick action nature.

In stage 1, the settings of legacy and cutting-edge voltage control devices have been calculated based on PV generation outputs and loads day-ahead forecasted values, stochastic analysis has been carried out hourly time scale with the aim of reduction of voltage violations and energy losses. In stage 2 (i.e. inner-day control stage), scheduling has been performed based on the short-term forecast deviations and day-ahead scheduling optimization results, model predictive control (MPC) based rolling optimization control model is employed to control the settings of advanced

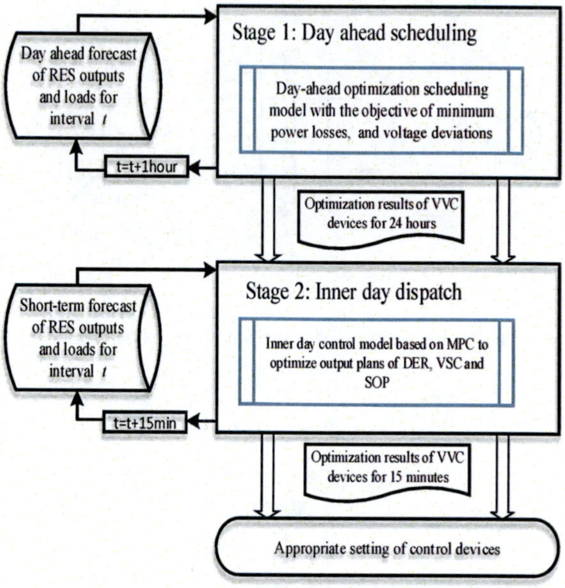

Fig. 2 Architecture of proposed two-stage coordination VVO method

voltage control devices with the aim of reduction of voltage violations and power loss in intra hour time scale.

4 Mathematical Formulation

The mathematical formulation of the two stages has been framed as follows.

4.1 Stage 1

In stage 1, a coordinated optimization model in view of legacy VVC devices, DER, VSC and SOP has been formulated. The aim is to lessen the power loss (f_{loss}) and voltage violations (f_{vd}) simultaneously with optimum scheduling of the remote-controlled switches, legacy and cutting-edge voltage control devices in every hour. The mathematical expression of objective functions as seen in (2)

$$\min .\{f_{loss}, f_{vd}\} \tag{2}$$

where

$$f_{loss} = \sum_{t=1}^{T} \left(\sum_{ij \in \Omega_{br}} R_{ij} I_{t,ij}^2 + \sum_{i'j' \in \Omega_{br}} R_{i'j'} I_{t,i'j'}^2 \right) \tag{3}$$

$$f_{vd} = \sum_{t=1}^{T} \left(\sum_{i \in \Omega_{nd}} \left| V_{t,i}^2 - v_{ref}^2 \right| \right) \tag{4}$$

In (3), the first term represents the power loss ($R_{ij} I_{t,ij}^2$) in the AC network for time interval t. Similarly, the second term represents the power loss $R_{i'j'} I_{t,i'j'}^2$ in the DC network for time interval t. Here, R_{ij},$I_{t,ij}$ represent the resistance and current flow linked among the AC buses i and j. Similarly, $R_{i'j'}$,$I_{t,i'j'}$ represent the resistance and current flow linked among the DC buses i and j. In (4), $V_{t,i}$ represents the voltage magnitude at bus i. v_{ref} represents the reference voltage.

AC power flow constraints: The distribution power flow model and big-M method [9] has been adopted for design of the DN and to integrate network reconfiguration constraints, respectively, and the constraints as follows:

$$\sum_{ij \in \Omega_{br}} \left(P_{t,ij} - R_{ij} I_{t,ij}^2 \right) + P_{t,j} = \sum_{jk \in \Omega_{br}} P_{t,jk} \tag{5}$$

$$\sum_{ij \in \Omega_{br}} \left(Q_{t,ij} - X_{ij} I_{t,ij}^2 \right) + Q_{t,j} = \sum_{jk \in \Omega_{br}} Q_{t,jk} \tag{6}$$

$$-M\left(1 - e_{t,ij}\right) \le \begin{pmatrix} V_{t,j}^2 - V_{t,i}^2 - \left(R_{ij}^2 + X_{ij}^2\right) I_{t,ij}^2 \\ +2\left(R_{ij} P_{t,ij} + X_{ij} Q_{t,ij}\right) \end{pmatrix} \le M\left(1 - e_{t,ij}\right) \tag{7}$$

$$-M e_{t,ij} \le P_{t,ij} \le M e_{t,ij} \tag{8}$$

$$-M e_{t,ij} \le Q_{t,ij} \le M e_{t,ij} \tag{9}$$

$$P_{t,j} = P_{t,j}^{DER} + P_{t,j}^{VSC_{AC}} + P_{t,j}^{SOP} - P_{t,j}^{L} \tag{10}$$

$$Q_{t,j} = Q_{t,j}^{DER} + Q_{t,j}^{VSC_{AC}} + Q_{t,j}^{SOP} + Q_{t,j}^{CB} - Q_{t,j}^{L} \tag{11}$$

$$I_{t,ij}^2 V_{t,i}^2 = P_{t,ij}^2 + Q_{t,ij}^2 \tag{12}$$

where Ω_{br} denotes the set of AC branches; $P_{t,ij}$, $Q_{t,ij}$ represents the active and reactive power flow in a branch linked among bus i and bus j, respectively, for time interval t. $P_{t,j}$, $Q_{t,j}$ represents the active and reactive power injection at bus j, respectively. $V_{t,i}$ is the magnitude of voltage at bus i for time interval t. $P_{t,j}^{DER}/Q_{t,j}^{DER}$ denotes the active and reactive power injected by DER, $P_{t,j}^{VSC_{AC}}/Q_{t,j}^{VSC_{AC}}$ symbolizes the active and reactive power provided by VSC at AC side, $P_{t,j}^{SOP}/Q_{t,j}^{SOP}$ symbolizes the active and reactive power provided by SOP at bus j, respectively. $P_{t,j}^{L}/Q_{t,j}^{L}$ signifies the active and reactive power consumption at bus j, respectively.

DC Power Flow Constraints

$$\sum_{i'j'\in\Omega_{br'}} \left(P_{t,i'j'} - R_{i'j'}I_{t,i'j'}^2\right) + P_{t,j'} = \sum_{j'k'\in\Omega_{br'}} P_{t,j'k'} \tag{13}$$

$$-M\left(1-e_{t,i'j'}\right) \le \left(\begin{array}{l} V_{t,j'}^2 - V_{t,i'}^2 - R_{i'j'}^2 I_{t,i'j'}^2 \\ +2R_{i'j'}P_{t,i'j'} \end{array}\right) \le M\left(1-e_{t,i'j'}\right) \tag{14}$$

$$P_{t,j'} = P_{t,j'}^{DER} + P_{t,j'}^{VSC_{DC}} - P_{t,j'}^{L} \tag{15}$$

$$I_{t,i'j'}^2 V_{t,i'}^2 = P_{t,i'j'}^2 \tag{16}$$

where $\Omega_{br'}$ denotes the set of DC branches; $P_{t,i'j'}$ represents the active power flow in a branch linked among bus i' and bus j' for time interval t. $P_{t,j'}$ symbolizes the active power injection at bus j', respectively. $V_{t,i'}$ symbolizes the voltage magnitude at bus i' for time interval t. $P_{t,j'}^{DER}$ symbolizes the active power injected by DER at bus j', $P_{t,j'}^{VSC_{DC}}$ symbolizes the active power injected by VSC at DC side. $P_{t,j'}^{L}$ symbolizes the active power consumption at bus j'.

5 MISOCP Model Conversion

In this paper, MISOCP method has been chosen to solve complex non-linear mixed-integer programming (MINLP) problem formulated in previous Sect. 4. To realize the linearization, the parameters $v_{t,i}$ and $l_{t,ij}$ are substituted with $V_{t,i'j'}^2$ and $I_{t,i'j'}^2$, respectively, in constraints (3)–(16) which are modified as follows:

$$f_{loss} = \sum_{t=1}^{T}\left(\sum_{ij\in\Omega_{br}} R_{ij}l_{t,ij} + \sum_{i'j'\in\Omega_{br}} R_{i'j'}l_{t,i'j'}\right) \tag{17}$$

$$f_{vd} = \sum_{t=1}^{T}\left(\sum_{i\in\Omega_{nd}} \left|v_{t,i} - v_{ref}^2\right|\right) \tag{18}$$

$$\sum_{ij\in\Omega_{br}} \left(P_{t,ij} - R_{ij}l_{t,ij}\right) + P_{t,j} = \sum_{jk\in\Omega_{br}} P_{t,jk} \tag{19}$$

$$\sum_{ij\in\Omega_{br}} \left(Q_{t,ij} - X_{ij}l_{t,ij}\right) + Q_{t,j} = \sum_{jk\in\Omega_{br}} Q_{t,jk} \tag{20}$$

$$\sum_{i'j'\in\Omega_{br'}}\left(P_{t,i'j'}-R_{i'j'}l_{t,i'j'}\right)+P_{t,j'}=\sum_{j'k'\in\Omega_{br'}}P_{t,j'k'} \tag{21}$$

$$-M\left(1-e_{t,ij}\right)\leq\begin{pmatrix}v_{t,j}-v_{t,i}-\left(R_{ij}^2+X_{ij}^2\right)l_{t,ij}\\+2\left(R_{ij}P_{t,ij}+X_{ij}Q_{t,ij}\right)\end{pmatrix}\leq M\left(1-e_{t,ij}\right) \tag{22}$$

$$-M\left(1-e_{t,i'j'}\right)\leq\begin{pmatrix}v_{t,j'}-v_{t,i'}-R_{i'j'}^2l_{t,i'j'}\\+2R_{i'j'}P_{t,i'j'}\end{pmatrix}\leq M\left(1-e_{t,i'j'}\right) \tag{23}$$

$$\left(\underline{V}\right)^2\leq v_{t,i}\leq\left(\overline{V}\right)^2 \tag{24}$$

$$l_{t,ij}\leq\left(\overline{I}\right)^2 \tag{25}$$

$$\left\|\left[2P_{t,ij}2Q_{t,ij}l_{t,ij}-v_{t,i}\right]^T\right\|_2\leq l_{t,ij}+v_{t,i} \tag{26}$$

$$\left\|\left[2P_{t,i'j'}2Q_{t,i'j'}l_{t,i'j'}-v_{t,i'}\right]^T\right\|_2\leq l_{t,i'j'}+v_{t,i'} \tag{27}$$

Constraints (26) and (27) are the relax constraints of (12) and (16), respectively, in the form of quadratic cone constraints.

Soft Open Point Operation Constraints

$$\left(P_{t,i}^{SOP}\right)^2+\left(Q_{t,i}^{SOP}\right)^2\leq 2\frac{P_{t,i}^{SOP,loss}}{\sqrt{2}A_i^{SOP}}\frac{P_{t,i}^{SOP,loss}}{\sqrt{2}A_i^{SOP}} \tag{28}$$

$$\left(P_{t,j}^{SOP}\right)^2+\left(Q_{t,j}^{SOP}\right)^2\leq 2\frac{P_{t,j}^{SOP,loss}}{\sqrt{2}A_j^{SOP}}\frac{P_{t,j}^{SOP,loss}}{\sqrt{2}A_j^{SOP}} \tag{29}$$

$$\left(P_{t,i}^x\right)^2+\left(Q_{t,i}^x\right)^2\leq 2\frac{S_{t,i}^x}{\sqrt{2}}\frac{S_{t,i}^x}{\sqrt{2}};\ x\in\{DER,VSC,SOP\} \tag{30}$$

Constraints (28)–(30) are the SOP operational relax constraints in the form of rotated quadratic cone constraint.

Capacitor Banks Operation Constraints

$$Q_{t,i}^{CB}=K_{t,i}^{CB}q_i^{CB} \tag{31}$$

$$K_{t,i}^{CB}\in\{0,1\}$$

Equation (31) represents the total reactive power provided by capacitor banks ($Q_{t,i}^{CB}$) at ith bus.

Radial Network Topology Constraints

$$\sum_{(i,j)\in\Omega_{nd}} e_{ij} = nd - 1 \tag{32}$$

$$e_{ij} = \beta_{ij} + \beta_{ij}; \forall(i,j) \in \Omega_{nd} \tag{33}$$

$$\sum_{j\in\Omega_{nd}} \beta_{ij} = 1; \forall i \in \Omega_{nd} and i \neq \Omega_{ss} \tag{34}$$

$$\beta_{ss,j} = 0; \forall i \in \Omega_{ss} \tag{35}$$

where Ω_{nd} and Ω_{ss} signifies the set of nodes and set of substation, respectively. Constraints from (32) to (35) are related to the distribution network radial topology and its connectivity [9]. Equation (32) represents the tree topology. Equation (33) specifies that branch (β_{ij}) when jth node is the parent of node ith or vice-versa (β_{ji}). Equation (34) represents that each bus has only one parent bus. Equation (35) represents that the substation bus has no parent bus.

ZIP Load Model

$$\left.\begin{aligned} P_{t,i}^{L} &= P_{t,i}^{L,nom}\left[\alpha_p^Z\left(\frac{V_{t,i}}{V^{nom}}\right)^2 + \alpha_p^I\left(\frac{V_{t,i}}{V^{nom}}\right) + \alpha_p^P\right] \\ Q_{t,i}^{L} &= Q_{t,i}^{L,nom}\left[\alpha_q^Z\left(\frac{V_{t,i}}{V^{nom}}\right)^2 + \alpha_q^I\left(\frac{V_{t,i}}{V^{nom}}\right) + \alpha_q^P\right] \end{aligned}\right\} \tag{36}$$

$$\left.\begin{aligned} \alpha_p^Z + \alpha_p^I + \alpha_p^P = 1 \\ \alpha_q^Z + \alpha_q^I + \alpha_q^P = 1 \end{aligned}\right\} \tag{37}$$

Equation (36) denotes the ZIP load, where $\{\alpha_p^Z, \alpha_p^I, \alpha_p^P\}$ and $\{\alpha_q^Z, \alpha_q^I, \alpha_q^P\}$ represents the percentage of constant impedance, current and power of active and reactive power, respectively. Equation (36) can be linearized by technique employed in [10] as follows:

$$\left.\begin{aligned} P_{t,i}^{L} &= P_{t,i}^{L,nom} + CVR^p\frac{P_{t,i}^{L,nom}}{2}\left(\frac{v_{t,i}}{(V^{nom})^2} - 1\right) \\ Q_{t,i}^{L} &= Q_{t,i}^{L,nom} + CVR^q\frac{Q_{t,i}^{L,nom}}{2}\left(\frac{v_{t,i}}{(V^{nom})^2} - 1\right) \end{aligned}\right\} \tag{38}$$

where $CVR^p = 2\alpha_p^Z + \alpha_p^I$, and $CVR^q = 2\alpha_q^Z + \alpha_q^I$.

OLTC Operation Constraints

$$v_{t,1} = A_t^{OLTC} v_0 \tag{39}$$

$$A_t^{OLTC} = \sum_{k=1}^{N_{tap}} st_k b_{t,k}^{oltc} \tag{40}$$

$$\sum_{k=1}^{N_{tap}} b_{t,k}^{oltc} = 1 \tag{41}$$

Equation (39) signifies the square of substation voltage. v_0 is square of nominal voltage. A_t^{OLTC} could be calculated by means of (40), here st_k varies from 0.90 to 1.10. $b_{t,k}^{oltc}$ is the binary control variable. Equation (41) guarantees that only one OLTC tap position is chosen.

5.1 Stochastic Optimization

To imitate the effect of forecast error on outcomes of the optimization problem, stochastic analysis has been performed. The beta and normal distribution function have been selected to exhibit the uncertainties in the solar irradiance and load demand, respectively [11]. Besides, MCS has been executed to generate N_s scenarios for load demand and solar irradiance. However, a greater number of scenarios would increase the complexity and computational time. Here, the K-means clustering scheme [7] has been implemented to reduce the number of scenarios (N_r). The reduced scenarios probability has been normalized (π_s^{norm}) as given below.

$$\pi_s^{norm} = \frac{\Pr obability\, of\, occurance\, of\, one\, of\, reduced\, scenario}{Sum\, of\, probability\, of\, occurance\, of\, one\, of\, reduced\, scenario} \tag{42}$$

Here, the objective function in (2) can be rewritten as given in (43) to examine the influence of achieved reduced scenarios (N_r)

$$\min\left\{\sum_{s\in N_r} \pi_s^{norm} \times (f_{loss}, f_{vd})\right\} \tag{43}$$

Subjected to the constraints from (19) to (41).

5.2 *Stage 2*

In stage 2 (i.e. inner-day dispatch stage) to come across the influence of forecast deviations, an inner-day rolling based on model predictive control (MPC) optimization has been developed. Figure 3 shows the MPC-based inner-day rolling scheduling, where the uncertain output power of load demand and solar generation output are forecast over a predicted horizon (T_p) period via MPC. Here, the cutting-edge voltage control devices (e.g. DER, VSC and SOPs) are planned over a (T_p) period, with the attention of constraints given from (19) to (41). Though, the first period (i.e. control horizon (T_c) period) optimized choices in the T_p, are executed whereas the remaining periods are discarded. Afterwards, the T_p is moved over to the following periods and the advanced voltage control devices are planned again with modernized prediction data. Here, the optimal settings of voltage control devices of the first time period would be handover to the real-time dispatch stage. The mathematical expression of the objective function can be expressed as given in (44)

$$\min \left\{ \sum_{t=t_k}^{t_k+T_p} (f_{loss}, f_{vd}) \right\} \tag{44}$$

subjected to the constraints from (19) to (41).

6 Results and Discussions

The 33-bus AC-DC hybrid distribution network (HDN) has been selected for the analysis [5]. Figure 4 depicts the 33 bus AC-DC HDN. CBs with a rating of 300 kVAR are mounted at node no. 3, 6, 11, 15, 23 and 31. The SOP rating of 500 kVA is mounted among bus no. 25 and bus no. 29. The VSC loss coefficient in SOP is considered as 0.02 [6]. Three PV-based DGs having a rating of 250 kVA each are installed at buses no. 10, 17, and 22. Three wind-based DGs rating of 500 kVA are mounted at buses no. 5, 24 and 29. Wind generation output, PV generation output

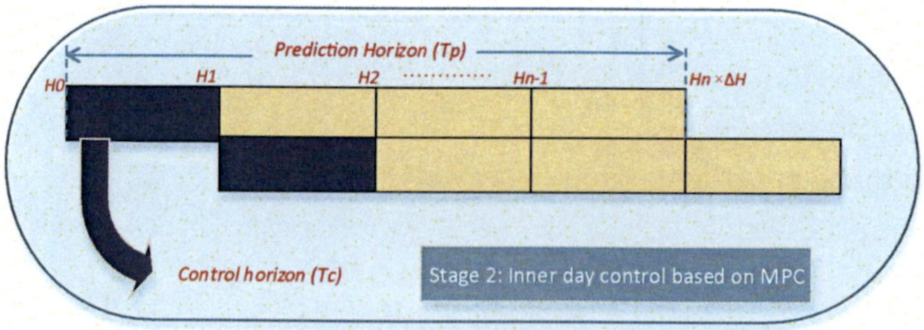

Fig. 3 Model predictive control based Inner-day rolling control model

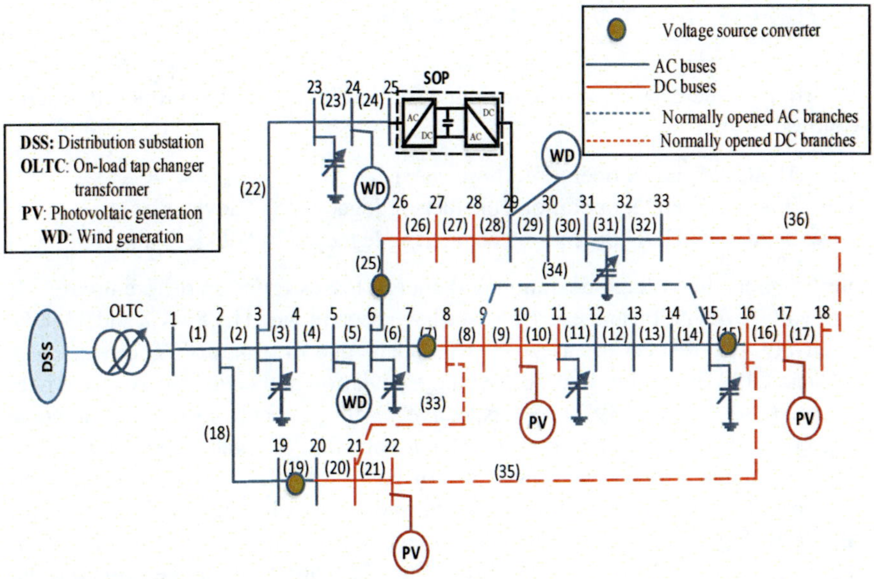

Fig. 4 33 bus AC and DC hybrid distribution network

and load over a typical day (i.e. 24 h) have been depicted in Fig. 5. The permissible range voltage limits are selected as 0.95 pu to 1.05 pu.

In this paper, simulations have been performed on MATLAB environment with GAMS toolbox [12] and CPLEX algorithm. The ε-constraint method has been used to determine Pareto solution and then the fuzzy decision method has applied to determine the best compromise solution [3].

Fig. 5 Typical data of load, PV and wind generation

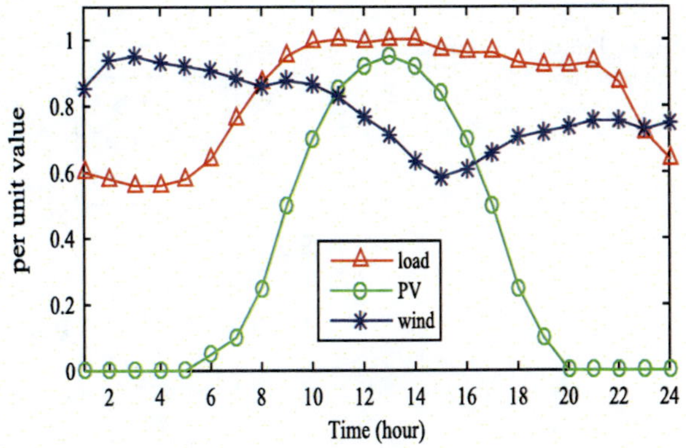

6.1 Discussions on Numerical Results

In this study, three cases have been studied for verification of the proposed scheme as follows.

- Case 1: AC-DC HDN operation (base case)
- Case 2: AC-DC HDN operation considering legacy VVC devices only
- Case 3: AC-DC HDN operation considering legacy VVC devices and SOP.

Table 1 depicts the voltage deviation and energy losses found in different cases. In case 2, voltage deviation and energy losses have been reduced by 8.94% and 33.09%, respectively with respect to case 1. This happens due to legacy VVC devices (e.g. OLTC and CBs) operation. In case 3, voltage deviation and energy losses have been decreased to 41.93% and 39.278%, respectively with respect to case 1. This happens owing to the SOP device operation with the association of legacy VVC devices.

Figure 6 depicts the active power loss of 33-bus AC-DC HDN under different cases. From Fig. 6, it can be observed higher power loss has been achieved in case 3. Similarly, Figs. 7, 8 and 9 depicts the voltage magnitude profile of AC-DC HDN under case 1 to case 3, respectively. From Figs. 7, 8 and 9, it can be observed that the voltage magnitude profile of HDN has been improved and maintained within permissible limit under case 3.

Table 1 Results in three different cases

Cases	Case 1	Case 2	Case 3
f_{loss} (MWh)	5.68	3.8	3.449
Δf_{loss} (%)	–	33.09	39.278
f_{vd}	90.358	82.28	52.463
Δf_{vd} (%)	–	8.94	41.93

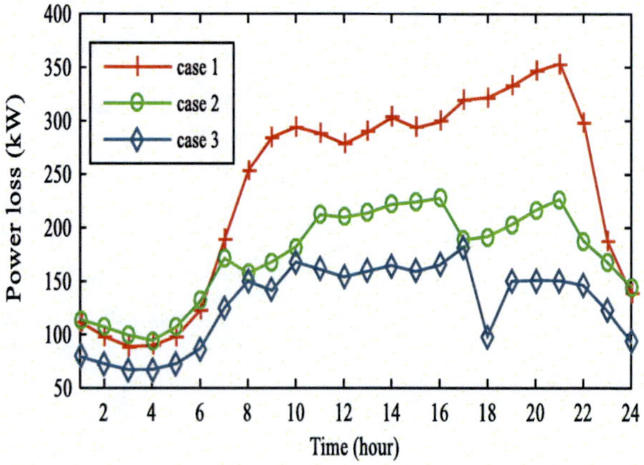

Fig. 6 Power loss under different cases

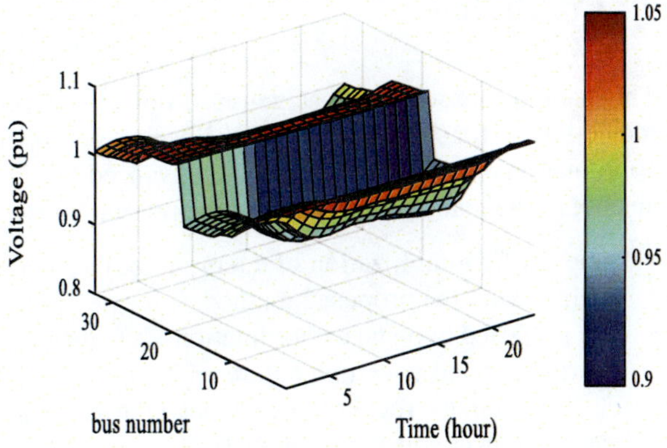

Fig. 7 Voltage magnitude profile AC-DC HDN under case 1

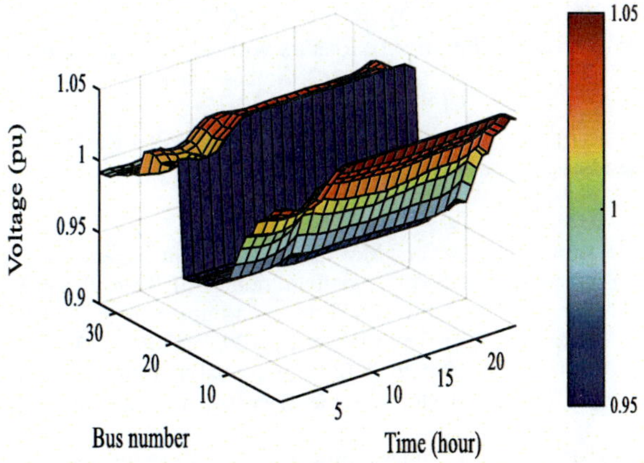

Fig. 8 Voltage magnitude profile of AC-DC HDN under case 2

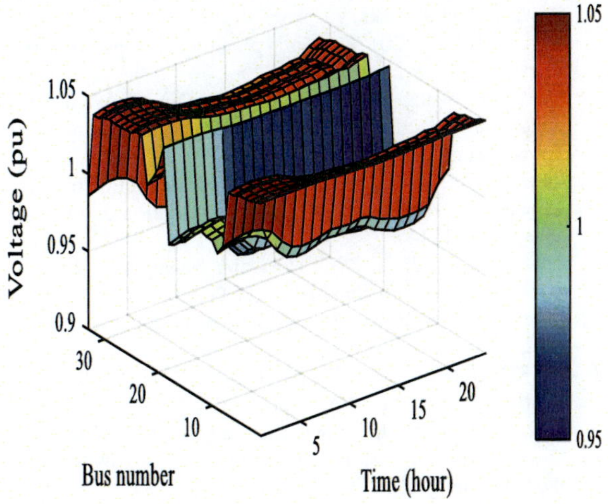

Fig. 9 Voltage magnitude profile of AC-DC HDN under case 3

7 Conclusion

This paper proposed a two-stage multi-objective stochastic Volt/VAR optimization methodology for AC-DC HDN considering soft open points (SOP) to realize the coordination operation of legacy and advanced VVC devices that have coexisted in an HDN. The main observations of the present investigations are:

- Noteworthy minimization of energy losses as well as voltage deviation has been attained with the proposed method.
- The proposed two-stage coordination VVC methodology is efficient to handle the intermittent and uncertain nature of RES.
- Further, voltage regulation and power loss minimization can be achieved by reactive power support through SOP.

Therefore, the deployment of the proposed scheme in AC-DC HDN is reliable and also reduces the energy loss as well as voltage deviations simultaneously.

References

1. Ahmed HM, Salama MM (2019) Energy management of AC–DC hybrid distribution systems considering network reconfiguration. IEEE Trans Power Syst 34(6):4583–4594
2. Wang Z, Wang J (2013) Review on implementation and assessment of conservation voltage reduction. IEEE Trans Power Syst 29(3):1306–1315
3. Pamshetti VB, Singh SP (2019) Optimal coordination of PV smart inverter and traditional Volt-VAR control devices for energy cost savings and voltage regulation. Int Trans Electr Energy Syst 29(7):e12042
4. Pamshetti VB, Singh S, Singh SP (2019) Combined impact of network reconfiguration and Volt-VAR control devices on energy savings in the presence of distributed generation. IEEE Syst J 14(1):995–1006
5. Ahmed HMAM (2017) Optimal planning and operation of AC-DC hybrid distribution systems. Thesis
6. Li P, Ji H, Wang C, Zhao J, Song G, Ding F, Wu J (2017) Coordinated control method of voltage and reactive power for active distribution networks based on soft open point. IEEE Trans Sustain Energy 8(4):1430–1442
7. Pamshetti VB, Singh S, Singh SP (2020) Reduction of energy demand via conservation voltage reduction considering network reconfiguration and soft open point. Int Trans Electr Energy Syst 30(1):e12147
8. Zhang L, Liang J, Tang W, Li G, Cai Y, Sheng W (2018) Converting AC distribution lines to DC to increase transfer capacities and DG penetration. IEEE Trans Smart Grid 10(2):1477–1487
9. Chen X, Wu W, Zhang B (2015) Robust restoration method for active distribution networks. IEEE Trans Power Syst 31(5):4005–4015
10. Jha RR, Dubey A, Liu CC, Schneider KP (2019) Bi-level volt-var optimization to coordinate smart inverters with voltage control devices. IEEE Trans Power Syst 34(3):1801–1813
11. Li R, Wang W, Wu X, Tang F, Chen Z (2019) Cooperative planning model of renewable energy sources and energy storage units in active distribution systems: a bi-level model and Pareto analysis. Energy 168:30–42
12. https://www.gams.com/download.

Effective Power Management in Renewable Energy Resources Based Power System Incorporating Electric Spring

Om Krishan and Sathans Suhag

Abstract The ever-increasing adoption of renewable energy resources (RERs) for fulfilling the electricity requirement led to some severe issues in the operation of modern power systems, such as frequency and voltage deviations. This problem is more significant in grid-isolated power systems, where grid inertia is not available to maintain frequency and voltage levels in specified limits. In this paper, an electric spring (ES) and its coordinated control are proposed for the dynamic active power compensation for frequency and voltage regulation in a grid-isolated RERs-based power system under different operating conditions. The phase angle and the amplitude control are adopted for frequency and voltage regulation, respectively. The simulation is performed in the MATLAB/Simulink platform.

Keywords Renewable energy resources · Electric spring · Power management · Voltage regulation · Battery energy storage

1 Introduction

The ever-increasing demand for electricity and worldwide government policies to minimize environmental pollution led to the more penetration of renewable energy resources (RERs) into electricity generation [1]. RERs provide a sustainable solution to meet increasing electricity demand and help transmission system deferral and reduce power loss. On the other hand, due to the high penetration level of RERs, there is always a mismatch between supply and demand, causing various issues like frequency and voltage deviations and power quality problems. These issues can lead to overheating electrical machines, false tripping of converters and generators, and power quality problems. At all levels, the mismatch between generation and demand can significantly challenge the power system's operation [2–4]. These challenges are well documented and tackled by different researchers. Demand-side management

O. Krishan (✉) · S. Suhag
National Institute of Technology Kurukshetra, Kurukshetra, India

S. Suhag
e-mail: sathans@nitkkr.ac.in

A. Kumar et al. (eds.), *Renewable Energy Towards Smart Grid*, Lecture Notes in Electrical Engineering 823, https://doi.org/10.1007/978-981-16-7472-3_24

has been extensively used to counter the active power mismatch between supply and demand caused by the intermittency of RERs [5]. Energy storage systems (ESSs) are also a solution to this problem, but the high cost of ESSs restricts their widespread use [6, 7]. Different types of direct load controllers are also proposed, but they also have some issues with the exact customer's load modeling, stimulation guidelines, etc. [8]. These methods have their own strengths and weaknesses, and one cannot entirely rely on a single method for active power balance in RERs-based power systems. The use of electric spring (ES) is recently suggested for active power balance in RERs-based power systems. ES is based on Hooke's Law and can be realized practically by using power-electronics-based circuits for active power balance between supply and demand. In this paper, a new control strategy is proposed for power management in RERs-based power system consisting of ES is presented.

2 Brief Overview of Electric Spring

The operating principle of ES draws an analogy from the conventional mechanical spring [9]. In RERs power system, ES can be implemented with an inverter circuit connected to the non-critical load in series connection as depicted in Fig. 1.

Non-critical loads are those types of load that can be operated on a wide range of voltage levels and can be switched on and off as per requirement. For example, refrigerators, heaters, air conditioners, etc., can be used as non-critical loads, and the voltage and power can be allowed to fluctuate in non-critical loads. Whereas critical load requires strict voltage and power, and variation in frequency and voltage cannot be afforded. Electricity supply to the security system, hospital loads, etc. Analogous to conventional mechanical spring, ES can be used for the following purposes:

- Store/release electricity according to the requirements.
- Provide active power support to the RERs-based power system.
- For limiting oscillations magnitude.

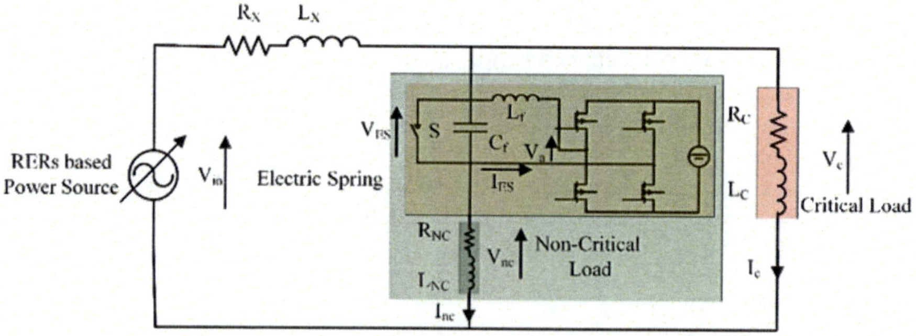

Fig. 1 Circuit diagram of electric Spring

Figure 1 shows the simplified diagram of RERs power system connected with ES with the help of an inverter to the rest of the system. The L-C filter filters the voltage of the inverter circuit as per Fig. 1. The non-critical load connected in series with the inverter circuit helps in making active power compensation. When ES is not connected to the system, the voltage across the critical load is equal to the voltage across the non-critical load and dependent upon the availability of the RERs-based power source.

$$\overrightarrow{V_c} = \overrightarrow{V_{nc}} = \overrightarrow{V_{in}} \tag{1}$$

In case the power availability of the RERs-based source is more than the load requirement, then the voltage across the critical load V_c is more than the reference voltage across the load V_{cref}. With ES connected to the system, the voltage across the critical load is given as follows:

$$\overrightarrow{V_c} = \overrightarrow{V_{es}} + \overrightarrow{V_{nc}} \tag{2}$$

The battery energy storage system connected on the input side of the inverter helps in bidirectional power flow.

2.1 *Modelling of Electric Spring*

The ES shown in Fig. 1 can be realized using an inverter circuit. In the mathematical modeling of ES, the effective resistance and inductance (R_f and L_f) of the filter circuit are neglected, and also the inverter is considered to be lossless. From Fig. 1, on applying KVL and KCL following equations are obtained:

$$\overrightarrow{V}_a - \overrightarrow{V}_{ES} = \overrightarrow{V}_{Lf} = L_f \frac{d\overrightarrow{I}_{ES}}{dt} \tag{3}$$

where V_a and I_{ES} are the voltage and current across inverter, V_{Lf} is the voltage across filter circuit.

Also,

$$\overrightarrow{V}_s = Z_c \overrightarrow{I}_c = Z_c\left(\overrightarrow{I}_{in} - \overrightarrow{I}_{nc}\right) \tag{4}$$

$$C_f \frac{d\overrightarrow{V}_{ES}}{dt} = \overrightarrow{I}_{ES} + \overrightarrow{I}_{nc} = \overrightarrow{I}_{ES} + \overrightarrow{I}_{in} - \frac{\overrightarrow{V}_s}{Z_c} \tag{5}$$

Here, C_f is the filter capacitance, Z_C is the critical load impedance, I_{nc} is the current through the non-critical load, and I_{in} is the current from RERs base power source.

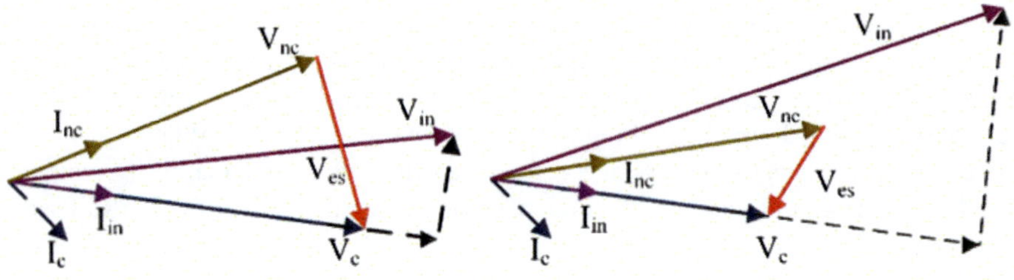

Fig. 2 Phasor diagram of Electric Spring for **a** under-voltage and **b** over-voltage

The filter circuit only allows fundamental frequency components of voltage V_{a1} and filters out high-frequency components, the Va can be written as

$$\vec{V}_a = \vec{V}_{a1} = \vec{m} \times V_{DC} \tag{6}$$

From Eqs. (3) and (6), the following equation is obtained:

$$L_f \frac{d\vec{I}_{ES}}{dt} = V_{DC}\vec{m}(t) - \vec{V}_{ES} \tag{7}$$

Here m is the modulation index.

Figure 2 shows the flexible operation of ES with both under-voltage and over-voltage scenarios.

In under-voltage case, the voltage, the capacitive and real power are injected by the ES for increasing the voltage V_c up to the level of the reference voltage. On the other hand, in the overvoltage case, the inductive and real power is injected to perform voltage regulation.

3 Power Management Strategy for Electric Spring in RERs-Based Power System

In this paper, ES is employed for active power compensation in a power system with high penetration of RERs, and its control strategy is shown in Fig. 3. The magnitude of the error difference between the V_c and V_{cref} decides the amount of active power compensated, and sign (±) difference determines whether active power is absorbed/injected from/to the system. A phased-locked loop (PLL) is employed for calculating the phase angle of I_{ES}.

PI controller compensates for the difference between Vc and Vcref and that signal is further passed through a scale factor 'K' for limiting the error in the range of ± 1. The signal is thus obtained, and the output of the phase-modulated sinusoidal signal

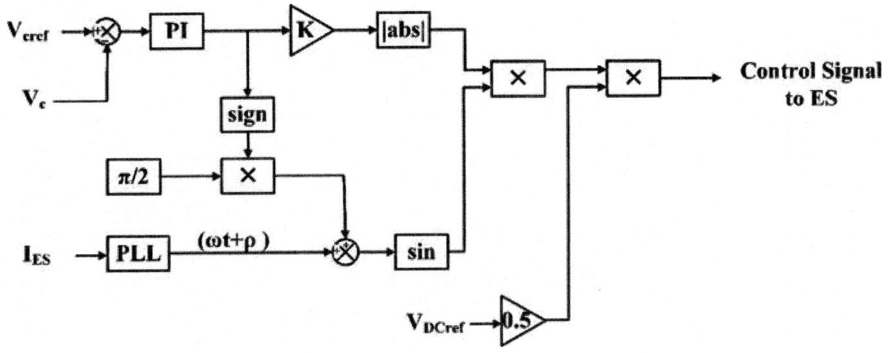

Fig. 3 Control strategy for electric spring

is multiplied, which is further used in the multiplication of the reference DC link voltage to generate control signals of the inverter circuit.

4 Simulation Results and Discussion

The RERs-based power system shown in Fig. 1 is simulated in MATLAB/Simulink environment. The RERs includes wind energy conversion system (WECS) and PV system. BESS balances the power and to protect BESS from deep charge and discharge, its upper and lower SoC limit is restricted to 20% and 80%, respectively. The output power of PV system and WECS is varied depending upon the availability of solar irradiance and wind speed as shown in Fig. 4a, d, respectively. The critical load is also varied at certain time intervals, as shown in Fig. 4g. The ES is controlled to absorb excess power as power consumed by the non-critical load when BESS is on full SoC level and able to maintain the DC link voltage and frequency in limits as shown in Fig. 5.

In order to confirm the efficacy of the control system, the grid isolated hybrid power system is subjected to variable operating conditions, viz. variable irradiance level, variable wind speed, and variable-connected electric load. The solar irradiance is varied at time intervals at 2 s and 4 s as shown in Fig. 5b. The corresponding current and power of the solar system are also varied and can be seen in Fig. 4a, d, respectively. The wind speed is changing from 8 m/sec to 10 m/sec and 10 m/sec to 12 m/sec at a time interval of 3 s and 5 s, respectively. The current and power obtained from WECS are also changed at same time interval which can be observed from Fig. 4d, g, respectively. The load is changed from 7.5 kW to 5 kW at a time interval of 6 s. To compensate for these variations, the BESS and ES come in to action. As the power generation from the solar system and WECS increased coupled with a reduction in load. So slow transients of excess power available go to BESS and fast transients to SCES. The BESS takes excess power whenever there is an

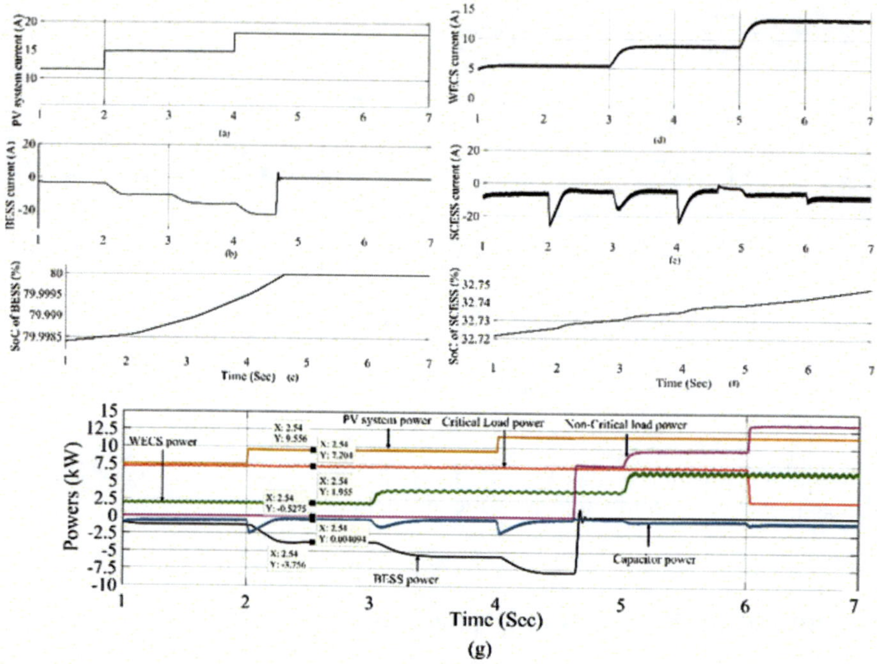

Fig. 4 Simulation results of RERs-based power system under different operating conditions

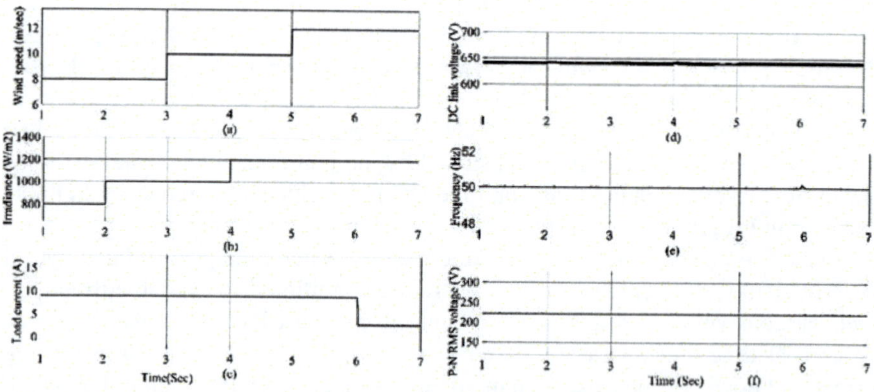

Fig. 5 Various waveforms of RERs-based system under different operating conditions

increase in generation at time intervals; 2 s, 3 s, and 4 s. The SoC of BESS reaches up to 80% at time 4.6 s, so that it cannot take more power for its charging. At this time, non-critical load gets connected to the system and takes the excess power and BESS gets disconnected. Meanwhile, the fast transients due to variable operating conditions are taken care of by ES as shown in Fig. 4d. From Fig. 5, it can be clearly seen that the voltage across the critical load and frequency level is not out of specified limits and good quality power is available across it. On the other hand, the power balance between different components of RERs-based power system components and V_{DC} are also maintained very effectively despite variations in wind speed, solar irradiance, and critical load.

5 Conclusions

In this paper, a power management strategy for the different components of RERs-based power system, consisting of WESC and PV system and BESS as a backup source and ES is proposed. ES is able to effectively to maintain the balance between supply and demand. In the case of excess power available from the RERs, the hybrid configuration of ES and BESS absorbs that excess power and thereby maintains power balance, which means DC link voltage and frequency are held within the specified limits. The system is subjected to different operating conditions like changes in wind speed, solar irradiance, and connected load. The frequency, V_{DC}, and the three-phase RMS voltage across the critical load, all are within the specified limits and the simulation results obtained verify the efficiency of the control scheme.

References

1. Krishan O, Suhag S (2018) An updated review of energy storage systems: Classification and applications in distributed generation power systems incorporating renewable energy resources. Int J Energy Res 1–40 (2018). https://doi.org/10.1002/er.4285
2. Vidyanandan KV, Senroy N (2016) Frequency regulation in a wind–diesel powered microgrid using flywheels and fuel cells. IET Gener Transm Distrib 10:780–788. https://doi.org/10.1049/iet-gtd.2015.0449
3. Krishan O, Sathans (2016) Frequency regulation in a standalone wind-diesel hybrid power system using pitch-angle controller. In: 2016 3rd International Conference on Computing for Sustainable Global Development (INDIACom). pp 1148–1152, New Delhi (2016)
4. Krishan O, Suhag S (2019) Techno-economic analysis of a hybrid renewable energy system for an energy poor rural community. J Energy Storage 23:305–319. https://doi.org/10.1016/j.est.2019.04.002
5. Groppi D, Pfeifer A, Garcia DA, Krajačić G, Duić N (2021) A review on energy storage and demand side management solutions in smart energy islands. Renew Sustain Energy Rev 135. https://doi.org/10.1016/j.rser.2020.110183
6. Krishan O, Suhag S (2020) Power management control strategy for hybrid energy storage system in a grid-independent hybrid renewable energy system: a hardware-in-loop real-time verification. IET Renew Power Gener 14:454–465. https://doi.org/10.1049/iet-rpg.2019.0578
7. Krishan O (2018) Sathans: design and techno-economic analysis of a HRES in a rural village. Procedia Comput Sci 125:321–328. https://doi.org/10.1016/j.procs.2017.12.043

8. Peyghami S, Mokhtari H, Blaabjerg F (2017) Decentralized load sharing in a low-voltage direct current microgrid with an adaptive droop approach based on a superimposed frequency. IEEE J Emerg Sel Top Power Electron 5:1205–1215. https://doi.org/10.1109/JESTPE.2017.2674300
9. Hui SY, Lee CK, Wu FF (2012) Electric springs—a new amart grid technology. IEEE Trans Smart Grid 3:1552–1561. https://doi.org/10.1109/TSG.2012.2200701

Bismuth Phosphate as an Efficient Electrode Material for Energy Storage Device Applications

Aman Joshi, Sunaina, and Prakash Chand

Abstract Bismuth phosphate nanostructures act as an active electrode material for next-generation energy storage device applications. Here, we report the effect of hexagonal phase, monoclinic phase, reaction time, and temperature on electrochemical properties of $BiPO_4$ nanostructures prepared with microwave technique. The X-ray diffraction (XRD) patterns of the synthesized materials confirmed the formation of pure phase formation $BiPO_4$ nanostructures. The Debye Scherrer formula calculated the size of the crystals. All the synthesized bismuth phosphate materials are in the nanometers range, which was also exposed to Field Effect Scanning Electron Microscopy (FESEM) assessment. Optical properties are explained by Raman spectra and Fourier Transform Infrared Spectroscopy (FTIR). Electrochemical properties are studied via cyclic voltammetry (CV), galvanostatic charge–discharge (GCD), and electrochemical impedance spectroscopy (EIS) techniques. The power law is used to confirm the behavior of the electrode, i.e., of battery type or supercapacitor type. This power law is applicable to CV curves of the samples. Further, the specific capacity was calculated from the GCD curve. The study reveals that bismuth phosphate can be used as an efficient electrode material for energy storage device applications.

Keywords Bismuth phosphate · Electrochemical · Energy storage

A. Joshi (✉) · Sunaina · P. Chand
Department of Physics, National Institute of Technology, Kurukshetra, Haryana 136119, India
e-mail: aman_6180070@nitkkr.ac.in

P. Chand
e-mail: prakash@nitkkr.ac.in

A. Joshi
Department of Physics, J.C. Bose University of Science and Technology, YMCA, Faridabad, Haryana 121006, India

A. Kumar et al. (eds.), *Renewable Energy Towards Smart Grid*, Lecture Notes in Electrical Engineering 823, https://doi.org/10.1007/978-981-16-7472-3_25

1 Introduction

The energy crisis and environmental degradation are currently two main concerns preventing the improvement in the quality of life of human society. As of today, the use of energy and environmental conservation is of great importance to our universe. Power is a critical asset to economic growth and environmental quality [1–3]. Our insatiable consumption of fossil fuels has contributed to global warming and changing the climate. The heavy use of these fossil resources has also caused a great deal of environmental pollution [4]. In the current energy situation, in everyone's life, portable devices such as laptops, mobiles, and all the other electronic gadgets use too much from morning to night. Clean energy requirements are more important than ever, so researchers have also concentrated on the production and use of renewable energy sources. For instance, solar cells, batteries, and supercapacitors (SCs) are useful technologies for energy conversion and storage to address the lack of natural resources and environmental pollution [5, 6]. Figure 1 shows the difference between battery and supercapacitor. In addition, much work has also been based on the photocatalyst's use of solar energy to extract organic contaminants from industrial waste water and the evolution of hydrogen by splitting water [7, 8]. SCs are supposed to be the best energy storage devices because of their advantages over conventional devices like batteries. SCs have great power density, safer, long charged/discharged lifecycle, and are highly reliable. Ultracapacitors or electrochemical capacitors are also called SCs. Classification of supercapacitors is depicted in Fig. 2. There are two types of SCs: pseudocapacitors and Electric Double-Layer Capacitors (EDLCs). The mechanism of charge storage in both forms of SCs is distinct. In EDLCs, there is pure physical charge accumulation during the charging and discharging process, and all carbon materials fall in this category. The materials for EDLCs are activated carbon,

Function	Supercapacitor	Battery
Charge time	1 – 10 seconds	10 – 60 Minutes
Cycle life	1 Million	500 or Higher
Cell Voltage	2.3 V – 2.75 V	3.6 V – 3.7 V
Specific Energy (Wh/kg)	5	100 - 200
Specific Power (W/kg)	Up to 10000	1000 - 3000
Service Life	10 to 15 years	5 to 10 years
Charge Temperature	-40 to 65^{0} C	0 to 45^{0} C

Fig. 1 Difference between supercapacitor and battery

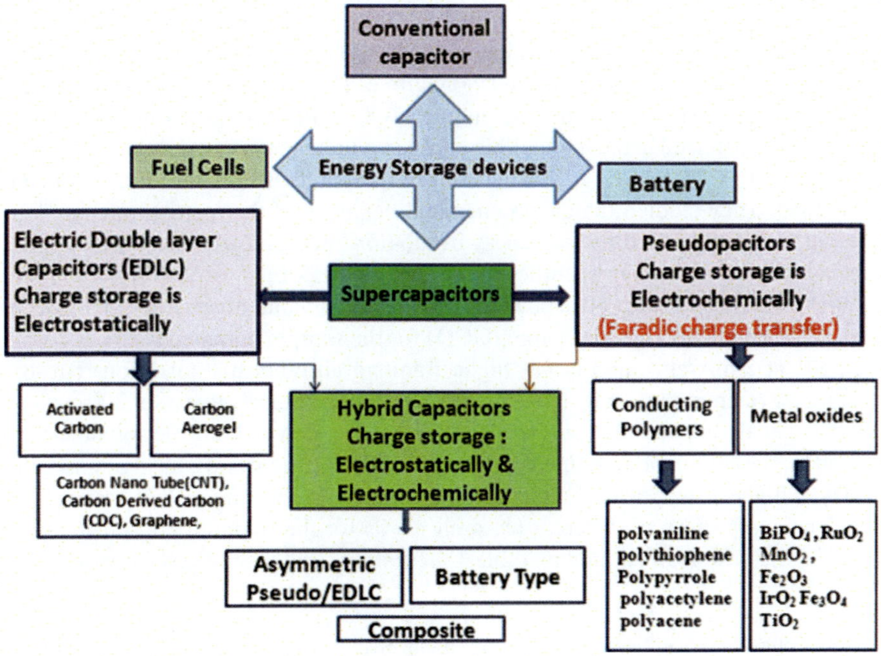

Fig. 2 Classifications of supercapacitors

carbon aero gel, Carbon Nano Tube (CNT), Carbon Derived Carbon (CDC), and graphene, while in the pseudocapacitor; redox (reduction and oxidation) reactions take place. Materials for pseudocapacitors are conducting polymers like polyaniline, polythiophene, polypyrrole, polyacetylene, polyacene, and various metal oxides like $BiPO_4$, RuO_2, MnO_2, Fe_2O_3, IrO_2, Fe_3O_4, and TiO_2, etc. Figure 3a–c shows the mechanism of EDLCs, Pseuodocapacitors, and hybrid supercapacitors, respectively. Among all the Bismuth-based compounds used for energy storage applications, the

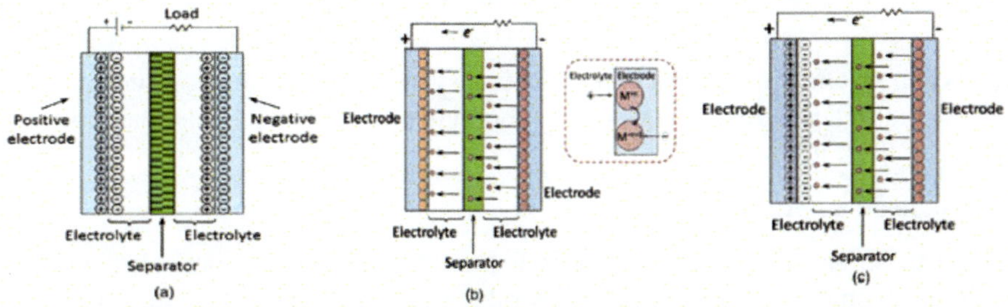

Fig. 3 (**a–c**) Schematic representation of supercapacitor Taxonomy: (**a**) EDLCs type (**b**) pseudocapacitor type (**c**) hybrid capacitors type. (Reproduced with permission from Ref. [17], © American Society of Civil Engineers 2013)

$BiPO_4$ gathered great attention due to its high thermal and chemical stability, eco-friendly nature, and abundantly available sources. The two crystal forms in which $BiPO_4$ mainly exists are hexagonal and monoclinic. The hexagonal phase is formed at low temperature and the co-ordination number is 8, while at relatively high temperature, it gets transferred to the monoclinic phase in which nine oxygen atoms surround Bi^{3+}. In 2015, V.D. Nithya et al. successfully prepared $BiPO_4$ by ultrasonication method [9]. They studied the electrochemical behavior of the material prepared at different irradiation conditions such as irradiation time and ultrasonic power and revealed that the maximum specific capacitance of 1052 Fg^{-1} is obtained when the reaction time was 2 h with 60% power. The calculated value of specific capacitance from galvanostatic charge–discharge (GCD) for the sample prepared with pH 7 was 302 Fg^{-1} at 2 mA/cm^2, and it was highest among all other pH conditions but the capacitance retention was only 64% of its starting capacitance value after 200 cycles. S. Vadivel et al. prepared the $BiPO_4$/MWCNT composite via the facile solvothermal method for the application of supercapacitors and photocatalysts [10]. This composite displayed highest specific capacitance value of 504 F g^{-1} when 5mVs^{-1} scan rate is used. The comparison of different research works for the Bismuth phosphate as an electrode material is illustrated in Table 1. The smart grid technology integrates the electrical network with digital communication systems. This technology can reduce the problems related to existing system by supplying renewable, reliable, and superior electricity to all users.

Table 1 Comparison of different research works for the Bismuth phosphate as an electrode material

Sr. No	Material	Method of fabrication	Electrolyte	Specific capacitance Fg^{-1}/ capacity(Cg^{-1})	References
1	$BiPO_4$	Microwave	2 M KOH	268 Fg^{-1} at 1Ag^{-1}	[11]
2	$BiPO_4$	Hydrothermal	4 M KOH	446 Fg^{-1} at 1Ag^{-1}	[4]
3	$BiPO_4$	Microwave	2 M KOH	1074 Cg^{-1} at 1Ag^{-1}	[12]
4	$BiPO_4$	Microwave	2 M KOH	610 Cg^{-1} at 1Ag^{-1}	[13]
5	$BiPO_4$	Microwave	2 M KOH	104 Fg^{-1} at 1Ag^{-1}	[14]
6	$BiPO_4$	Sonochemical synthesis	1 M KOH	302 Fg^{-1} at 2 $mAcm^{-2}$	[15]
7	$BiPO_4$	Hydrothermal	1 M KOH	480 Fg^{-1} at 1 $mAcm^{-2}$	[16]

Electrochemical Study

There can be any method of synthesis for the preparation of material, but after synthesis common techniques are used for the assessment of electrochemical properties of these materials. All the samples are characterized via CV, GCD, and EIS. Important properties of the material for the electrodes and their controlling parameters are shown in Figure 4.

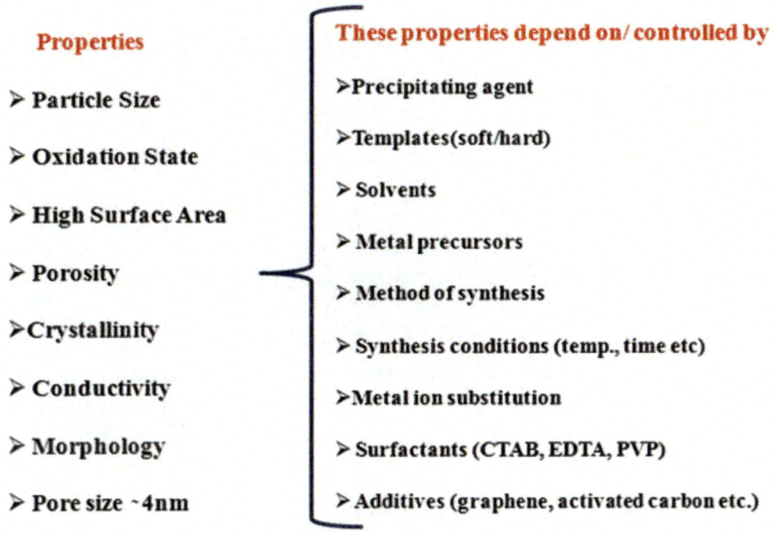

Fig. 4 Important properties of the electrode material and their controlling parameters

1.1 Fabrication of Electrodes

Working electrodes are fabricated from the synthesized material by adding synthesized material, activated carbon, Polyvinylidene fluoride (PVDF) in a particular ratio, i.e., 80:10:10. N-Methyl-2-pyrrolidone (NMP) is used as a solvent for mixed material slurry formation. Then around 1×1 cm electrodes are prepared on the nickel foil through the technique of the doctor blade or drop casting process. Before and after the slurry has been coated, the weight of the active material is measured. Figure 5 illustrates all the requirements of electrode materials for electrochemical supercapacitors (Fig. 6).

1.2 Cyclic Voltammetry (CV) Analysis

The CV technique is an essential electrochemical characterization used on the electrode/electrolyte interfaces to confirm the behavior of electrodes. CV analysis approves the Faradaic or non-Faradaic reactions that occur in the material. The affirmation of the electrode's potential window has a significant impact on the effective and safe utilization of energy storage devices. The potential window of the BiPO4 samples as prepared is -1.2 V to 0.2 V, inside which oxidation and reduction reaction happens completely in 2 M KOH solution as an aqueous electrolyte [11]. The oxidation–reduction conversion among various valence states of bismuth ions is shown by the observed CV peaks. The ions in the electrolyte are transported toward the

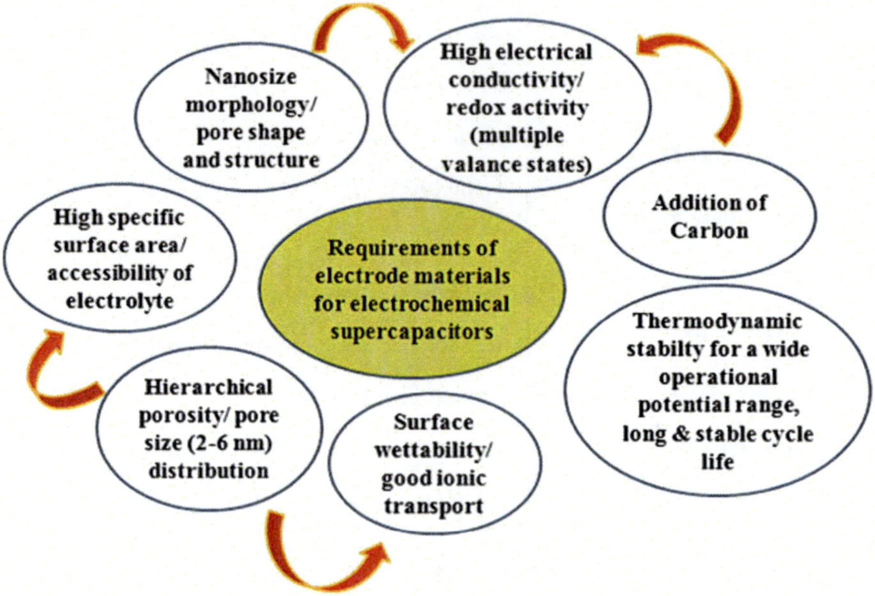

Fig. 5 Electrode material requirements for electrochemical supercapacitors

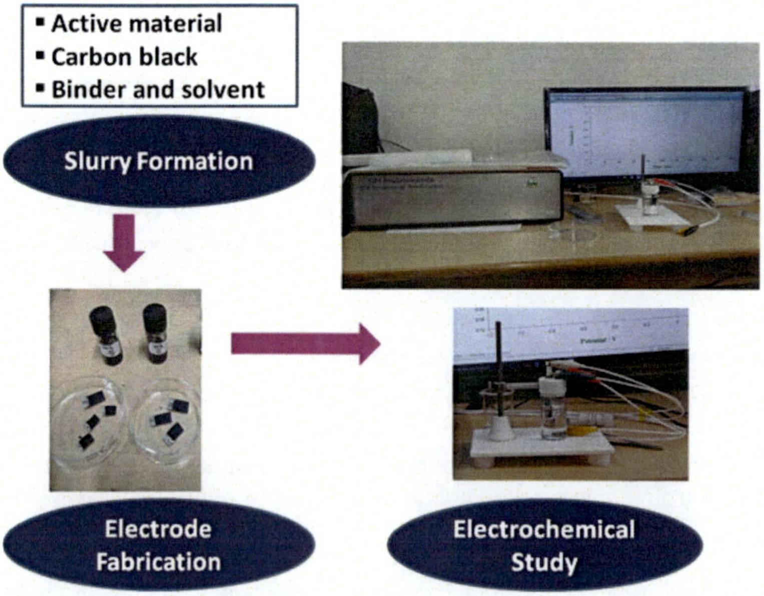

Fig. 6 Various steps involved in electrochemical study of electrodes

electrode's surface by oxidation–reduction reactions as a result of electrostatic interaction to counter-balance the charge on the electrode. The following Eq. 1 can be defined as the injection of K + ions through the electrochemical reaction at fabricated electrodes.

$$BiPO_4 + yK^+ + ye^- \leftrightarrow K_yBiPO_4Minimize(P_{LOSS}) \quad (1)$$

Here, 'y' is the mole fraction of the K^+ ions inserted. Power law is used to confirm the nature of the process occurring during CV. $i_p = a.v^b$, according to Power law, where i_p is cathodic peak response, v is the scan rate, and b is the log (i) – log (v) plot slope. If the value of b is equal to 0.5, then it is an electrode of the battery type, and if the value of b is equal to 1, it is an electrode of the supercapacitor type. For bismuth phosphate material, this value is nearly equal to 0.5, which represents a diffusion process, is taking place during reduction and oxidation, and confirms the electrode material's battery type behavior.

1.3 Galvanostatic Charge Discharge (GCD) Analysis

There are two types of discharge behavior in general, which are linear and non-linear discharge forms. In the first kind, i.e., rectilinear state, the storage of the charge is centered on the reaction of adsorption–desorption. The non-linear state, on other hand, designates the electrode's battery type performance. For bismuth phosphate material, a non-linear discharging behavior was observed at different current densities that specified the redox reactions in the synthesized material and confirmed the battery type behavior. It also agrees strongly with the result of the cyclic voltammetry curve. The GCD characterization technique is a suitable practice for approximating the rate capability and capacity of the electro-active synthesized material. Because of the diffusion process occurring during the electrochemical reaction, specific capacity is calculated according to Eq. 2 [14].

$$Cs = \frac{iX\Delta T}{M}(\text{in Cg} - 1) \quad (2)$$

Here, Cs is the value of calculated specific capacity, i is the applied current density, ΔT is the discharged time interval, and M is the active mass of the electrode material coated on the nickel foil.

With the increasing applied current density values, the value of calculated specific capacity is going to be declined. It can be seen that the ions will have ample time to travel at a low value of applied current density to store extra charge, resulting in the elevated value of specific capacity. Although at an upper value of functional current density, the ions would not have the right timing for the happening of faradic response, i.e., dissemination restricted behavior in the electro-active materials. Therefore the conclusion is the lower value of specific capacity.

1.4 Electrochemical Impedance Spectroscopy (EIS) Analysis

The material is analyzed by EIS with Nyquist plots and bode plot, which is recorded in the 0.1 Hz-100 kHz frequency range. The Nyquist plot can be depicted as the graph between real and imaginary portion of the impedance and the angle made by vector Z with X-axis is called the phase angle. The inclination of this angle describes the behavior of energy storage mechanism. Faradic reactions are negligible in ideal polarizable electrodes, and the current contributes to the double-layer capacitor being charged and discharged at the electrode electrolyte interface. CD is used instantaneously to demonstrate the impedance performance of numerous interface components existing in the supercapacitor device for a partly polarizable electrode, both the faradic charge transfer resistance, RF, and double-layer capacitance. In the bode plot, log of frequency is plotted on X-axis while both the log of absolute value of impedance and phase angle are plotted on Y-axis. This helps the systems to establish a relaxation time constant, and the real and imaginary component of the complex power vs. the frequency plots provide information on the frequency activity of the supercapacitor cells [15].

2 Conclusion

Bismuth phosphate material has immersed enormous deliberation for energy storage system applications as expectant active electrode material based on highly thermally and chemically stable, higher theoretical specific capacity, eco-friendly nature, safety, copious possessions cost-effectiveness. Additional, $BiPO_4$ is significant for improving the galvanic features of phosphate, reflecting futuristic curiosity as an ion-conducting, electronic, and rapid ion conductor. Such distinctive features invent it as a very inspiring electrode material for such applications.

References

1. Owusu KA et al (2017) Low-crystalline iron oxide hydroxide nanoparticle anode for high-performance supercapacitors. Nat Commun 8:14264
2. X. Shi et al., Recent advances of graphene-based materials for high-performance and new-concept supercapacitors, Journal of Energy Chemistry, 27 25–42(2018).
3. A. Muzaffar et al., A review on recent advances in hybrid supercapacitors: Design, fabrication and applications, Renewable and Sustainable Energy Reviews 101 123–145(2019).

4. A. Joshi, P. Chand, V. Singh "Optical and Electrochemical Performance of Hydrothermal Synthesis of $BiPO_4$ Nanostructures for Supercapacitor Applications" Materials Today: Proceedingsdoi:https://doi.org/10.1016/j.matpr.2020.02.716
5. Yu A, Chabot V, Zhang J (2013) Electrochemical supercapacitors for energy storage and delivery, Fundamentals and applications. CRC Press, Taylor & Francis Group
6. A. Joshi and P. Chand, "Electrochemical Properties of $Bi_{0.85}Mg_{0.15}PO_4$ Nanostructures for Supercapacitor Applications" AIP Proceedings 2220 (1) 020174 (2020).
7. Conway BE (1991) Transition from "supercapacitor" to "battery" behavior in electrochemical energy storage. J Electrochem Soc 138:1539–1548
8. D. Wei, M. R. J. Scherer, C. Bower, P. Andrew, T. Ryhanen, & U. Steiner, A Nanostructured ElectrochromicSupercapacitor, Nano Letters 12 1857–1862(2012).
9. V. D. Nithya, L. Vasylechko, R. Kalai Selvan, Phase and shape dependent electrochemical properties of $BiPO_4$ by PVP assisted hydrothermal method for pseudocapacitors, RSC Adv., 4(110) 65184–65194(2014).
10. S. Vadivel, A.N. Naveen, J. Theerthagiri, J. Madhavan, T. Santhoshini Priya, N. Balasubramanian, Solvothermal synthesis of $BiPO_4$ nanorods/MWCNT (1D–1D) composite for photocatalyst and supercapacitor applications, Ceramic International 42 14196–14205(2016).
11. Joshi A, Chand P, Singh V (2020) Electrochemical and Optical Study of $BiPO_4$ Nanostructures for Energy Storage Applications. Materials Today: Proceedings 28:302–307
12. Chand P, Joshi A, Singh V (2020) Impact of phase segregation on optical and electrochemical property of $BiPO_4$ nanostructures for energy storage applications. J Mater Sci: Mater Electron 31(19):16867–16881
13. P. Chand, A. Joshi, V. Singh, High performance of facile microwave-assisted $BiPO_4$ nanostructures as electrode material for energy storage applications, Materials Science in Semiconductor Processing 122 105472 (2021).
14. S. Vadivel, D. Maruthamania, M. Kumaravela, B. Saravanakumar, Bappi Paul, Siddhartha Sankar Dhar, K. Saravanakumar, V. Muthuraj, Supercapacitors studies on $BiPO_4$ nanoparticles synthesized via asimple microwave approach, Journal of Taibah University for Science 11 661–666 (2017).
15. V.D. Nithya, B. Hanitha, S. Surendran, D. Kalpana, R. Kalai Selvan, Effect of pH on the sonochemical synthesis of $BiPO_4$ nanostructures and its electrochemical properties for pseudocapacitors, UltrasonicsSonochemistry 22 300–310 (2015).
16. V.D. Nithya, R. Kalai Selvan, L. Vasylechko, Hexa methylene tetramine assisted hydrothermal synthesis of $BiPO_4$ and its electrochemical properties for supercapacitors, Journal of Physics and Chemistry of Solids 86 11–18 (2015).
17. Vangari M, Pryor T, Jiang L (2013) Supercapacitors : Review of materials and fabrication methods. Journal of Energy Engineering 139(2):72–79

Recent Advancement in Tungsten Oxide as an Electrode Material for Supercapacitor Applications

Sunaina Saini, Aman Joshi, and Prakash Chand

Abstract Clean energy innovations are currently gaining huge attention because of fossil fuel exhaustion and increased global warming. Among the various technologies for converting electricity to other forms of energy, electrochemical energy storage technology is widely used, which includes the battery, supercapacitor and their hybrids, and fuel cells. Due to the long cycle life, high power density, more reliability and performance, and less maintenance needed, supercapacitors are favorable among different energy storage devices. Supercapacitor technology is evolving by exploring new materials and concepts. Transition metal oxides are versatile materials with many advantages such as natural abundance, low cost, and negligible toxicity toward living organisms. Tungsten oxide has gained large interest in the field of electrochemical applications due to its wide negative potential window so that it can be used as a negative electrode for device fabrication. So, here we focus on the recent development of WO_3 nanostructures as electrode material for supercapacitor applications. To enhance the conductivity of tungsten trioxide, many researchers have also developed the composites of WO_3 with different carbon materials and other transition metal oxides. So this article presents the development of pristine WO_3 as well as their composites to provide insights into the rapidly growing field of energy storage that may inspire additional research.

S. Saini (✉) · A. Joshi · P. Chand
Department of Physics, National Institute of Technology, Kurukshetra 136119, Haryana, India

A. Joshi
e-mail: aman_6180070@nitkkr.ac.in

P. Chand
e-mail: prakash@nitkkr.ac.in

A. Joshi
Department of Physics, J.C. Bose University of Science and Technology YMCA, Faridabad 121006, Haryana, India

A. Kumar et al. (eds.), *Renewable Energy Towards Smart Grid*, Lecture Notes in Electrical Engineering 823, https://doi.org/10.1007/978-981-16-7472-3_26

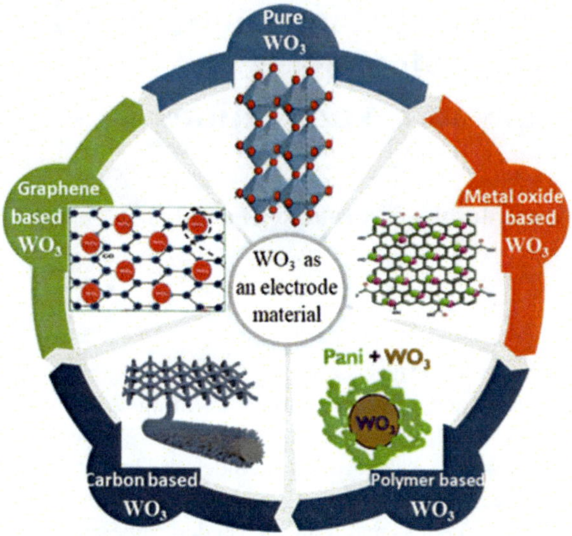

Keywords Transition metal oxides · Electrochemical · Supercapacitor · Tungsten oxide

1 Introduction

In order to meet the requirements of portable devices such as computers, watches, smartphones, and hybrid electric vehicles, there is a strong demand for advancement in the production of energy storage devices. People rely so heavily on modern technology that it is difficult to imagine their lives without them. One of the most commonly used electrochemical energy storage devices in the world is batteries, but due to some disadvantages like more charging time, low power handling capability, and short cycle life [1, 2], they cannot fulfill the high energy demands required for some applications. While on the other hand, supercapacitors possess elevated power density, quick charging capability, good cycle stability [3]. So, they are a promising alternative to batteries whenever high power demand is required. There are generally two types of charge storage mechanisms on the basis of which the supercapacitors are classified, as shown in Fig. 1. The first type, i.e., Electric Double Layer Capacitors (EDLCs), involves the electrostatic storage of energy by separating the charges at the interface of electrode and electrolyte without the transfer of electrons [4]. While the second type, i.e., pseudocapacitors, undergo fast redox reactions on the surface of the electrode during the charging and discharging process, this results in electrons being absorbed or released and utilizes electrochemical energy storage. The third type of supercapacitor is a hybrid capacitor, which is the combination of both EDLCs and pseudocapacitors, and the charge is stored electrostatically as well as electrochemically. Mainly the carbon-based materials such as CNTs, activated carbon, nanofibres and foams, graphene, and carbon aerogel are explored that show EDLC behavior

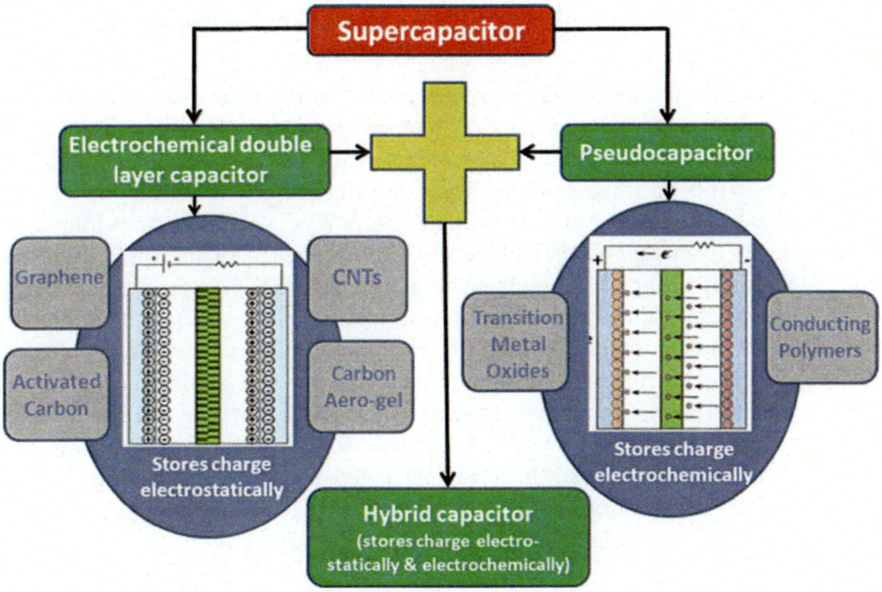

Fig. 1 Classification of supercapacitors

due to its high surface area, accessibility, and ability to exist in various ways. Pseudocapacitors include Transition Metal Oxides (TMOs) such as Manganese oxide (MnO_2), Ruthenium oxide (RuO_2), Tungsten oxide (WO_3), Nickel oxide (NiO), and conducting polymers such as polypyrrole (PPy), polyaniline (PANI), etc. The high theoretical specific capacitance with better rate capability and good electrical conductivity [5] has made the RuO_2 the best electrode material, but the toxicity and cost have limited its use. MnO_2 can replace the RuO_2 due to eco-friendliness, cost effectiveness, and high theoretical specific capacitance, but the only disadvantage is its poor ionic and electrical conductivities [6]. Among the various stoichiometric and non-stoichiometric of tungsten oxides, the tungsten trioxide (WO_3) has been extensively investigated as an electrode material for energy storage applications because of its unique tunneling structure, good corrosion resistance, and charge transport features. The energy density and specific surface area for pristine WO_3 are low, so it is necessary to also summarize the ongoing progress for WO_3-based material such as WO_3/carbon, WO_3/metal oxides from various perspectives to improve their performance.

2 WO_3-Based Materials for Supercapacitor Application

The stoichiometric WO_3 exhibits different crystal structures that are obtained during structural transformations by varying the temperature range. At lower temperature (<350 °C), it generally exists in the monoclinic or triclinic phase, while at a higher temperature (>700 °C), it gets transformed to orthorhombic or tetragonal phase. Each unit cell of tungsten trioxide contains eight WO_6 octahedra, and transition from one phase to another resulted in the W—O bond displacement, which shifts the tungsten atoms by tilting of WO_6 octahedra. Further, there is a hexagonal phase also which is not obtained during structural transformation, but it can only be produced in an aqueous medium, and it is found to be more efficient and advantageous for electrochemical properties. The hexagonal WO_3 is formed by sharing the corners of the WO_6 octahedron forming the six membered hexagonal rings and three membered trigonal rings. These rings are stacked together along the (001) axis to form the hexagonal and trigonal tunnels, which resulted in high specific capacitance due to the formation of accommodation sites for a huge number of ions during redox reactions. The schematic for the hexagonal WO_3 is shown in Fig. 2.

2.1 *Pristine WO_3*

WO_3 is a semiconductor material of electrochemically stable n-type having applications in different fields such as gas sensing, photocatalysis, electrochromic devices, and energy storage devices. In recent years, WO_3 has generated extensive research interest for supercapacitor application due to many advantages such as unique physiochemical properties, an outstanding performance due to tunneling structure, cost effective, and environment friendly. The effect of temperature, morphology, different crystal structures of WO_3 on the electrochemical performance of tungsten oxide is investigated by many researchers. Lokhande et al. have synthesized the four phases of tungsten oxide, i.e., monoclinic, hexagonal, tetragonal, and orthorhombic,

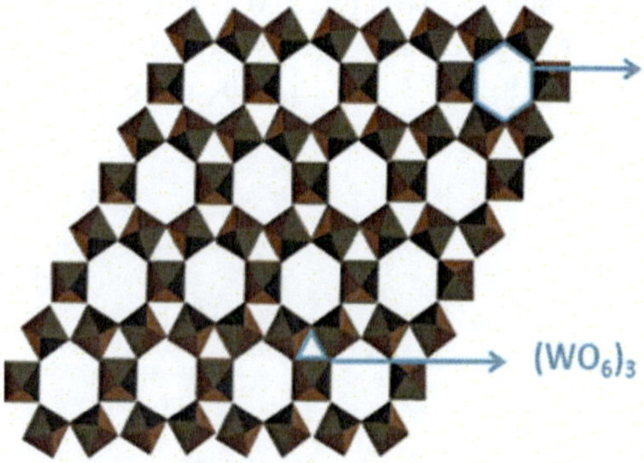

Fig. 2 Schematic of hexagonal WO_3

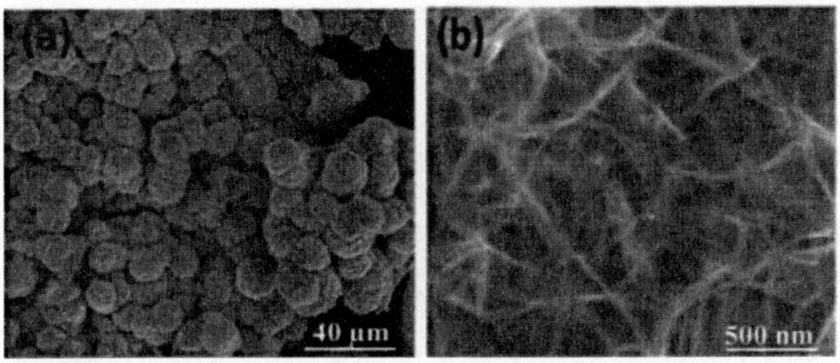

Fig. 3 **(a)** Low and **(b)** High magnification SEM images (Reprinted with permission from Ref. [9] © Elsevier (2015))

and reported that the hexagonal phase showed the highest specific capacitance of 377.5 Fg^{-1} at 2 mVs^{-1} with better rate capability and stability [7]. Shinde and co-workers have directly grown hexagonal WO_3 thin films on carbon cloth via hydrothermal method prepared at 180 °C, 160 °C, and 140 °C displaying nanorods-like, nanoplate-like, and nanogranule-like morphology, respectively [8]. The hexagonal nanorods exhibited a high specific capacitance of 694 F/g at 0.35 A/g with 87% retention after 2000 cycles. Xu et al. prepared the nanofibres assembled microspheres shown in Fig. 3 using H_2O_2 as inorganic additives and $K_2WO_4 \bullet 2H_2O$ as a source of tungsten [9]. The morphology of the as-prepared sample increased the effective electrolyte contact surface area and thus the overall specific capacitance of 795.05 F/g at 0.5 A g^{-1} in the 2 M electrolyte H_2SO_4.

2.2 *WO_3–Carbon-Based Material*

Synthesizing the WO_3 in composite with carbon is an efficient way to overcome the drawbacks associated with pristine WO_3, i.e., low electrical conductivity, poor rate capability, and chemical stability. Porous-activated carbon, graphene oxide, carbon aerogel, and carbon nanotubes are some of the carbonaceous materials that can enhance the electrochemical performance of tungsten oxide due to the synergistic effect that occurs between these two. Shinde et al. used a one-pot hydrothermal method to directly deposit WO_3 on the prepared Multiwalled Carbon Nanotube (MWCNT/CC) substrate [10]. The surface area of 67.54 m^2g^{-1} is obtained for the prepared sample that enhances the access of electrolyte ions to the active material by reducing the path length. These factors resulted in the improved electrochemical properties of the MWCNT–WO_3 hybrid electrode and achieved maximum specific capacitance of 429.6 Fg^{-1}, and when passed through 5000 cycles, then it retains 94.3% of capacitance, as compared to pristine WO_3 with 155.6 F/g capacity and 84.9% retention. Carbon aerogel has ordered and interconnected mesoporous texture, which is helpful to attain size selected nanoparticles via solvent immersion method.

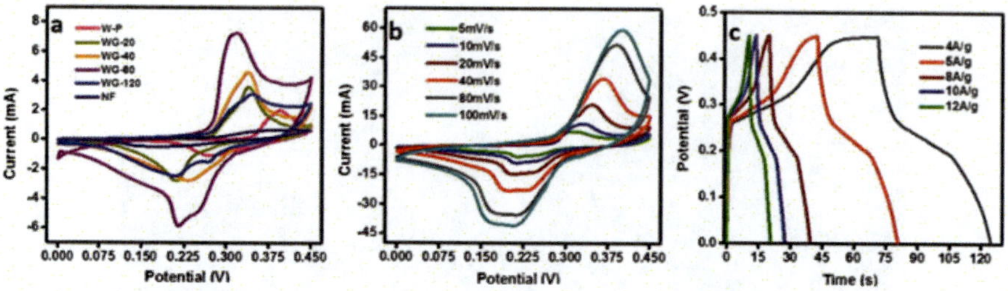

Fig. 4 **(a)** Cyclic voltammetry curves of all the composites at 5 mV/s, **(b)** Cyclic voltammogram of composite with 80 mg rGO at various scan rates, **(c)** GCD curves at various current densities (Reprinted with permission from Ref. [12] Copyright (2019) American Chemical Society)

So, in this direction, Liu et al. investigated the influence of the size distribution of WO_3 nanostructures in carbon aerogel and reported the exponential increase in specific capacitance with the decrease in size of nanoparticles [11]. They achieved the maximum specific capacitance of 1055 F/g at 5 mV/s with good cycle stability when the size of WO_3 nanostructures is decreased to 7.3 nm. The composite of graphene or reduced graphene oxide (rGO) with tungsten oxide facilitates the ion diffusion between WO_3 and the edges of the graphene sheet and also prevents the restacking of graphene and particle agglomeration. Samal and co-workers demonstrated WO_3–rGO composites by varying the concentration of GO and determined the high-specific capacitance of 801.6 F/g at a current density of 4 A/g in 3 M KOH electrolyte [12]. They also theoretically explained that this high value is achieved due to extra electronic states close to the Fermi level resulted from the hybridization of 2p orbitals of carbon atoms with the d state of tungsten and 2p state of oxygen. Figure 4 shows the various characteristics they used to study the electrochemical properties of the electrode material.

The electrochemical performance increased gradually with GO concentration and exhibited the best results for WG-80. When further increased to 120 mg then it leads to uneven growth, restacking, and uncontrollable aggregation and hence deteriorated the supercapacitive performance.

2.3 WO_3—Metal Oxide Based Materials

By mixing the two metal oxides, the synergistic effect increased the redox behavior and enabled the transfer of electrons and ions into the active material. The two-step atomic layer deposition method followed by annealing for the formation of TiO_2 nanoparticles—WO_3 films was proposed by Hai et al. [13]. The results confirmed that the surface functionalization of 2D WO_3 by TiO_2 nanoparticles exhibited the specific capacitance of 342.5 F/g at a current density of 1.5 A/g, which is 1.5 times better than pure WO_3. Zhou et al. successfully synthesized the Mo–W mixed oxide ($Mo_xW_{1-x}O_3$) due to some similarities between W^{6+} and Mo^{6+}, such as ionic radius,

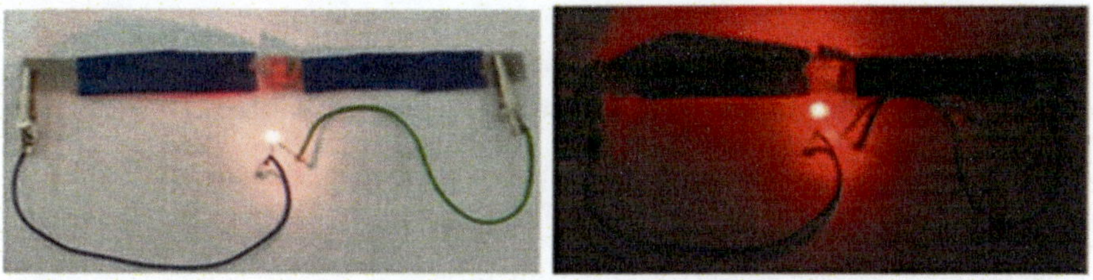

Fig. 5 Image for the lighting of a red LED by using two asymmetric devices (NiSe//WO_3@PPy) with series connection (Reprinted with permission from Ref. [15] Copyright (2019) American Chemical Society)

oxidation state, and electronegativity [14]. They deposited it on TiO_2 nanotube via co-electrodeposition method. The distortion in this mixed oxide leads to increased lattice space and reduced crystallite size, which provides an easy diffusion of ions in the solid. The outcomes revealed the specific capacitance of 517.4 F/g at a current density of 1 A/g with 89.3% retention when 10 A/g current density is maintained, indicating ultrahigh rate capability. Amit Kumar Das and his companions fabricated the solid-state asymmetric supercapacitor (ASC) by utilizing the WO_3@PPy composite and NiSe as the negative and positive electrodes, respectively [15]. The enhanced potential window of the device is 1.25 V that delivered the power density of 1249 W/kg with a maximum energy density of 37.3 Wh/kg at a current density of 2 Ag^{-1}. Also, the ASC device demonstrated good cyclic stability by holding 91% of its initial specific capacity when passed through 5000 GCD cycles (Fig. 5).

3 Conclusion

To meet the energy demands and prevent environmental risks, many efforts have been made in developing clean and green devices for storing energy. The performance of these devices largely depends upon the electrode material and the electrolyte used. This has lead to the exploration of novel materials as well as tuning the properties of existing materials via different methods. Tungsten oxide is a widely studied material in the last several years and has lead to many publications. Recently many researchers have boosted its performance as an electrode material for supercapacitor applications, which is briefly summarized in this article. They have developed the composites with carbon and other metal oxides to overcome the disadvantages of pristine WO_3 and to achieve high power density and cycle stability. Further studies should be focused on extending the working potential range by enhancing their energy density. This material has the potential to be employed as a negative electrode in the energy storage device.

References

1. Chen GZ (2017) Supercapacitor and supercapattery as emerging electrochemical energy stores. Int Mater Rev 62:173–202
2. Mukhopadhyay A (2014) Deformation and stress in electrode materials for Li-ion batteries. Prog Mater Sci 63:58–116
3. Muzaffar A (2019) A review on recent advances in hybrid supercapacitors: design, fabrication and applications. Renew Sustain Energy Rev 101:123–145
4. Yu A (2013) Electrochemical supercapacitors for energy storage and delivery. Fundamentals and applications. CRC Press, Taylor & Francis Group, Boca Raton
5. Hu CC (2006) Design and tailoring of the nanotubular arrayed architecture of hydrous RuO_2 for next generation supercapacitors. Nano Lett 6(12):2690–2695
6. Yu N (2016) High-performance fibre-shaped all-solid-state asymmetric supercapacitors based on ultrathin MnO_2 nanosheet/carbon fibre cathodes for wearable electronics. Adv Energy Mater 6(2). Article 1501458
7. Lokhande V (2019) Charge storage in WO_3 polymorphs and their application as supercapacitor electrode material. Results Phys 12:2012–2020
8. Shinde PA (2017) Temperature dependent surface morphological modification of hexagonal WO_3 thin films for high performance supercapacitor application. Electrochim Acta 224:397–404
9. Juan X (2015) Tungsten oxide nanofibers self-assembled mesoscopic microspheres as high-performance electrodes for supercapacitor. Electrochim Acta 174:728–734
10. Shinde PA (2019) Direct growth of WO_3 nanostructures on multi-walled carbon nanotubes for high-performance flexible all-solid-state asymmetric supercapacitor. Electrochim Acta 308:231–242
11. Liu X (2018) Dispersed and size-selected WO_3 nanoparticles in carbon aerogel for supercapacitor applications. Mater Des 141:220–229
12. Samal R (2019) Facile production of mesoporous WO_3-rGO hybrids for high-performance supercapacitor electrodes: an experimental and computational study. Sustain Chem Eng 7(2):2350–2359
13. Hai Z (2017) TiO_2 nanoparticles-functionalized two-dimensional WO_3 for high performance supercapacitors developed by facile two-step ALD process. Mater Today Commun 12:55–62
14. Zhou H (2017) Molybdenum-tungsten mixed oxide deposited into titanium dioxide nanotube arrays for ultrahigh rate supercapacitors. Appl Mater Interfaces 9(22):18699–18709
15. Das AK (2019) Highly rate capable nanoflower-like NiSe and WO3@PPy composite electrode materials toward high energy density flexible all-solid state asymmetric supercapacitor. Appl Electron Mater 1(6):977–990

GUPFC Impact in Managing the Congestion Using Generation Rescheduling

Charan Sekhar Makula and Ashwani Kumar

Abstract While overcoming the disadvantages of regulated monopoly power system, the restructured power system poses new issues to the system operator where the congestion management in the transmission lines is one among them where congestion in transmission line is known as violation of constraints of policies that are made in the market and constraints of physical model. The paper focuses on alleviation of congestion in hybrid electricity markets. The authors considered the generation rescheduling approach to manage the congestion in deregulated electricity markets with optimization of power flows with loadability factor. Secure bilateral transactions obtained from the optimization is considered for hybrid markets. Finally generalized unified power flow controller (GUPFC) is implemented to minimize the reschedule of generation and reduce the congestion cost. The proposed method is tested on IEEE 24—bus RTS for single line, two lines and three lines which are congested cases using MATLAB interfacing with GAMS software and the results are presented that show GUPFC implementation is more effective in reducing the congestion cost.

Keywords Hybrid electricity markets · Generation rescheduling · Congestion management · GUPFC

1 Introduction

The rapid growth in demand for electricity and desire to get maximum profit in the restructured power system causes the loading of transmission lines beyond policy constraints, physical and operational limits. The overload of line leads to line outage, overloading of other lines and subsequently their outages. Sometimes results in block out condition if proper action is not taken at the right time. Another problem is that electricity price spikes in the market. Thus congestion management (CM) is a critical issue that should be handled by ISO. To avoid these problems and to overcome the problem of congestion, various researchers proposed different techniques.

C. S. Makula (✉) · A. Kumar
National Institute of Technology Kurukshetra, Kurukshetra 136119, Haryana, India

A. Kumar et al. (eds.), *Renewable Energy Towards Smart Grid*, Lecture Notes in Electrical Engineering 823, https://doi.org/10.1007/978-981-16-7472-3_27

Christie et al. classified the congestion management methods based on the technique implemented by different markets around the world which is classified into three categories as optimal power flow (OPF) based, price area, and available transfer capability (ATC) methods [1]. Another way of classification of congestion management methods are market methods and non-market methods. Non-Market methods are as first come first serve, pro rata methods, whereas market methods include rescheduling of generation, load curtailment, zonal pricing, and nodal pricing. Technical methods include deployment of FACTS devices, tap changing transformer operation [2, 3]. Fang and David proposed the re-dispatch based congestion management in pool and hybrid electricity markets with modes of free operation, and either pool or bilateral mode transaction protection [4]. The authors proposed the generation rescheduling-based congestion management method where selection of generators is based on calculated values of Generation shift selectivity factors (GSSF) to increase or decrease their generation. The authors used relative electrical distance (RED) concept to determine the loading of generators [5]. Transmission congestion management using transmission congestion distribution factor (TCDF) based zonal/cluster method and real power rescheduling using generation contribution to overloading of lines is presented in [6].

The authors in [7] used generation rescheduling method to manage the congestion in hybrid electricity markets that are combination of pool market and bilateral contracts and compare the effectiveness of the FACTS devices that are STATCOM, IPFC and UPFC in alleviation of violation of transmission limits. Sumit et al. proposed the teaching learning based optimization technique for optimization of real power rescheduling in managing the congestion in pool electricity markets [8]. Ref. [9] presented the comprehensive survey on the optimization techniques used in congestion management and different approaches that are used to tackle the congestion in deregulated electricity markets. The congestion management based on generation rescheduling approach for the system consisting of zip loads is presented with impact of various FACTS devices in minimizing the congestion in [10]. A novel optimization technique refractor update based rider optimization algorithm (RU-ROA) which is combination of two rider optimization algorithm (ROA) and water wave optimization (WWO) is proposed to minimize the rescheduling cost of generation during congestion management [11]. Subhasish et al. proposed atom search optimization (ASO) technique for generation rescheduling to alleviate the congestion in the system [12]. Another artificial intelligence technique based on Bat algorithm to congestion cost minimization with optimized generation rescheduling is proposed in [13].

2 Model of GUPFC

The GUPFC is the combination of unified power flow controller (UPFC) and interline power flow controller (IPFC). In the simplest form GUPFC consists of two series converters each one in separate transmission line and one shunt converter which has capability to maintain the voltage at bus and each line real and reactive powers in transmission line [14]. The static power flow equation and power injections at buses with implementation of GUPFC is shown below [15, 16].

$$P_i = V_i^2(G_{ii} + G_{sh}) + \sum_h V_i V_h[G_{ih}\cos(\delta_{ih}) + B_{ih}\sin(\delta_{ih})] \\ + \sum_h V_i V_{se,ih}\left[G_{ih}\cos\left(\delta_i - \delta_{se,ih}\right) + B_{ih}\sin\left(\delta_i - \delta_{se,ih}\right)\right] \\ + V_i V_{sh}[G_{sh}\cos(\delta_i - \delta_{sh}) + B_{sh}\sin(\delta_i - \delta_{sh})] \tag{1}$$

$$Q_i = -V_i^2(B_{ii} + B_{sh}) + \sum_h V_i V_h[G_{ih}\sin(\delta_{ih}) - B_{ih}\cos(\delta_{ih})] \\ + \sum_h V_i V_{se,ih}\left[G_{ih}\sin\left(\delta_i - \delta_{se,ih}\right) - B_{ih}\cos\left(\delta_i - \delta_{se,ih}\right)\right] \\ + V_i V_{sh}[G_{sh}\sin(\delta_i - \delta_{sh}) - B_{sh}\cos(\delta_i - \delta_{sh})] \tag{2}$$

$$P_{hi} = V_h^2 G_{hh} + V_i V_h[G_{ih}\cos(\delta_{hi}) + B_{ih}\sin(\delta_{hi})] \\ + V_h V_{se,ih}\left[G_{ih}\cos\left(\delta_h - \delta_{se,ih}\right) + B_{ih}\sin\left(\delta_h - \delta_{se,ih}\right)\right] \tag{3}$$

$$Q_{hi} = -V_h^2 B_{hh} + V_i V_h[G_{ih}\sin(\delta_{hi}) - B_{ih}\cos(\delta_{hi})] \\ + V_h V_{se,ih}\left[G_{ih}\sin\left(\delta_h - \delta_{se,ih}\right) + B_{ih}\cos\left(\delta_h - \delta_{se,ih}\right)\right] \tag{4}$$

$$B_{ii} = \sum_h B_{ih};\; G_{ii} = \sum_h G_{ih} \tag{5}$$

The real power exchange constraint with GUPFC via DC link is zero in steady-state operation

$$P_{exchange} = Re\left[V_{sh}I_{sh}^* - \sum_h V_{se,ih}I_{hi}^*\right] = 0 \tag{6}$$

$$V_i^2 G_{sh} + V_i V_{sh}[G_{sh}\cos(\delta_i - \delta_{sh}) - B_{sh}\sin(\delta_i - \delta_{sh})] \\ + \left\{\sum_h \left[V_i V_{se}\left[G_{ih}\cos\left(\delta_i - \delta_{se,ih}\right) - B_{ih}\sin\left(\delta_i - \delta_{se,ih}\right)\right]\right.\right. \\ \left.\left. + V_h V_{se}\left[G_{ih}\cos\left(\delta_h - \delta_{se,ih}\right) - B_{ih}\sin\left(\delta_h - \delta_{se,ih}\right)\right]\right]\right\} = 0 \tag{7}$$

Where h = j, k etc.

The other constraints of the GUPFC as follows

$$V_{se,ihmin} < V_{se,ih} < V_{se,ihmax};\; \delta_{se,ihmin} < \delta_{se,ih} < \delta_{se,ihmax} \tag{8}$$

$$V_{shmin} < V_{sh} < V_{shmax};\ \delta_{shmin} < \delta_{sh} < \delta_{shmax} \tag{9}$$

3 Congestion Management Model

3.1 Base Case Real Power Output of Generators

With minimization of the marginal cost function of the generators, base case real power generation has been obtained, subject to their power generation limits, power flow limits, angle limits and voltage limits. After solving the OPF problem, the optimal generation obtained has been utilized as a base generation data for transmission line congestion management [17].

Minimize

$$obj = c_i P_g^2(i) + b_i P_g(i) + a_i \tag{10}$$

The power flow equations for real and reactive power P_{ij} and Q_{ij} are obtained as

$$P_{ij} = V_i^2 G_{ij} - V_i V_j \left[G_{ij}\, cos\left(\delta_i - \delta_j\right) + B_{ij}\, sin\left(\delta_i - \delta_j\right) \right] \tag{11}$$

$$Q_{ij} = -V_i^2 B_{ij} + V_i V_j \left[B_{ij}\, cos\left(\delta_i - \delta_j\right) - G_{ij}\, sin\left(\delta_i - \delta_j\right) \right] \tag{12}$$

Similarly P_{ji} and Q_{ji} are given by

$$P_{ji} = V_j^2 G_{ij} - V_i V_j \left[G_{ij}\, cos\left(\delta_i - \delta_j\right) - B_{ij}\, sin\left(\delta_i - \delta_j\right) \right] \tag{13}$$

$$Q_{ji} = -V_j^2 B_{ij} + V_i V_j \left[B_{ij}\, cos\left(\delta_i - \delta_j\right) + G_{ij}\, sin\left(\delta_i - \delta_j\right) \right] \tag{14}$$

Power injection balance equations at any bus-i is given by

$$P_{gi} - P_{di} = P_i \qquad i \in bus \tag{15}$$

$$Q_{gi} - Q_{di} = Q_i \qquad i \in bus \tag{16}$$

Where P_{gi}, P_{di} and P_i are real power generation, demand and losses, respectively at bus-i. Similarly Q_{gi}, Q_{di} and Q_i are reactive power generation, demand and losses, respectively, at bus-i.

Inequality constraints

Where the generation limits are taken as

$$P_{gimin} \leq P_{gi} \leq P_{gimax} \quad i \in bus \tag{17}$$

$$Q_{gimin} \leq Q_{gi} \leq Q_{gimax} \quad i \in bus \tag{18}$$

The voltage limits and voltage angle limits are

$$V_{imin} \leq V_i \leq V_{imax}\,;\ \theta_{imin} \leq \theta_i \leq \theta_{imax} \quad i \in bus \tag{19}$$

3.2 Congestion Management Model with Loadability Factor

The formulation of the objective functions and respective constraints for transmission line congestion management are same as for the case without loadability factor rescheduling of generator except the real power balancing equation. In this case while maximizing the loadability factor, the base power is obtained solving the fuel cost minimization problem described in the previous section. In every case, load is multiplied with loadability factor in the real power injection balancing equation. The real power balancing in each case will become as follows:

$$P_{gni} - ro * P_{di} = P_i \quad i \in bus \tag{20}$$

The objective function along with the fuel cost minimization is

Maximize.

$$\mathrm{ro} \tag{21}$$

Minimize.

$$CG = \sum_{g=1}^{ng}\left(R_g^{up}\Delta P_g^{up} + R_g^{down}\Delta P_g^{down}\right) \tag{22}$$

Up/down generation limits for ΔP_{gi}^{up} and ΔP_{gi}^{down} are given by

$$\Delta P_{gmin}^{up} \leq \Delta P_g^{up} \leq \Delta P_{gmax}^{up},\quad \Delta P_{gmin}^{down} \leq \Delta P_g^{down} \leq \Delta P_{gmax}^{down} \tag{23}$$

Power flow limits for congestion management

$$\mathrm{P_{ij}^2 + Q_{ij}^2 \leq S_{ij}^2} \tag{24}$$

Equality constraints

Change in generation limits

$$\sum_{\mathrm{g=1}}^{\mathrm{ng}} \Delta \mathrm{P_g^{up}} - \sum_{\mathrm{g=1}}^{\mathrm{ng}} \Delta \mathrm{P_g^{down}} = 0 \tag{25}$$

The power balance equation will become

$$\mathrm{P_{gni} - ro * P_{di} = P_i \quad i \in bus} \tag{26}$$

Where

$$\mathrm{P_{gn} = P_g + \Delta P_g^{up} - \Delta P_g^{down}} \tag{27}$$

3.3 Hybrid Market Model

The hybrid market model is the combination of pool model and bilateral model. In pool model, buyers and sellers submit their bid to pool based on amount of their will to purchase or sell. In bilateral model, ISO has to secure the bilateral contracts that are made between buyers and sellers based on negotiation of contracts [18].

The objective function for bilateral contracts is

$$\min \sum_i \sum_j b_{ij} \left(GD_{ij} - GD_{ij}^0 \right)^2 \tag{28}$$

The demand and generation equations are given by

$$P_{db} = \sum_i GD_{ij}; \; P_{gb} = \sum_j GD_{ij} \tag{29}$$

The total demand and generation in hybrid electricity markets are given by

$$P_g = P_{gb} + P_{gp}; \quad P_d = P_{db} + P_{dp} \tag{30}$$

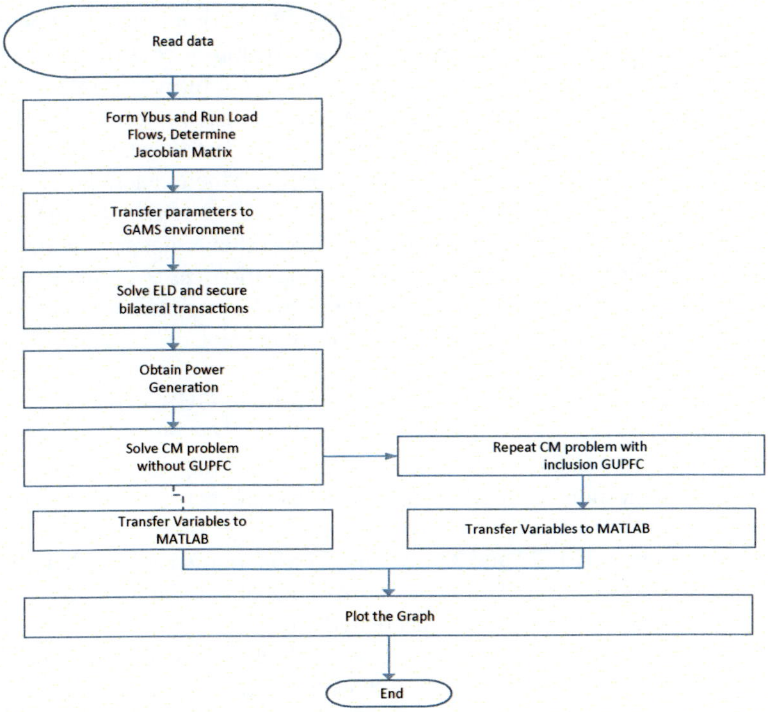

Fig. 1 Flowchart of the proposed method

The inequality constraints for the bilateral constraints are given by

$$0 \leq GD_{ij} \leq GD_{ij}^{max} \tag{31}$$

The sequence of execution of the program in the MATLAB and GAMS environment is represented in Fig. 1.

4 Results and Analysis

The results of the proposed method of generation rescheduling are obtaining with 50% of bilateral contracts in hybrid electricity market using MATLAB and GAMS environment on IEEE 24-bus RTS. The results include the cases of single line (SL) congestion, 2-lines (2L) congestion and 3-lines (3L) congestion. The results are obtained without inclusion and with inclusion of the GUPFC. Single line congestion is created by reducing the line-23 (between buses 14 and 16) power rating to 2.60 p.u. from 5.0 p.u. For 2-lines congestion case, line-18 maximum power flow rating is decreased to 2.25 p.u. from 5.0 p.u along with previous considered line-23 rating. Similarly 11th line (between bus-7 and bus-8) maximum power flow rating is decreased from 1.75 p.u. to 1.50 p.u. along with previously considered two lines.

In the case without GUPFC, from Table 1 we get the information that for single line congestion, Generator at bus 2 increased in generation while generators at buses 7, 22 decreased their generation. When 2-lines are congested, generators at bus-2 increase its generation and at buses-7, 13 and 16 decrease their generation. Similarly for 3-lines congested case generator at buses-2, 15, 22, 23 increased their generation and at buses-7, 13, 16, 21 decreased their generation. It shows that generator at bus-2 contribute economically dispatch power in all congestion cases by increasing the generation whereas generator at bus-7 participated by decreasing its generation. The graphical representation of numerical data for 3L congestion of Table 1 is presented in Fig. 2.

Comparing with case of without GUPFC, inclusion of GUPFC results in reduction of amount of power generation rescheduling. For SL congestion case, small change in generation has been obtained for congestion management. Similarly also for 2L and 3L congestion cases and at the same time number of generators that are to be rescheduled are also decreased. This is clearly observable from Figs. 2 and 3 that generators at buses 21, 22 and 23 are relieved from re-dispatching their generation with the implementation of GUPFC and also that at remaining generator buses amount of up/down generation is also decreased (Table 2).

Table 3 clearly indicates that the installation of GUPFC bring in substantial saving in the congestion cost especially in case if three lines are congested. The results are much more appreciable comparing with results from Ref. [7]

5 Conclusion and Future Works

In the present work, the congestion in transmission is relieved using generation rescheduling accompanied by optimal power flows achieved using MATLAB is interfacing with GAMS software. The FACTS device GUPFC which is more effective than other FACTS devices is implemented to reduce the deviations in generation in pool and bilateral contracts. This is clearly indicated in the results. At the same time GUPFC gave substantial saving in the congestion cost for the system operator. The work will be carried out for future considering optimal location for GUPFC using artificial intelligence technique.

Table 1 Generator up/down generation for congestion management without GUPFC

B. no	SL				2L				3L			
	P_g	P_{gn}	ΔP_g^{up}	ΔP_g^{down}	P_g	P_{gn}	ΔP_g^{up}	ΔP_g^{down}	P_g	P_{gn}	ΔP_g^{up}	ΔP_g^{down}
1	1.3524	1.3524	0	0	1.3524	1.3524	0	0	1.3524	1.3524	0	0
2	0.15	0.3186	0.1686	0	0.15	0.9343	0.7843	0	0.15	0.95	0.8	0
7	3	2.998349	0	0.001651	3	2.997052	0	0.002948	3	2.738548	0	0.261452
13	5.91	5.91	0	0	5.91	5.347494	0	0.562506	5.91	5.164931	0	0.745069
15	2.15	2.15	0	0	2.15	2.15	0	0	2.15	2.95	0.8	0
16	1.55	1.55	0	0	1.55	1.331181	0	0.218819	1.55	0.75	0	0.8
18	4	4	0	0	4	4	0	0	4	3.2	0	0.8
21	1.261351	1.261351	0	0	1.261351	1.261351	0	0	1.261351	0.667872	0	0.593479
22	3	2.83308	0	0.16692	3	3	0	0	3	3.8	0.8	0
23	6.6	6.6	0	0	6.6	6.6	0	0	6.6	7.4	0.8	0

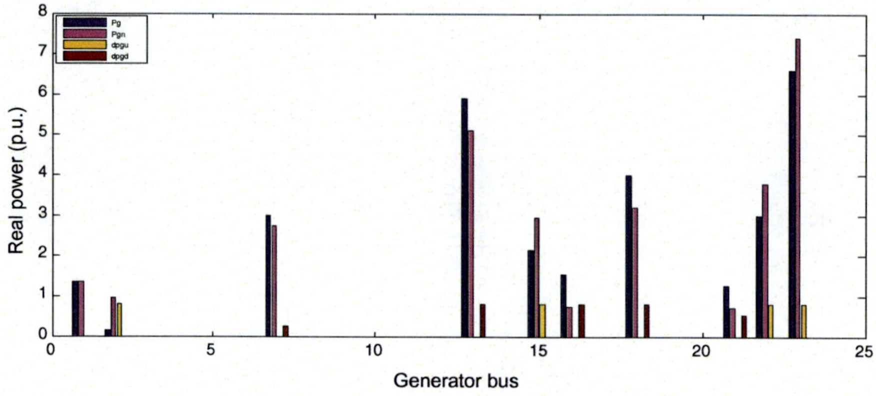

Fig. 2 Generators up/down generation without GUPFC for 3L congestion with loadability factor

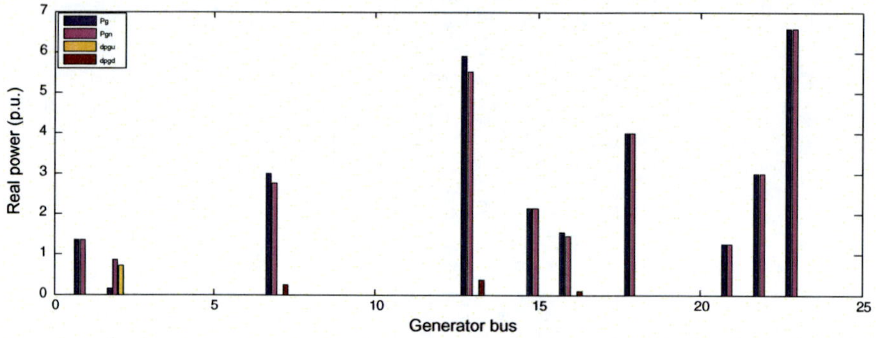

Fig. 3 Generator's up/down generation with GUPFC on line-13 for 3L congestion

Table 2 Generator up/down generation for congestion management with GUPFC

B. no	SL				2L				3L			
	P_g	P_{gn}	ΔP_g^{up}	ΔP_g^{down}	P_g	P_{gn}	ΔP_g^{up}	ΔP_g^{down}	P_g	P_{gn}	ΔP_g^{up}	ΔP_g^{down}
1	1.352	1.352	0	0	1.352464	1.352464	0	0	1.352464	1.352464	0	0
2	0.15	0.168	0.018008	0	0.15	0.624205	0.474205	0	0.15	0.873512	0.723512	0
7	3	3	0	0	3	3.002279	0.002279	0	3	2.758386	0	0.241614
13	5.91	5.91	0	0	5.91	5.528138	0	0.381862	5.91	5.525089	0	0.384911
15	2.15	2.15	0	0	2.15	2.15	0	0	2.15	2.15	0	0
16	1.55	1.55	0	0	1.55	1.455	0	0.095	1.55	1.453	0	0.097
18	4	4	0	0	4	4	0	0	4	4	0	0
21	1.261	1.261	0	0	1.261	1.261	0	0	1.261	1.261	0	0
22	3	2.982	0	0.018	3	3	0	0	3	3	0	0
23	6.6	6.6	0	0	6.6	6.6	0	0	6.6	6.6	0	0

Table 3 Congestion cost with and without GUPFC

Congestion		SL	2L	3L
Cost($)/hour	Without GUPFC	348.4286	594.7092	1561
	With GUPFC	288.2032	471.5936	570.4048

References

1. Christie RD, Wollenberg BF, Wangstien I (2000) Transmission management in the deregulated environment. Proc IEEE 88(2):170–195
2. Kumar A, Srivastava SC, Singh SN (2004) A zonal congestion management approach using real and reactive power rescheduling. IEEE Trans Power Syst 18(1):554–562
3. Bruno S, LaScala M (2004) Unified power flow controllers for security-constrained transmission management. IEEE Trans Power Syst 18(1):418–426
4. Fang RS, David AK (1999) Transmission congestion management in an electricity market. IEEE Trans Power Syst 14(2):877–883
5. Yesuratnam G, Pushpa M (2010) Congestion management for security oriented power system operation using generation rescheduling. In: IEEE 11th international conference on probabilistic methods applied to power systems, Singapore, pp 287–292
6. Mandala M, Gupta CP (2010) Comparative studies of congestion management in deregulated electricity market. In: 16th national power systems conference, December, pp 628–635
7. Sekhar C, Kumar A (2012) Congestion management in hybrid electricity markets with FACTS devices with loadability limits. Int J Electr Comput Eng 2(1):75–89
8. Verma S, Saha S, Mukherjee V (2018) Optimal rescheduling of real power generation for congestion management using teaching-learning-based optimization algorithm. J Electr Syst Inf Technol 5(3):889–890
9. Narain A, Srivastava SK, Singh SN (2020) Congestion management approaches in restructured power system: key issues and challenges. Electr J 33(3)
10. Kumar A, Mittapalli RK (2014) Congestion management with generic load model in hybrid electricity markets with FACTS devices. Int J Electr Power Energy Syst 57:49–63
11. Srivastava J, Yadav NK, Sharma AK (2020) A novel hybrid algorithm for rescheduling-based congestion management scheme in power system. Electr Eng (Springer) 102:1993–2010
12. Deb S, Datta S, Singh KR, Kumar R (2020) Real power rescheduling of generator for transmission line congestion management using atom search algorithm. In: IEEE 1st international conference for convergence in engineering (ICCE) 2020, Kolkata
13. Paul K, Kumar N, Dalapati P (2021) Bat algorithm for congestion alleviation in power system network. Technol Econ Smart Grids Sustain Energy 6
14. Fardanesh B, Shperling B, Uzunovic E, Zelingher S (2000) Multi-converter FACTS devices: the generalized unified power flow controller (GUPFC). In: IEEE PES summer meeting, Seattle, USA, July
15. Zhang X-P, Handschin E, Yao MM (2001) Modeling of the generalized unified power flow controller (GUPFC) in a nonlinear interior point OPF. IEEE Trans Power Syst 16(3):367–373
16. Kumar A, Sekhar C (2012) Congestion management based on demand management and impact of GUPFC. In: International conference on power, signals, controls and computation, Trissur, India. IEEE
17. Wood AJ, Wollenberg BF (1996) Power generation, operation and control. Wiley, New York
18. Hamoud G (2000) Feasibility assessment of simultaneous bilateral transaction in a deregulated environment. IEEE Trans Power Syst 15(1):22–26

Implementation of Black Widow Optimization Algorithm for Loss Minimization in an Unbalanced Radial Distribution System

Aliva Routray, Khyati D. Mistry, and Sabha Raj Arya

Abstract In this work, active power losses in the distribution lines are minimized by using Distributed Generation technology with improved contribution in power quality as well as efficiency of the system. The location and capacity of multiple DGs are determined by using Black Widow Optimization technique. For power flow analysis, Load Impedance Matrix method is used to find power losses in the lines, current flow and bus voltages in system. These multiple algorithms are implemented with standard IEEE-19 bus unbalanced radial distribution system and the overall system performance is studied.

Keywords Black Widow Optimization algorithm · Load Impedance Matrix · Loss minimization · Radial distribution system

1 Introduction

To match with the rising load demands from the consumer end, the usage of conventional as well as non-conventional energy sources are also growing in power industry applications. As distribution system is the most significant segment of power system, more Distributed Generation (DG) applications are implemented to fulfill the customer satisfaction without polluting the atmosphere. DGs can be a source of renewable or non-renewable energy like solar, wind, tidal, biomass, diesel, etc. [1]. Since the distribution section is near to the consumer side, the communication or transmission cast gets reduced with it. The other benefits of DGs can be active and reactive power loss minimization in the distribution lines, bus voltage improvement of the system, pollution free eco-friendly environment, reduced maintenance cost, optional system upgradation and enhanced system longevity [2]. However, due to technical development in machineries and increasing demands from the consumers,

A. Routray · K. D. Mistry · S. R. Arya (✉)
Department of Electrical Engineering, Sardar Vallabhbhai National Institute of Technology, Surat, India

K. D. Mistry
e-mail: kkp@eed.svnit.ac.in

A. Kumar et al. (eds.), *Renewable Energy Towards Smart Grid*, Lecture Notes in Electrical Engineering 823, https://doi.org/10.1007/978-981-16-7472-3_28

the DG placements are highly into the energy market. DG has been implemented with many methods of combined technologies of power electronics engineering and energy storing devices like battery or capacitors. Another important aspect of DG application is the right placement and capacity of DG in the system in order to fulfill the objective of any project planning [3]. The objective of the DG implementation in the system can be different like low line losses in the system, maximum profit, improving bus voltages, reducing cost, etc.

Most of the distribution system is radial in nature, hence there is unidirectional power flow. The distribution system can be balanced or unbalanced depending upon the magnitude of current flowing and phase angle between them. However, the negative reactance of line, higher R/X value and very high line reactance ratio of long to short line cause the distribution network to be unbalanced. There are numerous load flow studies implemented for transmission system but can't be used for distribution system because of the above causes [4]. Hence, few techniques have been developed to analyze the power flow in distribution system and mentioned in many literatures. Backward Forward Sweep (BFS) method has been traditionally implemented to distribution system as a load flow method [5]. Numerous literatures on this technique with few modifications are also implemented. However, in this work, a Load Impedance Matrix (LIM) method is executed in distribution system. This method comes with an advantage of single step analytical method and faster computational efficiency unlike other distribution system load flow techniques [6].

There are numerous optimization techniques that have been implemented to determine the DG location and capacity in any system. Individual techniques are nature inspired and solves the challenging real-world problems effectively. The optimization problems are not limited to electrical engineering domain but can be applied to any technical domain. There are several techniques used to obtain the optimum DG position and its capacity having objective function of loss minimization in distribution system. Out of all the popular heuristic techniques, Grey Wolf Optimization (GWO) [3], Teaching Learning Based Optimization (TLBO) [7], Particle Swarm Optimization (PSO) [8], Jaya [9], Grasshopper Optimization Algorithm (GOA) and Cuckoo Search (CS) [10], Quantum Behaved PSO (QPSO) [11] etc. are already used for the above problem formulation. These optimization methods individually use different nature-inspired ideas with algorithm specific parameters to reach the optimum solution of any problem formulation. For being an efficient technique, Black Widow Optimization (BWO) is implemented to determine the multiple DG locations and capacities in three phase unbalanced system. This algorithm is new and it has a strange technique of finding feasible solutions. It also has an advantage of early convergence and high accuracy for the optimum DG allotment solutions [12].

The objective of this work is to minimize the active power loss in the distribution lines for an unbalanced radial distribution system. In order to feed the additional power supply to the distribution grid, Distribution Generation sources are used in the system. However, the novelty of this paper is the implementation of Black Widow Optimization technique to locate the optimum position and capacity of multiple DGs in the test system. In this work, IEEE standard 19 bus system is studied along with its change in system parameters.

2 Algorithm in Three Phase Radial Distribution System

Traditional methods have been used for power flow analysis in any distribution system and can be applicable for balanced as well as unbalanced systems. The system carries line data, load data and line parameters for load flow study. In case of three phase unbalanced system, the line parameters are available in terms of matrix for individual phase of each node. The obtained matrix for each bus is having a 3 × 3 impedance matrix and a 3 × 1 load matrix. In this work Load Impedance Matrix (LIM) technique is implemented to find the current flow, power loss and bus voltage for every phase [6].

This LIM matrix is formed with the load data and impedance data for the entire system. This single step technique is having faster computational response and robust in nature which is an advantage over the traditional load flow methods. The Radial Distribution System under study is having 19 nodes and 18 branches, respectively. The LIM matrix for the system is expressed in Eq. (1). Here, Vxi is the nodal voltages at ith bus, Vx1 is the substation voltage which is 1 pu, Sxi is the active and reactive power in complex form of power at bus i and ØxBi is the branch path impedances (BPI) (i varies from 1 to 19).

$$\begin{bmatrix} V_{x1} \\ V_{x2} \\ V_{x3} \\ \vdots \\ \vdots \\ V_{x18} \\ V_{x19} \end{bmatrix} = \begin{bmatrix} V_{x1} \\ V_{x1} \\ V_{x1} \\ \vdots \\ \vdots \\ V_{x1} \\ V_{x1} \end{bmatrix} - \begin{bmatrix} S_1^{*}\sum(\varphi_{B1} \cap \varphi_{B1}) & S_2^{*}\sum(\varphi_{B2} \cap \varphi_{B1}) & \cdots & \cdots & S_{x18}^{*}\sum(\varphi_{xB18} \cap \varphi_{B1}) & S_{x19}^{*}\sum(\varphi_{xB19} \cap \varphi_{B1}) \\ S_{x1}^{*}\sum(\varphi_{xB1} \cap \varphi_{xB2}) & S_{x2}^{*}\sum(\varphi_{xB2} \cap \varphi_{xB2}) & \cdots & \cdots & S_{x18}^{*}\sum(\varphi_{xB18} \cap \varphi_{xB2}) & S_{x19}^{*}\sum(\varphi_{xB19} \cap \varphi_{B2}) \\ S_{x1}^{*}\sum(\varphi_{xB1} \cap \varphi_{xB3}) & S_{x2}^{*}\sum(\varphi_{xB2} \cap \varphi_{xB3}) & \vdots & \vdots & S_{x18}^{*}\sum(\varphi_{xB18} \cap \varphi_{xB3}) & S_{x19}^{*}\sum(\varphi_{xB19} \cap \varphi_{B3}) \\ \vdots & \vdots & & & \vdots & \vdots \\ \vdots & \vdots & \vdots & \vdots & \vdots & \vdots \\ S_{x1}^{*}\sum(\varphi_{xB1} \cap \varphi_{xB18}) & S_{x2}^{*}\sum(\varphi_{xB2} \cap \varphi_{xB18}) & \vdots & \vdots & S_{x18}^{*}\sum(\varphi_{xB18} \cap \varphi_{xB18}) & S_{x19}^{*}\sum(\varphi_{xB19} \cap \varphi_{xB18}) \\ S_{x1}^{*}\sum(\varphi_{xB1} \cap \varphi_{xB19}) & S_{x2}^{*}\sum(\varphi_{xB2} \cap \varphi_{xB19}) & \cdots & \cdots & S_{x18}^{*}\sum(\varphi_{xB18} \cap \varphi_{xB19}) & S_{x19}^{*}\sum(\varphi_{xB19} \cap \varphi_{xB19}) \end{bmatrix} \begin{bmatrix} \frac{1}{V_{x1}^{*}} \\ \frac{1}{V_{x2}^{*}} \\ \frac{1}{V_{x3}^{*}} \\ \vdots \\ \vdots \\ \frac{1}{V_{x18}^{*}} \\ \frac{1}{V_{x19}^{*}} \end{bmatrix} \tag{1}$$

The compressed form of the above equation is given in Eq. (2). Mathematically,

$$[V_x] = [V_{x1}] - [LIM]\left[\frac{1}{V_x^*}\right] \tag{2}$$

This technique is already executed with balanced, unbalanced and meshed distribution system and comes out to be the efficient analytical method with one step calculation unlike the traditional BFS method for the branch current, line losses and nodal voltage calculation. The implementation of the algorithm in this work is expressed in the flow chart as given in Fig. 1.

The 'ε' value is considered to be in order of 0.0001 so that the condition of the load flow convergence between two consecutive iterations of V_t and V_{t-1}. The LIM

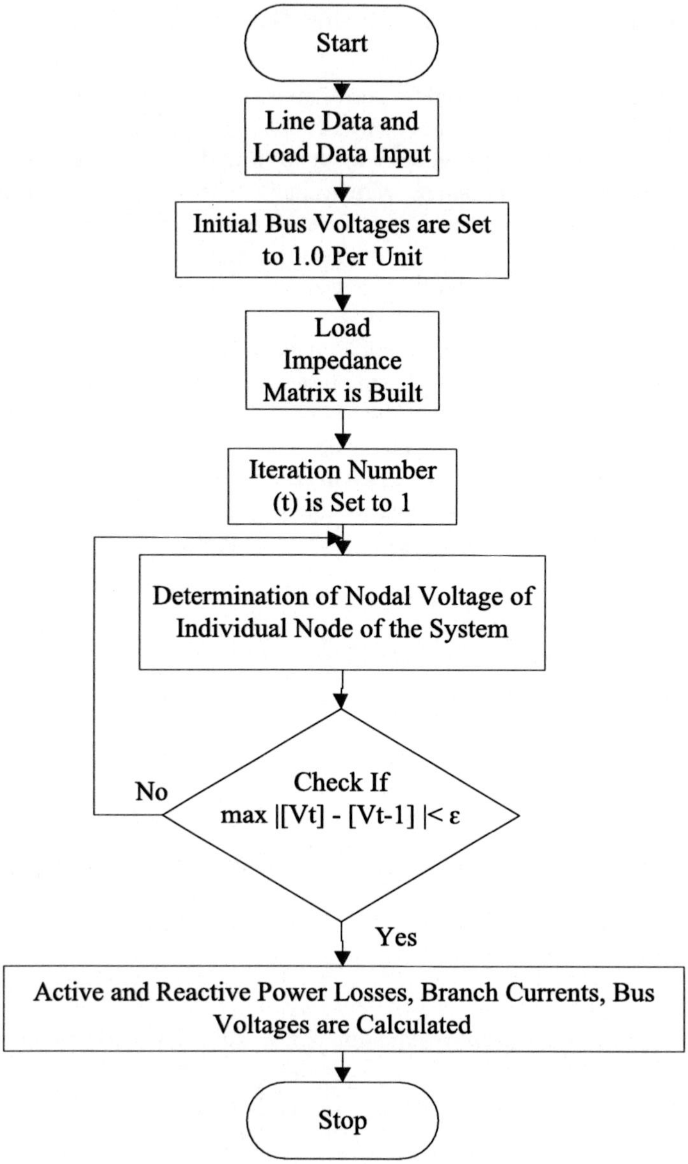

Fig. 1 Flow chart of LIM power flow algorithm

method of load flow analysis is executed with balanced, unbalanced and meshed distribution system. The explanation for above algorithm is best given in [6].

3 Problem Formulation

The objective function of the work is to determine the active power losses in the distribution system using DG. In order to find the power loss, the line currents in each branches is calculated and is given by Eq. (3).

$$[I_{xabc}]_{(nn\times 1)} = [Z_{xabc}]^{-1}_{(nn\times nn)} \times \left[V_{xdiff_{abc}}\right]_{(nn\times 1)} \tag{3}$$

Here, V_{xnn}–$V_{x(nn-1)}$) = V_{xdiff}, which is the potential difference between two nodes, I_{xabc} and Z_{xabc} are the three phase currents and branch impedances of the line and nn are the number of nodes.

The power loss for any branch $P_{xnnLOSS}$ and Active Power Loss (APL) of the system including all the branch losses can be determined using Eqs. (4–7).

$$P_{x(nn-1)} = V_{x(nn-1)} \times I_{x(nn-1)}{}^{*} \tag{4}$$

$$P_{xnn} = V_{xnn} \times (-I_{xnn})^{*} \tag{5}$$

$$P_{xnn_{Loss}} = P_{x(nn-1)} + P_{xnn} \tag{6}$$

$$Total\ \text{Active}\ Power\ Loss(APL_{xnn}) = \sum_{i=1}^{nn} P_{xi_{Loss}} \tag{7}$$

Here bus number varies from i to nn. Few operational constraints are to be satisfied in order to ensure the system integrity. The power balance equation at bus ‘i’ after DG is connected to the grid is given in Eq. (8).

$$P_{xDGi} = P_{xi} - P_{xDi} \tag{8}$$

Here, P_{xDGi} is the power generation from DG, P_{xi} is the bus power generation and P_{Di} is the power supplied to the loads connected at the nodes at bus ‘i’, respectively. Mathematically, the objective function in this work can be expressed as

$$\text{Total Power Loss} = \sum_{i=1}^{nn} P_{xi_{Loss}} \tag{9}$$

The constraints to the objective function are described below.

3.1 Constraints of Equality

The power losses in the lines and the power consumed by the consumers as loads must be equal to the sum of power generation from the grid as well as from the DG. The power balance equation is given by

$$P_{xG} + P_{xDG} = P_{xD} + P_{xLOSS} \tag{10}$$

Here, PxG is the power generated from the grid, P_{xD} is the consumed power by the consumers, P_{xDG} is the power generation from DG source and P_{xLOSS} is the losses in the distribution lines.

3.2 Constraints of Inequality

In every distribution lines, there is some limit to the current flow in the line due to the current carrying capacity. If I_{xmax} is the maximum current limit for the respective line and I_x is the flowing current, then the equation is given as in Eq. (11).

$$I_x \le I_{xmax} \tag{11}$$

Similarly, the nodal voltage of the system must be in its limitation or tolerance range. If Vx is the actual voltage at any node and Vxmin is the minimum and Vxmax is the maximum voltage limits, then the equation can be given as in (12).

$$|V_{xmin}| \le |V_x| \le |V_{xmax}| \tag{12}$$

3.3 Constraints of DG

DG is connected as an additional part to the existing system. Hence it also follows some limitation as the maximum capacity of DG should be less than its maximum limits. If P_{xDG} is the DG power generation at any time and P_{xDGmax} is the maximum power that can be generated using the DG, then the constraint is given in Eq. (13).

$$|P_{xDG}| \le P_{xDGmax} \tag{13}$$

This helps the system to stay stable and maximum benefits can be obtained to the system with consideration of DG constraints.

4 Computational Steps for Optimum Placement of DG

The radial distribution system is tested with load flow technique and optimization algorithm simultaneously so as to minimize the line losses in the distribution lines. The following are the steps implemented while incorporating both these algorithms together to justify the objective of the work.

Step-1: The unbalanced radial distribution system is run with power flow method and corresponding active power losses in the network is measured

Step-2: In optimization algorithm design program, the bus number is randomly generated out of all 19 buses apart from substation bus number-1

Step-3: The DG is placed at the random node with some upper and lower values of energy capacity depending on the load data. The optimization algorithm is carried out and the total active power losses are calculated

Step-4: The minimum power loss is determined with different value of DG location and capacity for all the three phases. These results are saved by the programmer

Step-5: The location of DG that gives most minimum power loss is picked and stored

Step-6: The DG capacities for all the three phases of the system is also found for the system that gives minimum losses

Step-7: Nodal Voltages and line losses for all the phases of the radial system are calculated with power flow analysis

5 Optimization Algorithm

BWO technique is implemented for DG location and capacity in a three phase test distribution system. It is a meta-heuristic optimization technique based on strange mating behavior of black widow spiders [12].

Initially, each spider is represented as population. The natural phenomena depends on few stages. It includes initial population, procreation, sibling cannibalism and mutation. The solution of the problem is represented by a widow which is an array of $1 \times NN$ or is given in Eq. (14).

$$\text{Black widow} = [b_1, b_2, \ldots, b_{NN}] \quad (14)$$

Now, the widow fitness can be determined by evaluating the fitness function Fb among the set of black widows ($b_1, b_2, \ldots, b_{NN}$). Mathematically,

$$\text{Fitness} = \text{Fb (blackwidow)} = \text{Fb}\,(b_1, b_2, \ldots, b_{NN}) \quad (15)$$

The initial population of the spider is generated with a matrix of population size and number of widow solutions. In procreation stage, an alpha is created which is further used to create offspring using parents. The process is expressed in Eq. (16–17).

$$c1 = \alpha \times b1 + (1 - \alpha) \times b2 \quad (16)$$

$$c2 = \alpha \times b2 + (1 - \alpha) \times b2 \quad (17)$$

Here, b1 and b2 are parents, c1 and c2 are offspring. This process is carried out for NN/2 times without duplication of randomly generated numbers. Now the widow offspring along with their mother widow are sorted by fitness values. During the cannibalism stage, the female black widow consumes the male one and transfers the eggs to the sock. The siblings obtained after hatching are also engaged in cannibalism process. The best fit and strong spider is considered to be the optimum solution of the objective function after going through above stages. The technique is well described in [12] with detailed explanation. The flow chart of model implementation is given in Fig. 2. This technique is considered because of its high performance in finding global optimum solution with accuracy and faster convergence rate.

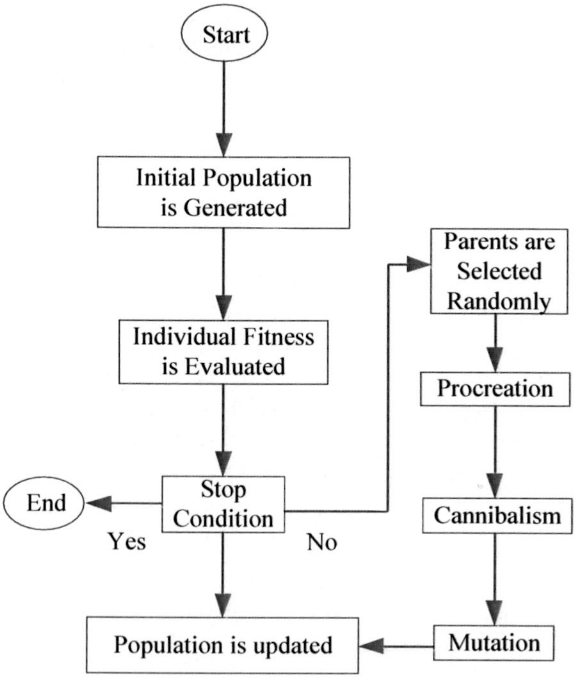

Fig. 2 Flowchart of BWO algorithm

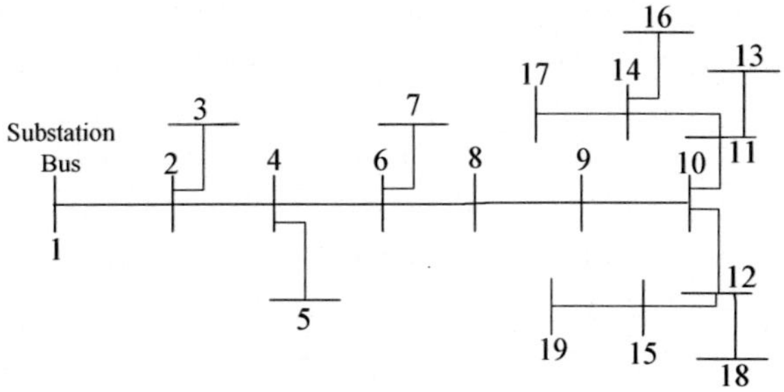

Fig. 3 IEEE-19 node radial distribution system

6 Test System

In this work, IEEE standard 19 bus unbalanced radial distribution system is studied. The test system is shown in Fig. 3 with base voltage and base MVA of 11 kV and 1 MVA, respectively [13]. The unbalanced radial system consists of load data and line data in terms of self-impedance, mutual impedance, conductor type and length of conductor. The final input data for the test system is converted into per unit and applied to the load flow program.

The self-impedances for three phases (Phase-A, Phase-B and Phase-C) are represented as Z_{XAA}, Z_{XBB} and Z_{XCC}, respectively. Similarly, the mutual impedances between the phases are expressed as Z_{XAB}, Z_{XBC} and Z_{XCA} and the sending or receiving end represents the branch from one node to another. The impedances and load data are tabulated in Tables 1, 2 and 3.

7 Proposed Methodology

The unbalanced test distribution system is subjected to undergo multiple algorithms implementation in this work. The additional power demand from the DG source improves the reduction in line losses and bus voltage enhancement. For this execution, the system is run with load impedance matrix method of power flow analysis as well as optimization algorithm of optimum objective function. However, the system is implemented with coronavirus optimization technique for best location and capacity of DG so that the system faces minimum line losses. These multiple algorithms are coded and executed in a software environment of MATLAB, version 20a with computer processor of i7, 8th generation, 500 SSD and 4 GB RAM.

Table 1 Self-impedance line data for IEEE-19 bus system

Sending/receiving end		Self-impedance (PU)		
Fr	To	Z_{XAA}	Z_{XBB}	Z_{XCC}
1	2	0.1161 + 0.0500i	0.1161 + 0.0500i	0.1161 + 0.0500i
2	3	0.2580 + 0.1110i	0.2580 + 0.1110i	0.2580 + 0.1110i
2	4	0.0290 + 0.0125i	0.0290 + 0.0125i	0.0290 + 0.0125i
4	5	0.0290 + 0.0125i	0.0290 + 0.0125i	0.0290 + 0.0125i
4	6	0.0129 + 0.0056i	0.0129 + 0.0056i	0.0129 + 0.0056i
6	7	0.0516 + 0.0222i	0.0516 + 0.0222i	0.0516 + 0.0222i
6	8	0.0806 + 0.0347i	0.0806 + 0.0347i	0.0806 + 0.0347i
8	9	0.1161 + 0.0500i	0.1161 + 0.0500i	0.1161 + 0.0500i
9	10	0.3225 + 0.1388i	0.3225 + 0.1388i	0.3225 + 0.1388i
10	11	0.0290 + 0.0125i	0.0290 + 0.0125i	0.0290 + 0.0125i
10	12	0.0290 + 0.0125i	0.0290 + 0.0125i	0.0290 + 0.0125i
11	13	0.3225 + 0.1388i	0.3225 + 0.1388i	0.3225 + 0.1388i
11	14	0.0129 + 0.0056i	0.0129 + 0.0056i	0.0129 + 0.0056i
12	15	0.3225 + 0.1388i	0.3225 + 0.1388i	0.3225 + 0.1388i
12	16	0.4644 + 0.1998i	0.4644 + 0.1998i	0.4644 + 0.1998i
14	17	0.1580 + 0.0680i	0.1580 + 0.0680i	0.1580 + 0.0680i
14	18	0.2064 + 0.0888i	0.2064 + 0.0888i	0.2064 + 0.0888i
15	19	0.2064 + 0.0888i	0.2064 + 0.0888i	0.2064 + 0.0888i

8 Result and Discussion

The generated power from DGs are fed to the grid with some limitation of power penetration. In this work, the percentage of DGs inflow into the system is assumed to be 30% of the total load connected to the test system [14]. In other words, the three DGs (DG-1, DG-2 and DG-3) are to be connected at the three phases of the system with optimum capacity in order to minimize the line losses with a constraint of total inflow DG power.

8.1 Line Loss Minimization

The system is executed with power flow technique integrated to the new optimization algorithm. The active and reactive power losses are found with best DG location and their capacities. However, in this work DG-1 is assumed to be connected to Phase-A, DG-2 to Phase-B and DG-3 to Phase-C, respectively. The optimum capacities for all the DGs are found and the results are mentioned in Table 4 and the graphical representation is shown in Fig. 4, respectively.

Table 2 Mutual-impedance line data for IEEE-19 bus system

Sending/receiving end		Mutual-impedance (PU)		
Fr	To	Z_{XAB}	Z_{XBC}	Z_{XCA}
1	2	0.0387 + 0.0167i	0.0387 + 0.0167i	0.0387 + 0.0167i
2	3	0.0860 + 0.0370i	0.0860 + 0.0370i	0.0860 + 0.0370i
2	4	0.0097 + 0.0042i	0.0097 + 0.0042i	0.0097 + 0.0042i
4	5	0.0097 + 0.0042i	0.0097 + 0.0042i	0.0097 + 0.0042i
4	6	0.0043 + 0.0019i	0.0043 + 0.0019i	0.0043 + 0.0019i
6	7	0.0002 + 0.0019i	0.0172 + 0.0074i	0.0108 + 0.0095i
6	8	0.0002 + 0.0019i	0.0269 + 0.0116i	0.0108 + 0.0095i
8	9	0.0387 + 0.0167i	0.0387 + 0.0167i	0.0387 + 0.0167i
9	10	0.1075 + 0.0462i	0.1075 + 0.0462i	0.1075 + 0.0462i
10	11	0.0097 + 0.0042i	0.0097 + 0.0042i	0.0097 + 0.0042i
10	12	0.0097 + 0.0042i	0.0097 + 0.0042i	0.0097 + 0.0042i
11	13	0.1075 + 0.0462i	0.1075 + 0.0462i	0.1075 + 0.0462i
11	14	0.0043 + 0.0019i	0.0043 + 0.0019i	0.0043 + 0.0019i
12	15	0.1075 + 0.0462i	0.1075 + 0.0462i	0.1075 + 0.0462i
12	16	0.1548 + 0.0666i	0.1548 + 0.0666i	0.1548 + 0.0666i
14	17	0.0527 + 0.0227i	0.0527 + 0.0227i	0.0527 + 0.0227i
14	18	0.0688 + 0.0296i	0.0688 + 0.0296i	0.0688 + 0.0296i
15	19	0.0688 + 0.0296i	0.0688 + 0.0296i	0.0688 + 0.0296i

8.2 *Increase in Nodal Voltages*

The additional power supply from DG minimizes the power loss in the distribution lines which indirectly improves the bus voltages of the system. The results are taken for all the phases of the system before DG and after DG penetration. The nodal voltage for the substation bus is always assumed to be constant which is 1.0 per unit. Hence, the bus voltages for nodes apart from substation node is determined. Here, the nodal voltage for Phase-B (minimum bus voltage) for all the buses are given in Table 5 and in Fig. 5.

8.3 *Optimal Location and Capacity of DG*

The optimization technique is validated with TLBO optimization technique for a 19-bus unbalanced radial distribution system [15]. The results are mentioned in Table 6.

In this work, BWO optimization technique is further implemented with test system for loss minimization in the lines. The best position of DG and optimum capacity

Table 3 Load data for IEEE-19 bus system

Bus no	Loads connected (PU)		
	Phase-A	Phase-B	Phase-C
1	0	0	0
2	0.01038 + 0.00501i	0.00519 + 0.00252i	0.01038 + 0.00501i
3	0.01101 + 0.00534i	0.00519 + 0.00252i	0.00972 + 0.00471i
4	0.00405 + 0.00195i	0.00567 + 0.00276i	0.00648 + 0.00315i
5	0.00648 + 0.00315i	0.00519 + 0.00252i	0.00453 + 0.00219i
6	0.0042 + 0.00204i	0.00309 + 0.0015i	0.00291 + 0.00141i
7	0.00972 + 0.00471i	0.0081 + 0.00393i	0.0081 + 0.00393i
8	0.00744 + 0.0036i	0.00534 + 0.00258i	0.00339 + 0.00165i
9	0.0123 + 0.00597i	0.01491 + 0.00723i	0.01329 + 0.00642i
10	0.00339 + 0.00165i	0.0042 + 0.00204i	0.00258 + 0.00126i
11	0.00744 + 0.0036i	0.00744 + 0.0036i	0.01101 + 0.00534i
12	0.00972 + 0.00471i	0.0081 + 0.00393i	0.0081 + 0.00393i
13	0.00438 + 0.00213i	0.00534 + 0.00258i	0.00648 + 0.00315i
14	0.00309 + 0.0015i	0.00309 + 0.00234i	0.00405 + 0.00195i
15	0.00438 + 0.00213i	0.00486 + 0.00234i	0.00696 + 0.00336i
16	0.00777 + 0.00378i	0.01038 + 0.00501i	0.00777 + 0.00378i
17	0.00648 + 0.00315i	0.00486 + 0.00234i	0.00486 + 0.00234i
18	0.00543 + 0.00258i	0.00534 + 0.00258i	0.00552 + 0.00267i
19	0.00876 + 0.00423i	0.01005 + 0.00486i	0.00714 + 0.00345i

Table 4 Active and reactive power losses for IEEE-19 bus

Active power losses (kW)				
Status	Phase A	Phase B	Phase C	Total losses
Without DG	4.4687	4.5893	4.4902	13.5482
With DG	2.1774	2.0485	2.1382	6.3641
Reactive power losses (kVAR)				
Status	Phase A	Phase B	Phase C	Total losses
Without DG	2.0155	1.7046	2.0495	5.7696
With DG	1.0029	0.7553	0.95406	2.7123

of DG are found after the execution of optimization algorithm with the system. The results are found with optimum DG location of node number 12 and optimum DG capacities at the specific node are 34.9 kW (DG-1), 39.1 kW (DG-2) and 35.9 kW (DG-3) respectively.

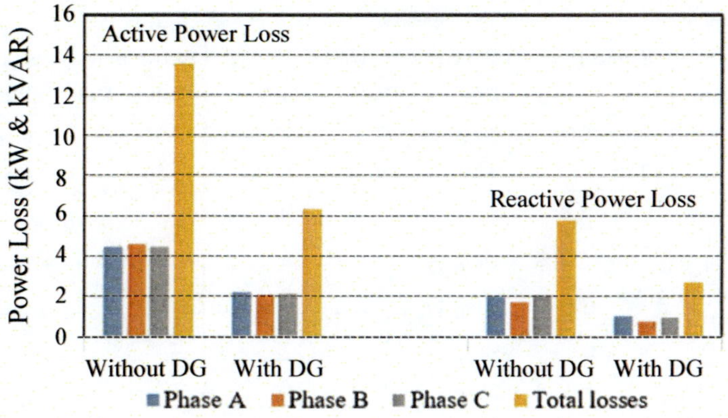

Fig. 4 Active and reactive power losses

Table 5 Bus voltage improvement in Phase-B

Bus no	2	3	4	5	6	7
No DG	0.9886	0.9875	0.9861	0.9859	0.9850	0.9846
With DG	0.9915	0.9905	0.9897	0.9896	0.9890	0.9886
Bus no	8	9	10	11	12	13
No DG	0.9782	0.9704	0.9530	0.9523	0.9521	0.9509
With DG	0.985	0.9801	0.9707	0.9700	0.9706	0.9686
Bus no	14	15	16	17	18	19
No DG	0.9522	0.9481	0.9480	0.9515	0.9512	0.9463
With DG	0.9699	0.9666	0.9666	0.9693	0.9690	0.9648

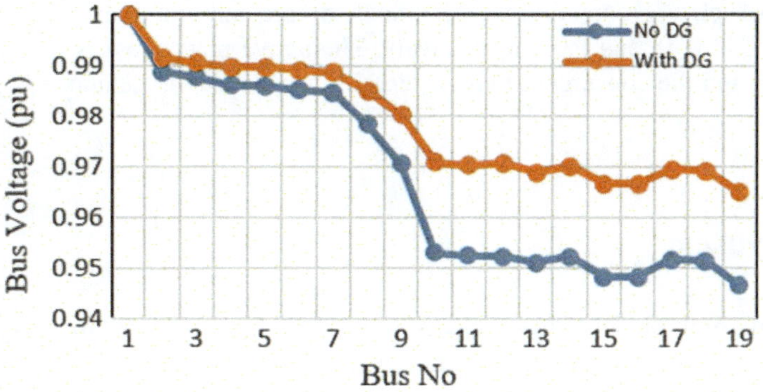

Fig. 5 Status of Phase-B nodal voltages

Table 6 Validation of DG location and capacity

Optimization technique	DG position	DG capacity (kW)			Active power loss (kW)	
		DG-1	DG-2	DG-3	Without DG	With DG
TLBO [15]	12	34.91	39.14	35.94	13.54	6.36
BWO	12	34.91	39.14	35.94	13.54	6.36

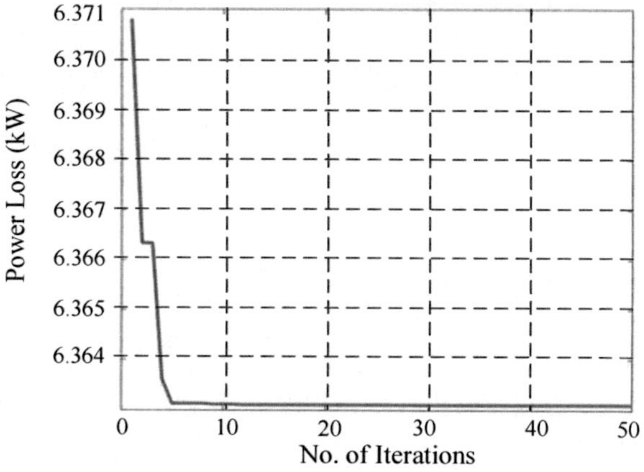

Fig. 6 Convergence characteristics of active power loss

8.4 *Optimization Algorithm*

In this optimization technique, best widow fitness represents the best fitness function. The number of iterations is set to 50 for the algorithm execution. The algorithm parameters such as procreation rate, cannibalism rate and mutation rate are considered to be 0.8, 0.45 and 0.02, respectively. The active power loss is minimized to 6.3641 kW and the convergence characteristics for the above technique is given in Fig. 6.

9 Conclusion

In this work, multiple Distributed Generations (DG) are connected to the system in order to minimize the power loss in the distribution lines. For the loss minimization objective function, the locations and capacities of DGs are determined using a meta-heuristic Black Widow Optimization (BWO) algorithm which is the novelty of this work. A power flow analysis using Load Impedance Matrix (LIM) method is implemented to the test system so as to monitor the system parameters like current flow, nodal voltages and active power losses in individual phases. The load flow algorithm along with optimization technique is executed with an IEEE-19 bus unbalanced Radial Distribution System (RDS). DGs with optimum capacities are placed in each phases of the optimum node which reduces the total line losses to 46.97% and the minimum bus voltage is significantly improved from 0.9463 pu to 0.9648 pu.

References

1. Bollen MHJ, Hassan F (2011) Integration of distributed generation in the power system. Wiley, Hoboken
2. Naik N, Vadhera S (2020) Power loss minimization and voltage improvement with small size distributed generations in radial distribution. In: Advances in electric power and energy infrastructure: proceedings of ICPCCI, vol 608, p 103
3. Routray A, Mistry KD, Arya S (2019) Wake analysis on wind farm power generation for loss minimization in radial distribution system. Renew Energy Focus 34:99–108
4. Bhimarasetti T, Kumar A (2014) Distributed generation placement in unbalanced distribution system with seasonal load variation. In: Proceedings of 2014 18th national power systems conference (NPSC), Guwahati, pp 1–5
5. Milovanović M, Radosavljević J, Perović B (2020) A backward/forward sweep power flow method for harmonic polluted radial distribution systems with distributed generation units. Int Trans Electr Energy Syst 30(5):e12310
6. Ghatak U, Mukherjee V (2017) A fast and efficient load flow technique for unbalanced distribution system. Int J Electr Power Energy Syst 84:99–110
7. Meena MK, Kumar Y, Kumar R, Kumar A (2020) Voltage stability improvement and loss minimization by optimal placement of STATCOM using teaching-learning based optimization technique. Available at SSRN 3575346
8. Karunarathne E, Pasupuleti J, Ekanayake J, Almeida D (2020) Multi-leader particle swarm optimization for optimal planning of distributed generation. In: 2020 IEEE student conference on research and development (SCOReD), pp 96–101
9. Routray A, Mistry KD, Arya S (2019) Loss minimization in a radial distribution system with DG placement using jaya optimization technique. In: 2019 IEEE international conference on sustainable energy technologies and systems (ICSETS), Bhubaneswar, India, pp 336–340
10. Suresh MCV, Edward JB (2020) A hybrid algorithm based optimal placement of DG units for loss reduction in the distribution system. Appl Soft Comput 10619
11. Hussain B, Amin A, Mahmood A, Usman M (2020) An optimal site selection for distributed generation in the distribution network by QPSO algorithm. In: 2020 international conference on engineering and emerging technologies (ICEET), Lahore, Pakistan, pp 1–6
12. Hayyolalam V, Kazem AAP (2020) Black widow optimization algorithm: a novel meta-heuristic approach for solving engineering optimization problems. Eng Appl Artif Intell 87:103249
13. Subrahmanyam JBV, Radhakrishna C (2009) A simple method for feeder reconfiguration of balanced and unbalanced distribution systems for loss minimization. J Electr Power Compon Syst 38:72–84
14. Duong MQ, Tran NTN, Sava GN, Scripcariu M (2017) The impacts of distributed generation penetration into the power system. In: Proceedings of 2017 international conference on electromechanical and power systems (SIELMEN), Iasi, pp 295–301
15. Routray A, Mistry KD, Aryaand S, Chittibabu B (2021) Applied machine learning in wind speed prediction and loss minimization in unbalanced radial distribution system. Energy Sources Part A: Recover Util Environ Eff 1–21

Bacterial Foraging Optimization Based Allocation of PV-DG in Power Distribution System

Ashish Verma and Atma Ram Gupta

Abstract PV is fastest growing renewable energy resource based DG for injecting real power into the system to fulfill the increasing load demand. DSTATCOM with energy storage can exchange both real and reactive power which is a more efficient and economical FACTS device. Most of the electrical loads are linear or constant type in the different types of connected loads. BFOA is one of the best optimization techniques for optimal allocation of DG and DSTATCOM because it is efficient in solving real-world optimization problems. In proposed paper, Solar PV and DSTATCOM are used to inject real and reactive power with different types of loads using BFOA for analysis of effect on voltage profile, real power losses, energy cost and savings in IEEE 33 bus RDS. Also, the behavior of voltage profile on 25% load variation with role of DNO for active as well as reactive power management is studied and analyzed.

Keywords PV · ZIP load · BFOA · DNO · DNR · DSTATCOM · LSF · Ancillary services

1 Introduction

Power flow in the electrical network depends on transmission line parameter such as impedance, phase angle and magnitude of the voltage of sending end and receiving end. Active power and reactive power flow can be controlled by controlling one or more of these parameters [1, 2]. As load in the power system is increasing day by day, it is required to increase load handling capability of power system. Transmission losses are a function of load. So to reduce the transmission losses, Distributed Network Operator (DNO) has responsibilities to manage the load demand in the power system efficiently. Efficient use of power system is a very complex task for DNO which depends on various factors. DNO can use Distributed Network Reconfiguration (DNR) and sitting of Distributed Generation (DG) for managing the

A. Verma (✉) · A. R. Gupta
Department of Electrical Engineering, NIT Kurukshetra, Kurukshetra, Haryana, India
e-mail: ashish_31910102@nitkkr.ac.in

A. Kumar et al. (eds.), *Renewable Energy Towards Smart Grid*, Lecture Notes in Electrical Engineering 823, https://doi.org/10.1007/978-981-16-7472-3_29

increased active as well as reactive load demand. Optimal allocation of DG at distribution system provides economic, environmental and technical advantages. DNO uses proper resources and Transmission & Distribution (T&D) system capacity for efficient allocation of DG. Solar photovoltaic energy is claimed to play a crescent role in the energetic system production. DNO should consider cost factors, technical efficiencies, carbon intensities, static and dynamic variation and growth of load demand [3]. Before allocation of DG and Distribution Static Compensator (DSTATCOM), basic concept of solar Photovoltaic (PV) power generation, advantages of solar PV as DG, mathematical modeling and its cost analysis are discussed in Sect. 2. In Sect. 3, basic concept of DSTATCOM and its advantages are discussed. In Sect. 4, solar irradiance with respect to different location is discussed. In Sect. 5, different types of loads for electrical power system are discussed. In Sect. 6, basic concept of Bacterial Foraging optimization algorithm (BFOA) and its steps for allocation of DG and DSTATCOM in IEEE 33 bus Radial Distribution System (RDS) are discussed.

2 Solar PV Power Generation Allocation

2.1 Solar PV Power Generation

Solar power generation is one of the largest source of renewable energy and there are different ways of solar power generation, i.e. PV and concentrating solar power (CSP) systems. PV technology includes thin film cells, crystalline, organic and inorganic dye-sensitized solar cells which are used as backup energy generation. The CSP technology includes linear Fresnel reflector, parabolic trough, solar dish and power tower which are used for large-scale solar power generation [4]. Factors affecting efficiency of solar PV generation are energy conversion efficiency of solar panel, control of maximum power point, environmental condition, operation and monitoring.

There are different types of algorithms of maximum power point tracking (MPPT) such as constant voltage control, improved constant voltage, interference observation method, conducting incremental method and fuzzy logic control. Energy generated by solar PV can be calculated by using monthly average radiation data according to the size of power plant. Average monthly radiation for solar PV is affected by cloudy days and also solar PV generates only for 10 h. Energy generated by 500 kW solar PV plant in a year (assume plant generate full power, i.e. 500 kW and average hour per day as 10) is given by Eq. (1).

$$Energy\ generated = 10 * 365 * 500\ \text{kWh} \tag{1}$$

Table 1 Various advantages of solar PV as DG

Economical	Environmental	Technical
Cost of losses reduces	Less air pollution	Voltage profile
Lower electricity bill	Using less water	Reduce T&D losses
Reduce operational cost	Slow climate change	System stability
Reduce maintenance cost	Reduce carbon emission	Improved power flow
Replacement cost is high		Developed technology
Less tax rate		

2.2 Solar PV as DG

Solar PV generation as DG injects real power to the micro-grid. It is required to operate at maximum power point for injection of maximum real power. Integration of solar PV as DG can provide different types of advantages as shown in Table 1.

2.3 Mathematical Modeling of Solar PV Power Plant

DC power delivered by a PV generator is given by Eq. (2).

$$P_{dc} = G_{eff}\eta_g A_g \tag{2}$$

Where G_{eff} is instantaneous incident irradiance, η_g is the efficiency of the generator which reduces continuously and A_g is the generator active surface which is constant. So Eq. (3) represents DC power at standard test conditions.

$$P_g^* = G^*\eta_g^* A_g \tag{3}$$

Where P_g^* is the rated peak power and $G^* = 1000\ \text{W/m}^2$. Combining these equations P_{dc} is given by Eq. (4).

$$P_{dc} = P_g^* \frac{G_{eff}\eta_g}{G^*\eta_g^*} \tag{4}$$

In proposed paper, 500 kW solar PV plant is used, suppose initial efficiency η_g^* 100% and average efficiency throughout the year (η_g) is 90%. So expression becomes as Eq. (5).

$$P_{dc} = 500\frac{G_{eff} * 90}{1000 * 100} \quad (5)$$

So final expression of DC power generated by solar panel is given by Eq. (6).

$$P_{dc} = 0.45 * G_{eff} \quad (6)$$

After that DC power generated by solar panel will convert into AC power by using inverter which can be given by Eq. (7).

$$P_{dc} = 0.45 * G_{eff} * \eta \quad (7)$$

η is the efficiency of inverter which is taken as 0.90 and expression of AC power is as Eq. (8).

$$P_{ac} = 0.45 * G_{eff} * 0.90 \quad (8)$$

2.4 Cost Analysis

Voltage profile, Energy losses and active and reactive power demand can be managed by DNR and allocation of solar PV at distribution side which helps for economic savings, long-life and efficiency of T&D equipment. Cost of Energy losses is given by Eq. (9).

$$\text{CEL} = (\text{Total Real Power loss}) * (\text{E}_\text{c}*\text{T})\ \text{INR} \quad (9)$$

where E_c = Energy cost rate (= 4.4664 INR/kWhr), T = Time duration (= 8760 h).

$$\text{Savings} = \text{CEL} - \text{CEL}' \quad (10)$$

Where CEL = Cost of Energy losses without solar PV, CEL' = Cost of Energy losses with solar PV.

Cost of solar PV depends on size and type of power plant (grid connected, off-connected and hybrid PV power plant) and as technology (thin film cells, crystalline, organic and inorganic dye-sensitized solar cells) is improving day by day, cost of solar PV reducing [5–7]. Savings due to Solar PV allocation can be given by Eq. (10).

2.5 *Load Sensitivity Factor for PV and DSTATCOM*

Loss sensitivity factor for sitting of PV and DSTATCOM as PV injects real power so loss sensitivity factor for sitting of PV is given by Eq. (11). DSTATCOM injects reactive power so loss sensitivity factor for optimal location of DSTATCOM is given by Eq. (12) [8, 9].

$$F_{PLS}(k, k+1) = \frac{dP_{lineloss}}{dP_{k+1,eff}} = \frac{2P_{k+1,eff}R_{k,k+1}}{|V_{k+1}|^2} \tag{11}$$

$$F_{QLS}(k, k+1) = \frac{dQ_{lineloss}}{dQ_{k+1,eff}} = \frac{2Q_{k+1,eff}X_{k,k+1}}{|V_{k+1}|^2} \tag{12}$$

3 DSTATCOM

DSTATCOM is a shunt connected Flexible AC Transmission Systems (FACTS) device which can be used in T&D lines for compensation of the reactive power continuously in the RDS, and it provides variable reactive power continuously which is not possible in case of shunt capacitors. DNO has to manage an extra cost of capacitors and for placing capacitors at optimal location in case of shunt capacitors. Also load balancing can't be possible with capacitor [10, 11].

DSTATCOM can be used to inject or absorb the reactive and active current. DSTATCOM can inject reactive power to compensate parameters of sensitive buses instantaneously and active power injection should be done by PV which can inject active power as per solar irradiance. Energy storage systems can be used to inject power for short period of time and it requires continuous charging [7]. The major components of a DSTATCOM device are DC-link capacitor, coupling transformer, PWM control strategy, inverter modules and AC filters. By changing voltage and current, reactive power can be control in two different modes such as voltage control mode (VCM) and current control mode (CCM). In proposed paper, DSTATCOM is used to inject reactive power into the micro-grid [12–15].

4 Solar Irradiance with Different Geographical and Environmental Conditions

Data of solar irradiance is taken from national renewable energy laboratory for each hour during a period of a year. In proposed paper, average value of solar irradiance for each day is taken for analysis. Solar irradiation varies with different geographical location [16].

5 Load Modeling

Analysis of impact of load on electrical parameter is increasing day by day. In power system, there are different types of loads such as static (linear, polynomial, exponential, induction motor and frequency-dependent load) and dynamic (exponential dynamic, dynamic induction motor and composite load model). In proposed paper, behavior of voltage and real power losses are determined.

In linear power load, power is independent of voltage ratio. In the proposed paper, it is base case when power load is taken as same as load (P_0, Q_0) for IEEE 33 bus RDS. Expression for linear power load can be given by Eqs. (13) and (14).

$$P = P_0 \tag{13}$$

$$Q = Q_0 \tag{14}$$

In polynomial load, power depends on voltage ratio of individual bus. In this proposed paper, polynomial power load at any bus is taken as actual power of that bus multiplied by voltage dependent polynomial as given in Eqs. (15) and (16).

$$P = P_0\left[n_1\left(\frac{V}{V_0}\right)^2 + n_2\left(\frac{V}{V_0}\right) + n_3\right] \tag{15}$$

$$Q = Q_0\left[n_1\left(\frac{V}{V_0}\right)^2 + n_2\left(\frac{V}{V_0}\right) + n_3\right] \tag{16}$$

Where V_0 is the reference voltage (= 1 pu), n_1, n_2 and n_3 are the ratio so that power depends on voltage ratio. In proposed paper, n_1, n_2 and n_3 are taken as 0.333.

In exponential load, the power is proportional to the exponential of voltage ratio of individual bus. In the proposed paper, the active and reactive power load at any bus is taken as actual power (P_0,Q_0) of that bus multiplied by exponential of voltage ratio of that bus which is given by Eqs. (17) and (18).

$$P = P_0\left(\frac{V}{V_0}\right)^{n_p} \tag{17}$$

$$Q = Q_0\left(\frac{V}{V_0}\right)^{n_q} \tag{18}$$

Where n_p and n_q are the exponential factor for voltage ratio which depends on the electrical load as given in Table 2.

In induction motor load, Most of the loads can be represented as induction motor type of load. As per analysis of equivalent circuit diagram of steady-state induction motor, active power depends on resistance, reactance, slip and voltage. For an

Table 2 Load component in electrical system

Load component	n_p	n_q
Air conditioner	0.50	2.50
Resistance space heater	2.00	0.00
Small industrial motor	0.10	0.60
Large industrial motor	0.05	0.50
Pump, fans	0.08	1.60
Fluorescent lightning	1.00	3.00
Average value	0.62	1.37

induction motor, all parameters are constant except supply voltage and power is proportional to the square of the voltage. In proposed paper, active power is used as proportional to the square of the voltage ratio which is given in Eqs. (19) and (20).

$$P = P_0\left(\frac{V}{V_0}\right)^2 \tag{19}$$

$$Q = Q_0\left(\frac{V}{V_0}\right)^2 \tag{20}$$

Sometimes electrical load depends on frequency of supply as shown in the Eqs. (19) and (20). But in proposed paper, frequency variation is taken as negligible so factor $(f - f_0)$ is zero. So power is constant which is same as base case.

$$P = P_0\left[1 + S_f(f - f_0)\right] \tag{19}$$

$$Q = Q_0\left[1 + S_f(f - f_0)\right] \tag{20}$$

In power system, due to continuous loading, load dependence on variable is also changing so these types of loading are called as dynamic loading. Expression for active power is given by Eqs. (21) and (22).

$$\mathrm{T_p}\frac{\mathrm{dP_r}}{\mathrm{dt}} + \mathrm{P_r} = \mathrm{P_0}\left(\frac{\mathrm{V}}{\mathrm{V_0}}\right)^{\mathrm{n_s}} - \mathrm{P_0}\left(\frac{\mathrm{V}}{\mathrm{V_0}}\right)^{\mathrm{n_t}} \tag{21}$$

$$\mathrm{P_i} = \mathrm{P_r} + \mathrm{P_0}\left(\frac{\mathrm{V}}{\mathrm{V_0}}\right)^{\mathrm{n_t}} \tag{22}$$

Where T_p is active load recovery time constant, P_r is active power recovery, P_0 is rated active power, n_s is the steady-state voltage dependence, n_t is the transient voltage dependence and P_i is the instantaneous power load response [17–19]. Load model currently used in industry is as shown in Fig. 1.

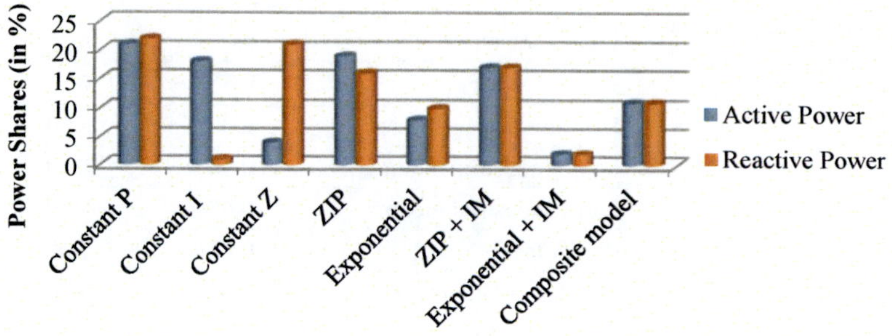

Fig. 1 Various load models currently used in industry

6 Allocation of Solar PV Power Plant Using BFOA

BFOA is based on the nature-inspired optimization algorithm. BFOA estimates objective function, and in the proposed paper real power loss function is used as objective function. Locomotion is achieved by a set of tensile flagella during foraging of real bacteria. There are two operations in this algorithm, first is 'tumble' and other is 'swim'.

After each iterative step of the program, the execution proceeds and leads to better progressively better fitness of power loss function. The parameters to be optimized represent coordinates of the bacteria. One bacterium is present at each point and after each progressive step the bacteria move to new positions. And at each position, power loss function is calculated and then with this calculated value of power loss function, further movement of bacteria is decided by decreasing direction of power loss function. So this will lead to a position with minimum power loss.

Before implementation of BFOA for calculation of optimal time of operation of solar generation with minimum power loss, it required to know about the buses for allocation of solar. And it can be calculated by using power Loss Sensitivity Factor (LSF) which can be given by Eqs. (10) and (11) [20–22].

LSFs can be calculated from the load flow analysis and values are arranged in a descending order for all buses of the given system. Descending order of LSFs decides the sequence in which the buses are to be considered for solar installation and DSTATCOM [12]. According to LSFs, most three sensitive buses are 25, 24 and 8 for IEEE 33 bus RDS for solar PV. Most three sensitive buses are 30, 24 and 25 for IEEE 33 bus RDS for DSTATCOM. The optimal size of solar generation at the candidate buses are calculated by using BFOA. Figure 2 shows flowchart of BFOA.

Where, i, j, k, l, m = variable for different loops, N_{el} = Elimination-dispersal steps, N_{re} = Reproduction steps, N_c = Chemotaxis steps, PL (I, j, k, l) = Min (Real Power Losses), i.e. Objective Function, N_s = Swim length, PL_{last} = Real Power Losses after last step.

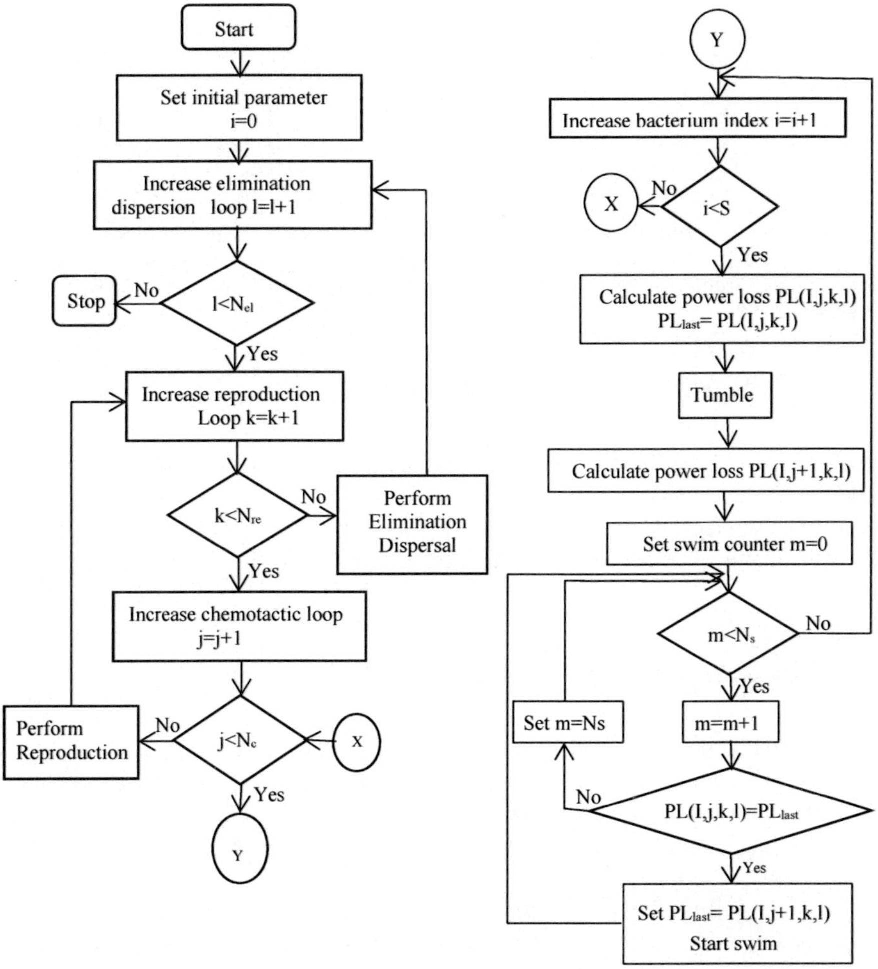

Fig. 2 Flowchart for BFOA

7 Results and Discussion

7.1 Input Data for Different Bus System

Proposed paper uses IEEE 33 bus RDS, and data for this system are taken from IEEE standard [23]. DGs have been allocated at most sensitive buses as per F_{PLS} and F_{QLS}.

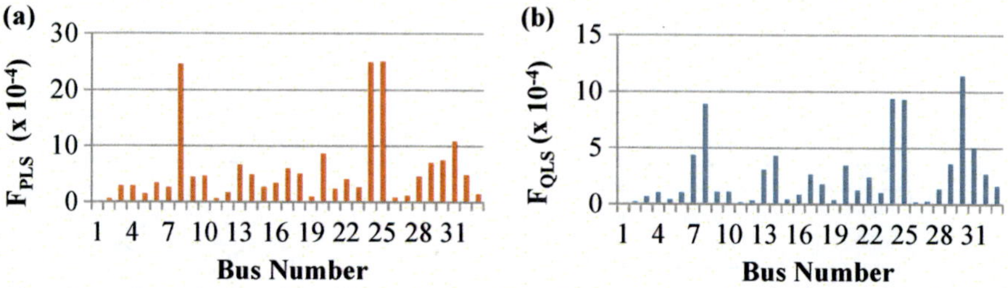

Fig. 3 **a** F_{PLS} for different bus for 33 bus system for PV **b** F_{QLS} for different bus for 33 bus system for DSTATCOM

7.2 Loss Sensitivity Factor at Different Buses

F_{PLS} and F_{QLS} helps to select sensitive buses that are suitable for allocation of PV and DSTATCOM. F_{PLS} of different buses is as shown in Fig. 3a which gives most sensitive buses for DG, i.e. 25, 24 and 8. F_{QLS} of different buses is as shown in Fig. 3b which gives most sensitive buses for DSTATCOM, i.e. 30, 24 and 25.

7.3 Bus Voltages at Different Loading

Load is variable throughout the year so proposed paper is also showing bus voltage for different loading (25% overloading, 25% under-loading) at all buses as shown in Fig. 4. Figure shows that voltage profile is better for 25% under-loading than 25% overloading. PV and DSTATCOM help to improve voltage profile and less power losses.

As shown in Fig. 4, there is very small difference between base case and 25% under-loading as compared to difference between base case and 25% overloading.

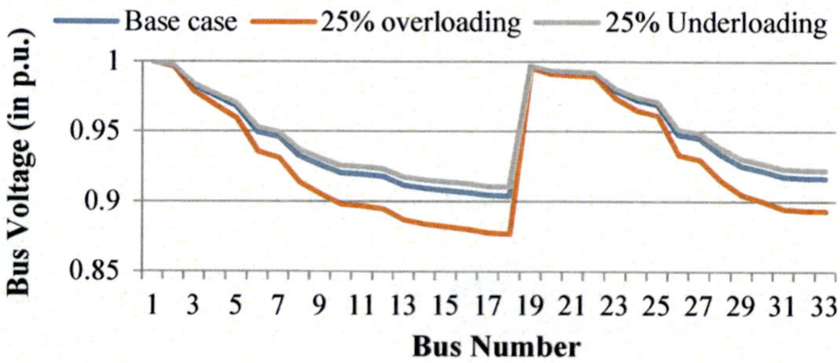

Fig. 4 Comparison of base case bus voltage with under/over loading

7.4 *Different Types of Loads*

As discussed earlier, there are different types of electrical loads in the power system. So Fig. 5a, b show bus voltage for different types of loads, real power loss for different cases without solar PV and with solar PV, respectively. Figure 6 shows bus voltage for different types of loads with DSTATCOM. Figure 7 shows real power losses for different types of loads in base case, with solar PV and with DSTATCOM.

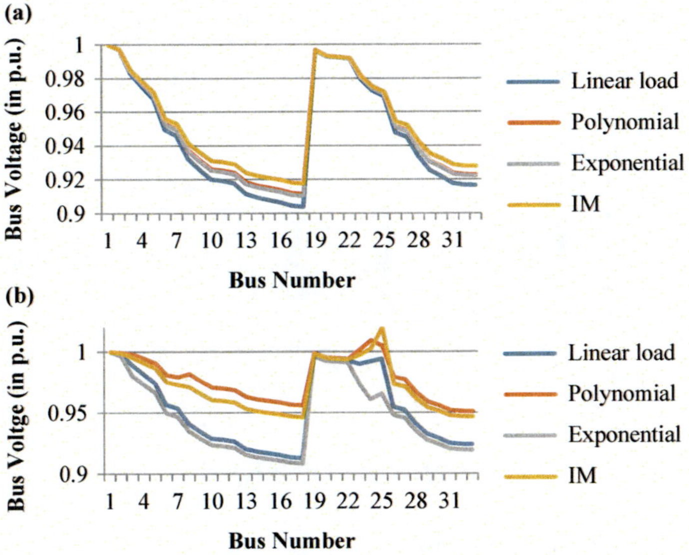

Fig. 5 **a** Comparison of bus voltage for different types of loads without solar **b** Comparison of bus voltage for different types of loads with solar PV allocation

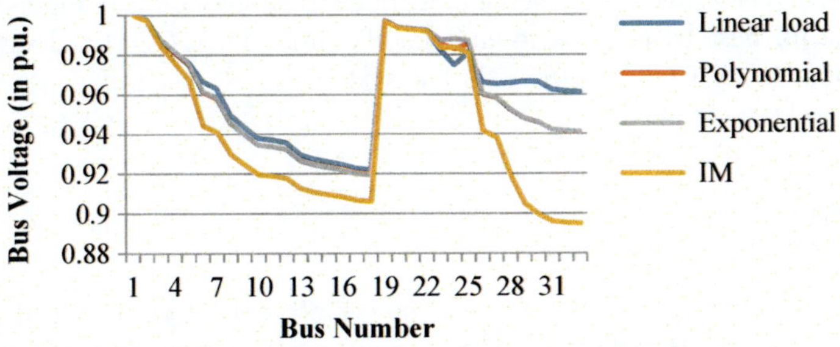

Fig. 6 Comparison of voltage for different types of loads with DSTATCOM

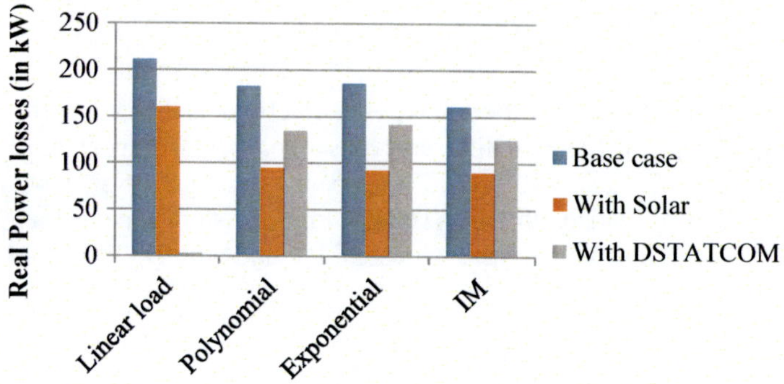

Fig. 7 Real power losses for different types of loads

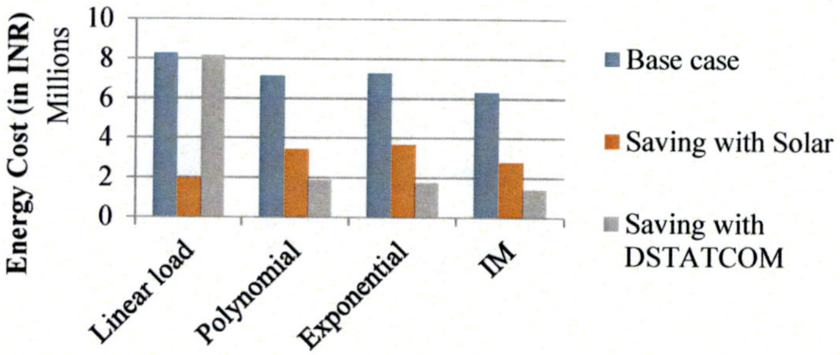

Fig. 8 Energy cost and savings with solar and with DSTATCOM

7.5 Energy Cost and Savings

Earlier proposed paper shows the better voltage profile and less energy losses with allocation of solar PV and DSTATCOM. After allocation of solar PV and DSTATCOM, DNO should be ensured that allocation is economical. Figure 8 shows energy cost for base case, i.e. without solar PV and DSTATCOM and savings for solar PV and DSTATCOM for IEEE 33 bus system.

8 Conclusions

In this proposed paper, modeling of solar power plant, DSTATCOM and different types of loads with role of DNO has been discussed. Sensitive buses for allocation of PV and DSTATCOM are selected from the value of F_{PLS} which are 24, 25, 8 for solar PV and 30, 24, 25 for DSTATCOM. Then impact 25% overloading and under-loading at each bus-on-bus voltage profile is discussed. For overloading bus voltage reduces, and for under-loading bus voltage increase. Then, bus voltage for different types of electrical load is discussed. Then, optimal allocation of solar PV and DSTATCOM for linear, polynomial, exponential and induction motor load using BFOA have been discussed. Solar PVs are important for minimum power losses. By placing the PV and DSTATCOM, Voltage profile will improve and real power losses will reduce as shown in Figs. 5a, b, 6 and 7. The proposed paper shows large amount of savings (in INR) with PV and DSTATCOM which is important for DNO.

References

1. Tashackori A, Hosseini SH, Sabahi M (2015) Power quality improvement using a power electronic transformer based DVR. In: 2015 23rd Iranian conference on electrical engineering, Tehran, pp 1597–1601. https://doi.org/10.1109/IranianCEE.2015.7146474
2. Jadhav S, Jangle N (2018) Improvement in power quality performance using S-transform based D-STATCOM. In: 2018 IEEE international conference on system, computation, automation and networking (ICSCA), Pondicherry, pp 1–6. https://doi.org/10.1109/ICSCAN.2018.8541210
3. Theo WL, Lim JS, Ho WS, Hashim H, Lee CT (2017) Review of distributed generation (DG) system planning and optimisation techniques: Comparison of numerical and mathematical modelling methods. Renew Sustain Energy Rev (Elsevier) 67(C):531–573
4. Sa'ed JA, Quraan M, Samara Q, Favuzza S, Zizzo G (2017) Impact of integrating photovoltaic based DG on distribution network harmonics. In: 2017 IEEE international conference on environment and electrical engineering and 2017 IEEE industrial and commercial power systems Europe (EEEIC / I&CPS Europe), Milan, pp 1–5. https://doi.org/10.1109/EEEIC.2017.7977786
5. Ahmed IM, Kamel S, Abdel-Mawgoud H, Yu J (2018) Optimal allocation of PV based DG in distribution networks at different load levels. In: 2018 20th international middle east power systems conference (MEPCON), Cairo, Egypt, pp 649–654. https://doi.org/10.1109/MEPCON.2018.8635122
6. Bhuyan S, Das M, Bhuyan KC (2017) Particle swarm optimizations based DG allocation in local PV distribution networks for voltage profile improvement. In: 2017 international conference on computer, electrical & communication engineering (ICCECE), Kolkata, pp 1–4. https://doi.org/10.1109/ICCECE.2017.8526201
7. Khasanov M, Kamel S, Tostado-Véliz M, Jurado F (2020) Allocation of photovoltaic and wind turbine based DG units using artificial ecosystem-based optimization. In: 2020 IEEE international conference on environment and electrical engineering and 2020 IEEE industrial and commercial power systems Europe (EEEIC / I&CPS Europe), Madrid, Spain, pp 1–5. https://doi.org/10.1109/EEEIC/ICPSEurope49358.2020.9160696
8. Dehkordi BM, Moallem M, Rezazadeh SJ, Amanifar O, Keivanfard M (2012) Optimal capacitor placement and sizing in TABRIZ distribution system using loss sensitivity factors and particle swarm optimization (PSO). In: 2012 Proceedings of 17th conference on electrical power distribution, Tehran, pp 1–6
9. Paul S, Jewell W (2012) Optimal capacitor placement and sizes for power loss reduction using combined power loss index-loss sensitivity factor and genetic algorithm. In: 2012 IEEE power and energy society general meeting, San Diego, CA, pp 1–8. https://doi.org/10.1109/PESGM.2012.6345635
10. Amin A, Kamel S, Yu J (2018) Developed analytical technique for optimal placement and sizing of DG and DSTATCOM in radial distribution systems. In: 2018 20th international middle east power systems conference (MEPCON), Cairo, Egypt, pp 1087–1091. https://doi.org/10.1109/MEPCON.2018.8635168

11. Samal P, Mohanty S, Ganguly S (2016) Effect of DSTATCOM allocation on the performance of an unbalanced radial distribution systems. In: 2016 IEEE international conference on engineering and technology (ICETECH), Coimbatore, pp 927–931. https://doi.org/10.1109/ICETECH.2016.7569383
12. Sanam J, Ganguly S, Panda AK, Panigrahy D (2016) Forecasting of AELC and TESC of distribution systems with the optimal allocation of DSTATCOM. In: 2016 IEEE innovative smart grid technologies - Asia (ISGT-Asia), Melbourne, VIC, pp 1100–1103. https://doi.org/10.1109/ISGT-Asia.2016.7796539
13. Sanam J, Ganguly S, Panda AK (2016) Allocation of DSTATCOM and DG in distribution systems to reduce power loss using ESM algorithm. In: 2016 IEEE 1st international conference on power electronics, intelligent control and energy systems (ICPEICES), Delhi, pp 1–5. https://doi.org/10.1109/ICPEICES.2016.7853724
14. Ghatak SR, Sannigrahi S, Acharjee P (2018) Comparative performance analysis of DG and DSTATCOM using improved PSO based on success rate for deregulated environment. IEEE Syst J 12(3):2791–2802. https://doi.org/10.1109/JSYST.2017.2691759
15. Gupta AK, Kumar A (2019) Deployment of distributed generation with D-FACTS in distribution system: a comprehensive analytical review. IETE J Res 1–18. https://doi.org/10.1080/03772063.2019.1644206
16. Bais R (2016) On the occurrence of Common Baron (Lepidoptera: Nymphalidae: Limenitidinae: Euthalia aconthea Cramer, 1777) in the Delhi area and analysis of abiotic factors affecting its distribution in India. J Threat Taxa 8:9418. https://doi.org/10.11609/jott.3112.8.12.9418-9433
17. Satyanarayana S, Thiruveedula R, Sivanagaraju S, Rao G (2007) An efficient load flow solution for radial distribution network including voltage dependent load models. Electr Power Compon Syst 35:539–551. https://doi.org/10.1080/15325000601078179
18. Dharmakeerthi CH, Mithulananthan N, Saha TK (2013) Impact of electric vehicle load on power system oscillatory stability. In: 2013 Australasian universities power engineering conference (AUPEC), Hobart, TAS, pp 1–6. https://doi.org/10.1109/AUPEC.2013.6725401
19. Arif A, Wang Z, Wang J, Mather BA, Bashualdo H, Zhao D (2017) Load modeling – a review. United States. https://doi.org/10.1109/TSG.2017.2700436
20. Das S, Biswas A, Dasgupta S, Abraham A (2009) Bacterial foraging optimization algorithm: theoretical foundations, analysis, and applications. https://doi.org/10.1007/978-3-642-01085-9_2
21. Chen H, Zhu Y, Hu K (2011) Adaptive bacterial foraging optimization. Abstr Appl Anal. https://doi.org/10.1155/2011/108269
22. Teng J-H (2003) A direct approach for distribution system load flow solutions. IEEE Trans Power Deliv 18(3):882–887. https://doi.org/10.1109/TPWRD.2003.813818
23. Vita V (2017) Development of a decision-making algorithm for the optimum size and placement of distributed generation units in distribution networks. Energies 10:1433. https://doi.org/10.3390/en10091433

Effect of Penetration Levels on Thermal Demands with V and f Regulation in Multi-energy Microgrid

Ankit Uniyal, Saumendra Sarangi, and Mahiraj Singh Rawat

Abstract With the integration of different energy carriers namely power, heat, cooling, storage, etc. at a single place, the efficacy of the microgrids has improved. The multi-energy microgrid however may face challenges in voltage (V) and frequency (f) control and meeting the thermal demands in near future due to rising distributed generation (DG) penetrations. Thus, it is therefore significant to analyze the impacts of DG penetration levels on the economics of power and thermal demands in such systems. The present work aims at analyzing the impact of various DG penetration levels on hot water supply economics in a multi-energy distribution-based autonomous microgrids (ME-DAM) considering minimized hourly frequency deviations and regulated voltage limits.

Keywords Distributed generation (DG) · Load flow · Multi-energy microgrid · Optimization algorithm

1 Introduction

In the present times, the conventional power microgrids are getting transformed as multi-energy systems owing to integration of diverse energy carriers, apart from electric power, such as thermal power in form of heat and cooling demands of the consumers. This has resulted in a more efficient energy network wherein cogeneration of different cooling, heat, and power units are set up as combined cooling, heat, and power units (CCHP) [1]. The CCHP units along with renewable power generation units and energy storage units can be used as multi-energy microgrid [1, 2].

The high penetration level of DGs causes voltage (V) and frequency (f) issues in the autonomous microgrid. The dumping of excess power as heat energy is one of the better solutions to solve this issue as it not only provides V and f control

A. Uniyal (✉) · M. Singh Rawat
National Institute of Technology Uttarakhand, Srinagar Garhwal, India
e-mail: msrawat@nituk.ac.in

S. Sarangi
Motilal Nehru National Institute of Technology Allahabad, Prayagraj, India

A. Kumar et al. (eds.), *Renewable Energy Towards Smart Grid*, Lecture Notes in Electrical Engineering 823, https://doi.org/10.1007/978-981-16-7472-3_30

but also contributes to thermal demand of the multi-energy microgrid. The proper management of the power dumping exercise during low load and high generation hours is essential to get optimized V and f in the system [3]. This analysis assumes more complications when an hourly power and thermal schedule of the microgrid is to be prepared from the forecasted demands of the network.

In a microgrid apart from non-dispatchable DGs, i.e., the DGs which are not converter controlled such as renewable-based DGs without energy storage facility, there are dispatchable DGs which are converter controlled and exhibit droop characteristics as active power (P) versus f and reactive power (Q) versus V as shown in Fig. 1. Due to these characteristics the power flow for distribution systems (or microgrids) is modified which involves the frequency parameter also in the power flow equations [3, 4]. Thus, the new power flow method in coordination with a suitable optimization method can aid in getting the optimized values of both V and f during any hour of the day.

There have been some recent works related to multi-energy systems such as authors in [1] have prepared an optimal dispatch coordination of multiple energy carriers such as fuel cell, electric boiler, CCHP, electric chiller, heat and ice storages, etc. in grid connected and islanded modes for a microgrid. Authors in [5] have done simultaneous power and heat scheduling of microgrids considering operational uncertainties using a new stochastic p-optimization optimization approach, Authors in [6] have done an optimal stochastic deployment of power and thermal energy storage in a residential multi-energy microgrid incorporating demand side management. Authors in [7] have obtained optimal placement of heterogenous DGs in a grid connected multi-energy microgrid having uncertainties.

All these recent works have done planning of power and thermal demands along with storages but not have studied the impact of penetration level of DGs. As the penetration level is rising with time, its impact on multi-energy microgrid in near future will be of grave concern for the network operator.

Moreover, the above works have not considered the droop characteristics of the converter-based DGs which are significant in autonomous droop regulated microgrids.

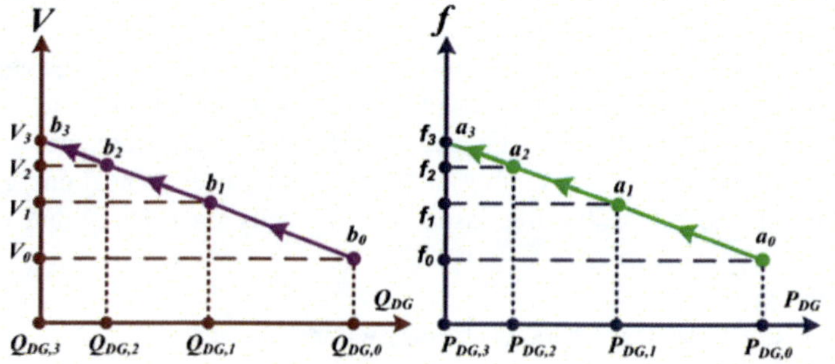

Fig. 1 Droop characteristics of converter-based DGs

Thus, to fill above research gaps the present work deals with conducting a cost-benefit analysis to know as how the DG penetration level impacts the hot water economics for 24 h of a day in multi-energy droop-based autonomous microgrid (ME-DAM) considering minimized hourly frequency deviations with regulated bus voltage limits. The excess power dumping due to high renewable generations is used as a mechanism for meeting the daily hot water generation as compared to the conventional gas boiler (GB) system. The hourly dumping of energy is calculated and corresponding hot water generation using electric boiler (EB) is noted down. The hourly hot water demand is known in the microgrid. The excess hot water demand which is not met by the dumped thermal generation has to be met using GB system. The analysis is conducted on an IEEE 33-bus weakly meshed test system modified as a multi-energy autonomous droop-based microgrid. The optimization problem of obtaining hot water generation corresponding to minimum hourly f deviations is solved using a meta-heuristic technique named as Modified Whale Optimization (MWO) algorithm. The analysis has been done using MATLAB® [8]. Finally, the results are compared and significant conclusions are drawn.

2 Problem Formulation

In the present analysis an optimization problem is formulated where hourly dumped power is calculated for a day considering hourly variations in renewable and converter-based DG powers.

The objective is to minimize the hourly frequency deviations by allocating optimized active and reactive dumped loads, i.e., P_{dp} and Q_{dp} respectively. The P_{dp} in kW is required by the electric boiler to generate hot water for the microgrid consumers, i.e., a multi-energy microgrid.

The locations of DGs and DLs are predefined. A technoeconomic analysis is further conducted to study the effect of different penetration levels on hourly dumped energy which is used for hot water generation. The hot water demand which was not generated using dumped energy is assumed to be generated using GBs.

The optimization problem is formulated as follows:

$$\text{Min } (delF)$$

St.

$$|V_l| \leq |V| \leq |V_u| \tag{1}$$

$$|IB_i| \leq |I_{i,u}| \tag{2}$$

$$|P_{DL,l}| \leq |P_{DL}| \leq |P_{DL,u}| \tag{3}$$

$$|Q_{DL,l}| \le |Q_{DL}| \le |QL_{DL,u}| \tag{4}$$

$$|f_l| \le |f| \le |f_u|, \tag{5}$$

where V, V_l, and V_u are the voltage at each node, lower and upper voltage limits taken as 0.9 pu and 1.05 pu, respectively. IB_i and $IB_{i,u}$ are the branch current and upper limit of branch current, respectively. $P_{DL,l}$ and $P_{DL,u}$ are lower and upper values of active dump loads taken as 100 kW and 1200 kW, respectively.

$Q_{DL,l}$ and $Q_{DL,u}$ are low and up limits of reactive dump loads taken as 100 kVAR and 1200 kVAR, respectively.

f is the frequency of network.

f_l and f_u are the low and up frequency limits taken as 0.999 pu and 1.001 pu, respectively.

3 Methodology

This work requires two subroutines which are included in the coding platform. First subroutine (sub-1) is a load flow analysis for droop controlled autonomous microgrid called Droop-based Distribution System Load Flow (DB-DLF) and second subroutine (sub-2) is an meta-heuristic optimization algorithm called modified whale optimization (MWO) algorithm. An iterative process is followed while having mutual transfer of some variables between both the subroutines as shown in Fig. 2. During each iterative process, the three DL sizes for each thermal group located at nodes 3, 10, and 28 are optimized in sub-2 and send sub-1 where they are used in the load flow analysis, and objective $delF$ is calculated. The power flow variables are sent to sub-1 once again where they are compared with the ones obtained in the previous iterations and the best one is retained while rest are discarded. This process is repeated till convergence is reached.

Fig. 2 Proposed methodology

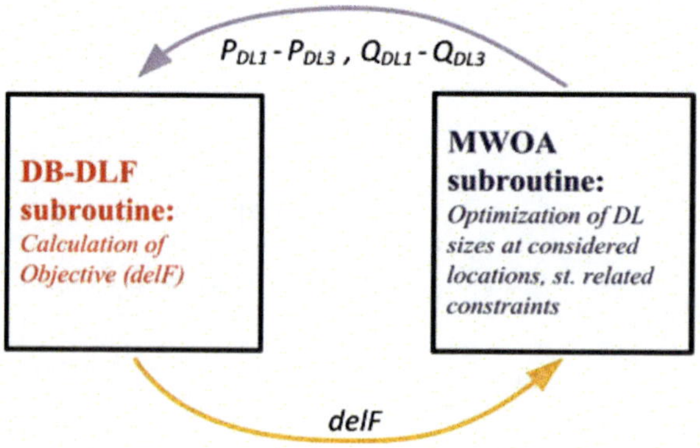

3.1 DB-DLF

The converter-based DGs have power electronic interfaces, by virtue of which the V and f are related to the Q and P at the DG buses by the droop characteristics [3, 9]. The normal distribution system load flow (DLF) thus has to be modified so as to incorporate these characteristics also in the load flow analysis.

The flowchart in Fig. 3 shows the various steps implemented in the droop-based distribution load flow (DB-DLF) analysis. As per the flowchart, after the initial variables are decided, a DLF analysis [10] is conducted and tol_1 is calculated as per Eq. (6)

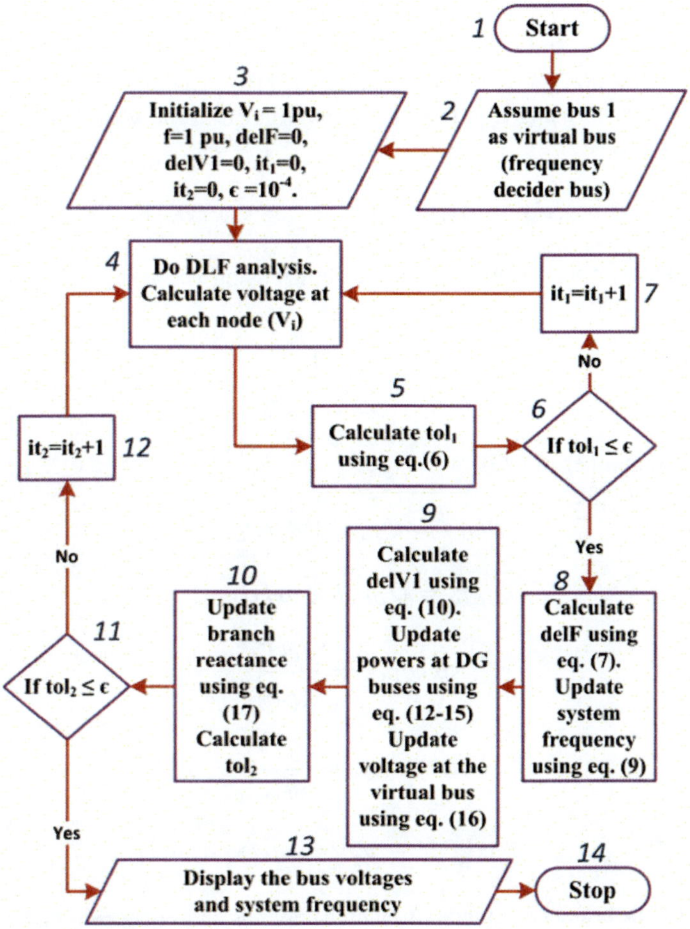

Fig. 3 DB-DCLF flowchart

$$tol_1 = Max.|V_i^{iter_1} - V_i^{iter_1 - 1}|. \tag{6}$$

The tol_1 is checked at Step 6 and if it is obeyed, then delF is calculated using Eq. (7), frequency update at substation bus (virtual bus) is done using Eq. (9)

$$delF = \alpha_{pe} \times \left[P_{DG,1} - \Re\left(V_1 \cdot IB_1^*\right)\right]. \tag{7}$$

The real part of complex term is shown using $\Re$.
α_{pe} is calculated as

$$\alpha_{pe} = \frac{1}{\sum_{M=1}^{n_{dg}} \frac{1}{\alpha_{p,M}}} \tag{8}$$

$$f^{iter_2+1} = f^{iter_2} + delF. \tag{9}$$

In Step 9, the $delV1$ is obtained using Eq. (10). The active and reactive DG powers are updated using Eqs. (12)–(15) and voltage at the virtual bus is updated using Eq. (16).

$$delV_1 = \alpha_{qe} \times \left[Q_{DG,1} - \Im\left(V_1 \cdot IB_1^*\right)\right]. \tag{10}$$

The imaginary part of complex term is shown using $\Im$.
α_{qe} is calculated as

$$\alpha_{qe} = \frac{1}{\sum_{M=1}^{n_{dg}} \frac{1}{\alpha_{q,M}}} \tag{11}$$

Here M is the number of DGs.

$$P_{DG_M} = P_{DG_M 0} + delP_{DG_M} \tag{12}$$

$$Q_{DG_M} = Q_{DG_M 0} + delQ_{DG_M} \tag{13}$$

$$delP_{DG_M} = \frac{delF}{\alpha_{p,M}} \tag{14}$$

$$delQ_{DG,M} = \frac{delV}{\alpha_{q,M}} \tag{15}$$

$$V_1^{iter_2+1} = V_1^{iter_2} + delV_1. \tag{16}$$

In Step 10, the branch reactance (X) which depends on frequency is updated using Eq. (17).

$$X_{se,re} = X_{se,re} \times \frac{f^{iter_2+1}}{f_{iter_2}}. \tag{17}$$

Here se and re are sending and receiving end nodes, respectively. If Step 11 is true then load flow is converged, otherwise Steps 4–11 are followed again till convergence is reached.

3.2 MWO Algorithm

Authors in [11] presented a meta-heuristic method inspired from hunting pattern of Blue Whales. The trajectory of their hunting pattern was very efficient and thus authors in [11] used it to obtain global optimum in an optimization problem. This algorithm is again adapted and improved by [12] and renamed as Modified Whale Optimization (MWO) algorithm. In MWO, Mantegna's algorithm for Levy flight was incorporated to improve global search tendency of WOA. The detailed algorithm of MWOA is given in [12].

4 Results and Discussions

The IEEE 33-bus microgrid in island/autonomous mode has base values of KVA and KV as 500 and 12.66. The load and line data has been taken from [13]. The system processor is Intel I5 and RAM is 8 GB. The locations for solar PV and wind-based DGs have been assumed at nodes 3, 6, 10, 13, 17, 19, 25, and 31 as per the composite sensitivity index for a distribution network [6, 7, 14]. The converter-based DGs for each thermal group are located at nodes 1, 10, and 28. The electric boiler (dump load) is located at nodes 3, 10, and 28 to supply the thermal demand for each group.

There are three cases assumed, each has a different penetration level of DGs for transition season (season between summer and winter):

Case 1: When renewable DG penetration level is low, i.e., 80% of case 2 and the nominal value of converter-based DGs are kept at 0.8 pu for base (1 pu = 500 kVA).
Case 2: When renewable DG penetration is as shown in Fig. 4, as given in [6] and converter-based DGs are kept at 1 pu.
Case 3: When renewable DG penetration is 1.2 times of case 2 and converter-based DGs are kept at 1.2 pu.

The hot water tanks are assumed to be located at three thermal groups ($TG1$–$TG3$) as shown in Fig. 5. The hourly hot water generated is stored in these tanks and circulated to fulfill the hot water requirements of consumers at TG1–TG3 [6]. The lower limit of hot water in tank is kept as 5 kL and upper limit as 200 kL. It means that tank always has 5 kL of minimum level maintained. In such times if hot water is required, then GBs are used as second preference for hot water generation. In the

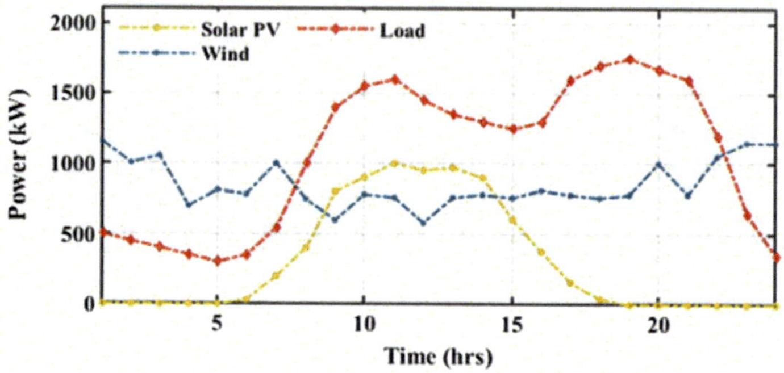

Fig. 4 Solar PV, wind, and load power in the microgrid during transition season

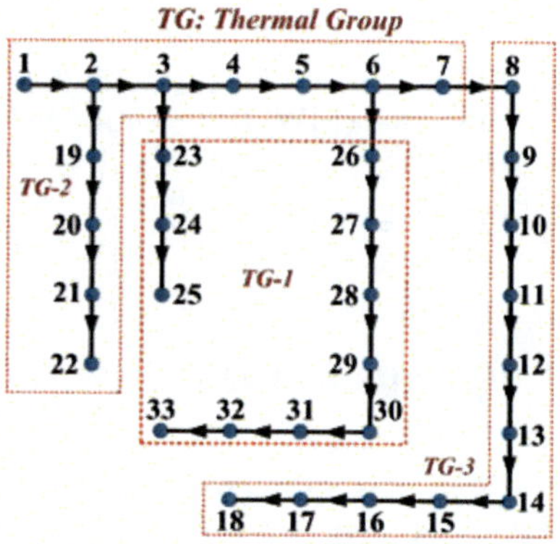

Fig. 5 The multi-energy IEEE 33-bus microgrid with thermal groups

present case it is assumed that before the start of first hour, there was 50 kL of hot water in the tank. The active and reactive droop coefficients for converter-based DGs are assumed as −0.05, −1, and −0.1.

The unit cost of operation of converter-based DGs is 85.6 \$/h + 4.37 \$/kWh, wind DG is 0.115 \$/kWh, PV DG is 0.68 \$/kWh and GB is 0.039 \$/kWh [1, 2, 5–7].

Figure 6 shows the 24 h power profile from DGs, i.e., solar PV (P_{pv}), wind (P_w), converter-based DG (P_{dg}), and active dump load (P_{dp}). It is evident from Fig. 5 that with increase in DG penetrations the power to be dumped (P_{dp}) has increased.

Figure 7 shows the 24 h volume of hot water generated (V_{gen}), hot water stored in the tank (V_{tank}), hourly demand of hot water (V_{dem}) and extra hourly hot water demand which is not met by DL based electric boilers. From Fig. 6, due to increase in penetration the volume of hot water generated per hour is more, i.e., case 3 > case 2 > case 1. The hot water demand (V_{dem}) during low DG penetrations could not be met using dumped power, thus GB's need to be employed to provide extra demand (V_{ext}). During high DG penetrations, the hot water demand is mostly met using dumped power so GBs are used to somewhat less extent for extra volume of hot water demand.

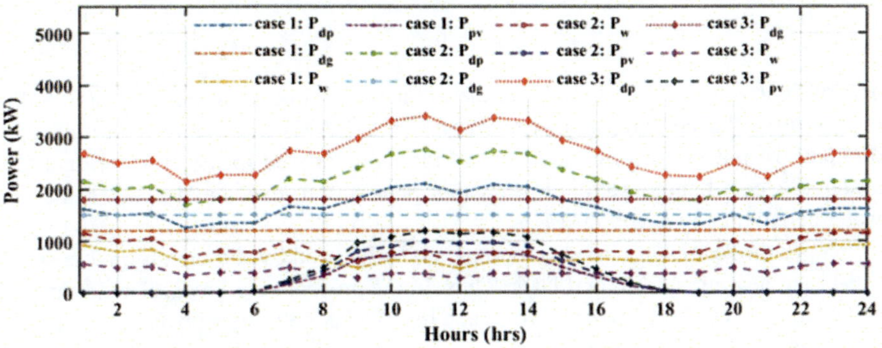

Fig. 6 24 h power profile from DGs and DLs

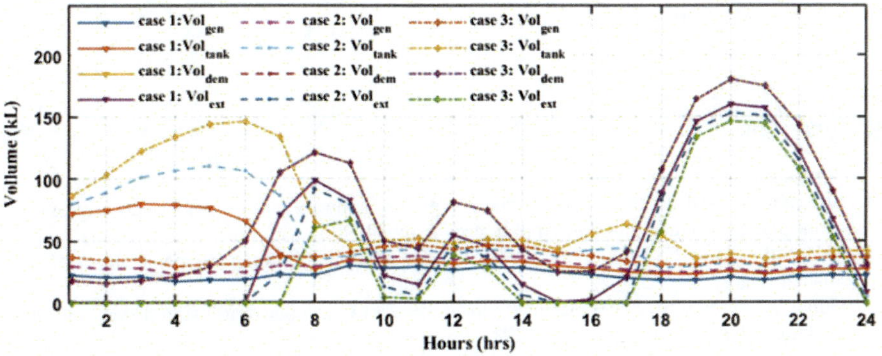

Fig. 7 24 h volume profile of hot water generation

Table 1 24 h power cost of DGs and DLs

Cases	$C_{P_{dg}}$ (\$k)	C_{P_w} (\$k)	$C_{P_{pv}}$ (\$k)	C_{GB}(\$k)	Net cost (\$k)
Case 1	125.89	1.89	0.4	3.339	131.519
Case 2	157.42	2.36	0.5	2.851	163.131
Case 3	188.92	2.83	0.6	2.391	194.741

From Table 1, the 24 h power costs of DGs and DLs are shown for all three cases. It is shown that net operational costs, i.e., sum of all DGs operation costs and cost of water generated from GB for a day (24 h) are \$131.52 k for case 1, \$163.13 k for case 2 and \$194.74 k for case 3. The less net cost in case 1 is due to less unit cost of GB system over DG setups. Thus it could be seen that with increase in DG penetration GB's can be utilized effectively in meeting hot water demands in the multi-energy system. However it depends on each unit cost of DGs and heating systems.

5 Conclusion

The present work aimed at studying the impact of DG penetration level in a multi-energy droop-based autonomous microgrid (ME-DAM) to observe the efficacy of dumping of power for hot water generation as compared to the conventional GB system. A cost-benefit study showed that penetration level has impacted the hot water economics in a ME-DAM.

As per the results obtained, economics is influenced by the operational costs of GB-based heating system and DGs. Thus, in the present case low DG penetrations, provided most cost-effective solutions and high DG penetrations provided least cost-effective solutions. It was due to the domination of converter-based DG unit costs over the operational cost of GB heating system that has increased the net costs in high penetration scenario. Moreover, during the entire analysis the optimal regulation of V and f was maintained in the network.

References

1. Li Z, Xu Y (2018) Optimal coordinated energy dispatch of a multi-energy microgrid in grid-connected and islanded modes. Appl Energy 210:974–986
2. Li Z, Shi Y, Fang F (2014) Combined cooling, heating and power systems: a survey. Renew Sustain Energy Rev 35:1–22
3. Uniyal A, Sarangi S (2019) Optimal allocation of ELC in microgrid using droop controlled load flow. IET GTD 13(20):4566–4578
4. Nutkani NU, Loh PC, Weng P, Blaabjerg F (2017) Decentralized economic dispatch scheme with online power reserve for microgrids. IEEE Trans Smart Grid 8(1):139–148
5. Mohammadreja M, Rezaei N, Ghaderi A (2019) Simultaneous power and heat scheduling of microgrids considering operational uncertainties: a new stochastic p-robust optimization approach. Energy 185:239–253
6. Li Z, Xu Y, Feng X, Wu Q (2020) Optimal stochastic deployment of heterogenous energy storage in a residential multi-energy microgrid with demand side management. IEEE Trans Ind Inform 17(2):991–1004
7. Li Z, Xu Y, Fang S, Mazzoni S (2019) Optimal placement of heterogenous distributed generators in a grid-connected multi-energy microgrid under uncertainties. IET Renew Power Gener 13(14):2623–2633
8. MATLAB R2018a 1994-2018 The math works
9. Díaz G, Aleixandre JG, Costo J (2016) Direct backward/forward sweep algorithm for solving load power flows in AC droop-regulated microgrids. IEEE Trans Smart Grid 7(5):2208–2217
10. Teng JA (2003) Direct approach for distribution system load flow solutions. IEEE Trans Power Deliv 18(3):882–887
11. Mirjalili S, Lewis A (2016) The whale optimization algorithm. Adv Eng Softw 95:51–67
12. Sun Y (2018) A modified whale optimization algorithm for large-scale global optimization problems. Expert Syst Appl 114:563–577
13. Baran ME, Wu FF (1989) Network reconfiguration in distribution systems for loss reduction and load balancing. IEEE Trans Power Deliv 4(2):1401–1407
14. Zhang C, Xu Y, Dong Z, Ma J (2016) A composite sensitivity factor-based method for network distributed generation planning. In: IEEE power systems computation conference (PSCC), pp 1–6

Modeling and Performance Analysis of a PV System Under Mismatch Scenarios

Shahida Khatoon, Mohd Faisal Jalil, and Rahma Aman

Abstract This work will present a comprehensive performance analysis of the solar PV array configurations under mismatch scenarios. The analysis done so far aims to study the mismatch losses of power in a solar array under irradiations that are not uniform in nature. The irradiance mismatching effects also exhibit many peak power points on current-voltage and power-voltage characteristics and non-linearity because of which various traditional MPPT algorithms are not able to track global peak. The extracted power in the PV array depends on PV array configuration and mismatch scenarios. There are numerous solar array available configurations. The examination is based on making the optimal solar modules configuration in an array to get maximum instantaneous power under mismatch scenarios. The PV array configurations modeling that is considered would be done in MATLAB/Simulink environment. The most optimal configuration will reduce the occurrence of multiple local peaks and avoid having a complex maximum power point tracking algorithm. The additional advantage of this configuration would be a co'st-effective and simple photovoltaic array structure.

Keywords Array arrangements · Partial shading conditions · Shading patterns

1 Introduction

As the demand for electrical energy increases day by day, one cannot rely completely on conventional sources, so renewable sources are being looked upon as they are a clean form of energy. A PV system has drawn most of the attention because of its efficiency and various boons. The effect of shading caused by clouds, buildings, trees, bird litters, etc., can diversely affect the PV array because of which the current and voltage mismatch reduces the power produced. The loss of power in a PV array is because of the array's shading pattern and arrangement. The PV array power-voltage

S. Khatoon · R. Aman
Jamia Millia Islamia, New Delhi, India

M. F. Jalil (✉)
Aligarh Muslim University, Aligarh, India

A. Kumar et al. (eds.), *Renewable Energy Towards Smart Grid*, Lecture Notes in Electrical Engineering 823, https://doi.org/10.1007/978-981-16-7472-3_31

curve shows multiple peaks; only one is the global peak. These multiple local peaks make it difficult in getting the maximum power point. Numerous techniques of MPPT have been developed to get the maximum of local peaks, after getting the GMPP, the problem of extracting a significant amount of output power is present as the bypass diodes short-circuited the modules under the shading. Also, the hotspot problem occurred due to the mismatch problems, which further degrades the performance of PV panels and makes the cell parameter more consistent. So to overcome the losses, many different array configurations are being developed which are Bridge Linked (BL), Series-Parallel (SP), Honey Comb (HC), and Total Cross Tied (TCT). These interconnections are compared for parameters like losses, optimum power, fill factor, and reliability [9]. According to the study, we can consider TCT configuration as the finest solution that reduces the mismatch losses caused by partial shading of arrays. To reduce the losses and increase the generated power, there are numerous static and dynamic reconfigurations; all these techniques will be discussed in this paper.

2 Literature Review

This review will discuss and investigate different papers on an array configuration of the PV system. The paper's study would be done in terms of FF, accuracy, optimum power point, etc. We will discuss the positive and weak points of each technique and do a complete analysis. Numerous research papers [1–21] are taken to study the various PV module reconfiguration schemes in an array for finding the best scheme that can perform better under all the shading conditions.

2.1 PV Array Configuration

The panels can be connected in a series and also in parallel according to get the voltage and power parameters. To get good performance various conventional methods that are series-parallel series(S-P-S), series-parallel(SP), bridge linked(BL), honey comb(HC), total cross-tied(TCT), multi-string are being introduced. The researchers' main concern is of making the performance of the PV system better and to meet the demand numerous other configurations are being looked upon like hybrid configuration, reconfigured/modified Configuration, and the puzzle-based advanced configurations. These are the PV array configurations that will be compared and discussed below.

2.1.1 Traditional Configuration

(a) *Series PV array configuration.*

Arrays connected in series are influenced by partial shading more easily because the shading imposes a limit to the flow of current on the panel connected in strings. When the affected panel operates at its maximum power point, and the bypassing diode is not active, the unaffected solar PV panels will not operate effectively. The mismatch is further increased as the shade intensity gets further increased [1]. In [2], the author shows how the series-connected array gives multiple peaks because of which it is difficult to track maximum power and lowers the power. In [3], the author investigates how we can control the power interface in a better way and can minimize the operation of the overall system. This paper deals firstly on how the performance of MPPT is diversely affected by the shading conditions and power generated by the solar PV system, partial shading has greatly affected the panels connected due to which its performance gets poor. The efficiency of the panel can be increased by some popular inverter connections of photovoltaic modules connecting in series. In the paper proposed here for each photovoltaic module, an individual power interface is done. Accordingly, the structure is recommended for the solar photovoltaic cell features and for minimizing the reduced performance caused by the partial shading conditions MPPT is being used. Lastly, in this paper, to get the correct converter topology for DC/DC MPPT module applications, a comparative study is done. There is some advantage of the boost converter over the buck converter as they have better implementations and dynamic response. In [4], simulation of the effect of shading in arrays with various string configurations is being studied. Four different strings size: 6 × 18, 4 × 27, 3 × 36, and 2 × 54; all of them having 108 cells in total are connected in series. More deformation in the I–V characteristic is caused as the cell per string is increased, it will displace the working point to lower voltages which causes an increase in power losses. Series configuration shows multiple peaks which results in the problem of tracking optimum power point. There is a difficult task in tracking multiple peaks caused by shading on the series interconnection, and research effort is greatly dedicated to developing power electronic interfaces having a control system which are better to reduce the difficulty, some schemes have been developed to reduce the losses. This is done by rearranging the connection of solar photovoltaic modules that will make uniform irradiation. However, these strategies have some advantages and disadvantages. In [5], modeling of PV characteristics which is experimentally based is studied, and the challenges seen during the solar PV model development are being discussed in this paper. Moreover, the previous method's limitations and how to overcome them are described here. This paper discusses a novel modeling technique in PSCAD software and illustrates the results under various shading conditions. PSCAD software is used for achieving any configuration of the PV modules under any type of conditions that could be shaded and unshaded. The model can be done together with power conversion systems, where such systems' behavior is analyzed and minimized under various conditions.

(b) *Parallel PV array configuration*

In a parallel connection, the parallel arrangement of the solar PV array to analyze electrical performance a MATLAB/Simulink is being implemented. In this type of configuration, there is no current limitation under shaded conditions as the total of individual current in the panel is the current in the string. The voltage does not vary much for that reason in this arrangement; the individual panels operate near to optimum power points. This helps in getting maximum power from the panel easily under all environmental conditions [1]. In [6], the author has compared the performance of series and parallel using translucent shading and opaque shading spots. The parallel configuration gave a better response compared to the series configuration. In [7], Brune's conditions are being used to analyze the solar PV panel's performance. A set of matrix equations characterizes the parameter of the equivalent circuits. Further in this paper, a LabView application for proving the theoretical model is being implemented. The effect different irradiance conditions have on the performance of the PV array is reviewed in this paper; this is done to achieve MPPT. Further in this paper, a real-time emulator is developed whose model structure is based on the closed-loop. With this scheme, MPPT techniques and PV system behavior under partial shading can easily be analyzed in the laboratory.

(c) *Series-parallel-series PV array configuration*

In [8], the behavior of the interconnection of solar cells is being implemented through simulation. The hotspot heating mechanism in various interconnections such as series-parallel-series, series-parallel, and series circuits are being briefly explained, and its protection is also explained in this paper. To forecast the behavior of photovoltaic cells, I-V characteristics must be considered, and modification is required for much better performance under PSCs in the configuration that already exists. The S-P-S configuration of parallel-configured PV array having a series arrangement and when the performance comparison is made in a series-parallel-series(S-P-S)) configuration it gave the best response.

(d) *Series-parallel, total cross-tied, bridge-link and honey comb PV array configuration*

In [9], the author developed an effectual algorithm for simulating the PV array efficiency, as in the model simulated earlier at every junction of the array Kirchhoff's law is applied to get the currents and voltages equations in the network. There is a need to make the junction current disappear, but mathematically it is not possible as in actual the size of the array is enormous. The nodes and meshes in the network cause difficulty in implementation as they are varying and the other drawback is that the time of execution is not fast. The algorithm for its efficiency uses time and space as the two main measures that make it faster and make a more capable algorithm to get an equilibrium state. To reduce the current in the array of the order of 10^5, the proposed technique uses a linear programming technique. The interconnected configurations, such as TCT and BL, take less time for simulation. In this paper, further investigation of arithmetical calculation of the faults in conditions of fields in the simple total

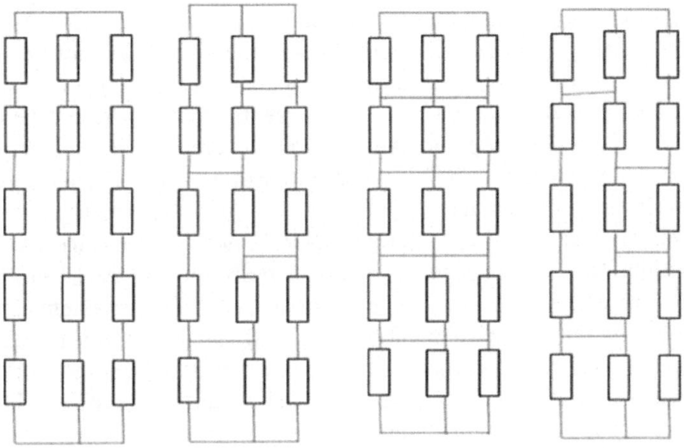

Fig. 1 Topologies [9]

crossed-tied, series-parallel, and bridge-linked solar cell interconnection arrangements are done. The conclusion shows that to the electrical mismatch, the BL and TCT are less liable. In [10], the author studies the interconnection schemes among the modules, and in this paper, the author proposes a study of all the different partial shading conditions possible. Here the internal mismatch is not considered for study purposes; rather, it focuses on external mismatch. Two setups for simulation are used to establish the relationship between shadow, interconnections, and the maximum power point. These are the shadow study and optimization study. In shadow study, the study is done for various shadow patterns on various interconnections such as SP, HC, BL, TCT, where TCT shows the best response. The Figure below (Fig. 1) presents various PV array connection schemes used for the simulation study. These laws are only valid for a 3 × 3 PV array, and to overcome this drawback, research is still being done. The topologies discussed above are shown in Fig. 1.

2.1.2 Hybrid PV Array Configurations

(a) *SP-TCT, BL-TCT, and BL-HC PV array configuration.*

In [12], Su-Do-Ku puzzle-based study is done for various shading patterns named as- 1, 2, and 3 are compared with the efficiency of TCT and SP-TCT, where patterns 1,2 characterized by four different cases and pattern 2 by two different cases and the effect caused by these patterns on solar array characteristics is discussed here and Su-Do-Ku shows the most required response. In [18], a detailed investigation of extraction of power with the help of number placement reconfiguration method under various faults like partial shading, issues due to wiring, module degradation, etc., is done. Here in this paper extraction of power is carried out of various interconnections like series-parallel, bridge linked, TCT and compared with relocation techniques. It is found considering the relocation method can help in increasing power generation of a shaded PV array that is not uniform. This is done for the faults in consideration. In[17], the author develops the auto-rearrangements scheme wherein the electrical connections of a 3 × 3 solar photovoltaic (PV) array are interchanged, from series-parallel (SP) to total cross-tied (TCT) configuration under PSCs. Further in this paper, an experimental system is discussed that comprises a relay used in the module of size 3 × 3. A relay circuit based on Arduino is made to modify the electrical connection of solar module. Here in this paper, a discussion based on wireless UEOS(User End Operating System) Performance analysis is done. A PV array mathematical model is made to validate in an environment of MATLAB simulation. It is concluded that power is high in TCT when compared to SP and form factor(FF) is more in TCT than SP and power loss is less when compared to SP. In [19], an analysis of fault in solar PV arrays is significantly concerned because of the safety and efficiency risk in a solar PV system. It shortens the life of the PV system. For the complete analysis of the fault, we have to consider the impact of different materials and solar PV configuration. In this paper, a detailed analysis of faults in different PV array arrangements, non-uniform shadings such as series-parallel (SP), honey comb (HC), and total-cross-tied (TCT). Here, under non-uniform irradiance, a complete investigation is done for multiple fault conditions for analyzing different faults. Monocrystalline and thin-film solar PV is considered, and a simulation in MATLAB/Simulink is done. Comparative study of results is done, and lastly, it is concluded that thin-film PV with an appropriate PV array arrangement is best suitable for improving the performance under various fault conditions.

2.1.3 PV Array Configurations Based on Puzzle

Su-Do-Ku PV array configuration

In [12], the author presents a method that configures the TCT arrangement to improve the generated power under the solar array's mismatch scenarios. The arrangement is made according to the Su-Do-Ku array configuration in which no changes are required in the electrical connections in Fig. 2. The advantage of the reconfiguration method over the newly proposed technique is that the connections of the modules are made in such a way that it dynamically changes the shading conditions. Practically implementing the reconfiguration is expensive. The disadvantage of the newly proposed puzzle technique is that it is applied to the 9 × 9 PV module and for different array voltages the difference in row current is less when compared to the TCT interconnection. Its performance can be improved further if the difference is not minimum. To get better output, losses caused by the mismatch in the solar cell can be further reduced. To remove such drawbacks array reconfiguration technique based on GA configuration is proposed in [13]. In this, the authors have employed an experimental analysis that is hardware based for a 9th row and 9th column and 6th row and 6th column size of the array to examine the decrement in line and mismatch

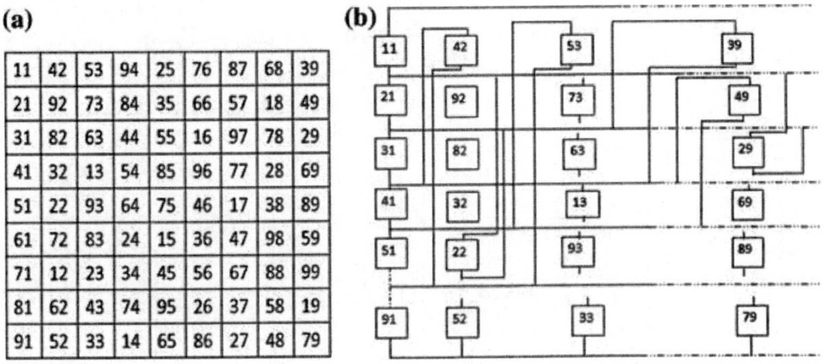

11	42	53	94	25	76	87	68	39
21	92	73	84	35	66	57	18	49
31	82	63	44	55	16	97	78	29
41	32	13	54	85	96	77	28	69
51	22	93	64	75	46	17	38	89
61	72	83	24	15	36	47	98	59
71	12	23	34	45	56	67	88	99
81	62	43	74	95	26	37	58	19
91	52	33	14	65	86	27	48	79

Fig. 2 **a** Chosen SU-Do-Ku pattern [12]. **b** SuU-Do-Ku arrangement [12]

losses caused by different conditions of partial shading. Using a genetic algorithm, comparison is done between TCT and the puzzle technique is done and concluded that the improved techniques show much better GMPP for various shading cases and give a better power output. In [20], the author has considered the line loss and developed an improved Su-Do- Ku arrangement which will reduce the length of the wire that is the cause for line loss. The system is studied for various patterns and the performance has been simulated and the obtained result confirms that the proposed structure has improved the PV Array power output when compared to any randomly selected Sudoku pattern.

2.1.4 PV Array Configuration Done in an Odd-Even Structure

In [14], a new technique was developed to improve the output of the solar PV array. The shading is distributed over the complete array, and by doing this shade, dispersion minimizes the mismatch losses. According to the requirement of current and voltage decisions, rows and columns are taken. The technique proposed is for any number of array sizes. In the odd–even technique, reconfiguration of an array of size 4 × 4 is done. The structure is presented in Fig. 3. Here many case studies are being done for the better-enhanced power and lesser mismatch losses and from the comparison of the odd-even technique with conventional techniques and it was shown that the proposed odd-even shows a better response.

3 Conclusion

The Study of various interconnection schemes for a PV system is done so that the effect due to partial shading is reduced. Here, each configuration study gives its benefits, inadequacies, and significant features, which help develop better performance techniques. TCT schemes have better performance when compared to other

(a)

M_{11}	M_{13}	M_{31}	M_{33}
M_{42}	M_{22}	M_{44}	M_{24}
M_{12}	M_{32}	M_{14}	M_{34}
M_{41}	M_{21}	M_{43}	M_{23}

(b)

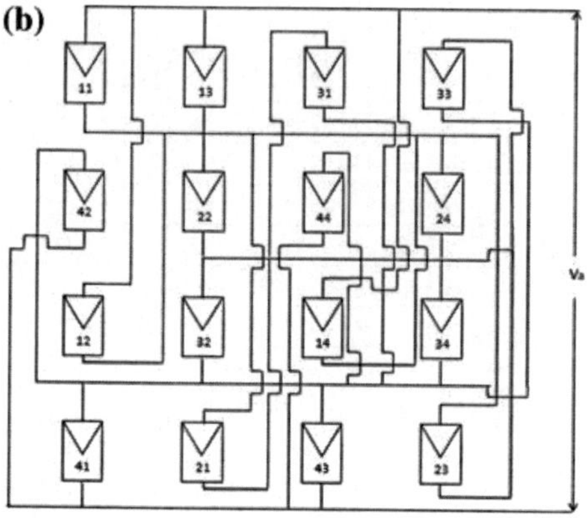

Fig. 3 **a** An odd-even structure [14]. **b** An odd–even structure connection [14]

conventional configurations. In hybrid configuration, the conventional configuration was developed by different modeling approaches, and further, the performance is analyzed and compared to find which configuration is better. A puzzle-based configuration is being reviewed here, where different types of puzzle patterns are developed for studying its performance, and the TCT configuration is reconfigured by most of the researchers according to the puzzle pattern. In most of the proposed techniques, shade dispersion is done on the entire module. As a result, the power output is being enhanced.

References

1. Malathy S, Ramaprabha R (2018) Reconfiguration strategies to extract maximum power from the photovoltaic array under partially shaded conditions. Renew Sustain Energy Rev Elsevier 81(P2):2922–2934
2. Khatoon S, Ibraheem, Jalil MF (2014)Analysis of solar photovoltaic array under partial shading conditions for different array configurations. In: 2014 innovative applications of computational intelligence on power, energy and controls with their impact on humanity (CIPECH), Ghaziabad, 2014, pp 452-456. https://doi.org/10.1109/CIPECH.2014.7019127
3. Xiao W, Ozog N, Dunford WG (2007) Topology study of photovoltaic interface for maximum power point tracking. IEEE Trans Ind Electron 54(3):1696–1704
4. Alonso-Garcia MC, Ruizb JM, Herrmann W (2006) Computer simulation of shading effects in photovoltaic arrays. Renew Energy 31:1986–1993
5. Moballegh S, Jiang J (2011) Partial shading modeling of photovoltaic system with experimental validations. In: Proceedings of IEEE conference on power and energy society general meeting, pp 1–9
6. Gao L, Dougal RA (2009) Parallel connected solar PV system to address partial and rapidly fluctuating shadow conditions. IEEE Trans Ind Electron 56(5):1548–1556
7. Kadri R, Andrei H, Gaubert JP, Ivanovici T, Champenois G, Andrei P (2012) Modeling of the photovoltaic cell circuit parameters for optimum connection model and real-time emulator with partial shadow conditions. Energy 42:57–67

8. Bishop JW (1988) Computer simulation of the effects of electrical mismatches in photovoltaic cell interconnection circuits. Solar Cells 25:73–89
9. Gautam NK, Kaushika ND (2002) An efficient algorithm to simulate the electrical performance of solar photovoltaic arrays. Energy 27:347–361
10. Villa LFL, Picault D, Raison B, Bacha S, Labonne A (2012) Maximizing the power output of partially shaded photovoltaic plants through optimization of the interconnections among its modules. IEEE J Photovolts 2(2):154–163
11. Yadav S, Pachauri RK, Chauhan YK (2015) Comprehensive investigation of PV arrays under different shading patterns by shade dispersion using puzzled pattern-based Su-do- Ku puzzle configuration. In: Proceedings of IEEE conference on next generation computing technologies, pp 824–830
12. Rani BI, Ilango GS, Nagamani C (2013) Enhanced power generation from PV array under partial shading conditions by shade dispersion using Su Do Ku configuration. IEEE Trans Sustain Energy 4(3):594–601
13. Deshkar SN, Dhale SB, Mukherjee JS, Babu TS, Rajasekar N (2015) Solar PV array reconfiguration under partial shading conditions for maximum power extraction using genetic algorithm. Renew Sustain Energy Rev 43:102–110
14. Nasiruddin S, Khatoon MF, Jalil, Bansal RC (2019) Shade diffusion of partial shaded PV array by using odd-even structure. Sol Energy 181:519–529
15. Khatoon S, Ibraheem, Jalil MF (2015) Feasibility analysis of solar photovoltaic array configurations under partial shading conditions. In: Annual IEEE India conference (INDICON), pp 1–6
16. Gautam NK, Kaushika ND (2002) An efficient algorithm to simulate the electrical performance of solar photovoltaic arrays. Energy 27(4):347–361
17. Pachauri R, Singh R, Gehlot A, Samakaria R, Choudhury S (2019) "Experimental analysis to extract maximum power from PV array reconfiguration under partial shading conditions. Eng Sci Technol Int J 22:109–130
18. Haq AU, Alammari R, Iqbal A, Jalal M, Gul S (2020) Computation of power extraction from photovoltaic arrays under various fault conditions. IEEE Access 8:47619–47639
19. Gul S, Haq AU, Jalal M, Anjum A, Khalil IU (2020) A unified approach for analysis of faults in different configurations of PV arrays and its impact on power grid. Energies 13:1–23
20. Rao PS, Dinesh P, Ilango GS, Nagamani C (2015) Optimal Su-Do-Ku based interconnection scheme for increased power output from PV array under partial shading conditions. Front Energy 9(2):199–210
21. Jalil MF, Khatoon S, Nasiruddin I, Bansal RC (2020)An improved feasibility analysis of photovoltaic array configurations and reconfiguration under partial shading conditions, electric power components and systems

An Analytical Study to Determine Performance and Economic Benefits of the Grid-Interactive Rooftop Solar Power Plants

Bhaskarjyoti Das and Ashwani Kumar

Abstract Fossil fuel resources are limited in nature and also have adverse effects on the environment due to harmful emissions. With the increasing energy demand, it has therefore become of utmost importance to exploit alternate energy sources such as renewable energy sources to the fullest. Renewable energy sources are abundant in nature and cause minimal harm to the environment. Rooftop Solar Photovoltaic system is one of the most prominent and fast-growing in the market today. The awareness amongst people, non-requirement of additional ground space, supportive government policies, ease of installation, economic benefits, etc., have resulted in the fast adaption of these systems in the society. As the rooftop solar PV plants are increasing, it is important to study the generation pattern, economic viability and the impact on the distribution utility. This would also help in planning the network well and scheduling energy storage systems with the increasing renewable energy penetration to the grid. In this paper, the annual recorded data of a particular electrical circle under the distribution utility of Assam state is studied. The solar generation pattern and the export–import from the solar PV plants throughout the year are studied. Also, the monthly savings pattern due to the energy injected into the grid is studied. The result gives a comparative study and the maximum and minimum of the parameters throughout the year.

Keywords Photovoltaic (PV) system · Economic benefits · Solar energy generation · Distribution company (DISCOM) · Export–Import metering

---

B. Das (✉)
New and Renewable Energy Wing, Assam Power Distribution Company Ltd, Guwahati 781001, India

A. Kumar
Department of Electrical Engineering, NIT Kurukshetra, Kurukshetra 136119, India
e-mail: ashwani.k.sharma@nitkkr.ac.in

A. Kumar et al. (eds.), *Renewable Energy Towards Smart Grid*, Lecture Notes in Electrical Engineering 823, https://doi.org/10.1007/978-981-16-7472-3_32

1 Introduction

The energy demand is increasing day by day. The fossil fuel sources which still are the major source of energy for the energy needs are limited in nature. They are also a major contributor to environmental pollution leading to global warming and other problems. It is high time that the alternative energy resources such as renewable energy sources are exploited as much as possible. Renewable energy sources such as solar energy, wind energy, etc., are natural sources and also cause minimal or no environmental pollution at all. These abundant resources of energy are the most promising alternatives for energy generation in the coming future as these will not only provide a cleaner source of energy but will also help reduce gradually the dependence on fossil fuels for the energy requirements [1].

Energy generation through renewable energy sources is very fast catching up worldwide. Many nations around the world have nearly 50 percent of their energy supply through renewable with few nations like Norway and Iceland which are generating 100 percent electricity through renewable energy sources [2]. The government of India too has set ambitious renewable energy goals; the total RE target set by the year 2022 is 175 GW. The national target is segregated state-wise. The state-wise target set by the Ministry of New and Renewable Energy for Assam state is 0.688 GW. Out of which target for rooftop solar is 0.250 GW in Assam [3].

1.1 *Rooftop Solar PV System*

Without requiring additional ground space for installation, rooftop solar is one of the best solutions to tap into solar energy for our energy needs. The unutilized space on rooftops of the individual households, industrial buildings, commercial buildings, or any other type of building provides a large potential for generating solar power.

Rooftop solar can be an off-grid system with battery backup or a grid interactive one in which the power generated can be used to fully/partially fulfill the requirements of the building occupants and surplus generation, if any, can be fed to the grid for which the owner is benefitted. A Grid interactive rooftop solar PV system may or may not have an energy storage facility [4, 5].

In India awareness amongst the people about the benefits of using renewable energy especially solar energy has increased significantly over the years and with fast-growing solar business in the country, advancement in technology, government subsidies for setting up solar PV systems, etc., have led to the increased deployment of solar PV systems in the country [6].

In Assam too, energy generation through solar PV systems is fast catching up. Large MW scale solar projects are already being set up while a few more are in progress; Solar microgrids, standalone solar PV systems in households are also installed in different parts of the state. Grid interactive rooftop solar PV systems in residences, commercial buildings are also seen fast growing. As of date, for the rooftop solar PV system in Assam, the total cumulative capacity installed is 20.6753 MW under the DISCOM area.

With the increasing penetration of solar PV systems to the grid, it is essential to perform an extensive study to know the impact of the integration of solar PV systems to the DISCOM, its economic viability to further strengthen the network and gear up for future large-scale integration of renewable energy sources to the grid. As a bid to analyze this aspect, in this paper, annual recorded data of the grid-connected solar rooftop plants installed under a specific electrical circle under the DISCOM is studied. A comparative study is done based on the data. The Export-Import metering compensation, benefits to the consumer and impact on DISCOM are studied in this paper.

2 Solar Generation Data

The annual recorded data of the grid-connected solar rooftop plants installed under Guwahati Electrical Circle (GEC)-II, Assam is considered for the study. There are a total of 24 numbers of rooftop solar plants installed in different locations under GEC-II with an aggregate capacity of 2.859 MWp. The plants are connected to the grid and have proper metering utilities installed. Brief details about these rooftop solar plants are shown in Table 1.

2.1 Analyzing the Database

The annual generation data of the grid-connected rooftop solar PV plants is collected from the respective electrical subdivisions under the DISCOM. A comparative study of the data is done to basically understand the following:

- Solar generation pattern throughout the year,
- Energy export–import,
- Benefits to the Consumers.

The monthly solar generation data of Sl. No. 4 and Sl. No. 5 from Table 1 is plotted and the generation pattern throughout the year is studied. To observe the pattern of energy export–import from the grid, the monthly export–import data of the same consumers are used. The percentage of energy exported to the grid and energy imported from the grid round the year is calculated.

In the bi-directional export–import metering facility in a grid-interactive rooftop solar PV plant, the energy imported by the consumer during a particular month from the grid is billed as per the prevailing tariff applicable for the consumer category. The energy exported (if any) to the grid is reckoned at the Average Pooled Power Cost (APPC) rate fixed by the State Electricity Regulatory Commission, which is

Table 1 Details of Consumers with grid-connected rooftop solar plants under GEC-II

Sl. No	Name and address of the consumer	Consumer No.	Solar plant capacity (in KWp)
1	Amiya Kr. Das	11,000,018,008	3
2	Rajendra Kr. Das	51,000,065,715	3
3	Phanidhar Das	51,000,053,738	3
4	Bimal Das	60,000,001,101	10
5	Kamala Devi	52,000,001,397	5
6	Faculty Higher Secondary School	61,000,002,159	50
7	Managing Trustee, Faculty H.S School	61,000,002,137	50
8	Arpana Nath	51,000,378,565	3
9	Ajit Kr Nath	51,000,378,423	3
10	Arya Nursing College	61,000,002,148	30
11	National Law University	63,000,001,606	285
12	Principal SBMS college	63,000,001,304	28
13	M/S CENTPLY	67,000,001,768	1000
14	Purbanchal Timber Industries	67,000,002,633	500
15	HITESH BARUA	61,000,003,836	80
16	Secy, Shalom SVD House	30,000,051,131	10
17	Pavitra Atma Sevak	64,000,002,133	10
18	Father Joseph, Don Bosco	61,000,003,835	160
19	Srimanta Shankar Academy	63,000,002,832	230
20	Divisional Engineer BSNL	61,000,003,826	66
21	RTTC BSNL	61,000,003,856	60
22	Director, NIC Pig	64,000,002,134	240
23	Ruben Ramesh Choudhury	30,000,006,486	10
24	Holy Home Charitable Trust	61,000,005,313	20

at present Rs. 3.78/KWh [7]. Considering the monthly export data of the selected consumers, the savings are calculated for round the year and the pattern is studied.

3 Results and Observations

3.1 Solar Energy Generation Pattern

The monthly solar energy generation record for Consumer no. 60000001101 (Sl. No. 4) and Consumer no. 52000001397 (Sl. No. 5) from Table 1 is plotted and is shown in the Figs. 1 and 2, respectively.

It is observed that in both the figures that the maximum solar generation is recorded in the month of April. The minimum on the other hand is recorded in the duration of July–September for Consumer 4 and in the duration of December–February for Consumer 5.

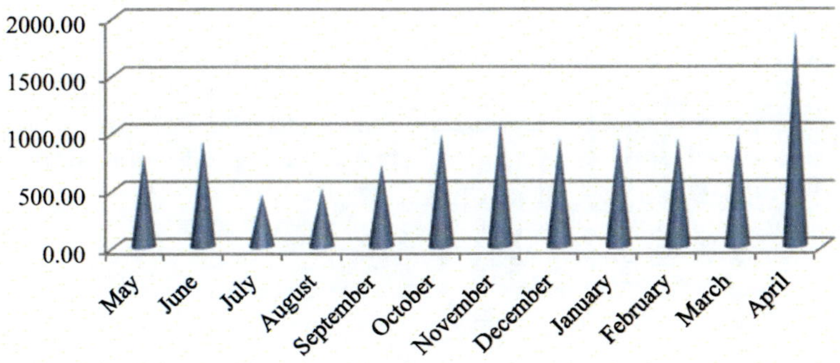

Fig.1 Monthly solar generation (in KWh) for Sl. No. 4

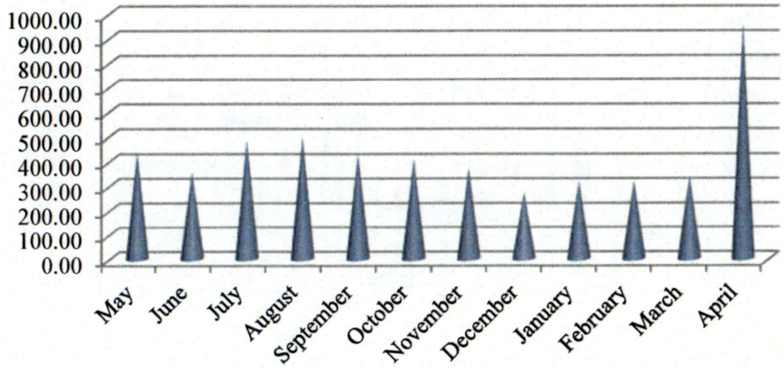

Fig.2 Monthly solar generation (in KWh) for Sl. No. 5

3.2 *Energy Export–Import from Grid*

To observe the pattern of energy export–import from the grid, the monthly export–import data of the same consumers are considered and plotted in the figures below.

It is found that the duration of maximum and minimum energy exported to the grid in both cases coincides with their solar energy generation pattern. The energy imported from the grid reaches a maximum in the month of July–September in Fig. 3, whereas in Fig. 4 for Sl. No. 5, it is seen that the maximum is reached in the month of April. The comparison of the total annual solar generation, energy export–import from the grid for the two consumers considered here is depicted in Figs. 5 and 6.

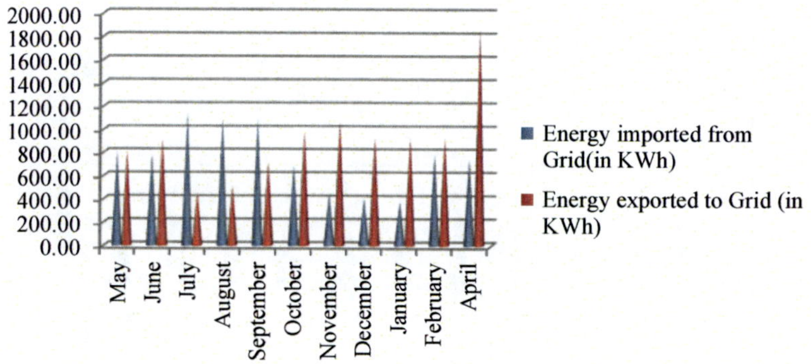

Fig. 3 Monthly export–import data (in KWh) for Sl. No. 4

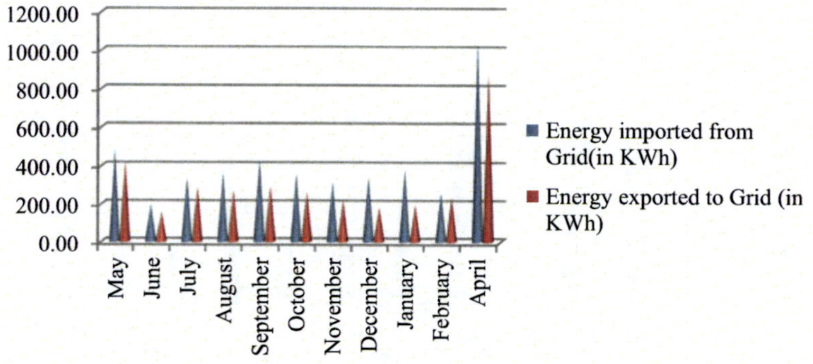

Fig. 4 Monthly export–import data (in KWh) for Sl. No. 5

Fig. 5 Depiction of total annual solar generation, energy export–import for Sl.No. 04

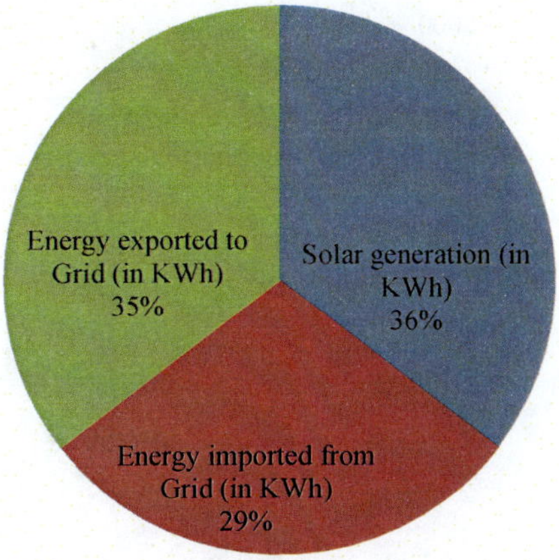

Fig. 6 Depiction of total annual solar generation, energy export–import for Sl.No. 05

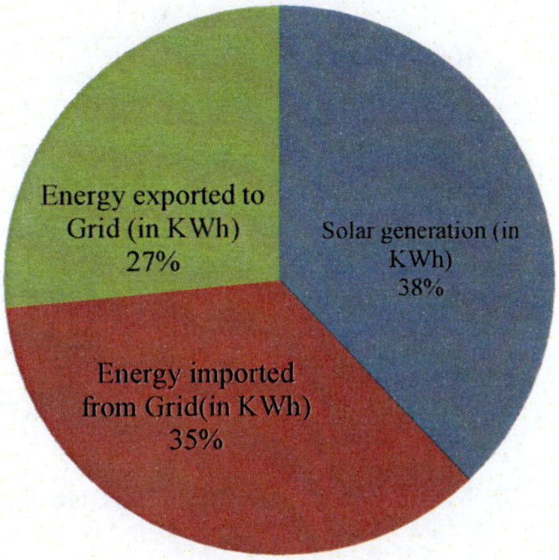

3.3 Benefits to Consumers

The amount due to the consumer on account of injection of solar energy to the grid shall be reckoned at APPC rate, i.e., Rs. 3.78/KWh. This amount shall be adjusted from the monthly bill. Thus for building owners, rooftop solar installations will help in cutting down electricity bills. To understand the pattern of savings of the selected consumers the monthly export data is utilized to calculate the monthly savings. The savings hence calculated is plotted and shown in the figures below (Figs. 7 and 8).

It is observed that the monthly savings are maximum in the month of April for both the cases. The minimum is seen during July–September for the first case. Whereas, the minimum for the second case is observed during June and in the months of November–December.

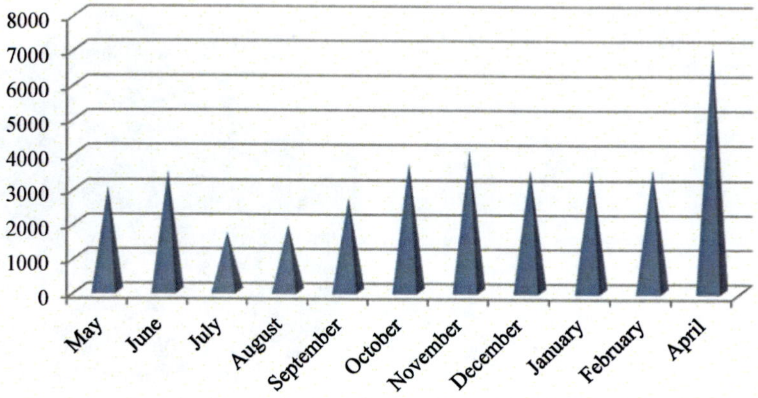

Fig. 7 Monthly savings (in Rs.) for Sl. No. 04

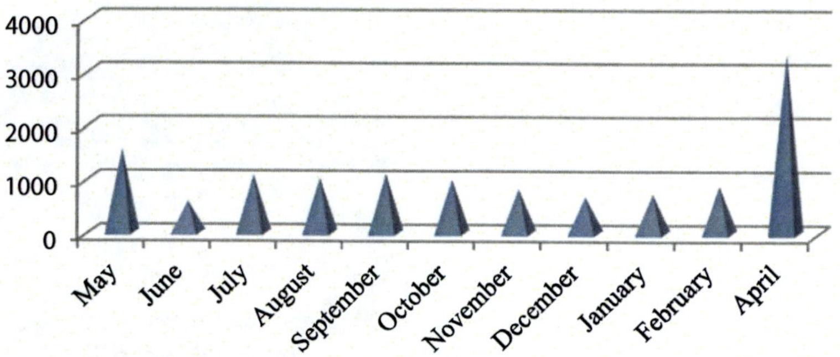

Fig. 8 Monthly savings (in Rs.) for Sl. No. 05

4 Conclusion

The annual recorded data of the solar rooftop plants under a particular electrical circle under the DISCOM in the state of Assam are studied. The solar generation of two random consumers under the circle, Sl.no. 4, 5 from Table 1 is considered. It is seen that the maximum solar generation is recorded in the month of April. The minimum on the other hand is recorded in the duration of July–September for Consumer 4 and in the duration of December–February for Consumer 5. Similarly, the export and import data of the two consumers are studied to observe the pattern. It is seen that in the first case, generation from solar energy is 36%, energy imported from the grid is 29%, and the energy exported to the grid is 35%. In the second case, the generation from solar energy is 38%, energy imported from the grid is 35%, and the energy exported to the grid is 28%.

The energy injected into the grid by the consumer is reckoned at the APPC rate, i.e., Rs. 3.78/KWh. This amount will be saved from the monthly electricity bill of the consumer. To understand the savings pattern, the savings of the two consumers considered in the previous cases are considered. The maximum and minimum points are observed. It is also seen that during April for the first case, the exported energy to the grid is much higher than the imported energy particularly because of the higher solar energy generation.

As a whole, the solar generation is good enough round the year giving ample savings to the consumers. The DISCOM is also benefitted from the rooftop solar PV plants. It helps to meet the Renewable Purchase Obligations. A good amount of energy export to the grid may help in the reduction of power procurement, improve the life of the assets in use due to reduction of loading of the network; solar rooftop also helps in deferring new distribution infrastructure which assumes added importance in congested localities. However, due to the intermittent nature of renewable energy generation, it is a must to consider ways to mitigate the effects that might occur to the grid such as grid instability, power quality issues, etc., with increasing penetration of renewable energy sources. In that case, energy storage plays a vital role by storing the surplus energy and releasing the energy in the required quantum whenever required. The energy storage systems may either be located centrally or decentralized storage systems may also be a solution. The solar energy generation pattern, export–import data can be studied to optimally schedule the energy storage system in the case of decentralized energy storage systems [8].

References

1. Gielen D, Boshell F, Saygin D, Brazilian MD, Wagner N (2019) Ricardo Gorini "The role of renewable energy in the global energy." J Energy Strat Rev 24:38–50
2. "Performance of Solar Power Plants in India (2011). http://cercind.gov.in
3. Ministry of New and Renewable Energy (MNRE). Accessed 16 Aug 2012
4. Al Ali M, Emziane M (2013) Performance analysis of rooftop PV systems in Abu Dhabi. Energy Proc 42:689–697
5. Cucumo M, De Rosa A, Ferraro V, Kaliakatsos D, Marinelli V (2006) Performance analysis of a 3kW grid-connected photovoltaic plant. Renew Energy 31(8):1129–1138
6. Ramadhan M, Naseeb A (2011) The cost benefit analysis of implementing photovoltaic solar system in the state of Kuwait. Renew Energy 36:1272–1276
7. Assam solar energy policy (2017)
8. Das B, Kumar A (2018) A NLP approach to optimally size an energy storage system for proper utilization of renewable energy sources. Proc Comput Sci 125:483–491
9. Metayer M, Breyer C, Fell HJ (2015) The projections for the future and quality in the past of the world energy outlook for solar PV and other renewable energy technologies, energy watch group, Berlin, Germany. http:// energywatchgroup.org/wp-content/uploads/2015/09/EWG_WEO-Study_2015.pdf.
10. Nguyen Hk, Song JB, Han Z (2015) Distributed Demand side management with energy storage in smart grid. IEEE Trans Parallel Distrib Syst 26:3346–3357
11. Rao S, Parulekar BB (2012) Energy technology non-conventional, renewable & conventional, 3rd edn. Khanna Publication
12. Khanna M, Mundra K, Ullah A (2011) Parametric and semi-parametric estimation of the effect of firm attributes on efficiency: electricity generating industry in India. J Int Trade Econ Dev 8(4):419–436

Demand-Side Management Approach Using Heuristic Optimization with Solar Generation and Storage Devices for Future Smart Grid

Neeraj Kumar Mahto, Smriti Jaiswal, and Dulal Chandra Das

Abstract Nowadays utilities are offering an incentive for direct control over the loads present on the consumer side to implement demand-side management, where utilities have greater control over consumer-side consumption but it lacks the application of suitable criteria and objectives independently. This paper aims to focus on the problem of power and energy management in the consumption side of the grid by decision-making and load management as per the dynamic price of electricity. It discusses the load shifting technique to handle a large number of controllable devices present on the distribution side using a heuristic-based evolutionary algorithm to handle complex operations and constraints in MATLAB/Simulink. As per the survey conducted in a residential area of Karnal City, Haryana, simulation studies are carried out for different types of consumers with a variety of control loads offering energy management in the distribution of loads as well as energy dynamic pricing.

Keywords Demand-Side Management (DSM) · Evolutionary Algorithm (EA) · Smart grid

1 Introduction

In recent years, India has remarkably strengthened its position under the Technology Collaboration Programme (TCP) globally by focusing on energy supply technologies, system integration technologies such as smart grids and Demand-Side Management (DSM) with investments in storage systems and distribution networks, expanding its photovoltaic power systems, solar heating, and cooling and storage devices along with fusion power systems. Demand-side management is an involuntary traditional interrupting service or schedule often controlled by distribution companies, for example, demand shedding at peak hour whereas Demand Response

N. K. Mahto · S. Jaiswal (✉) · D. C. Das
Department of Electrical Engineering, NIT Silchar, Silchar, Assam, India
e-mail: smriti_rs@ee.nits.ac.in

D. C. Das
e-mail: dulal@ee.nits.ac.in

A. Kumar et al. (eds.), *Renewable Energy Towards Smart Grid*, Lecture Notes in Electrical Engineering 823, https://doi.org/10.1007/978-981-16-7472-3_33

(DR) is a voluntary responsive approach to a price signal driven by customer's needs and interests [1]. Smart grid integrates all the branches of consumers and distribution service providers for maintaining grid stability, improving efficiency, and overall operational cost reduction [2].

Considering the present potential of solar generation and storage devices in the residential and commercial area of Karnal, Haryana, this paper highlights the fulfillment of objectives of an evolutionary DSM approach in a smart grid managing energy on-site [3, 4].

DSM possess the communication infrastructures and mathematical techniques in the real-time scenario, where the major interest of smart grid infers to smart metering, smart appliance, home automation system, to allow pervasive control and monitoring [5, 6]. This up-gradation is smart and can be achieved by: (1) Maximization of the utilization of renewable energy sources, (2) Monetary incentive to the consumer employing technologies of DSM techniques like building automation and smart metering, targeting the results of cost function and peak load shaving [2, 7–11].

The major challenges encountered in DSM are to stabilize the power system frequency under extreme climate conditions and direct load control along with generation control while considering real-time data for Renewable Energy Resources (RES) units like solar/wind and Bioenergy based Generating (BEGS) units.

In the literature, many works have been done in this regard employing PI/PID-type controllers parameters using various algorithms like firefly (FF) and particle swarm optimization (PSO), etc. [12, 13]. In order to optimally allocate resources and the distributed co-generation such that a dual objective of energy-waste management can also be served [14, 15]. Then another feature of DSM in focus is cost–benefit analysis for optimal sizing, charging/discharging of an energy storage system by different techniques like time series and feedforward (FF), neural network techniques, a Mixed Linear Integer Problem (MLIP), Genetic Algorithm (GA), etc. [16, 17]. A Battery Energy Storage System (BESS) is considered as a spinning reserve for frequency regulation to compensate for the forecast inefficiency and the line loss of distribution systems. As per the scope of this paper, instead of a spinning reserve, it is considered to act as an emission-free ancillary service in the microgrid to ensure the short-term power availability as backup within a few seconds to minutes in times of need at the user end.

Among various DSM approaches that have been proposed in this area of research, this paper aims to contribute as another 'smart' solution for distribution companies focusing on the grid-connected solar power generation along with an efficient energy storage system with an inclusive and replicable model of sustainable development aiding to consumer awareness. The primary objective of the proposed consumer-driven DSM approach in this paper is to assess and promote maximum user end participation and control over local appliances in order to minimize the utility bill in the real distribution network of Karnal city, Haryana is simulated in MATLAB/Simulink.

This paper is organized as follows: Sect. 2 describes the DSM Architecture, Sect. 3 proposes the suitable DSM strategy, Sect. 4 deals with problem formulation, Sect. 5 refers to the test data set and case study, Sects. 6 and 7 conclude the paper with results of simulation in MATLAB/Simulink.

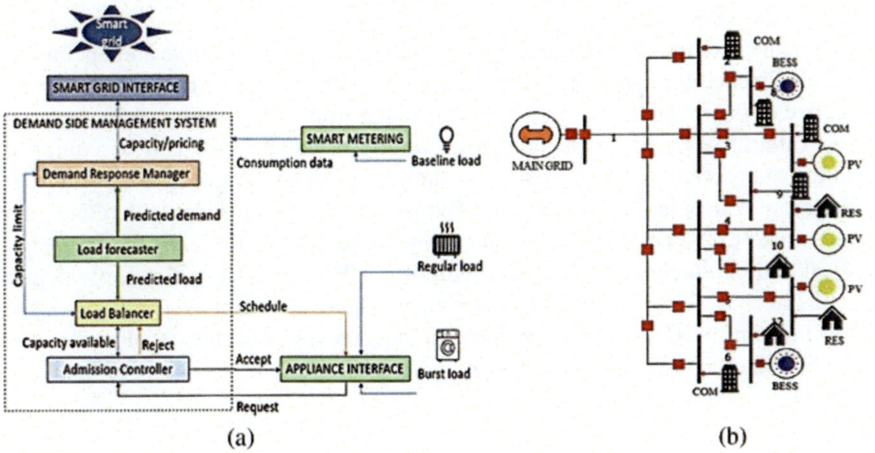

Fig. 1 DSM architecture and schematic diagram of the smart city area

2 DSM Architecture

This section presents the layer-wise architecture of DSM as shown in Fig. 1a. It includes subparts: (1) Admission Controller (AC), (2) Load Forecaster (LF), (3) Load Balancer (LB), (4) Demand Response Manager (DRM) [11]. DSM techniques need to handle a variety of loads of large quantities in the residential and commercial areas of smart grid as shown in Fig. 1b.

AC is accountable for peak load operation as per the suggested DSM strategy. A load balancer communicates AC along with smart appliances for operation depending on consumption requirement, e.g., the DSM technique shifts burst loads such that the energy cost can be minimized. D/R manager trades energy with the grid and the load forecaster analyzes the expected energy requirement of the system [9]. Load profile of the industry, domestic, commercial consumers can be altered by six broad methods: peak clipping, load shifting, valley filling, flexible load shape, static load growth, strategic conversion [10, 18, 20].

3 Proposed Demand-Side Management Approach

The proposed DSM approach employs the load shifting technique with an evolutionary algorithm [11, 19]. As per the objective curve referred in the loads may have different characteristics depending upon the variation over a period of time. So, the proposed methodology must encounter all those possible control durations for the variety of controllable loads mentioned in Sect. 5 [18, 20].

Here, the DSM control strategy fetches the objective curve as input then the mathematical computation is carried out targeting the expected curve and operation-schedule for each appliance in residential and commercial areas of Karnal city with an objective of the reduction in operational cost. The grid-tied domestic Solar Rooftop Photovoltaic (SPV) power plant with or without a battery ought to be equipped with the appropriate features to support the grid during daylight hours in order to incorporate consumers request for the operation of any appliances, the requisition is forwarded to the DSM control system which responds on the basis of the mathematical output which was calculated beforehand. The response will be provided to the appliance is either allowed or issued a new connection time slot. The proposed algorithm aims to achieve much efficiency and resiliency in further development and contribution of a cost-effective solution to the problem.

4 Problem Formulation

This algorithm allows the implementation of demand patterns of different types of consumption lifestyle so that inconvenience to the consumers can be minimized is carried out as follows:

Step 1: Types of load, tariffs from distribution utility, consumption pattern, and duration of the device are gathered.

Step 2 Targeted load curves based on respective pricing structures are prepared.

Step 3: Heuristic Optimization (HO) EA is applied to inputs from steps 1, 2.

Step 4: Handling information of the load curve with the device usage shifting guidelines are sent to the customers based on the objective of reduction of utility bills of consumers. The nomenclature and symbols used are mentioned below in Table 1.

The objective curve of the proposed DSM, which is based on load shifting technique can be mathematically formulated as [7]

$$\text{Minimize: } \sum_{t=1}^{N} (P_{Load}(t) - Target(t))^2 \tag{1}$$

$$\begin{aligned} \text{Wherer, } Pload(t) = (&Forecast(t) + Connect(t) - Disconnect(t) \\ &+ Charging(t) - Generating(t)) \end{aligned} \tag{2}$$

Here, Connect(t) is the combination of load enhancement at time t by shifting the load from other time intervals and another due to the device connections scheduled for times that proceed [11]. Similarly, Disconnect(t) refers to the reduction in load due to delay in the connection of devices which were originally supposed to start

Table 1 Nomenclature and symbols used

Symbol	Nomenclature
P Load (t)	Load consumption function at time(t)
Target(t)	Targeted curve at time(t)
Connect (t)	Loads connected at time t during shifting,
Disconnect (t)	Loads disconnected at time t
Charging (t)	Battery charging at time t
Discharging(t)	Battery discharging at time t
Generating(t):	Amount of solar generation at time t
X_{kit}	Devices of type k shifted from time i to t
f_{loss}	Loss factor due to dirt, temperature, etc.
$P_{sol(t)}$	Energy generation due to PV panel
P_T^S	Standard energy generation of PV panel
G h(t)	Expected hourly solar insolation at the surface of solar PV panel (W/m^2)
GS	Standard incident radiation
SOC(t)	State of Charge at time t
$P_b(t)$	Battery's input/output power
η_{batt}	Round trip efficiency of the battery
V_{bus}	Voltage of bus
M	Maximum permissible delay
RES	Residential area
COM	Commercial area

at time t and another which start their consumption at time steps of t subjected to following constraints:

1. The devices shifted must be positive:

$$X_{kit}(t) > 0 \forall \text{i, t, k.} \quad (3)$$

2. The devices shifted cannot be more than the controllable device present at the particular time k_{kit}:

$$\sum\nolimits_{t=1}^{N} X_{kit}(t) \text{ controllable (k)} \quad (4)$$

4.1 Solar Generation

Solar PV output depends on solar radiation [8]. Mathematically, it can be represented as,

$$P_{sol}(t) = P_T{}^S * f_{loss} * \frac{Gh(t)}{Gs} \tag{5}$$

4.2 Battery Storage

Energy measurement system uses the information about the state of charge (SOC) during peak demand and/or renewable generation is not sufficient to cater to the load [3, 12]. The state of charge (SOC) is given as

$$\frac{SOC(t)}{SOC(t-1)} = \int_{t-1}^{t} \frac{P_b(t) * \eta_{batt}}{V_{bus}} dt \tag{6}$$

where battery is in charging mode if $P_b(t) > 0$, otherwise it is in discharging mode. In this paper, different test cases are considered where different types of controllable devices are given by

$$X_{kit} = 0 \; where \; i > t \tag{7}$$

Considering a maximum of 12 h. as allowable delay for all the devices which is denoted by 'm', i.e.

$$X_{kit} = 0 \; where \; (i - t) > m \tag{8}$$

It is randomly generated between the limits of the controllable devices. Then, the fitness function can be given as follows:

$$fitness = \frac{1}{1 + \sum_{t=1}^{24} (P_{load}(t) - Target(t))^2} \tag{9}$$

5 Test Data and the Case Study

5.1 Test Data

For heuristic optimization, the suggested DSM approach is operated at 410 V,500KVA grid capacity with area-wise hourly forecasted load demand data listed in Table 2, [11, 18]. Residential area consists of 2604 controllable load devices with 14 different types; (b) Commercial area held 800 controllable devices of 8 different varieties given in Table 3. Indigenous solar crystalline PV modules with solar PV cells with the minimum 16% performance rate and the loading factor more than 70% with the capping of solar generation is 90% of the sanctioned load. The

Table 2 Load demands, forecasting, and energy prices in real time

Time (h)	Price (INR/MWh)	Hourly forecasted residential load (KW)	Hourly forecasted commercial load (KW)
08–09	3426.98	729.4	923.5
09–10	3502.58	713.5	1154.4
10–11	3626.36	713.5	1443
11–12	3814.67	808.7	1558.4
12–13	3706.47	824.5	1673.9
13–14	3427.02	761.1	1673.9
14–15	3292.88	745.2	1673.9
15–16	3502.38	681.8	1587.3
16–17	3904.75	666	1558.4
17–18	3910.00	951.4	1673.9
18–19	3647.25	1220.9	1818.2
19–20	3900.37	1331.9	1500.7
20–21	4016.13	1363.6	1298.7
21–22	3703.21	1252.6	1096.7
22–23	3296.34	1046.5	923.5
23–24	2870.57	761.1	577.2
00–01	2399.33	475.7	404
01–02	2164.57	412.3	375.2
02–03	2040.74	364.7	375.2
03–04	1999.81	348.8	404
04–05	2055.89	269.6	432.9
05–06	2397.83	269.6	432.9
06–07	2689.10	412.3	432.9
07–08	3229.49	539.1	663.8

Table 3 Details of hourly consumption in the residential and commercial area

	Device Type	Hourly consumption (KW)			No. of devices
		1st	2nd	3rd	
RES	Blender	0.3	0	0	66
	Coffee Maker	0.8	0	0	56
	Dish Washer	0.7	0	0	288
	Dryer	1.2	0	0	189
	Fan	0.2	0.2	0.2	288
	Frying Pan	1.1	0	0	101
	Hair Dryer	1.5	0	0	58
	Iron	1	0	0	340
	Kettle	2	0	0	406
	Oven	1.3	0	0	279
	Rice Cooker	0.85	0	0	59
	Toaster	0.9	0	0	48
	Vacuum Cleaner	0.4	0	0	158
	Washing Machine	0.5	0.4	0	268
COM	Air Conditioner	4	3.5	3	56
	Coffee Maker	2	2	0	99
	Dryer	3.5	0	0	117
	Fan	3.5	3	0	93
	Kettle	3	2.5	0	123
	Lights	2	1.75	1.5	87
	Oven	5	0	0	77
	Water Dispenser	2.5	0	0	156

overall solar PV array capacity must not be below the capacity (kWp) of the residential solar power modules of at least 300 Wp including 72 cells. The Lithium Ferro Phosphate (LiFePO4) battery has to have a working temperature range (both for charging and unloading) of 20–60 Deg. C with a minimum capacity of each prismatic cell 3.2 V 20Ah [21]. Based on the above technical requirement of rooftop solar panel installation, the rating of a solar panel ranges from 200 to 400 watts and battery rating 100 Ah / 1 KW is considered here.

5.2 *Case Study*

As per a survey conducted in a residential area of Karnal City, Haryana, the data for various controllable loads present in the consumer end has been collected accounting for all 316 connections comprises of the total load of 1276 (KW) with 70% of which are residential load and rest are commercial load, and total solar generation in smart city area about 250 (KW) details are given in Table 4. When burst loads like dishwashers, dryers, washing machines are accumulated along with the regular load then peak consumption is created at the consumer end. Such appliances can be interrupted intermittently to run the burst load in Table 5. Mainly, the DSM technique shifts these loads such that the energy cost is minimized. The case study is further divided into three scenarios to analyze the impact of two different heuristic

Table 4 Load detail of smart city area of Karnal

Sr. No	Location of feeder pillar	Location of service pillar	No. of connections
1	Feeder Pillar inside KCGMC	Service Pillar near RTA Office	23
2	Feeder Pillar near Ambedkar Chowk	(1) Service Pillar near sport House (2) Service Pillar near Amul Dairy (3) Service Pillar near Ashoka Theater	71
3	Feeder Pillar near Karnal Club	(1) Service Pillar near Diwan Colony, Main Gate	54
		(2) Service Pillar inside Diwan Colony	57
4	Feeder Pillar Minar Road	(1) Service Pillar Minar Road (2) Service Pillar near Civil line Quarter Gate	56
5	Feeder Pillar near Emergency Gate of KCGMC	Service Pillar Opp. Union Bank, Ambedker Chowk	55

Table 5 Details of controllable load

Device type	Hourly consumption (KW)			No. of devices
	1st	2nd	3rd	
Blender	0.3	0	0	276
Coffee Maker	0.8	0	0	82
Dish Washer	0.7	0	0	55
Dryer	1.2	0	0	225
Hair Dryer	1.5	0	0	85
Iron	1	0	0	289
Kettle	2	0	0	176
Oven	1.3	0	0	169
Toaster	0.9	0	0	115
Vacuum Cleaner	0.4	0	0	137
Washing Machine	0.5	0.4	0	225

approaches of demand management in the distribution system in terms of economic advantages:

Scenario 1: Initial stage of employing heuristic DSM strategies on residential and commercial controllable loads as listed in Table 5. of smart city area of Karnal separately without any power storage or solar generation.

Scenario 2: Second stage of employing DSM with HO strategies on residential and commercial area load separately with SPV generation at feeder pillar no. 2, 3, and 4, during the day time period between 10 a.m. and 6 p.m.

Scenario 3: Third stage of employing DSM with HO strategies on controllable residential and commercial area load individually with SPV generation utilizing battery storage system in radial distribution feeder during evening hours, i.e., 6 p.m. onwards.

6 Results

The comparative analysis of operational cost obtained as per the proposed algorithm considering three different cases of smart city of Karnal area including both Residential (Res) and Commercial (Com) are listed in Table 6.

The achievement of the primary objective of the optimization technique is to demonstrate the consumer-driven load shifting DSM process and to assess the local electricity usage in the distribution network while minimizing the expense of the user in the above cases are discussed here as follows:

Scenario 1: Aims to leap forward the objective demand curve where x-axis represents 24-h slots as given in Table 2 versus power in KW on y-axis and study the implication of heuristic DSM strategies in demand response with residential and commercial controllable load shifting in smart city area of Karnal for 24 h starting at 08.00 h separately without any power storage or solar generation installation is shown in Figs. 2 and 5 along with results listed in Table 6. The total cost without application of DSM in Karnal city area as the base case is (INR) 34,296.75 and with support of

Table 6 Operational cost reduction

Sr. No	Cases	Area type	Total cost with DSM (Rs.)		% Cost reduction
			GA	PSO	
1	DSM with HO	Res	216,202	213,795	7.16%
		Com	323,521.32	315,724.4	12.90%
2	DSM with HO and Solar Generation	Res	210,641	208,234	9.58%
		Com	316,106.12	308,309.2	14.95%
3	DSM with HO, Solar Generation and storage facility	Res	203,314	202,185	12.20%
		Com	306,295.12	297,034.18	18.06%

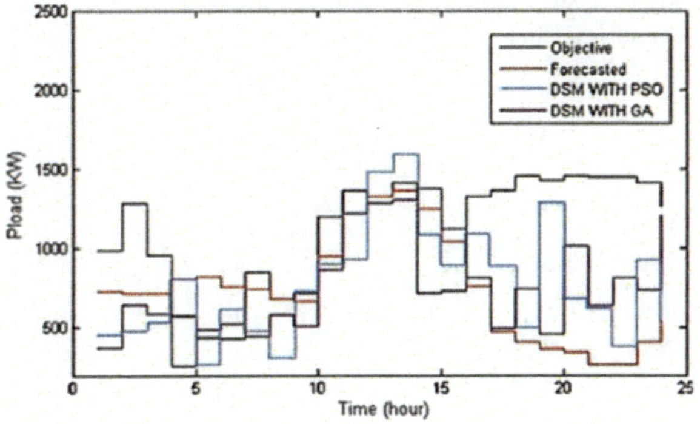

Fig. 2 DSM results of the residential area

DSM heuristic optimization technique, a significant reduction in overall utility bill of 11% is observed in the area (Figs. 3 and 4).

Scenario 2: An overall reduction of 12.07% in utility bill is obtained from the simulation results on employing DSM based on PSO on residential and commercial area separately with SPV generation during the daytime period between 10 a.m. and 6 p.m.

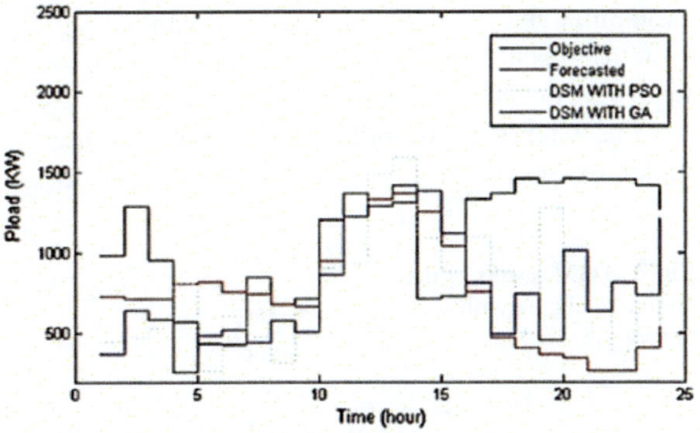

Fig. 3 DSM results of the residential area with solar

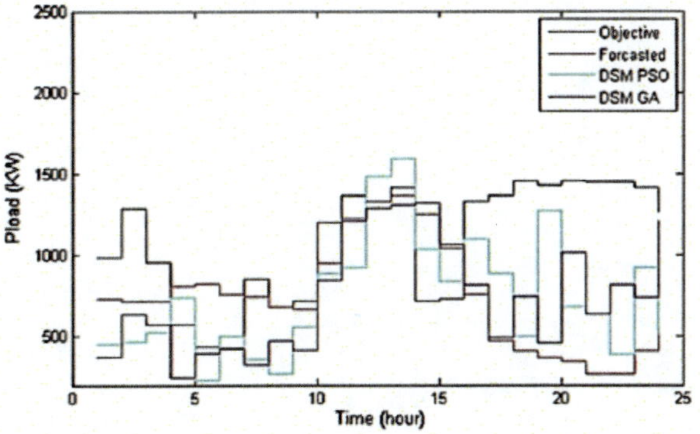

Fig. 4 DSM results of the residential area with solar and storage

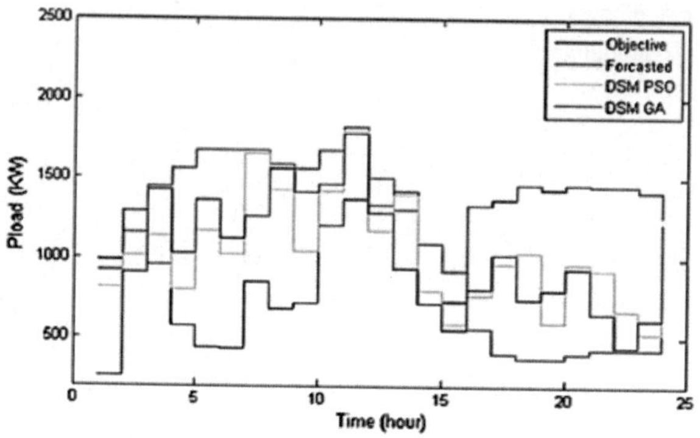

Fig. 5 DSM result of the commercial area

at designated feeder pillar near Ambedkar Chowk, Karnal Club, and Minar Road for which the graphs are shown in Figs. 3, 6 between 10 a.m. and 6 p.m. is listed in Table 6. From 6 p.m. to 11 p.m. for residential area and 10 a.m. to 7 p.m. for commercial area, it operates at high loading hours and without any efficient power storage and management system economic advantages are aligned purely to the elasticity of the operating loads with a permissible delay of 5 min at that instance.

Scenario 3: Simulation results of employing DSM with HO strategies on controllable residential and commercial area load with SPV generation utilizing energy storage system as backup near a feeder at KCGMC and emergency gate during evening hours, i.e.,6 p.m. onwards in addition to scenario 2 are shown in Figs. 4 and 7 separately. The rise in saving on utility bills of the Karnal city area is obtained as 14.96% compared to the initial stage results with uncurtailed load forecast on the consumer side.

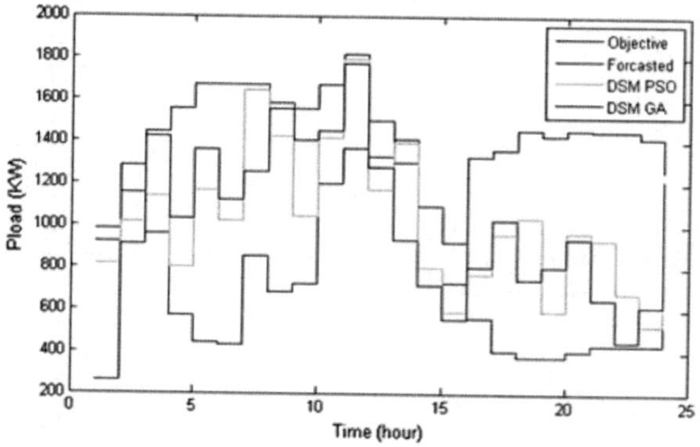

Fig. 6 DSM results of the commercial area with solar generation

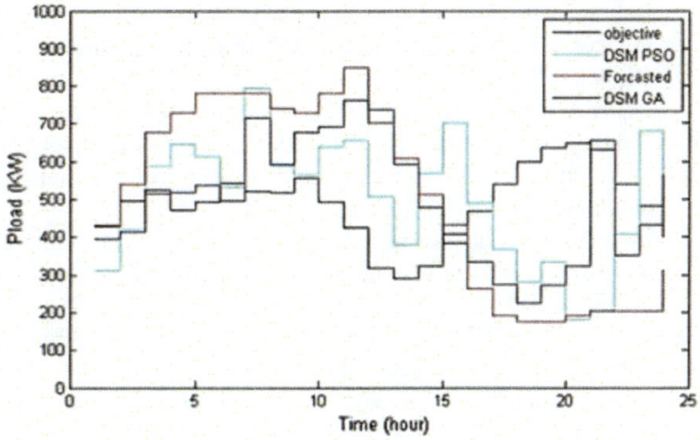

Fig. 7 DSM results of the commercial area with solar generation and storage

7 Conclusion

The simulation results as obtained justify that the DSM technique is very close to the objective curve in all three cases of heuristic optimization such that the suggested approach handles the controllable devices and their pattern efficiently. The comparative analysis of results highlights the significant drop in operational cost of the system in residential as well as in commercial areas with respect to different DSM approaches including GA and PSO. Demand-side management encourages the efficient utilization of existing power system infrastructure instead of expanding the generation capacity by selecting by mitigation of peak load demand and shifting the load during off-peak hours such a system can be implemented by proper communication between the utilities and the consumers.

References

1. IEA (2020) India 2020 energy policy review. https://www.iea.org/reports/india-2020. Accessed 10 Jan 2020
2. Ullah MN, Javid N, Khan I, Mahmood A, Farooq MU (2013) Residential energy consumption controlling techniques to enable autonomous demand side management in future smart grid communications. In: 8th International conference on broadband, wireless computing, communication and applications, pp 545–550
3. Patterson T (2012) DC, come home: Dc micro grids and the birth of the ethernet. IEEE Power Energy Mag 10(6):60–69
4. Xu Y, Milanovic JV (2016) Day-ahead prediction and shaping of dynamic response of demand at bulk supply points. IEEE Trans Power Syst 31(4):3100–3108
5. Farhangi H (2010) The path of the smart grid. IEEE Power Energy Mag 8(1):18–28
6. Nguyen HK, Song JB, Han Z (2015) Distributed demand side management with energy storage in smart grid. IEEE Trans Smart Grid 26(12)
7. Logenthiran T, Srinivasan, Shun TZ (2012) Demand side management in smart grid using heuristic optimization. IEEE Trans Smart Grid 3:1244–1252
8. Atzeni I, Ordonez LG, Scutari G, Palomar DP, Fonollosa JR (2013) Demand –side management via distributed energy generation and storage optimization. IEEE Trans Smart Grid 4(2):866–876

9. Soliman HM, Leon-Garci A (2014) Game theoretic demand-side management with storage devices for future smart grid. IEEE Trans Smart Grid 5(3)
10. Kinhekar N, Padhy NP, Li F, Gupta HO (2016) Utility oriented demand side management using smart AC and Micro DC Grid cooperative. IEEE Trans Power Syst 31(2):1151–1160
11. Balakumar P, Sathiya S (2017) Demand side management in smart grid using load shifting technique. In: IEEE international conference on electrical, instrumentation and communication engineering (ICEICE), Karur, pp 1–6
12. Logenthiran T, Srinivasan D, Khambadkone AM, Aung HN (2012) Multiagent system for real-time operation of a microgrid in real-time digital simulator. In: IEEE Trans Smart Grid 3(2):925–933
13. Latif A, Das DC, Biswas K, Kumar K, Kumar R, Hussain SI (2020) Non-critical demands managed load frequency stabilization of dish-stirling-biodiesel based islanded microgrid system using FF optimized controller In: Dawn S, Balas V, Esposito A, Gope S (eds) Intelligent techniques and applications in science and technology. ICIMSAT 2019. Learning and analytics in intelligent systems, vol 12. Springer, Cham, pp 188–196
14. Singh S, Singh M, Chandra KS (2016) Feasibility study of an islanded microgrid in rural area consisting of PV, wind, biomass and battery energy storage system. Energy Conver Manag 128:178–190
15. Barik AK, Tripathy D, Das DC, Sahoo SC (2020) Optimal load-frequency regulation of demand response supported isolated hybrid microgrid using fuzzy PD+I controller. In: Dawn S, Balas V, Esposito A, Gope S (eds) Intelligent techniques and applications in science and technology. ICIMSAT 2019. Learning and analytics in intelligent systems, vol 12. Springer, Cham, pp 798–806
16. Chen SX, Gooi HB, Wang Q (2012) Sizing of energy storage for microgrids. IEEE Trans. Smart Grid 3(1):142–151
17. Teng JH, Luan SW, Lee DJ, Huang YQ (2013) Optimal charging/discharging scheduling of battery storage systems for distribution systems interconnected with sizeable PV generation systems. IEEE Trans Power Syst 28(2):1425–1433
18. Mohsenian-Rad AV, Wong WS, Jatskevich J, Schoberr R, Leon-Garcia A (2010) Autonomous demand-side management based on game-theoretic energy consumption scheduling for future smart grid. IEEE Trans Smart Grid 1(3)
19. Ramachandran B, Ramanathan A (2015) Decentralized demand side management and control of PEVs connected to a smart grid. In: Power systems conference (PSC), 2015 Clemson University, Clemson, SC, pp 1–7
20. Deng R, Yang Z, Chen J, Asr NR, Chow M (2014) Residential energy consumption scheduling: a coupled-constraint game approach. IEEE Trans Smart Grid 5(3):1340–1350
21. Technical Specification: Grid Connected Solar Rooftop Photo Voltaic (SPV) power plant-with/without battery bank, Annexure-I. (2020). https://hareda.gov.in/. Accessed 27 June 2020

Optimal Placement of Micro-Phasor Measurement Units in Active Distribution Systems Using Mixed-Integer Programming

V. V. S. N. Murty, P. Ramakrishna, Vijay Babu Pamshetti, and Ashwani Kumar

Abstract The network operators are unable to detect the cause for grid disturbances, oscillations in the grid, outage of electrical equipment, and sudden change in grid parameters due to the non-availability of real-time information. Therefore, phasor measurement units shall be installed at appropriate locations to get real-time data of grid operation, tripping/failure details of grid elements, behavior of electrical machines, pre-incidents and post incidents etc. The main aim of the μPMUs placement problem is to ensure system observable with a minimum number of μPMUs. This paper presents the optimal location of Micro-Phasor Measurement Units (μPMUs) in microgrids to achieve fault observability (based on complete observability of the system) using mixed-integer programming. Further, aspects of line outage, with and without zero injection nodes and μPMU outage on μPMU placement problem are also discussed in this work. In this work, the μPMU placement problem is solved using mixed integer programming to minimize total cost and maximize system observability. Numerical results are obtained for 33-bus and 69-bus active distribution test systems [25].

Keywords Phasor Measurement Unit (PMU) · Optimal location · Active distribution system · Observability · Mixed-integer programming

V. V. S. N. Murty (✉)
Engineers India Limited, New Delhi, India

P. Ramakrishna
Department of Electrical Engineering, NIT Andhra Pradesh, Tadepalligudem, India

V. B. Pamshetti
Department of Electrical Engineering, B V Raju Institute of Technology, Medak, India

A. Kumar
Department of Electrical Engineering, NIT Kurukshetra, Kurukshetra, India
e-mail: ashwani.k.sharma@nitkkr.ac.in

A. Kumar et al. (eds.), *Renewable Energy Towards Smart Grid*, Lecture Notes in Electrical Engineering 823, https://doi.org/10.1007/978-981-16-7472-3_34

1 Introduction

Synchro phasors and Wide Area Monitoring System [1] is a promising technology for wide-area measurement and monitoring of power systems from a centralized control center. For many years, Supervisory Control and Data Acquisition (SCADA) is the most dominant tool for real-time operation, grid monitoring, and management of power systems. Synchrophasor measurements using phasor measurement units (PMUs) are deployed over a wide area, facilitating dynamic state measurement and visualization of a power system, which are useful in monitoring the safety and security of the grid. This fast-monitoring system detects the events and trends in grids with fluctuating load flows or highly loaded lines that conventional systems cannot detect at all or can detect too late. The synchrophasor technology has brought about a paradigm shift from state estimation to state measurement. Utilities in many countries are adopting PMU devices that assist system operators in real time even post-incidents. Due to the observability problem, low X/R value, communication issues, high penetration of renewable energy sources, high accuracy state estimation in microgrid-based distribution systems is a challenging task for the system operator [2].

A wide range of applications are available with PMUs [3] in the power system as shown in Fig. 1. Synchrophasor measurement and data transfer are defined in IEEE Std C37.118 [4, 5]. This standard describes measurements of synchro phasors, frequency, and df/dt in real-time. Micro-phasor measurement units were installed in distribution systems for complete system observability [6, 7]. Micro-PMU is coordinated with the most accurate GPS clock to measure synchro phasors. Accurate measurement of synchronized voltage and current phasors is helpful to the system operator for monitoring and controlling of the microgrid in real-time.

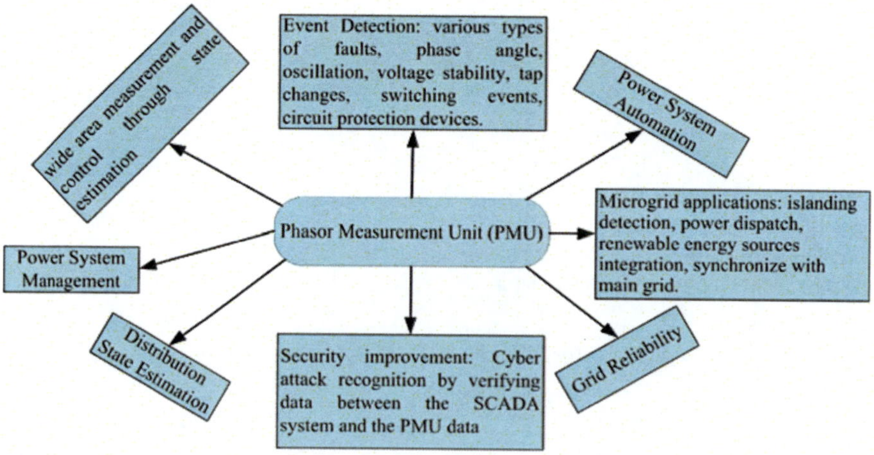

Fig. 1 Applications of PMU [3]

A lot of research work already available pertains to micro-PMU placement to ensure the entire system is observable. The schematic of micro-PMU integration into the grid is shown in Fig. 2. A centralized GPS server is allocated for several micro-PMUs to retrieve real-time information. In [8–10], integer linear program was used to solve the μPMU placement problem in reconfigured distribution systems. Graph-theoretic approach [11], genetic algorithm [12], and matrix manipulation algorithm [10] were applied to solve μPMU placement problems in distribution systems. Minimum number of PMUs were installed using PSAT in a standalone microgrid for complete system observable [13]. Practical implementation of PMUs in the Northern Region in India by Power grid Corporation of India Ltd for monitoring and control of large power grids [14]. In [15], PMU-based robust state estimation was formulated as a quadratic programming problem. Reliability index was introduced in PMU placement in distribution systems, solved using robust optimization [16]. The effect of component failure in PMUs was taken care of in the optimization problem. Voltage stability index was derived based on phasor data from micro-PMU in distribution systems [17].

In [19, 20], voltage and current phasor data received from PMUs was used to identify faults in microgrids. The compensation theorem in network theory was used to develop an equivalent circuit to represent the event by using voltage and current phasors data from micro-PMUs [21]. In [22], the rate of change of voltage phase

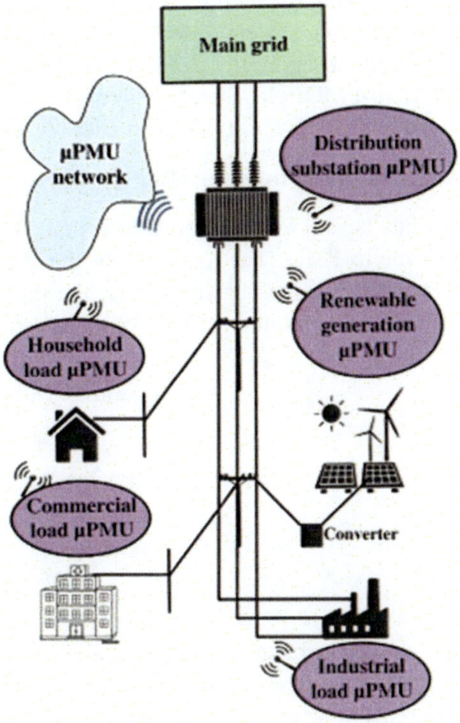

Fig. 2 Micro-phasor measurement unit in active distribution system [18]

angle difference between the point of common coupling and the bus closest to the fault point was utilized for fault detection.

In this paper, comparative analysis has been carried out for the optimal placement of micro-PMUs in active distribution systems. The objective of the micro-PMUs placement problem is to ensure complete system observability at minimum cost. Further, the micro-PMU placement problem is solved considering the loss of micro-PMU devices and loss of line. Moreover, the impact of feeder reconfiguration and network topology is also included in the optimization problem.

The paper is organized as follows: Sect. 2 discusses the optimal placement of micro-PMUs. Results and discussions are presented in Sect. 3. Finally, conclusions of the paper are given in Sect. 4.

2 Optimal Placement of Micro-PMUs

Following case studies are investigated to improve system observability by optimal placement of micro-PMUs. A detailed mathematical model for each case is explained below:

- Minimum number of micro-PMUs to maintain system observable,
- Minimizing the cost of micro-PMUs,
- Minimizing the number of micro-PMUs considering zero injection buses,
- Maximizing system observability with a minimum number of micro-PMUs,
- System observability against loss of micro-PMU,
- Micro-PMUs placement for system observability under line outage, and
- System observability under feeder reconfiguration.

(a) **Minimum number of micro-PMUs to maintain system observable**

The bus is said to be observable if that particular bus or adjacent bus is equipped with micro-PMU. The total cost of the system would be very high if micro-PMU is installed at each bus. Therefore, the minimum number of micro-PMUs shall be determined to maintain system observable as defined in Eq. (1).

$$MinimizeOF_{Xi} = \sum_{i \in nb} X_i \tag{1}$$

$$X_i + \sum_{j \in nb_{ij}^l} X_j \geq \alpha_i \forall i \in nb \tag{2}$$

$$\alpha_i = 1 \tag{3}$$

where X_i is the binary variable equal to 1 if the bus has micro-PMU and equal to 0 if the bus has no micro-PMU, nb is the set of number of buses, α_i is the binary variable represents network observability, nb_{ij}^l represents set of buses connected to ith bus.

(b) **Minimizing the cost of micro-PMUs to maintain system observable**

The cost of micro-PMU device is depending on the number of channels as described in Eq. (4).

$$MinimizeOFC_{Xi} = \sum_{i \in nb} W_i X_i \tag{4}$$

$$W_i = (1 + 0.1 * n) * C \tag{5}$$

Subject to Eqs. (2) and (3).

Where C is the cost of each micro-PMU and n is the number of used channels in PMU.

(c) **Minimum number of micro-PMUs considering zero injection buses**

In this case, the total network buses nb divided into zero injection buses nza and normal buses nr.

$$MinimizeOF_{Xi} = \sum_{i \in nb} X_i \tag{6}$$

$$nb = nza + nr \tag{7}$$

$$\sum_{a \in n_{ia}^{l}} (X_a + \sum_{j \in n_{aj}^{l}} X_j) \geq |n_{za}| - 1 \forall i \in n_{za} \tag{8}$$

Subject to Eqs. (2) and (3).

(d) **Maximizing system observability with the minimum number of micro-PMUs**

The minimum number of micro-PMUs deployed to ensure system observable.

$$MaxOF_{Xi,\alpha i} = \sum_{i \in nb} \alpha_i \tag{9}$$

$$\sum_{i \in nb} X_i \leq N_{micro-PMU} \tag{10}$$

$$1 \leq \alpha_i \tag{11}$$

Subject to Eq. (2).

Where $N_{micro-PMU}$ is total number of micro-PMUs.

(e) **System observability against loss of micro-PMU**

To maintain system observability in case of a single micro-PMU outage, each bus should be monitored by at least two micro-PMUs.

$$MinimizeOF_{Xi} = \sum_{i \in nb} X_i \tag{12}$$

$$X_i + \sum_{j \in nb^l_{ij}} X_j \geq 2\alpha_i \forall i \in nb \tag{13}$$

Subject to Eq. (3).

(f) **Micro-PMUs placement for system observability under line outage**

To maintain system observability in case of a single line outage, each bus should have a micro-PMU or each bus should be monitored by at least two other micro-PMUs.

$$MinimizeOF_{Xi,\alpha i} = \sum_{i \in nb} X_i \tag{14}$$

$$X_i + \alpha_i \geq 1 \forall i \in nb \tag{15}$$

3 Results and Discussions

The optimal placement of the micro-PMUs problem is solved in the GAMS environment using mixed integer programming (MIP). In this paper, the following cases are implemented: (a) minimum number of micro-PMUs to maintain system observable, (b) minimizing of cost of micro-PMUs, (c) minimizing of the number of micro-PMUs considering zero injection nodes, (d) maximizing system observability with a minimum number of micro-PMUs, e) system observability against loss of micro-PMU, and (f) micro-PMUs placement for system observability under line outage. Initially, results are obtained for system observability. Consequently, the effect of network topology and feeder reconfiguration is also analyzed on the micro-PMUs placement problem. Standard benchmark distribution test systems 33-bus [23] and 69-bus [24] are used to demonstrate the simulation results.

3.1 Results for 33-Bus System

As depicted in Fig. 3, the 33-bus radial system has 32 lines and sectionalizing switches are in the open position. Distributed generation is deployed at 13th, 24th, and 30th buses. For the 33-bus radial system, numerical results for each case are given in Table 1. It can be observed that the locations and the number of micro-PMUs are different for each objective function.

In this case, 5 loops are considered in a 33-bus mesh system to demonstrate the impact of network topology, which is also investigated on optimal placement of

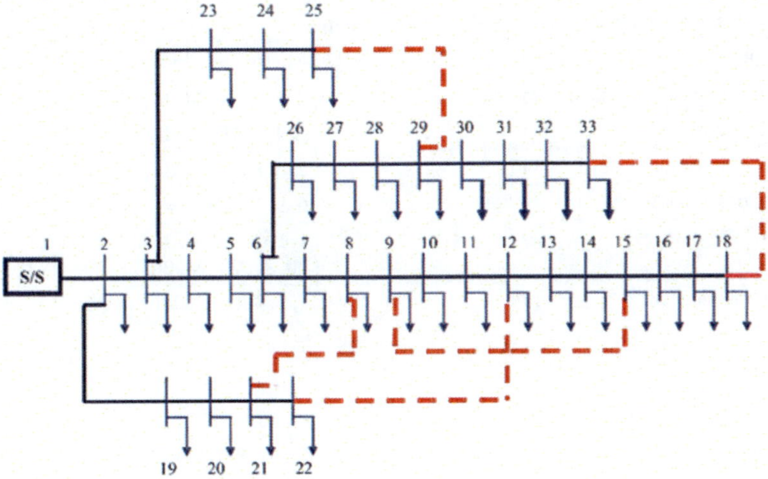

Fig. 3. 33-bus distribution system

Table 1 Simulation results for the 33-bus radial system

	Number of micro-PMUs	Micro-PMU locations
Without zero injection nodes	11	2, 5, 8, 11, 14, 17, 21, 24, 26, 29, 32
Minimum cost	11	2, 5, 8, 11, 14, 17, 21, 24, 27, 30, 33
With zero injection nodes	11	2, 5, 8, 11, 14, 17, 21, 24, 26, 29, 32
System observability	4	3, 6, 17, 28
Single PMU outage	24	1, 2, 3, 5, 6, 8, 9, 11, 12, 14, 15, 17, 18, 20, 21, 22, 24, 25, 27, 28, 29, 30, 32, 33
Single line outage	18	1, 4, 6, 8, 10, 12, 14, 16, 18, 19, 20, 22, 23, 25, 27, 29, 31, 33

micro-PMUs. The 33-bus mesh system has 37 lines and the sectionalizing switches: 33, 34, 35, 36, and 37 are closed positions. Numerical results for the test system for various cases is given in Table 2. From Table 2, it can be noticed that the optimal location of micro-PMUs is changed from radial to mesh topology. Further, the number of micro-PMUs is reduced in mesh topology for Case: E and Case: F.

During feeder reconfiguration, lines 7, 9, 14, 32, and 37 are kept open to reduce voltage drop and line losses and improve voltage profile subject to maintaining radial structure. Optimal locations of micro-PMUs in the reconfigured distribution system are given in Table 3 and observed that the location of micro-PMUs is different compared to radial system. The simulation results given in Tables 1–3 highlight the importance of objective function considered for micro-PMUs placement in distribution systems. This study may help the distribution network operator while planning synchronous phasors in active distribution systems.

Table 2 Simulation results for the 33-bus mesh system

Description	Number of micro-PMU	Micro-PMU locations
Without zero injection nodes	11	2, 5, 8, 11, 14, 17, 21, 24, 27, 29, 32
Minimum cost	11	2, 4, 7, 10, 13, 16, 21, 24, 27, 30, 33
With zero injection nodes	11	2, 5, 8, 11, 14, 17, 21, 24, 27, 29, 32
System observability	4	6, 9, 12, 29
Single PMU outage	22	1, 2, 3, 5, 6, 8, 9, 11, 12, 13, 15, 17, 18, 20, 21, 23, 24, 26, 28, 29, 31, 32
Single line outage	17	1, 3, 4, 6, 8, 10, 12, 14, 16, 17, 19, 21, 24, 27, 29, 31, 33

Table 3 Simulation results for the 33-bus reconfigured system

Description	Number of micro-PMU	Micro-PMU locations
Without zero injection nodes	11	2, 3, 6, 11, 13, 15, 18, 21, 25, 27, 31
Minimum cost	11	2, 3, 6, 11, 14, 15, 18, 21, 25, 28, 31
With zero injection nodes	11	2, 3, 6, 11, 13, 15, 18, 21, 25, 27, 31
System observability	4	3, 6, 12, 31
Single PMU outage	25	1, 2, 3, 4, 6, 7, 9, 10, 11, 12, 13, 14, 15, 16, 18, 20, 21, 24, 25, 27, 28, 30, 31, 32, 33
Single line outage	19	1, 3, 5, 7, 8, 10, 12, 14, 15, 17, 19, 21, 23, 25, 26, 28, 30, 32, 33

3.2 Results for the 69-Bus System

As illustrated in Fig. 4, the 69-bus distribution system has 68 lines and sectionalizing switches are in open position. Distributed generation is deployed at 61st, 18th' and 11th buses. For the 69-bus radial system, numerical results for each case are given in Table 4. It can be observed from Table 4 that optimal locations and the number of micro-PMUs are different for different objective functions.

In this case, a 69-bus mesh distribution system with 5-loops is considered to demonstrate the impact of network topology is also investigated on optimal placement of micro-PMUs. The 33-bus mesh system has 68 lines and the sectionalizing switches: 69, 70, 71, 72, and 73 are in closed positions. Numerical results for the 69-bus mesh system for various cases are given in Table 5. From Table 5, it can be noticed that the optimal location of micro-PMUs is changed from radial to mesh topology. Further, number of micro-PMUs is reduced in mesh topology for Case: E and Case: F.

During feeder reconfiguration, lines 14, 55, 61, 69, and 70 are kept open to reduce voltage drop and line losses subject to maintaining radial structure. Optimal locations of micro-PMUs in the reconfigured distribution system are given in Table 6 and

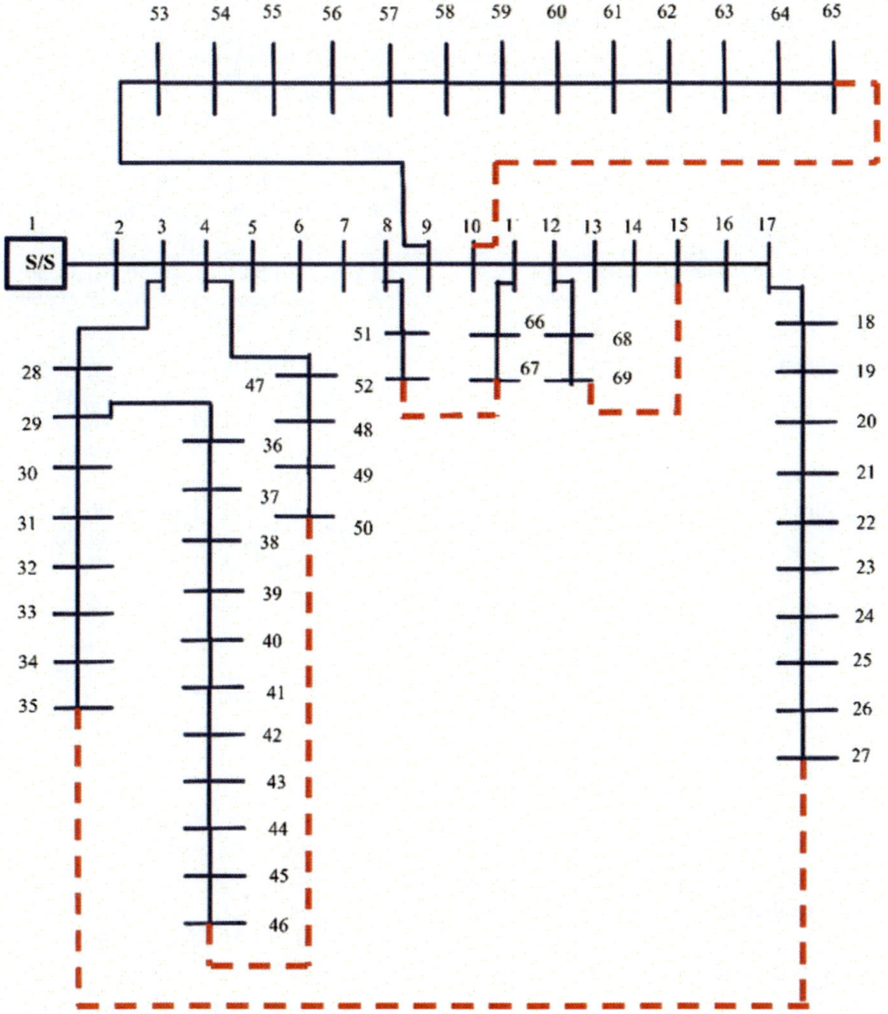

Fig. 4. 69-bus distribution system

observed that the location of micro-PMUs is different compared to the radial system. The simulation results given in Tables 4, 5 and 6, highlight the importance of objective function considered for micro-PMUs placement in distribution systems.

Table 4 Simulation results for the 69-bus radial system

Description	Number of micro-PMU	Micro-PMU locations
Without zero injection nodes	24	2, 4, 6, 9, 14, 17, 20, 23, 26, 28, 31, 34, 37, 39, 42, 45, 49, 51, 55, 58, 61, 64, 66, 68
Minimum cost	24	2, 6, 9, 14, 17, 20, 23, 26, 29, 32, 35, 37, 40, 43, 46, 47, 50, 52, 55, 58, 61, 64, 66, 68
With zero injection nodes	23	3, 6, 9, 14, 17, 21, 23, 26, 30, 31, 34, 36, 39, 42, 46, 49, 51, 55, 58, 61, 65, 66, 68
System observability	4	3, 9, 12, 30
Single PMU outage	50	1, 2, 4, 6, 7, 9, 10, 13, 14, 16, 17, 18, 20, 21, 23, 24, 26, 27, 28, 29, 31, 32, 34, 35, 36, 37, 39, 40, 42, 43, 45, 46, 47, 49, 50, 51, 52, 54, 55, 57, 58, 60, 61, 62, 64, 65, 66, 67, 68, 69
Single line outage	38	1, 3, 4, 6, 8, 9, 11, 12, 14, 16, 18, 19, 21, 23, 25, 27, 29, 31, 33, 35, 37, 39, 40, 42, 44, 46, 48, 50, 52, 54, 56, 57, 59, 61, 63, 65, 67, 69

Table 5 Simulation results for the 69-bus mesh system

Description	Number of micro-PMU	Micro-PMU locations
Without zero injection nodes	24	2, 4, 6, 9, 14, 17, 20, 23, 26, 29, 31, 34, 37, 39, 42, 45, 49, 51, 55, 58, 61, 64, 66, 68
Minimum cost	24	1, 6, 9, 13, 17, 20, 23, 26, 28, 31, 34, 37, 40, 43, 46, 47, 50, 52, 55, 58, 61, 64, 66, 69
With zero injection nodes	22	3, 6, 9, 14, 17, 21, 25, 26, 31, 34, 36, 39, 42, 46, 49, 51, 55, 58, 61, 64, 66, 68
System observability	4	3, 9, 12, 25
Single PMU outage	50	1, 2, 3, 4, 6, 7, 9, 10, 13, 14, 16, 17, 19, 20, 22, 23, 25, 26, 27, 29, 31, 32, 34, 35, 37, 38, 40, 41, 42, 43, 45, 46, 47, 49, 50, 51, 52, 54, 55, 57, 58, 60, 61, 62, 64, 65, 66, 67, 68, 69
Single line outage	37	1, 3, 4, 6, 8, 9, 11, 12, 14, 16, 18, 20, 23, 25, 27, 29, 31, 33, 35, 37, 39, 40, 42, 44, 46, 48, 50, 52, 54, 56, 57, 59, 61, 63, 65, 67, 69

Table 6 Simulation results for the 69-bus reconfigured system

Description	Number of micro-PMU	Micro-PMU locations
Without zero injection nodes	25	2, 4, 6, 9, 13, 15, 18, 21, 24, 27, 28, 31, 34, 37, 40, 43, 44, 49, 51, 54, 57, 60, 63, 66, 68
Minimum cost	25	1, 6, 10, 13, 16, 19, 22, 25, 29, 32, 35, 36, 39, 42, 45, 47, 50, 51, 54, 57, 61, 62, 65, 67, 69
With zero injection nodes	23	3, 6, 9, 13, 15, 17, 21, 25, 27, 28, 31, 34, 36, 40, 44, 49, 51, 54, 57, 60, 63, 66, 68
System observability	4	3, 8, 11, 59
Single PMU outage	50	1, 2, 3, 4, 6, 7, 9, 10, 13, 14, 15, 17, 18, 20, 21, 22, 24, 25, 27, 28, 29, 31, 32, 34, 35, 37, 38, 40, 41, 43, 44, 46, 48, 49, 51, 52, 54, 55, 56, 57, 59, 60, 61, 62, 63, 65, 66, 67, 68, 69
Single line outage	38	1, 3, 4, 6, 8, 9, 11, 12, 14, 15, 17, 19, 21, 23, 25, 27, 29, 31, 33, 35, 37, 39, 41, 43, 45, 47, 49, 52, 53, 55, 56, 58, 59, 61, 62, 64, 67, 69

4 Conclusions

In this paper, the micro-PMUs placement problem is solved using mixed integer programming for complete system observability. Minimizing the number of micro-PMUs and cost, zero injection buses, outage of line, and micro-PMU are considered in the optimization problem. Based on the numerical results, it is observed that the number of micro-PMUs in the meshed system is more than the radial structure for case E and case F. Further, it is noticed that micro-PMUs locations in the radial network are different than reconfigured network and mesh systems.

References

1. Agarwal PK, Sarma NDR (2016) Synchro phasors and WAMS – an Indian experience. IFAC-Papers Online 49(27):66–72
2. JMcDonald J (2008) Adaptive intelligent power systems: active distribution networks. Energy Policy 36(12):4346–4351
3. Advancement of Synchro phasor Technology (2016) U.S. Department of Energy. https://www.energy.gov/sites/prod/files/2016/03/f30/Advancement%20of%20Sychrophasor%20Technology%20Report%20March%202016.pdf
4. IEEE C37.118.1-2011, Standard for synchro phasor measurements for power systems
5. IEEE C37.118.2-2011, Standard for synchro phasor data transfer for power systems
6. Liu J, Tang J, Ponci F, Monti A, Muscas C, Pegoraro PA (2012) Trade-offs in PMU deployment for state estimation in active distribution grids. IEEE Trans Smart Grid 3(2):915–924

7. Von Meier A, Culler D, McEachen A, Arghandeh R (2014) Micro-synchro phasors for distribution systems. In: IEEE PES innovative smart grid technologies conference (ISGT), pp 1–5
8. Dua GS, Tyagi B, Kumar V (2019) MILP based deployment of micro-PMU in reconfigurable active distribution network, North American power symposium (NAPS), Wichita, KS, USA, USA, pp 1–6
9. Chauhan K, Sodhi R (2019) A comparative analysis of μPMU placement for active distribution network's observabilit. In: 8th international conference on power systems (ICPS), Jaipur, India, pp 1–6
10. Abdul-Aziz Fish S, Chowdhury SP, Chowdhury (2011) Optimal PMU placement in a power network for full system observability. In: IEEE power and energy society general meeting, Detroit, MI, USA, pp 1–8
11. Tahabilder A, Ghosh PK, Chatterjee S, Rahman N (2017) Distribution system monitoring by using micro-PMU in graph-theoretic way. In: 4th International conference on advances in electrical engineering (ICAEE), Dhaka, Bangladesh, pp1–5
12. Khanjani N, Moghaddas-Tafreshi SM (2020) Analysis of observability constraint on optimal feeder reconfiguration of an active distribution network with μPMUs. In: 28th Iranian conference on electrical Engineering (ICEE), Tabriz, Iran, Iran, pp 1–5
13. Ghosh S, Das JK, Chanda CK (2019) Placement of phasor measurement unit for complete observability of an isolated microgrid system. Microsyst Technol 25:4671–4674
14. Agrawal VK, Agarwal PK, Kumar R (2010) Experience of commissioning of PMUs pilot project in the northern region of India. In: 16th National power systems conference, 15–17 December, pp 249–253
15. Lin C, Wu W, Guo Y (2020) Decentralized robust state estimation of active distribution grids incorporating microgrids based on PMU measurements. IEEE Trans Smart Grid 11(1):810–820
16. Gholami M, Abbaspour A, Fattaheian-Dehkordi S, Lehtonen M, Moeini-Aghtaie M, Fotuhi M (2020) Optimal allocation of PMUs in active distribution network considering reliability of state estimation results. IET Gener Transm Distrib 14(18):3641–3651
17. Kumar DS, Savier JS, Biju SS (2020) Micro-synchrophasor based special protection scheme for distribution system automation in a smart city. Prot Control Mod Power Syst 5(9):1–14
18. Dutta S, Sadhu PK, Reddy MJB, Mohanta DK (2020) Role of micro phasor measurement unit for decision making based on enhanced situational awareness of a modern distribution system, pp 181–199
19. Sharma NK, Samantaray SR (2020) PMU assisted integrated impedance angle-based microgrid protection scheme. IEEE Trans Power Deliv 35(1):183–193
20. Sanitha G, Savier JS (2019) Faulty phase and fault location identification using micro-PMU data. In: International conference on power electronics applications and technology in present energy scenario (PETPES), pp 1–6, Mangalore, India, 29–31 August 2019
21. Farajollahi M, Shahsavari A, Stewart EM, Mohsenian-Rad H (2018) Locating the source of events in power distribution systems using micro-PMU data. IEEE Trans Power Syst 33(6):6343–6354
22. Sharma NK, Samantaray SR (2019) Assessment of PMU-based wide-area angle criterion for fault detection in microgrid. IET Gener Transm Distrib 13(19):4301–4310
23. Baran ME, Wu FF (1989) Network reconfiguration in distribution systems for loss reduction and load balancing. IEEE Trans Power Del 4(2):1401–1407
24. Baran ME, Wu FF (1989) Optimal sizing of capacitors placed on a radial distribution system. IEEE Trans Power Deliv 4(1):735–743
25. Soroudi A (2017) Power system optimization modeling in GAMS. Springer International Publishing, pp 1–295

Binary Metal Oxide Spinel-$NiCo_2O_4$ as Electrode Material for Supercapacitor Application

Manpreet Kaur, Hardeep Anand, and Prakash Chand

Abstract In the era of modern technology, electronic portable devices have become a strong need today. Conventional energy storage devices, that is, batteries and capacitors are inadequate to fulfill the high demand of energy supply. Hence, supercapacitor comes into the picture as a fast emerging energy storage device due to its high power density, high energy density, fast charging, and good cyclic stability. A supercapacitor consists of two electrodes separated with an electrolyte. This implies that the electrode material and electrolyte play a key role in the efficiency of a supercapacitor. This book chapter explains the utility of nickel cobaltite ($NiCo_2O_4$) as a working electrode material in a pseudocapacitor. A redox reaction mechanism to explain the charge transfer in charging–discharging is also highlighted in this chapter. The method of synthesis of electrode material is another important factor that greatly affects the specific capacitance of the material. Hence, some important methods are also discussed briefly.

Keywords Spinel transition metal oxide · Nickel Cobaltite · Morphology · Mechanism · Supercapacitor

1 Introduction

Today we can't imagine our life without portable devices. From morning to evening, kitchen to office and restaurant to the gym, we depend so much on electricity and energy. After the exhaustion of conventional energy sources like coal and petroleum, we start focusing on non-conventional sources like solar cells, wind energy, fuel cells, etc., to fulfill the growing energy demand [1]. In this direction, conventional energy storage devices like batteries and capacitors play a noteworthy role. But for these rapid technological modern days, we need such a fast device which can overcome

M. Kaur (✉) · H. Anand
Department of Chemistry, Kurukshetra University Kurukshetra, Haryana 136119, India

P. Chand
Department of Physics, National Institute of Technology, Kurukshetra, Haryana 136119, India
e-mail: prakash@nitkkr.ac.in

A. Kumar et al. (eds.), *Renewable Energy Towards Smart Grid*, Lecture Notes in Electrical Engineering 823, https://doi.org/10.1007/978-981-16-7472-3_35

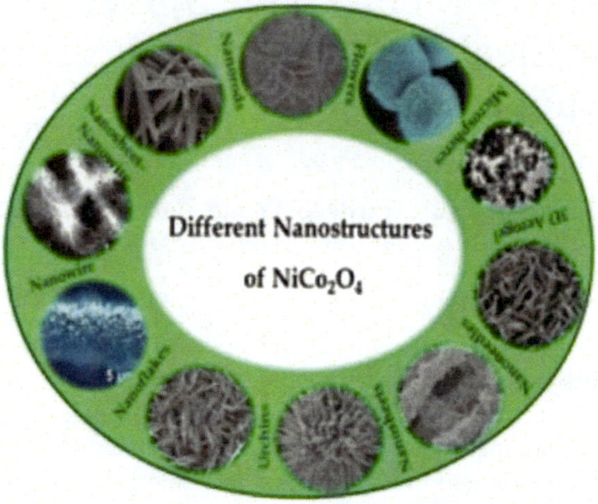

Fig. 1 Different morphologies of $NiCo_2O_4$ from different research works. [Reprinted with the permission from Ref. [4], © Elsevier 2015]

the slow charging of batteries and fast discharging of capacitors. The drawback of a battery is low power density and of a capacitor is low energy density. Supercapacitor overcomes these limitations because its characteristics lie in the midway of batteries and capacitors, that is, it has high power density and energy density [2]. The capacitance of the supercapacitors majorly depends upon the characteristics of the electrode material, temperature of the synthesis, pH, surfactants, electrolyte, capping reagents, and method of formation of the electrode material.

In the past couple of decades, metal oxides like RuO_2, MnO_2, CuO, Cu_2O_3, Fe_2O_3, NiO, spinel structures like $NiCo_2O_4$, Co_3O_4, $ZnCo_2O_4$, $MgCo_2O_4$, $CaCo_2O_4$, etc., have become a source of attraction to be worn as an electrode material for the ultracapacitor. Out of all the metal oxides mentioned above, $NiCo_2O_4$ comes out to be most appropriate. Reason being, its large surface area, variable oxidation states, and suitable distribution of pore size [3]. The porous nature of the metal oxide lets the electrode–electrolyte contact be very effective for charge storage. It makes room for the carrying ions and electrons at the electrode–electrolyte surface and in the bulk of the electrolyte. $NiCo_2O_4$ emerges as a gifted material for the synthesis of electrode material due to its ground cost, environment-friendly nature, good electrochemical properties, and catalytical properties (Fig. 1).

1.1 Nickel Cobaltite ($NiCo_2O_4$)

Nickel Cobaltite ($NiCo_2O_4$) has been developed into a new class of energy storage material for electrochemical supercapacitors due to its excellent electrochemical properties, low cost, and environmental friendliness. It reduces the pressure of energy crisis and environmental contamination. $NiCo_2O_4$ possesses richer electroactive sites

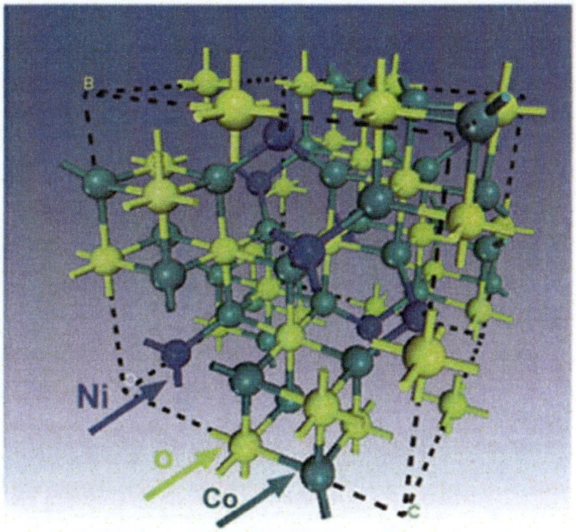

Fig. 2 Spinel structure of $NiCo_2O_4$. [Reprinted with the permission from Ref. [7], © Royal Society of Chemistry 2012]

and has at least two periods of elevated electrical conductivity than that of corresponding monometallic oxides, NiO and Co_3O_4. Along with higher power density, $NiCo_2O_4$ also possesses higher energy density. Many studies reveal that the resistance of $NiCo_2O_4$ is lesser than that of NiO and Co_3O_4 [5]. It is also assumed that redox reactions offered by the nickel cobaltite, including the contribution from both nickel and cobalt ions in different oxidation states, are richer than that of nickel oxide and cobalt oxide. Due to the spinel structure, it can store more charge. Another reason is the easy availability of nickel and cobalt making nickel cobaltite a very fascinating electrode material for scientists to work and explore more properties of the material [6] (Fig. 2).

1.2 Mechanism of $NiCo_2O_4$ as a Supercapacitor Electrode

$NiCo_2O_4$ electrode is used in the pseudocapacitor and the energy storage mechanism in the pseudocapacitor is the Redox reaction. Reversible and fast Faradaic reactions occur in the charge–discharge process. Due to variable oxidation states of nickel and cobalt, i.e., Co^{3+}/Co^{4+}, as well as Ni^{2+}/Ni^{3+} on the surface of the electrode materials, make nickel cobaltite facilitate a fast redox reaction. Following reactions takes place in the charging–discharging process when KOH is taken as an electrolyte [8].

$$NiCo_2O_4 + H_2O + OH^- \leftrightarrow NiOOH + 2CoOOH + e^- \quad (1)$$

.

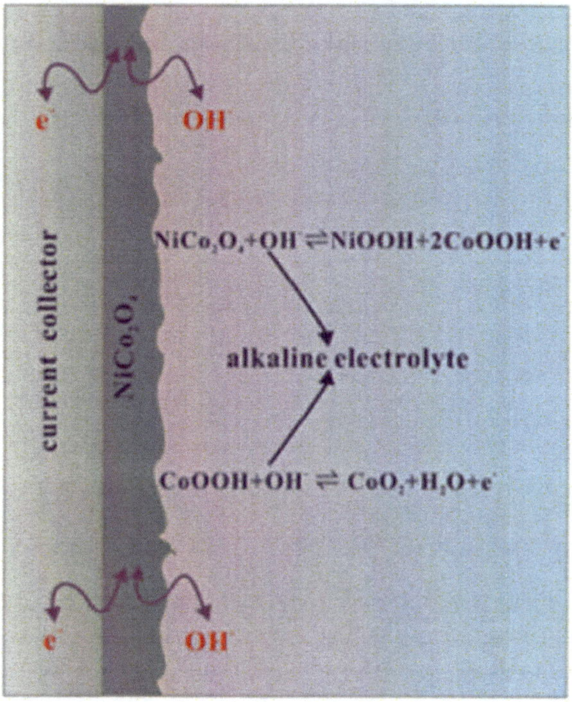

Fig. 3 The schematic diagram for the mechanism of $NiCo_2O_4$ as an electrode [Reproduced with the permission from Ref. [10], © Royal Society of Chemistry 2014]

$$CoOOH + OH^- \leftrightarrow CoO_2 + H_2O + e^- \quad (2)$$

$NiCo_2O_4$ is an example of an inverse spinel in which O^{2-} ions are present at the centers of an FCC close-packed structure. Ni^{2+} ions are present at the octahedral voids and Co^{3+} ions are present at the Tetrahedral as well as octahedral voids [9]. There are 4 octahedral voids and 8 tetrahedral voids present in a crystal lattice (Fig. 3).

2 Methods of Synthesis of Electrode Materials

There are several methods of the formation of $NiCo_2O_4$. These are the hydrothermal method, co-precipitation method, sol–gel method, electrode deposition method, chemical vapor deposition method, solid-state method, etc. A descriptive explanation of these methods is given below.

2.1 Hydrothermal Method

Hydrothermal is the simplest way for the fabrication of electrode materials. Almost every type of electrode material can be synthesized by this method. In this method, requisite salt is dissolved in a particular solvent to make precursor solutions in separate beakers with continuous stirring. Then, one precursor solution is added to the other in a dropwise manner slowly with constant stirring. Further, the reaction mixture is subjected to stirring for about another half hour after the addition of all the solutions. Transfer the reaction mixture into an autoclave and place it into a microwave oven for a specific time. Remove the autoclave from the oven, let it cool to room temperature, open it up, and filter the precipitates, dry at 60 °C.

$NiCo_2O_4$ is synthesized by the hydrothermal method with precursors $Ni(NO_3)0.6H_2O$, $Co(NO_3)0.6H_2O$, and urea. In the step of the reaction mechanism, urea undergoes thermal decomposition and yields melamine, ammonia, and carbon dioxide. The following reaction sequences take place in the autoclave for the formation of nickel-cobalt hydroxide [11].

$$6CO(NH_2)^2 \rightarrow C_3H_6N_6 + 3CO^2 + 6NH_3 \quad (3)$$

$$NH_3 + H_2O \rightarrow NH_4^+ + OH^- \quad (4)$$

$$Ni^{2+} + 2Co^{2+} + 6OH^- \rightarrow NiCo_2(OH)_6 \quad (5)$$

Nickel–cobalt hydroxide is then further calcined to give the nickel-cobaltite product under the air atmosphere. The calcination reaction can be written as follows:

$$2NiCo_2(OH)_6 + O_2 \rightarrow 2NiCo_2O_4 + 6H_2O \quad (6)$$

2.2 Co-precipitation Method

Just like the hydrothermal method, co-precipitation is another simple technique for the formation of metallic nanoparticles. It is a cost-effective, easy to perform, and fast method in comparison to other methods. For this method, first of all, precursor solutions are prepared in separate beakers with steady stirring. Then, one precursor is added to the other in a dropwise manner slowly with uninterrupted stirring. The reaction mixture is subjected to stirring for about another half hour after the addition of all the solutions. Precipitates formed are filtered on filter paper or collected by centrifugation. Washed several times, dried in the oven, and undergo calcination, if required [12].

$NiCo_2O_4$ is synthesized by the co-precipitation method with precursors $Ni(NO_3)0.6H_2O$, $Co(NO_3)0.6H_2O$, and KOH. The chemical reaction that takes place under co-precipitation is just similar to the reactions that happened in the hydrothermal method. Dipositive nickel ions with dipositive cobalt ions and hydroxide ions from potassium hydroxide yield the formation of nickel-cobalt hydroxide. The reaction can be shown as follows:

$$Ni^{2+} + 2Co^{2+} + 6OH^- \rightarrow NiCo_2(OH)_6 \quad (7)$$

$NiCo_2(OH)_6$ is then further calcined to give the nickel-cobaltite product under air atmosphere. The calcination reaction can be written as follows:

$$2NiCo_2(OH)_6 + O_2 \rightarrow 2NiCo_2O_4 + 6H_2O \quad (8)$$

2.3 *Sol–Gel Method*

One of the easy, cost-effective, and precise methods is the sol–gel method. In this method, we usually need to make a solution of the requisite precursors in the first step. In the next step, this precursor solution is heated to make a sol. After that, the sol is again stirred for several hours to make a gel. The gel is then heated to get the solid powder. The solid powder is calcined in the final step to get the product $NiCo_2O_4$. The following process shows the gelation mechanism [13].

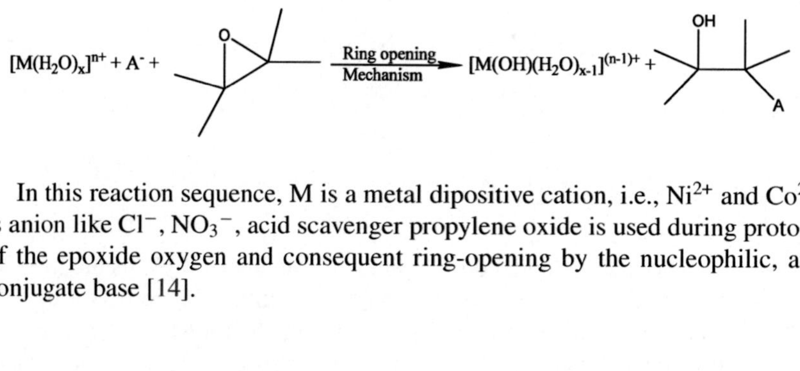

(9)

In this reaction sequence, M is a metal dipositive cation, i.e., Ni^{2+} and Co^{2+}, A^- is anion like Cl^-, NO_3^-, acid scavenger propylene oxide is used during protonation of the epoxide oxygen and consequent ring-opening by the nucleophilic, anionic conjugate base [14].

2.4 *Electrode Deposition Method*

It is one of the inexpensive and time-efficient methods for the formation of $NiCo_2O_4$. A three-electrode system is used in the electrode deposition method for the preparation of the nickel-cobalt hydroxide complex, which is then processed under calcination to get the final nickel-cobaltite product. The following reaction shows the chemical changes occurring in the system when we use nickel nitrate and cobalt nitrate as initial precursors [15].

$$NO_3^- + 2e^- + H_2O \rightarrow NO_2^- + 2OH^- \tag{10}$$

$$NO_2^- + 6e^- + 6H_2O \rightarrow NH_4^+ + 8OH^- \tag{11}$$

$$xNi^{2+} + 6xOH^- + 2xCo^{2+} \rightarrow Ni_xCo_{2x}(OH)_{6x} \tag{12}$$

$$Ni_xCo_{2x}(OH)_{6x} + 0.5xO_2 \rightarrow xNiCo_2O_4 + 3xH_2O \tag{13}$$

3 Results and Discussion

Chen fabricated $NiCo_2O_4$ using metal acetates as precursors to achieve flower-like morphology [16]. Moderate values of specific capacitance are reported as 677, 673, 664, 650, 638, 623, 687, 562, 539, and 520 F/g at various scan rates of 1, 2, 3, 5, 7, 10, 20, 30, 40, and 50 mV/s, respectively, which are multiples than the specific capacitances values of NiO and Co_3O_4. Guragain et al. (2020) recently studied the effect of urea on the electrochemical properties and the morphology of $NiCo_2O_4$ and reported the highest value of specific capacitance of 3143.451 F/g at a 2 mV/s scan rate [17]. It is also reported that the morphology of the material changes significantly with differing the quantity of urea. GCD measurements are taken in 3 M KOH electrolyte for electrochemical studies and to calculate the specific capacitance of the nickel cobaltite. The reported value of specific capacitance for urea content 4.44, 2.99, 1.49, 0.37, and 0.22 g are 1227, 1107.9, 3143.5, 837.6, and 902.7 F/g at a scan rate of 2 mV/s. Cao et al. (2020) reveal that poor crystallinity in the sample generates oxygen vacancies which in turn improves the electrical conductivity of the material [18]. At the current density of 0.5 A/g, the value of specific capacitance reported is 1076 F/g. Similarly, Yin et al. (2021) fabricated metal–organic framework (MOF) derived hollow $NiCo_2O_4$ nano-cages (NCs) [19]. Excellent electrochemical performance with a large specific capacitance of 1377.6 F/g at a current density of 1 A/g, good rate capability of 68.8%, capacitance retention at 20 A/g, and excellent cycling stability of 88.3% and capacitance retention after 6000 cycles is reported.

Researchers from different areas of the world are working on the performance of nickel cobaltite. Some are making composites of nickel cobaltite with other metal oxides, whereas others are trying to merge it with activated carbon. Methods of fabrication, electrolyte, aging period, the temperature of synthesis, time and temperature of calcination, etc., are a few variations in the manufacturing of the $NiCo_2O_4$ as electrode material. These variations highly affect the morphology which leads to the change in specific capacitance of the material (Fig. 4).

Following is the list of different research works carried out recently on the high performance of nickel cobaltite as a working electrode for the supercapacitor application. This table compares the year-wise progression in the performance of $NiCo_2O_4$ as a supercapacitor electrode (Table 1).

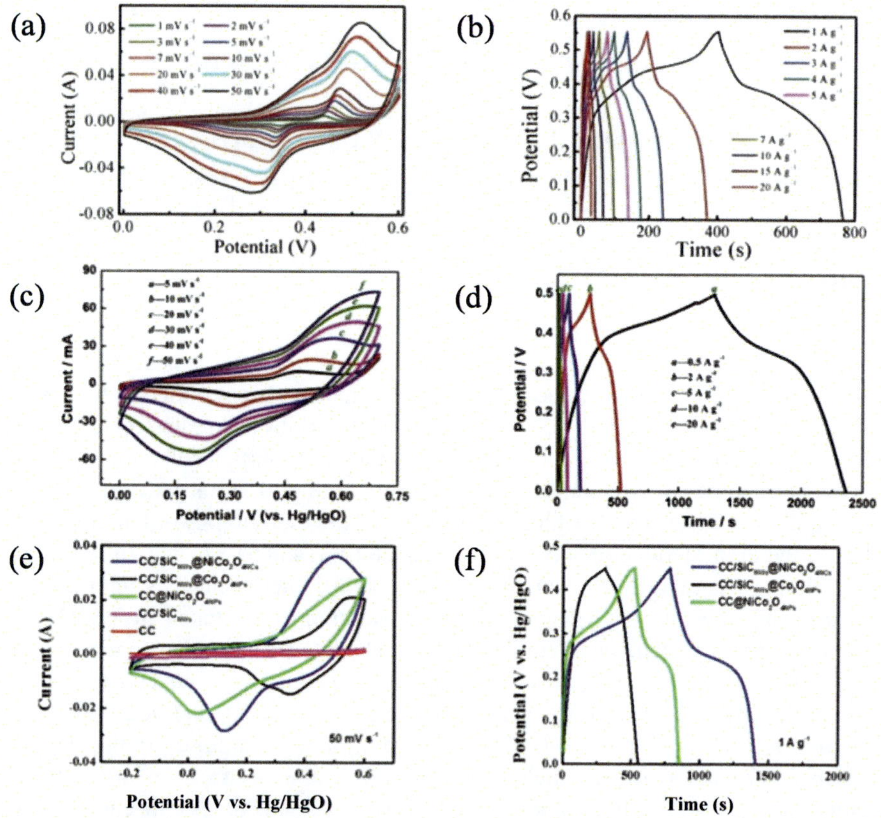

Fig.4 **a** CV for $NiCo_2O_4$ at different scan rates ranges from 1 to 50 mV/s **b** GCD for the at various current densities ranges from 1 to 10 A/g [Reprinted with the permission from Ref. [16], © Elsevier 2014] **c** CV for $NiCo_2O_4$/NF electrode at different scan rates ranges from 5 to 50 mV/s **d** GCD at various current densities [Reprinted with the permission from Ref. [18], © Elsevier 2020] **e** CV for different samples of $NiCo_2O_4$ at a scan rate of 50 mV/s **f** GCD of various samples [Reprinted with the permission from Ref. [19], © Elsevier 2021]

Table 1 Comparison of different research works for the $NiCo_2O_4$ as electrode material

SN	Material	Method of fabrication	Specific capacitance	Refs.
1	$NiCo_2O_4$ at Ni foam	Co-electrode deposition followed by calcination	2010, 1859, 1694, 1596, and 1450 F/g at current densities of 2, 4, 8, 12, and 20 A/g resp.	[20]
2	$NiCo_2O_4$	Hydrothermal method followed by calcination	294 F/g at 1 A/g	[21]
3	$NiCo_2O_4$	Sol–gel method	1254 F/g at 2 A/g using oxalic acid as a chelating ligand	[22]
4	Ni foam supported $NiCo_2O_4$-graphene oxide	Electrochemical deposition method	1078, 999 F/g at current density 1 and 5 mA resp.	[23]
5	$NiCo_2O_4$ and $NiCo_2O_4$@PANI nanocomposite	Hydrothermal and electrode deposition method	901, 865, 800, 763, and 725 F/g at current densities of 1, 2, 4, 8, and 10 A/g resp.	[24]
6	$NiCo_2O_4$	Solvothermal route	2118 F/g at 2 mV/s	[25]
7	Reduced $NiCo_2O_4$ electrode	Chemical reduction method	1590, 1586, 1560, 1548, 1496 F/g at current densities of 1, 2, 5, 10, and 20 A/g resp.	[26]
8	$NiCo_2O_4$ and $NiCo_2O_4$/C composite	Hydrothermal method	1480.9 and 995.2 F/g at 1 and 10 A/g resp.	[27]
9	$NiCo_2O_4$	Hydrothermal method	2876 and 1290 F/g at 1 and 10 A/g resp.	[28]
10	$NiCo_2O_4$	Co-precipitation method followed by calcination	1525 F/g at 1 A/g	[29]

4 Conclusion

Upon reviewing several research works, we can conclude that nickel cobaltite is emerging as one of the most promising electrode materials for supercapacitors. Due to the spinel crystal lattice of nickel cobaltite, it shows a great tendency of a fast redox reaction, and hence a good value of specific capacitance is expected. Many research works can be seen for pure nickel cobaltite and activated carbon. But doping the lattice of nickel cobaltite with the other s-block or first transition metals can further improve the electrochemical properties of this material. More research is expected in this field with the synthesis of spinel-type cobaltites due to their high theoretical specific capacitance. In recent years, nanocomposites of nickel cobaltite with other metal oxides, polymers, and activated carbons are fabricated with great interest. In this direction, more work is expected worldwide with some novel nickel cobaltite composites.

References

1. Wang Y (2015) Mesoporous transition metal oxides for supercapacitors. Nanomaterials 5(4):1667–1689. https://doi.org/10.3390/nano5041667
2. Borenstein A (2017) Carbon-based composite materials for supercapacitor electrodes: a review. J Mater Chem A 5(25):12653–12672. https://doi.org/10.1039/C7TA00863E
3. Zhang L (2018) A review of supercapacitor modeling, estimation, and applications: a control/management perspective. Renew Sustain Energy Rev 81(2):1868–1878. https://doi.org/10.1016/j.rser.2017.05.283
4. Dubal DP (2015) Nickel cobaltite as an emerging material for supercapacitors: an overview. Nano Energy 11:377–399. https://doi.org/10.1016/j.nanoen.2014.11.013
5. Wang C (2017) $NiCo_2O_4$-based supercapacitor nanomaterials. Nanomaterials 7(2):41. https://doi.org/10.3390/nano7020041
6. Li Y (2019) Review and prospect of $NiCo_2O_4$-based composite materials for supercapacitor electrodes. J Energy Chem 31:54–78. https://doi.org/10.1016/j.jechem.2018.05.010
7. Wang Q (2012) Morphology evolution of urchin-like $NiCo_2O_4$ nanostructures and their applications as pseudocapacitors and photoelectrochemical cells. J Mater Chem 22(40):21647–21653. https://doi.org/10.1039/C2JM34705A
8. Zhang Y (2014) Selective synthesis of hierarchical mesoporous spinel $NiCo_2O_4$ for high-performance supercapacitors. Nanoscale 6(8):4303–4308. https://doi.org/10.1039/C3NR06564B
9. Liu XY (2013) Self-assembled porous $NiCo_2O_4$ hetero-structure array for electrochemical capacitor. J Power Sources 239:157–163. https://doi.org/10.1016/j.jpowsour.2013.03.106
10. Wu Z (2014) $NiCo_2O_4$-based materials for electrochemical supercapacitor. J Mater Chem A 2(36):14759–14772. https://doi.org/10.1039/C4TA02390K
11. Shen L (2013) Mesoporous $NiCo_2O_4$ nanowire arrays grown on carbon textiles as binder-free electrodes for energy storage. Adv Func Mater 24(18):2630–2637. https://doi.org/10.1002/adfm.201303138
12. Karmakar S (2018) Investigation of structural and electrical transport properties of nano-flower shaped $NiCo_2O_4$ supercapacitor electrode materials. J Alloy Compd 757:49–59. https://doi.org/10.1016/j.jallcom.2018.05.056
13. Cui H (2005) Sol-Gel synthesis of nanoscale spinels using propylene oxide as a gelation agent. J Sol-Gel Sci Technol 35:175–181. https://doi.org/10.1007/s10971-005-4165-0
14. Liu MC (2012) A sol-gel process for fabrication of $NiO/NiCo_2O_4/Co_3O_4$ composite with improved electrochemical behavior for electrochemical capacitors. Appl Mater Interfaces 4(9):4631–4636. https://doi.org/10.1021/am301010u
15. Liu WW (2014) A three-dimensional vertically aligned multiwall carbon nanotubes/$NiCo_2O_4$ core/shell structure for novel high-performance supercapacitors. J Mater Chem A 2(14):5100–5107. https://doi.org/10.1039/C4TA00107A
16. Chen H (2014) Facilely synthesized porous $NiCo_2O_4$ flower-like nanostructure for high-rate supercapacitors. J Power Sources 248:28–36. https://doi.org/10.1016/j.jpowsour.2013.09.053
17. Guragain D (2020) Influence of urea on the synthesis of $NiCo_2O_4$ nanostructure: morphological and electrochemical studies. J Nanosci Nanotechnol 20(4):2526–2537. https://doi.org/10.1166/jnn.2020.17366
18. Cao Z (2020) Oxygen-vacancy-rich $NiCo_2O_4$ nanoneedles electrode with poor crystallinity for high energy density all-solid-state symmetric supercapacitors. J Power Sources 449:227571. https://doi.org/10.1016/j.jpowsour.2019.227571
19. Yin X (2021) Hierarchical self-supporting sugar gourd-shape MOF-derived $NiCo_2O_4$ hollow nanocages@SiC nanowires for high-performance flexible hybrid supercapacitors. J Colloids Interface Sci 586:219–232. https://doi.org/10.1016/j.jcis.2020.10.086
20. Yuan C (2012) Ultrathin Mesoporous $NiCo_2O_4$ nanosheets supported on Ni foam as advanced electrodes for supercapacitors. Adv Func Mater 22(21):4592–4597. https://doi.org/10.1002/adfm.201200994

21. Pu J (2013) Porous hexagonal $NiCo_2O_4$ nanoplates as electrode materials for supercapacitors. Electrochemica Acta 106:226–234. https://doi.org/10.1016/j.electacta.2013.05.092
22. Zhu Y (2014) Spinel $NiCo_2O_4$ for use as a high-performance supercapacitor electrode material: understanding its electrochemical properties. J Power Sources 267:888–900. https://doi.org/10.1016/j.jpowsour.2014.05.134
23. Mitchell E (2015) Ultrathin porous hierarchically textured $NiCo_2O_4$-graphene oxide flexible nanosheets for high-performance supercapacitors. New J Chem 39(3):2181–2187. https://doi.org/10.1039/C4NJ02110J
24. Jabeen N (2016) Unique core-shell nanorod arrays with PANI deposited into mesoporous $NiCo_2O_4$ support for high-performance supercapacitor electrodes. Appl Mater Interfaces 8(9):6093–6100. https://doi.org/10.1021/acsami.6b00207
25. Kumar V (2017) Pseudocapacitance of mesoporous spinel-type MCo_2O_4 (M=Co, Zn, and Ni) rods fabricated by a facile solvothermal route. ACS Omega 2(9):6003–6013. https://doi.org/10.1021/acsomega.7b00709
26. Yan D (2018) $NiCo_2O_4$ with oxygen vacancies as better performance electrode material for supercapacitor. Chem Eng J 334:864–872. https://doi.org/10.1016/j.cej.2017.10.128
27. Yang G (2018) Facile hydrothermal synthesis of $NiCo_2O_4$-decorated filter carbon as electrodes for high-performance asymmetric supercapacitors. Electrochem Acta 285:405–414. https://doi.org/10.1016/j.electacta.2018.08.013
28. Hao C (2018) Preparation of hierarchical spinel $NiCo_2O_4$ nanowires for high-performance supercapacitors. Ind Eng Chem Res 57(7):2517–2525. https://doi.org/10.1021/acs.iecr.7b04412
29. Bhagwan J (2019) Rapid synthesis of hexagonal $NiCo_2O_4$ nanostructures for high-performance asymmetric supercapacitors. Electrochem Acta 299:509–517. https://doi.org/10.1016/j.electacta.2018.12.174

Impact of Storage Energy on Operation and Control of Smart Grid

V. P. Meena, P. K. Meena, Surjeet Choudhary, Nitin Mathur, and V. P. Singh

Abstract In an attempt to reduce the CO_2 emission and the dependence on fossil fuels for generation of energy, power systems are moving towards the integration of a number of Renewable Energy Resources (RESs). Intermittent nature of wind and solar power introduces complexity with the grid and impacts negatively on power system reliability, stability, and power quality. This creates the problem of large-scale integration of RES with grid. Battery Energy Storage System (BESS) can mitigate these problems and help in increasing the penetration of RES into the power grid. The paper presents the role of BESS and its impact on the smart grid. An idea of Virtual Power Plant (VPP) to operate a smart grid along with BESS is presented.

Keywords Battery Energy Storage System (BESS) · Battery Management System (BMS) · Distributed energy resources penetration · Dy namic voltage regulator · Grid integration · Transmission congestion relief · Virtual power plant

1 Introduction

Our vision of having a grid that is mostly supplied by carbon-free energy can be realized by integrating more renewable energy resources into our grid. The variable nature of solar and wind generation due to environmental conditions poses a problem of fluctuating power output. This power needs to be regulated and should be made available at times when required. Energy storage helps us to smooth out intermittencies which is an issue with RES. The wind can stop blowing or the sun can stop shining and we need to instantaneously have another source of energy to take up the

V. P. Meena (✉) · S. Choudhary · N. Mathur · V. P. Singh
MNIT, Jaipur, India
e-mail: vmeena1@ee.iitr.ac.in

V. P. Singh
e-mail: vinay.ee@mnit.ac.in

P. K. Meena
IISER, Bhopal, India
e-mail: pavan19@iiserb.ac.in

A. Kumar et al. (eds.), *Renewable Energy Towards Smart Grid*, Lecture Notes in Electrical Engineering 823, https://doi.org/10.1007/978-981-16-7472-3_36

deficit. If we can store it after it is produced, then we can call upon that energy to feed the grid at any moment even when the sun and wind are absent.

In this paper, the development in battery technologies and a comparison of different batteries is presented with a focus on Li-ion batteries. The challenges with the use of batteries and the role of Battery Management System (BMS) that takes care of parameters that affect the operation and life of the batteries. BESS stabilizes the power grid and balances the supply and demand by integrating renewables in real time. The paper presents the impact of BESS on the smart grid in terms of renewable integration, penetration level, voltage and frequency regulation, power quality, transmission congestion relief. The intermittent nature of solar and wind generation limits us to penetrate more power into the grid. To increase the level of penetration of renewables, an idea of the virtual power plant is presented.

The remainder of the paper is arranged as follows. Section 2 describes the technology development in batteries. Utility-scaled energy storage system is derived in Sect. 3. Challenges and opportunities are described in Sect. 4 and finally, the work is concluded in Sect. 5.

2 Technological Development in Batteries

The electrical energy is stored in the form of chemical energy in rechargeable batteries. The technology currently used in the power system industry are lead-acid batteries, Lithium-Ion Batteries (LIB), Nickel-Metal Hydride (NiMH) batteries, Vanadium Redox Flow Batteries (VRFB), and sodium-sulfur batteries (NaS). The comparison is made in Table 1 [1].

Table 1 Comparison of other types of Batteries with LIB

Battery type	Lead-Acid	Ni-Cd	Ni-MH	Zn-Br	Fe-Cr	Li-Ion
Energy density	30–50	45–80	60–120	35–54	20–35	110–160
Power density	180	150	250–1000	-	70–100	1800
Operating voltage	2 V	1.25 V	1.25 V	1.67 V	1.18 V	3.6 V
Operating temperature	−20–60	−40–60	−20–60	−20–60	−40–60	−20–60
Cycle life	200–300	1500	300–500	> 2000	–	1000
Charge efficiency	79	–	–	–	–	–
Energy efficiency	70	60–90	75	80	66	80
Voltage efficiency	–	–	–	–	82	–
Overcharge	High	Moderate	Low	High Moderate		Very
Tolerance					Low	
Self-discharge	Low	Moderate	High	Low	HighVery Low	
Thermal stability	Least	Least	Least	Least	StableMost	

Table 2 Characteristic of different Li-ion Batteries

Type	LMO	LPF	LNMC	LTO	Li–s
Energy Density (Wh/Kg)	160	120	200	70	500
Power Density	200	200	200	1000	–
Cycle Life (Wh/Kg) (100% Depth of Discharge)	≥2000	≥2500	≥2000	≥10,000	~100
Cost (US$/kWh)	~360	~360	~360	~860	–
Safety	Good	Good	Good	Good	Good
Maturity	Commercial	Commercial	Commercial	Commercial	Commercial

The driving factor for the development of batteries has been their requirement for smart grid and Electric Vehicle (EV) application. Li-Ion Batteries (LIB) offer high energy density (high output voltage), long life cycle, high efficiency, and eco-friendliness. The main characteristics of different types of Li-ion batteries are listed in Table 2, [2].

Li-ion, NaS, and Vanadium flow batteries are the major battery technologies for large-scale energy storage. RFB possesses a small response time and is therefore suitable for balancing highly variable renewables but the cost competitiveness makes it less popular. A schematic diagram of BS interfaced with a system showing key components. The battery management system is a key component in the design of a storage system as it affects the life and performance of the battery.

Battery management plays a crucial role in the integration of battery energy storage system with the utility grid as far as performance, reliability, economy, and safety is concerned. The important features of Li-ion batteries are their size (energy density), long life (life cycle, capacity), charging and discharging characteristics, cost, their performance in a wide range of temperature, leakage and self-discharge profile, and poisonous impact [2]. The Li-ion batteries have good charging and discharging characteristics, as shown in Fig. 1 [1].

The components of BMS are shown in Fig. 2 [1].

Li-ion battery cells must be connected in series and parallel combination to meet the desired voltage level and power requirement forming a pack. Despite cells being similar and having the same specification, this may result in cell variances that accentuate with battery usage. For multi-cell pack, unequal cell voltages increase the risk of overcharge and undercharge. The overall efficiency and life span of the battery storage can be enhanced by optimizing and quality charging and discharging of the Li-ion batteries. It is proposed in [1] to charge the LIB by the Constant Current-Constant Voltage (CC-CV) charge-optimized PI (proportional-integral) controller. The discharging of LIB is done by the Discontinuous Current Mode (DCM) discharge control. The charging of battery is controlled by the battery properties and the State of Charge (SOC) of the battery. The load demand and the charge available in the battery systems control the discharging of the batteries [1]. The various component of BMS

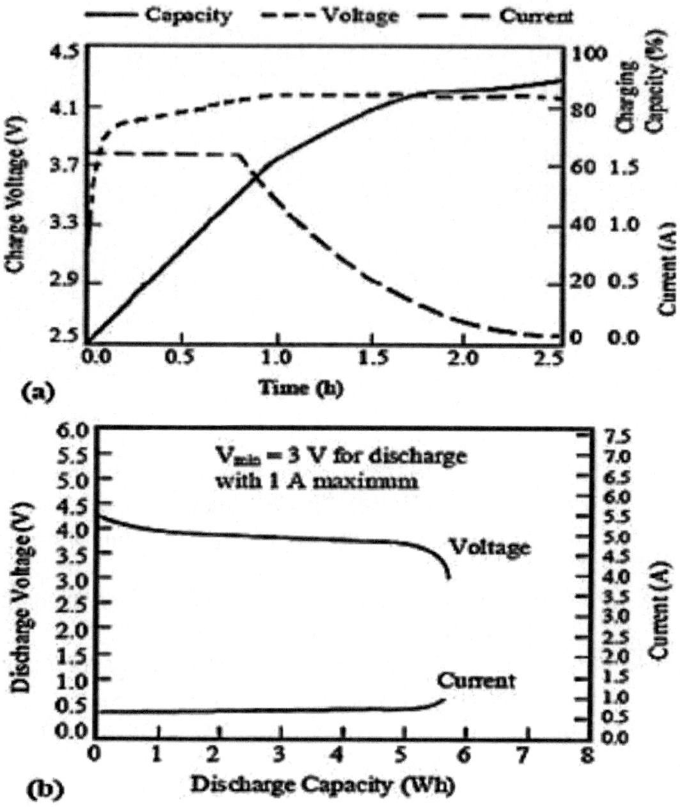

Fig. 1 **a** Charging. **b** Discharging

are shown in Fig. 2. The BMS continuously monitors and detects an unbalance in the current, voltage, and temperature of battery cells. The equalization is done by transferring the excess of charge from the overcharged cell to the undercharged cell. An active and passive equalization scheme is employed for this. In passive scheme, the excess charge is dissipated as heat in a resistor which increases the temperature which is not desired. In active equalization scheme, the circuit transfers energy from pack to cell, from cell to pack, or between cells to make it efficient and faster [3]. The battery needs hardware setup for avoiding overcharge/discharge and overheating. A battery cell operating out of its operating range loses its capacity with time. It affects the life of the battery [1]. The individual load pattern for a particular application has unique requirements for the ESS, especially the performance parameters of the Li-ion cells [4]. For different applications, different LIBs are used with performance parameters given in Table 2.

Various degradation mechanisms that lead to the aging of the battery include decomposition of the electrolyte, formation of a thin passive film, particle cracking, and dissolution of electrode material. These effect on the material and battery cell

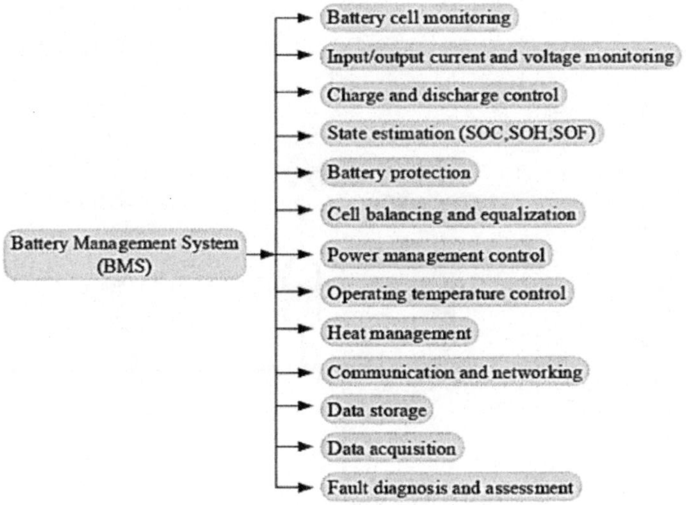

Fig. 2 Overview of BMS

level leads to reduced capacity retention, increase in resistance, and risk of an unsafe battery condition.

3 Utility-Scaled Energy Storage System

Renewable Energy Resources (RES) are sustainable and growing forms of electrical energy. The problem with solar and wind is that they are not there all the time. It's not sunny at night and there is no wind on a calm day. To implement RES at a large scale on a grid, we got to have Energy Storage (ES). Energy Storage helps us to smooth out intermittency [4].

The various components are grouped under battery, power electronic converter and grid connection interface. Each component plays an important role and involves extensive study and integrates them to utilize their potential in different applications.

Grid Application of BESS is illustrated in Fig. 3

BESS can be used by heavy industries for altering their demand-side response and use energy batteries during peak hours. The energy storage supports grid stability by frequency and voltage control. Energy storage helps in EV-Grid integration by relieving the grid from supplying power. The demand is met by stored energy in the battery. Some of the impacts on the smart grid are qualitatively been explained in this paper.

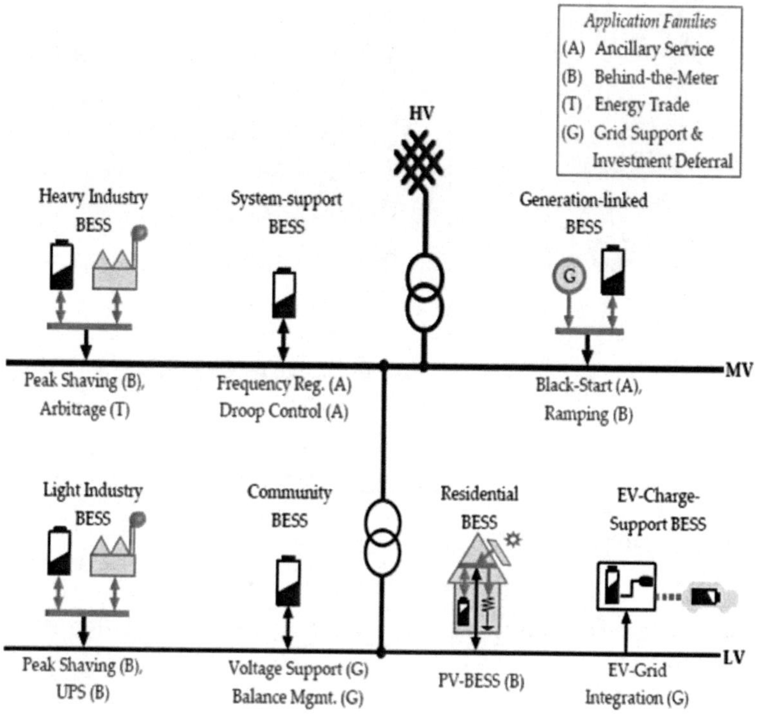

Fig. 3 Application of BESS

A. **Load Following and Peak Shaving**

Batteries charge when there is more renewable generation than demand and supply that energy to a load during peak hours. The PV farms integrated with ES can meet the changing load demand and help smoothen out fluctuation in power generated by the solar PV system by charging the battery when excess energy is available and vice versa. A typical load following and peak shaving application are shown in Fig. 4 [5].

In (b), we can see that the power from the electric grid can be used to charge the batteries at night during off-peak periods. This can later be used to increase the total output of the plant in times of peak load or generation shortfall. The capacity and size of battery storage are estimated based on power produced by PV plants and load as per system requirement.

B. **Increased Levels of Distributed Energy Resources Grid Pene tration**

Increased levels of renewable energy injection in the grid may reduce its stability due to intermittency in renewable power generation. A Virtual Power Plant (VPP) which can emulate the characteristic of a traditional generator with voltage and frequency regulation can help mitigate this issue. It allows more number of DER to be integrated with the grid. A VPP composed of DER with

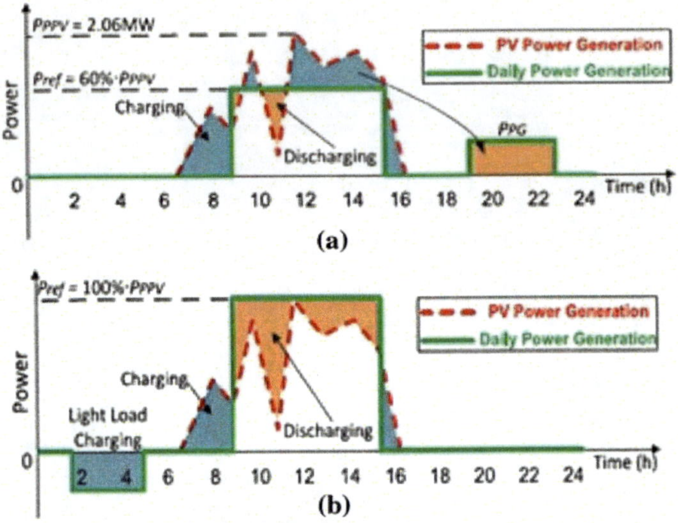

Fig. 4 Load Following and peak shaving

dedicated tools for load-generation forecasting. The VPP has two components, namely forecast and analytics (Fig. 5), and electrical systems control (Fig. 6).

Depending on load and network configuration, the forecasting component prepares the dispatch schedule for all sources in the VPP. This is achieved by taking historical weather and load data and the VPP sources model as inputs. This dispatch schedule is used by the electrical system component to regulate power flows between individual power sources and the loads.

The architecture and functional diagram representing various components of a VPP [6] are shown in Figs. 5 and 6.

(a) **Forecast and Analytics**

The weather data for the next 24 h in terms of temperature, solar insulation, and wind speed is known in advance. The controller has historical weather records through which it generates generation patterns. Load forecasting is also done by the controller. The objective is to prepare an economic dispatch schedule depending on weather and load forecasting.

(b) **Electrical System and Control**

Each VPP source such as PV solar generation and wind plants, battery energy storage, and demand-side response systems have their dedicated controllers, which regulate the flow of power from the source to the loads. The VPP dispatch schedule is made by utilizing generation and load schedule from the first component, viz. forecast and analytics tool. The controller's reference for real and reactive power is set using the droop control characteristics as shown in Fig. 7. The local loads are met by the renewable and the remaining power is supplied to the grid according to

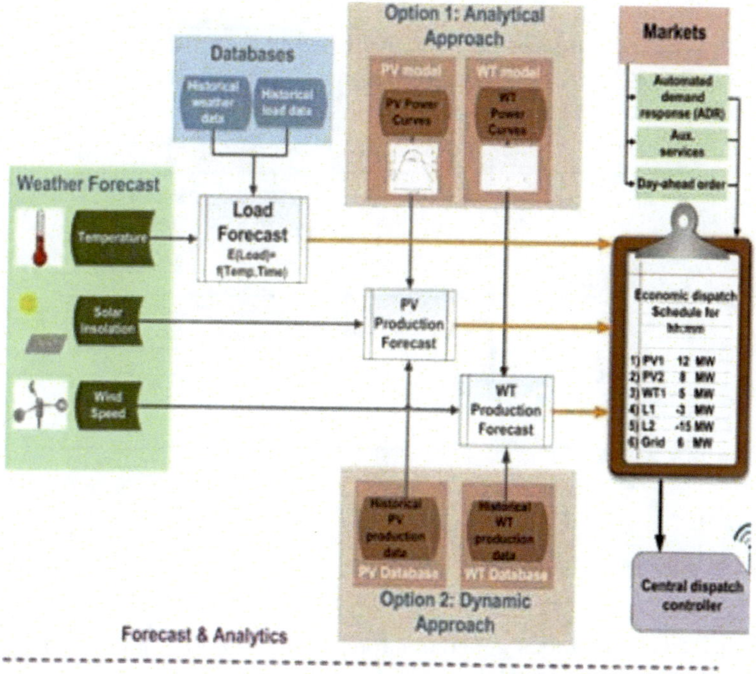

Fig. 5 Forecast and analytics

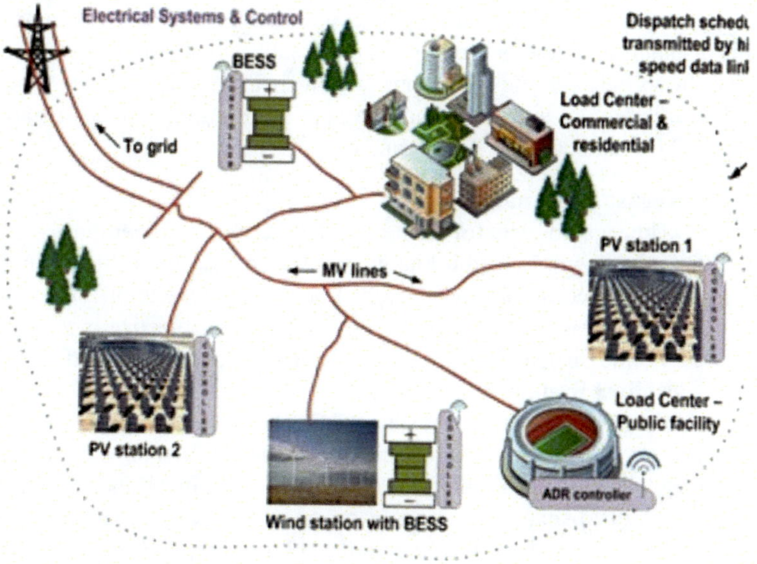

Fig. 6 Electrical system and control

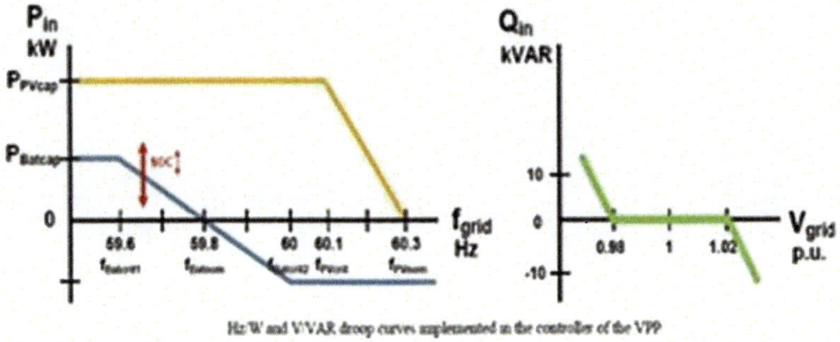

Fig. 7 Droop control Characteristic

the scheduled dispatch. If the local load is more than VPP generation, the utility grid supplies the shortfall. The mechanism for power flow control, i.e., real and reactive, features like voltage and frequency control that support grid, and VPP emulating characteristic of the traditional generator are part of the individual DER control unit. [6]

Droop control equations

$$P = \frac{V_{DG}V_{grid}}{X}\sin(\delta) \approx \frac{V_{DG}V_{grid}}{X}\delta \quad (1)$$

$$P = \frac{V_{grid}}{X}\left(V_{grid} - V_{DG}\cos(\delta)\right) \approx \frac{V_{grid}}{X}\left(V_{grid} - V_{DG}\right) \quad (2)$$

$$V - V_{ref} = -m_Q\left(Q - Q_{ref}\right) \quad (3)$$

$$\omega - \omega_{ref} = -m_P\left(P - P_{ref}\right) \quad (4)$$

C. **Transmission Congestion Relief**

Power limitations in the transmission network lead to bottlenecks and they impose a lot of stress on power lines and substation. The unpredictability of wind and solar power combined with the above scenario may lead to the need to curtail scheduled generation [7]. A BESS can provide congestion relief to such a grid. It allows the system planner and operator to add advantages such as an increase in reserve margin and avoiding curtailment of renewables.

Congestion happens mainly due to the outages of transformers and feeders in an interconnected power system network. The location of the outage and the system parameters decides the severity of congestion. [8] With more renewables coming up, there are chances that the power generation injection at certain nodes reach their crucial limit and then we

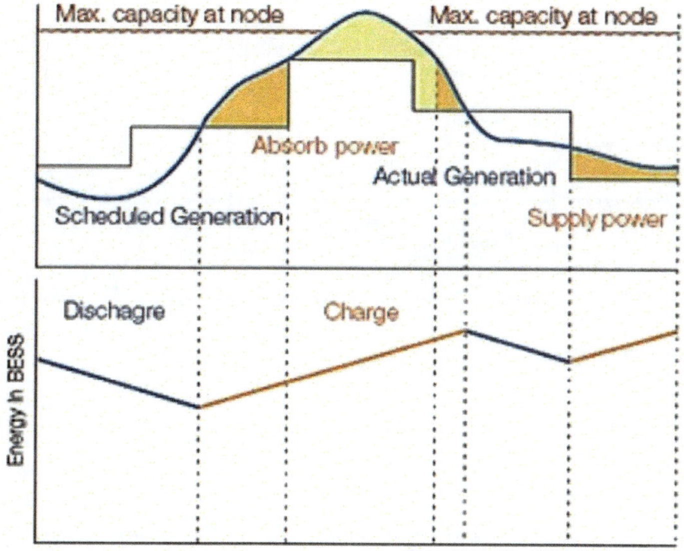

Fig. 8 BESS response to scheduled generation at a node

have to curtail our generation and increase it elsewhere to be transmitted through networks which are underloaded.

Transmission-line congestion is one of the critical issues that need to be investigated for ensuring the stable and reliable operation of the power system. A battery energy storage system can provide a solution to system planners and operators to solve line congestion problems as shown in Fig. 8. The load at a node is kept below its maximum capacity. The battery charging and discharging is coordinated with schedule generation and load mismatch. Any increase in load is met locally by battery storage without causing line congestion.

Rather than making upgrades in the transmission and distribution network, the system planners may defer these investments. These congestions are there only for peak hours and investing on an upgrade is not economic unless the power levels involved are high [7].

Figure 9 shows a simplified line diagram of a RES with a battery storage scheme connected at a node to support it at the time of distress.

What is important in implementing BESS for transmission congestion relief is to identify the bottlenecks in the network. Identify the nodes and the lines which get congested during peak hours. This is done by analyzing the past data. The load flow can help in selecting the nodes which require active and reactive power support. Algorithms are available in the literature that aims at economic sizing and siting of BESS that results in relief from transmission congestion. Locating a DER at suitable locations in the system reduces system losses and provides relief from the

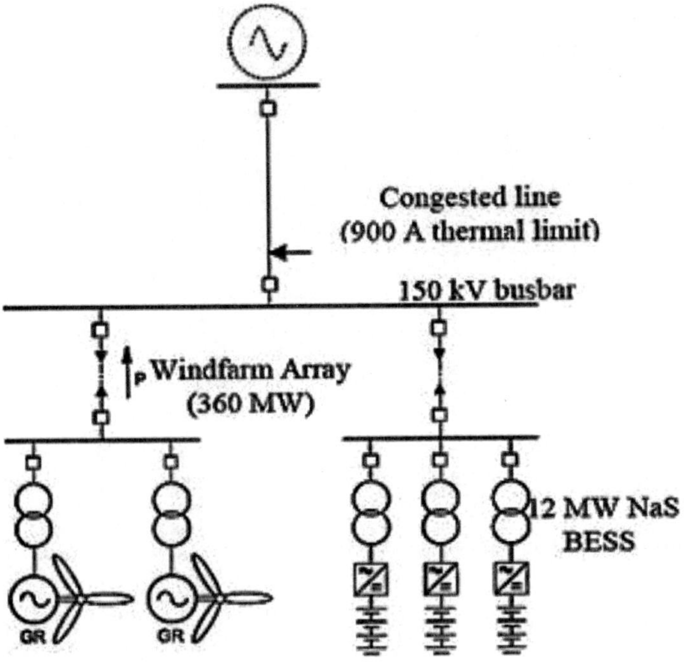

Fig. 9 Simplified single-line diagram

congestion by providing power near the load in the transmission network [9].

D. **Power Quality Improvement**

Most of the loads in the network are nonlinear and they inject harmonic currents in the system. The presences of harmonics result in system resonance, overloading of capacitors, changes in voltage level, and decrease in efficiency. Power quality is the cause of concern to power utilities and consumers as it affects the performance of load [10].

The power quality issues can be rectified by connecting a Static Compensator (STATCOM) at a point of common coupling with a battery energy storage system. A system to mitigate some of the power quality issues at PCC with wind energy system coupled with BESS is shown in Fig. 10. The battery energy storage maintains constant real power from changing wind power. The excess power generated charges the batteries during low energy demand hours. The combination of wind energy generation with a battery storage system is switched such that it synthesizes an output waveform required by the load by absorbing or delivering reactive power and facilitates the real power flow [9].

In the control scheme [10], the DC link capacitor couples the wind energy storage system with the AC grid. The excess power from the wind energy source is absorbed by the battery. The current-controlled voltage source inverter injects

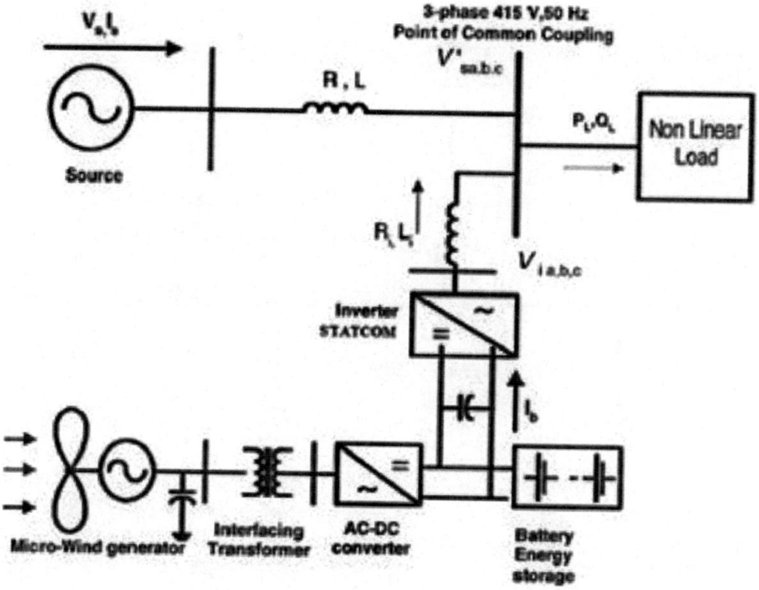

Fig. 10 Scheme of wind generation with battery storage

a fundamental component of current and harmonic currents required by the load. The harmonic currents do not come from the source, and therefore the voltage at PCC is maintained sinusoidal. It improves power quality by reducing the total harmonic distortion in the supply. It also relieves the grid from supplying load partially or fully taking up the load.

4 Challenges and Opportunities

The lack of regulation, technologies, safety, concern, and cost competitiveness prevent large-scale implementation of BESS in the network. The lifetime of the batteries is a major cause of concern because a number of factors like charging and discharging cycle, depth of discharge, SOC, temperature affects it. The inability to predict the life of batteries accurately imposes a problem in planning and executing BESS on large scale. The large-scale BESS unit is composed of series and parallel connected cells forming a battery pack with power reaching a few hundred kilowatts. Any imbalance in voltage/charge between these parallel-connected battery packs can cause a high circulating current among them [10]. The prominent cost drivers for Li-Ion-based BESS are battery aging and overall system losses. To improve system design and performance, a better understanding of the cell internal degradation mechanisms is required [4]. The placement of BESS, sizing of storage, and the dispatch strategy need to be optimized for exploiting the full potential of the energy storage

system [4]. The most challenging part is to bring down the cost of the entire system to make it economic and competitive with conventional power in the energy market.

5 Conclusion

Renewable Energy Sources (RES) is a way toward a more sustainable electricity supply. The recent developments and various projects on energy storage worldwide have cleared our vision about RES playing a crucial role in energy supply in a sustainable energy system in the future [4]. The increase in the number of photo voltaic (PV) and wind-driven electric generators in modern demand are getting challenging. It is important to equalize distributed and intermittent power generation with load demand within the grid for all times. Energy Storage Systems (ESS) are found to be capable of equalizing fluctuations and compensating a mismatch of power generation and utilization by coordinating power supply and energy time-shift.

References

1. Hannan MA, Hoque MM, Hussain A, Yusof Y, Ker PJ (2018) State-of- the-art and energy management system of lithium-ion batteries in electric vehicle applications: issues and recommendations. IEEE Access 6:19362–19378
2. Lee S-W, Lee K-M, Choi Y-G, Kang B (2018) Modularized design of active charge equalizer for li-ion battery pack. IEEE Trans Ind Electron 65(11):8697–8706
3. Hesse HC, Schimpe M, Kucevic D, Jossen A (2017) Lithium-ion battery storage for the grid—a review of stationary battery storage system design tailored for applications in modern power grids. Energies 10(12):2107
4. Yang Y, Ye Q, Tung LJ, Greenleaf M, Li H (2017) Integrated size and energy management design of battery storage to enhance grid integration of large-scale PV power plants. IEEE Trans Ind Electron 65(1):394–402
5. Essakiappan S, Shoubaki E, Koerner M, Rees J-F, Enslin J (2017) Dispatchable virtual power plants with forecasting and decentralized control, for high levels of distributed energy resources grid penetration. In: 2017 IEEE 8th international symposium on power electronics for distributed generation systems (PEDG), IEEE, pp 1–8
6. Hegstad RS (2014) Smarte løsninger i distribusjonsnett med høy grad av distribuert produksjon. Master's thesis, Institutt for elkraftteknikk
7. Palone F, Rebolini M, De Simone M, Gentili S, Giannuzzi G (2015) Operating strategies for congestion management of HV lines using NAS batteries. In: 2015 AEIT international annual conference (AEIT), IEEE, pp 1–6
8. Varghese JP, Ashok S, Kumaravel S (2017) Optimal siting and sizing of DGS for congestion relief in transmission lines. In: 2017 IEEE PES Asia-Pacific power and energy engineering conference (APPEEC), IEEE, pp 1–6

9. Dharavath R, Raglend IJ, Manmohan A (2017) Implementation of solar PV—battery storage with DVR for power quality improvement. In: 2017 Innovations in power and advanced computing technologies (i-PACT), IEEE, pp 1–5
10. Mundackal J, Varghese AC, Sreekala P, Reshmi V (2013) Grid power quality improvement and battery energy storage in wind energy systems. In: 2013 annual international conference on emerging research areas and 2013 international conference on microelectronics, communications and renewable energy, IEEE, pp 1–6